T0401392

Bioinformatics

METHODS IN MOLECULAR BIOLOGY™

John M. Walker, SERIES EDITOR

METHODS IN MOLECULAR BIOLOGY™

Bioinformatics

Volume II
Structure, Function and Applications

Edited by

Jonathan M. Keith, PhD

*School of Mathematical Sciences, Queensland University of Technology,
Brisbane, Queensland, Australia*

 Humana Press

Editor
Jonathan M. Keith
School of Mathematical Sciences
Queensland University of Technology
Brisbane, Queensland, Australia
j.keith@qut.edu.au

Series Editor
John Walker
Hatfield, Hertfordshire AL10 9NP
UK

ISBN: 978-1-60327-428-9   e-ISBN: 978-1-60327-429-6
ISSN 1064-3745   e-ISSN: 1940-6029
DOI: 10.1007/978-1-60327-429-6

Library of Congress Control Number: 20082922946

Cover illustration: Fig. 1A, Chapter 23, "Visualization," by Falk Schreiber (main image); and Fig. 4, Chapter 5, "The Classification of Protein Domains," by Russell L. Marsden and Christine A. Orengo (surrounding images)

Printed on acid-free paper

9 8 7 6 5 4 3 2 1

springer.com

Preface

Bioinformatics is the management and analysis of data for the life sciences. As such, it is inherently interdisciplinary, drawing on techniques from Computer Science, Statistics, and Mathematics and bringing them to bear on problems in Biology. Moreover, its subject matter is as broad as Biology itself. Users and developers of bioinformatics methods come from all of these fields. Molecular biologists are some of the major users of Bioinformatics, but its techniques are applicable across a range of life sciences. Other users include geneticists, microbiologists, biochemists, plant and agricultural scientists, medical researchers, and evolution researchers.

The ongoing exponential expansion of data for the life sciences is both the major challenge and the *raison d'être* for twenty-first century Bioinformatics. To give one example among many, the completion and success of the human genome sequencing project, far from being the end of the sequencing era, motivated a proliferation of new sequencing projects. And it is not only the quantity of data that is expanding; new types of biological data continue to be introduced as a result of technological development and a growing understanding of biological systems.

Bioinformatics describes a selection of methods from across this vast and expanding discipline. The methods are some of the most useful and widely applicable in the field. Most users and developers of bioinformatics methods will find something of value to their own specialties here, and will benefit from the knowledge and experience of its 86 contributing authors. Developers will find them useful as components of larger methods, and as sources of inspiration for new methods. Volume II, Section IV in particular is aimed at developers; it describes some of the "meta-methods"—widely applicable mathematical and computational methods that inform and lie behind other more specialized methods—that have been successfully used by bioinformaticians. For users of bioinformatics, this book provides methods that can be applied as is, or with minor variations to many specific problems. The Notes section in each chapter provides valuable insights into important variations and when to use them. It also discusses problems that can arise and how to fix them. This work is also intended to serve as an entry point for those who are just beginning to discover and use methods in bioinformatics. As such, this book is also intended for students and early career researchers.

As with other volumes in the Methods in Molecular Biology™ series, the intention of this book is to provide the kind of detailed description and implementation advice that is crucial for getting optimal results out of any given method, yet which often is not incorporated into journal publications. Thus, this series provides a forum for the communication of accumulated practical experience.

The work is divided into two volumes, with data, sequence analysis, and evolution the subjects of the first volume, and structure, function, and application the subjects of the second. The second volume also presents a number of "meta-methods": techniques that will be of particular interest to developers of bioinformatic methods and tools.

Within Volume I, Section I deals with data and databases. It contains chapters on a selection of methods involving the generation and organization of data, including sequence data, RNA and protein structures, microarray expression data, and functional annotations.

Section II presents a selection of methods in sequence analysis, beginning with multiple sequence alignment. Most of the chapters in this section deal with methods for discovering the functional components of genomes, whether genes, alternative splice sites, non-coding RNAs, or regulatory motifs.

Section III presents several of the most useful and interesting methods in phylogenetics and evolution. The wide variety of topics treated in this section is indicative of the breadth of evolution research. It includes chapters on some of the most basic issues in phylogenetics: modelling of evolution and inferring trees. It also includes chapters on drawing inferences about various kinds of ancestral states, systems, and events, including gene order, recombination events and genome rearrangements, ancestral interaction networks, lateral gene transfers, and patterns of migration. It concludes with a chapter discussing some of the achievements and challenges of algorithm development in phylogenetics.

In Volume II, Section I, some methods pertinent to the prediction of protein and RNA structures are presented. Methods for the analysis and classification of structures are also discussed.

Methods for inferring the function of previously identified genomic elements (chiefly protein-coding genes) are presented in Volume II, Section II. This is another very diverse subject area, and the variety of methods presented reflects this. Some well-known techniques for identifying function, based on homology, "Rosetta stone" genes, gene neighbors, phylogenetic profiling, and phylogenetic shadowing are discussed, alongside methods for identifying regulatory sequences, patterns of expression, and participation in complexes. The section concludes with a discussion of a technique for integrating multiple data types to increase the confidence with which functional predictions can be made. This section, taken as a whole, highlights the opportunities for development in the area of functional inference.

Some medical applications, chiefly diagnostics and drug discovery, are described in Volume II, Section III. The importance of microarray expression data as a diagnostic tool is a theme of this section, as is the danger of over-interpreting such data. The case study presented in the final chapter highlights the need for computational diagnostics to be biologically informed.

The final section presents just a few of the "meta-methods" that developers of bioinformatics methods have found useful. For the purpose of designing algorithms, it is as important for bioinformaticians to be aware of the concept of *fixed parameter tractability* as it is for them to understand NP-completeness, since these concepts often determine the types of algorithms appropriate to a particular problem. *Clustering* is a ubiquitous problem in Bioinformatics, as is the need to *visualize* data. The need to interact with massive data bases and multiple software entities makes the development of *computational pipelines* an important issue for many bioinformaticians. Finally, the chapter on *text mining* discusses techniques for addressing the special problems of interacting with and extracting information from the vast biological literature.

Jonathan M. Keith

Contents

SECTION IV: ANALYTICAL AND COMPUTATIONAL METHODS

Contributors

BISSAN AL-LAZIKANI • *Biofocus DPI, London, United Kingdom*

FÁTIMA AL-SHAHROUR • *Department of Bioinformatics, Centro de Investigación Príncipe Felipe (CIPF), Valencia, Spain*

JENS AUER • *Department of Life Science Informatics, Bonn-Aachen International Center for Information Technology (B-IT), Rheinische Friedrich-Wilhelms-University Bonn, Bonn, Germany*

JÜRGEN BAJORATH • *Professor and Chair of Life Science Informatics, Department of Life Science Informatics, Bonn-Aachen International Center for Information Technology (B-IT), Rheinische Friedrich-Wilhelms-University Bonn, Bonn, Germany*

RICHARD W. BEAN • *ARC Centre of Excellence in Bioinformatics, and Institute for Molecular Bioscience, The University of Queensland, Brisbane, Queensland, Australia*

REGINA BERRETTA • *Centre of Bioinformatics, Biomarker Discovery and Information-Based Medicine, The University of Newcastle, Callaghan, New South Wales, Australia*

DARIO BOFFELLI • *Children's Hospital Oakland Research Institute, Oakland, CA*

ANDREW B. CLEGG • *Institute of Structural Molecular Biology, School of Crystallography, Birkbeck College, University of London, London, United Kingdom*

WAGNER COSTA • *School of Electrical Engineering and Computer Science, The University of Newcastle, Callaghan, New South Wales, Australia*

SHAILESH V. DATE • *PENN Center for Bioinformatics, Department of Genetics, University of Pennsylvania School of Medicine, Philadelphia, PA*

JOAQUÍN DOPAZO • *Department of Bioinformatics, Centro de Investigación Príncipe Felipe (CIPF), Valencia, Spain*

HANNA ECKERT • *Department of Life Science Informatics, Bonn-Aachen International Center for Information Technology (B-IT), Rheinische Friedrich-Wilhelms-University Bonn, Bonn, Germany*

RICHARD D. EMES • *Department of Biology, University College London, London, United Kingdom*

MARIO FALCHI • *Twin Research and Genetic Epidemiology Unit, King's College London School of Medicine, London, United Kingdom*

NICOLAS GOFFARD • *Research School of Biological Sciences and ARC Centre of Excellence for Integrative Legume Research, The Australian National University, Canberra, Australian Capital Territory, Australia*

MARK HALLING-BROWN • *Institute of Structural Molecular Biology, School of Crystallography, Birkbeck College, University of London, London, United Kingdom*

NICHOLAS HAMILTON • *ARC Centre of Excellence in Bioinformatics, Institute for Molecular Bioscience and Advanced Computational Modelling Centre, The University of Queensland, Brisbane, Queensland, Australia*

EMMA E. HILL • *The Journal of Cell Biology, Rockefeller University Press, New York, NY*

MOU'ATH HOURANI • *Newcastle Bioinformatics Initiative, School of Electrical Engineering and Computer Science, The University of Newcastle, Callaghan, New South Wales, Australia*

THOMAS HUBER • *School of Molecular and Microbial Sciences and Australian Institute for Bioengineering and Nanotechnology, The University of Queensland, Brisbane, Queensland, Australia*

FALK HÜFFNER • *Institut für Informatik, Friedrich-Schiller-Universität Jena, Jena, Germany*

OLE N. JENSEN • *Department of Biochemistry and Molecular Biology, University of Southern Denmark, Odense, Denmark*

JONATHAN M. KEITH • *School of Mathematical Sciences, Queensland University of Technology, Brisbane, Queensland, Australia*

DENNIS KOSTKA • *Max Planck Institute for Molecular Genetics and Berlin Center for Genome-Based Bioinformatics, Berlin, Germany*

INSUK LEE • *Center for Systems and Synthetic Biology, Institute for Molecular Biology, University of Texas at Austin, Austin, TX*

CLAUDIO LOTTAZ • *Max Planck Institute for Molecular Genetics and Berlin Center for Genome-Based Bioinformatics, Berlin, Germany*

EDWARD M. MARCOTTE • *Center for Systems and Synthetic Biology, and Department of Chemistry and Biochemistry, Institute for Molecular Biology, University of Texas at Austin, Austin, TX*

NICHOLAS R. MARKHAM • *Xerox Litigation Services, Albany, NY*

FLORIAN MARKOWETZ • *Max Planck Institute for Molecular Genetics and Berlin Center for Genome-Based Bioinformatics, Berlin, Germany*

RUSSELL L. MARSDEN • *Biochemistry and Molecular Biology Department, University College London, London, United Kingdom*

RUNE MATTHIESEN • *CIC bioGUNE, Bilbao, Spain*

GEOFFREY J. MCLACHLAN • *ARC Centre of Excellence in Bioinformatics, Institute for Molecular Bioscience, and Department of Mathematics, The University of Queensland, Brisbane, Queensland, Australia*

ALEXANDRE MENDES • *Centre of Bioinformatics, Biomarker Discovery and Information-Based Medicine, The University of Newcastle, Callaghan, New South Wales, Australia*

VERONICA MOREA • *National Research Council (CNR), Institute of Molecular Biology and Pathology (IBPN), Rome, Italy*

GABRIEL MORENO-HAGELSIEB • *Department of Biology, Wilfrid Laurier University, Waterloo, Ontario, Canada*

PABLO MOSCATO • *ARC Centre of Excellence in Bioinformatics, and Centre of Bioinformatics, Biomarker Discovery and Information-Based Medicine, The University of Newcastle, Callaghan, New South Wales, Australia*

SHU-KAY NG • *Department of Mathematics, The University of Queensland, Brisbane, Queensland, Australia*

ROLF NIEDERMEIER • *Institut für Informatik, Friedrich-Schiller-Universität Jena, Jena, Germany*

CHRISTINE A. ORENGO • *Biochemistry and Molecular Biology Department, University College London, London, United Kingdom*

JOSÉ M. PEREGRÍN-ALVAREZ • *Hospital for Sick Children, Toronto, Ontario, Canada*

ALBIN SANDELIN • *The Bioinformatics Centre, Department of Molecular Biology and Biotech Research and Innovation Centre, University of Copenhagen, Copenhagen, Denmark*

FALK SCHREIBER • *Leibniz Institute of Plant Genetics and Crop Plant Research (IPK) Gatersleben, Germany and Institute for Computer Science, Martin-Luther University Halle-Wittenberg, Germany*

ADRIAN J. SHEPHERD • *Institute of Structural Molecular Biology, School of Crystallography, Birkbeck College, University of London, London, United Kingdom*

RAINER SPANG • *Max Planck Institute for Molecular Genetics and Berlin Center for Genome-Based Bioinformatics, Berlin, Germany*

GEORG F. WEILLER • *Research School of Biological Sciences and ARC Centre of Excellence for Integrative Legume Research, The Australian National University, Canberra, Australian Capital Territory, Australia*

SEBASTIAN WERNICKE • *Institut für Informatik, Friedrich-Schiller-Universität Jena, Jena, Germany*

MICHAEL ZUKER • *Mathematical Sciences and Biology Department, Rensselaer Polytechnic Institute, Troy, NY*

Contents of Volume I

Section I

Structures

Chapter 1

UNAFold

Software for Nucleic Acid Folding and Hybridization

Nicholas R. Markham and Michael Zuker

Abstract

The UNAFold software package is an integrated collection of programs that simulate folding, hybridization, and melting pathways for one or two single-stranded nucleic acid sequences. The name is derived from "Unified Nucleic Acid Folding." Folding (secondary structure) prediction for single-stranded RNA or DNA combines free energy minimization, partition function calculations and stochastic sampling. For melting simulations, the package computes entire melting profiles, not just melting temperatures. UV absorbance at 260 nm, heat capacity change (C_p), and mole fractions of different molecular species are computed as a function of temperature. The package installs and runs on all Unix and Linux platforms that we have looked at, including Mac OS X. Images of secondary structures, hybridizations, and dot plots may be computed using common formats. Similarly, a variety of melting profile plots is created when appropriate. These latter plots include experimental results if they are provided. The package is "command line" driven. Underlying compiled programs may be used individually, or in special combinations through the use of a variety of Perl scripts. Users are encouraged to create their own scripts to supplement what comes with the package. This evolving software is available for download at http://www. bioinfo.rpi.edu/applications/hybrid/download.php.

Key words: RNA folding, nucleic acid hybridization, nucleic acid melting profiles (or DNA melting profiles), free energy minimization, partition functions, melting temperature.

1. Introduction

The UNAFold software predicts nucleic acid foldings, hybridizations, and melting profiles using energy-based methods and a general computational technique known as dynamic programming (1). Early software for RNA folding predicted minimum

Jonathan M. Keith (ed.), *Bioinformatics, Volume II: Structure, Function and Applications, vol. 453*
© 2008 Humana Press, a part of Springer Science + Business Media, Totowa, NJ
Book doi: 10.1007/978-1-60327-429-6 Springerprotocols.com

free energy foldings only *(2–6)*. It became clear early on that such methods were unreliable in the sense that many different foldings, with free energies close to the computed minimum, could exist. Although constraints deduced from experiments or phylogenetic analyses could be applied to reduce uncertainty, a method to compute a variety of close to optimal foldings was needed.

The mfold software *(7–10)* computes a collection of optimal and suboptimal foldings as well as a triangular-shaped plot called an energy dot plot (EDP). The EDP contains a dot or other symbol in row i and column j ($i < j$) to indicate that the base pair between the ith and jth nucleotides can occur in a folding within some user prescribed free energy increment from the minimum. The *Vienna RNA Package* (Vienna RNA) *(11–13)* differs fundamentally from mfold because the underlying algorithm computes partition functions, rather than minimum free energies. This leads naturally to the computation of base pair probabilities *(14)* and what they call "boxplots." We call these triangular shaped plots probability dot plots (PDPs). In this case, all base pairs with probabilities above a certain threshold are plotted as boxes (or other symbols) whose area is proportional to the probability of that base pair. This software can compute all possible foldings close to optimal. The *sfold* package *(15–17)* also computes partition functions, but it uses a simpler algorithm than the Vienna RNA package because base pair probabilities are not computed directly (exactly). Instead, it computes a "statistically valid sample" (Gibbs sample) that permits the estimation of not only base pair probabilities, but probabilities of any desired motif(s). UNA-Fold encompasses all three methods by computing minimum and suboptimal foldings in the mfold style, and full partition functions that allow both exact base pair computations as well as stochastic sampling. In addition, stochastic sampling may be used to compute both ensemble enthalpy and ensemble heat capacity for single sequence folding.

For practical applications, it is common to use crude methods, often ad hoc, to compute melting temperatures for dimers. It is usual to assume that hybridized strands are perfectly complementary. When mismatches are permitted, they are few and isolated. There is no question about base pairs and no computations are required to determine the hybridization. For example, the simple case:

$$5'\text{-ACGGtCCAGCAA-}3'$$

$$3'\text{-TGCCgGGTCGTT-}5'$$

shows the dimerization of two almost complementary 12-mers. Note the single T·G wobble pair. A computer is not needed to compute the hybridization. A calculator would be helpful in adding up nearest neighbor free energies and enthalpies from published tables *(18–20)*, and a computer would be needed to process thousands of these dimers. In any case, free energy at 37°C, ΔG

(or DG_{37}) and enthalpy, ΔH, are easily computed. From this, the entropy, ΔS, is easily computed as:

$$\Delta S = 1000 \times \frac{\Delta H - \Delta G}{T},$$

where T is the hybridization temperature (K) and the factor of 1000 expresses ΔS in e.u. (entropy units; 1 e.u. = 1 cal/mol/K). The terms "free energy," "enthalpy," and "entropy" are really changes in free energy, enthalpy, and entropy with respect to the "random coil state." In reality, "random coil" is a large ensemble of states. The free energy and enthalpy of this ensemble of states may be (and is) set to zero. Only entropy cannot be arbitrarily assigned. Thus a ΔS term is $R\ln(N/N_{ref})$, where R is the universal gas constant, 1.9872 e.u.; N is the number of states consistent with hybridization; and N_{ref} is the number of states, always larger, in the unconstrained "random coil."

The simplest model for hybridization assumes that there are two states: hybridized, where the structure is unique and known, and "random coil." It is then a simple matter to deduce a melting temperature, T_m, at which half of the dimers have dissociated. The formula is given by:

$$T_m = 1000 \times \frac{\Delta H}{\Delta S + R\ln(C_t/f)}, \qquad [1]$$

where ΔH, ΔS (expressed in e.u.) and R have already been defined, C_t is the total strand concentration, and $f = 4$ when the two strands are different and $f = 1$ when self-hybridization takes place. This formula implicitly assumes that both strands are present in equal concentrations. Competition with folding and with other possible hybridizations, such as homodimer formation, is not considered at all.

UNAFold offers a number of increasingly sophisticated ways to compute melting temperatures and also computes entire melting profiles: UV absorbance at 260 nm, heat capacity (C_p), and mole fractions of various single and double-stranded molecular species as a function of temperature. Even when Equation [1] is used, the hybridization is computed by minimizing free energy. The various levels of complexity and choice of methods are described in the following.

It is important to issue a warning to those who wish to apply these methods to microarrays. Hybridization on microarrays is complicated by the fact that one of each hybridizing pair is immobilized. It is difficult to compute the effective "solution concentration" for such molecules, and diffusion is probably an important factor in slowing the time necessary to reach equilibrium. However, even for hybridization in solution, kinetic simulations that treat different probe-target pairs independently can lead to totally incorrect results for the time required to reach equilibrium as well as the equilibrium concentrations themselves (21).

It is important to emphasize that the computations used by the UNAFold software are based on a number of assumptions.

1. The simulations are for molecules in solution. For microarrays, computing an effective concentration of the immobilized oligos is necessary. One needs to estimate the "surface phase concentration," c, in units such as "molecules/cm^2," and the solution volume per cm^2 of surface area, v (L/cm^2) (21).

2. The system contains one or two different molecules. If, for example, a true system contains three different oligos, A, B and C, then some common sense must be used. If A and B do not hybridize with each other over the temperature range of interest, then A and C could be simulated separately from B and C if, in addition, the concentration of C is a couple of orders of magnitude greater than that for both A and B.

3. The computations assume that the system is in equilibrium. Thus, the melting profile predictions assume that temperature changes are slow enough so that the system is always at equilibrium. When this is not true, observed melting profiles depend on the rate of temperature change. In particular, a hysteresis effect can be observed. For example, even if the temperature is raised at a uniform rate from T_0 to T_1, and then brought down to T_0 at the same rate, the measured UV absorbance profiles for melting and cooling may differ.

2. Materials: The UNAFold Software

2.1. Supported Platforms

The UNAFold package compiles and runs on many platforms. **Table 1.1** lists a few operating system/architecture combinations under which the package is known to work.

In fact, the UNAFold package is not known *not* to run on any platform. The software should function on any system having the following basic tools:

- A Unix-style shell.
- Make.
- A C compiler.
- A Perl interpreter.

2.2. Dependencies

As noted, UNAFold requires only some very basic tools to build and run. However, there are several optional libraries and programs that provide additional functionality when installed:

- gnuplot: If gnuplot is detected, the hybrid2.pl script will use it to produce several plots in Postscript format.
- OpenGL/glut: If OpenGL and the glut library (22) are detected, the hybrid-plot program will be built.

Table 1.1
Some platforms supported by the UNAFold package

Operating system	Architecture
Linux	x86
Linux	x86_64
FreeBSD	x86
SunOS	sparc
IRIX	mips
IRIX64	mips
AIX	powerpc
MacOS	powerpc
MS Windows	x86

- gd: If the gd library is detected, the hybrid-plot-ng program will use it to create images in GIF, JPEG, and/or PNG formats directly. (With or without this library, hybrid-plot-ng creates Postscript files which can be converted to virtually any other image format.)

2.3. Downloading and Installing

The UNAFold software is available for download at http://www.bioinfo.rpi.edu/applications/hybrid/download.php. Binaries for Linux (in RPM format) and Windows (in EXE format) are available along with the source code.

After downloading and unpacking, building the software consists of three steps:

- Configuration: Usually this is as simple as typing ./configure. Running ./configure --help lists options that may be used to configure installation locations, specify locations of libraries, and set compiler options.

- Compilation: A single command—make—compiles and links all programs.

- Installation: Typing make install copies the programs, scripts, documentation, and data files to the location set with the configure script.

2.4. Core Programs

The core programs in UNAFold are written in C and are optimized when compiled since they are computationally intensive. Man pages exist for all of these programs. In addition, invoking any program with the --help option will generate an abbreviated set of instructions. Some programs have counterparts formed by adding the suffix -same. When simulating a one-sequence ensemble (in which the dominant dimer is a homodimer rather than a heterodimer) the -same version replaces the regular one.

- Folding:
  — hybrid-ss: Computes full partition functions for folding RNA or DNA. It may be run using the --energyOnly option, in which case the probabilities of base pairs and of single-stranded nucleotides and dinucleotides will not be computed, saving significant time and memory. It may also be run with the --tracebacks option to generate a stochastic sample of foldings. UNAFold also includes two simplified versions of hybrid-ss: hybrid-ss-simple and hybrid-ss-noml. hybrid-ss-simple assigns fixed entropic costs to multibranch loops and ignores single-base stacking, while hybrid-ss-noml does not allow multibranch loops at all.

  — hybrid-ss-min: Computes minimum energy foldings. It can predict a single, minimum free energy folding or, using an extended algorithm, generate an mfold-like collection of foldings and an EDP. For fast folding of many sequences with the same parameters, the --stream option reads sequences, one at a time, from standard input and writes free energies to standard output.

- Hybridization:
  — hybrid: Computes full partition functions for hybridizing RNA or DNA (without intramolecular base pairs). The --energyOnly and --tracebacks options function as in hybrid-ss.

  — hybrid-min: Computes minimum energy hybridizations. The --mfold and --stream options function as in hybrid-ss-min.

- Ensemble computations:
  These programs rely on the output of the core programs.

  — concentration(-same): Computes the mole fractions of different molecular species using the already computed free energies for individual monomer (folded) and dimer (hybridized) species.

  — ensemble-dg(-same): Computes ensemble free energies using species free energies and computed mole fractions.

  — ensemble-ext(-same): Computes UV absorbance at 260 nm using computed probabilities for single-strandedness of nucleotides and (optionally) dinucleotides, computed mole fractions, and published extinction coefficients (23).

2.5. Auxiliary Programs

- ct2rnaml: Converts structures in .ct format to RNAML *(24)*. ct2rnaml supports files with multiple structures on a single molecule as well as multiple molecules.

- ct-energy: Evaluates the free energy of structures in .ct format, with detailed free energy information for each loop if desired.

- ct-ext: Evaluates the UV absorbance of structures in .ct format.

- ct-prob: Estimates base pair probabilities from a stochastic sample in .ct format. ct-prob can also compare these probabilities to the "correct" probabilities if they are available.

- ct-uniq: Selects the unique structures from a set in .ct format. ct-uniq is typically used to remove duplicates structures from a stochastic sample.

- dG2dH, dG2dS, dG2Cp: Compute enthalpy, entropy and heat capacity, respectively, by differentiating free energy. Enthalpy and entropy are given by $\Delta H = \Delta G + T \dfrac{\partial \Delta G}{\partial T}$ and $\Delta S = \dfrac{\partial \Delta G}{\partial T}$; heat capacity is given by $C_p = -T + \dfrac{\partial^2 \Delta G}{\partial T^2}$. dG2Cp also determines the melting temperature(s) $T_m(C_p)$, the maximum or local maxima of the heat capacity curve.

- hybrid-plot(-ng): Displays PDPs created with hybrid or hybrid-ss. hybrid-plot is an interactive program based on glut, while hybrid-plot-ng produces static images in Postscript and GIF/JPEG/PNG formats.

2.6. Linking Software: Perl Scripts

We describe our scripts that combine a number of basic programs. This set is expected to grow, and we welcome both suggestions from users and user-generated scripts.

- concentrations(-same).pl: Produces normalized mole fractions (between 0 and 1) from the concentrations calculated by concentration(-same). concentrations(-same).pl also computes $T_m(\text{Conc})$, the temperature at which one half of the strands are single-stranded and unfolded.

- ct-energy-det.pl: Converts the verbose output of ct-energy to energy details in plain text or HTML.

- h-num.pl: Computes h-num values from an EDP. h-num is a measure of well-definedness for a helix.

- hybrid(-ss)-2s.pl: Simulates two-state hybridization or folding. hybrid-2s.pl uses hybrid-min to compute a minimum energy hybridization and extrapolates it over a range of temperatures. hybrid-2s.pl can be used in place of hybrid to obtain a two-state model instead of a full ensemble. Likewise, hybrid-ss-2s.pl uses hybrid-ss-min and can replace hybrid-ss.

- hybrid2.pl: Uses hybrid, hybrid-ss, concentration, ensemble-dg, and ensemble-ext to produce a full melting profile. The "flagship" script, hybrid2.pl also employs gnuplot to produce Postscript plots of mole fractions (from concentrations.pl), heat capacity (from dG2Cp) and UV absorbance. hybrid2.pl operates in several different modes depending on how it is invoked:

  — If hybrid2.pl is invoked with only one sequence (or with two copies of the same sequence), it automatically replaces concentration with concentration-same, ensemble-dg with ensemble-dg-same, and so on.

  — If hybrid2.pl is invoked as hybrid2-min.pl, hybrid-min and hybrid-ss-min replace hybrid and hybrid-ss so that the entire computation is performed using only an optimal structure for each species at each temperature, instead of a full ensemble. Likewise, invoking hybrid2-2s.pl replaces hybrid(-ss) with hybrid(-ss)-2s.pl to obtain a two-state model for each species.

  — Finally, if hybrid2.pl is invoked as hybrid2-x.pl, hybrid2-min-x.pl or hybrid22s-x.pl, the homodimers and monomers are excluded from consideration, so that only the heterodimer is simulated. hybrid2-2s-x.pl thus implements the classical two-state model, although it is still enhanced in that strand concentrations may be unequal.

- hybrid-select.pl: Folds or hybridizes sequences using only selected base pairs. hybridselect.pl runs two folding or hybridization programs; the PDP or EDP from the first run is used to select base pairs to be considered in the second run. If the first run is an energy minimization (EM), base pairs that appear in foldings with energies in a user-specified percentage of the minimum energy are selected; if it is a partition function (PF), base pairs with at least a user-specified probability are selected. hybrid-select.pl may be used for foldings (running hybrid-ss and/or hybrid-ss-min) and for hybridizations (running hybrid and/or hybrid-min). Since there are four possible methods—EM followed by EM, EM followed by PF, PF followed by EM, and PF followed by PF—there are a total of eight ways to use hybrid-select.pl. Additionally, the --mfold option may be used if the second stage is an EM, whereas --tracebacks is acceptable when the second stage is a PF.

- ITC-plot.pl: Produces isothermal titration calorimetry (ITC) plots (25). In an ITC plot, the concentration of one strand is varied while the other is held constant, and enthalpy change is plotted as a function of the changing strand concentration.

- melt.pl: Quickly computes two-state melting for monomers, homodimers or heterodimers. Only ΔG, ΔH, ΔS, and

T_m are output, but melt.pl can process many sequences very quickly.

- plot2ann.pl: Creates probability annotations for a structure in .ct format. Paired bases are annotated with the probability of the pair, whereas unpaired bases are annotated with the probability of being single-stranded. The resulting .ann file can be used by sir_ graph (part of the mfold_util package) to display an annotated structure graph.

- ss-count.pl: Estimates the probability of each base being single-stranded from a stochastic sample in .ct format. Like ct-prob, ss-count.pl can also compare these probabilities to the "correct" probabilities if they are available.

- UNAFold.pl: Folds a sequence with detailed textual and graphical output. The package's namesake script, UNAFold. pl, uses hybrid-ss-min to fold a sequence and then uses components of mfold_util to expand on the output. boxplot_ng is used to produce an EDP in Postscript format and, if possible, as a PNG, JPEG, or GIF. sir_graph_ng creates a structure plot for each structure in the same formats; the --ann option allows these structures to be automatically annotated with ss-count or p-num information. Finally, UNAFold.pl can produce an HTML file with hyperlinks to all of the plots and other information, unifying the folding results.

- vantHoff-plot.pl: Produces van't Hoff plots. In a van't Hoff plot, the concentrations of both strands are varied together, and the inverse of the melting temperature is plotted as a function of total strand concentration. A linear van't Hoff plot may be taken as evidence that the transition follows a two-state model.

3. Methods

This section presents a number of examples of using the UNA-Fold software.

3.1. Energy Data

The UNAFold software has two built in sets of energy files for DNA and RNA, respectively. For DNA, we use free energies at 37°C and enthalpies from the SantaLucia laboratory at Wayne State University in Detroit, MI (19). For RNA, we use equivalent parameters from the Turner laboratory at the University of Rochester in Rochester, NY (26). In both cases, it is possible to generate free energies at different temperatures in order to compute melting profiles. These DNA and RNA energy parameters are referred to as versions 3.1 and 2.3, respectively. More up-to-date free energies for RNA are available, also from the Turner laboratory, but they are for 37°C only *(27)*. These are denoted as version 3.0 energies. The built in energies are the default. A suffix flag, --suffix, is required to use any other set of (free) energies. Thus, the flag, --suffix DAT, tells any folding or hybridization program that version 3.0 RNA free energy parameters should be used.

The transition from minimum free energy (mfe) folding to partition function computations forced us to examine the energy rules with great care to avoid over-counting various states. For example, for version 3.0 RNA parameters, single wobble pairs (G·T or T·G) are allowed, but tandem wobble pairs are prohibited. Such motifs may still occur, but only as tandem mismatches in 2×2 symmetric interior loops. The result is that the number of valid base pairs varies slightly when using the RNA version 3.0 rules. In this context, three sets of zero energies are available and may be used in conjunction with partition function programs to count the number of valid hybridizations or foldings. They are NULD, NUL, and NULDAT, and refer to DNA version 3.2, RNA version 2.3 and RNA version 3.0 parameters, respectively. The first parameter sets are currently identical. Only RNA version 3.0 rules alter permitted base pairs, 1×1 loops and 2×2 loops; and these changes always involve wobble pairs.

3.2. Extended ct File Format

A ct file is a text file that defines a nucleic acid sequence together with its secondary structure. It may contain multiple foldings of a single sequence or even multiple foldings of different sequences. The original format was described by Zuker *(8)*, although it originated years earlier.

The first line (record) of a ct file contains an integer, N, which is the length of the nucleic acid sequence, followed by a "title." The title comprises all characters that occur on this line after N. For example, suppose that:

$$dG = -244.9 \text{ AvI5 Group 2 intron}$$

is the first line of a ct file. This intron contains 565 bases. The title is:

$$dG = -244.9 \text{ AvI5 Group 2 intron}$$

In this example, the title contains a predicted free energy and a description of the RNA. However, any text is valid in the title. Unfortunately, this format is not well understood or else it is simply ignored. For example, the RnaViz program *(28, 29)*, requires "ENERGY =" in the title line. Certain programs in the mfold package expect "dG =".

The following N lines (records) contain eight columns. The last two columns were introduced for UNAFold. The ith record following the title contains:

1. i, the index of the ith base.

2. r_i, the ith base. It is usually A, C, G, T, or U (upper or lower case), although other letters are permitted.

3. $5'(i)$. The 5′, or upstream connecting base. It is usually $i-1$, and is 0 when r_i is the 5′ base in a strand. If $5'(1) = N$, then the nucleic acid is circular.

4. $3'(i)$. The 3′, or downstream connecting base. It is usually $i+1$, and is 0 when r_i is the 3′ base in a strand. If $3'(N) = 1$, then the nucleic acid is circular.

5. bp(i). The base pair, $r_i \cdot r_{\mathrm{bp}(i)}$, exists. It is 0 if r_i is single-stranded.

6. The "historical numbering" of r_i. It may differ from i when a fragment of a larger sequence is folded or to indicate genomic DNA numbering in spliced mRNAs. The UNAFold software currently folds entire sequences only, but this feature is used when two sequences hybridize.

7. 5′stack(i). If this is 0, then r_i does not stack on another base. If it is $k \neq 0$, then r_i and r_k stack. In our applications, $k = i - 1$ when not zero, indicating that the base 5′ to r_i stacks on r_i. When bp(i) = $j > 0$, then r_k stacks 5′ on the base pair $r_i \cdot r_j$.

8. 3′stack(i). If this is 0, then r_i does not stack on another base. If it is $k \neq 0$, then r_i and r_k stack. In our applications, $k = i + 1$ when not zero, indicating that the base 3′ to r_i stacks on r_i. When bp(i) = $j > 0$, then r_k stacks 3′ on the base pair $r_i \cdot r_j$.

The UNAFold software uses the last two columns of the ct file only to indicate stacking on adjacent base pairs. Such stacking is assumed in base pair stacks. That is, if two adjacent base pairs, $r_i \cdot r_j$ and $r_{i+1} \cdot r_{j-1}$, both exist, then 5′stack($i + 1$) = i, 5′stack(j) = $j - 1$, 3′stack(i) = $i + 1$ and 3′stack($j - 1$) = j.

When 5′stack(i) = $i - 1$ and bp($i - 1$) = 0, then r_{i-1} is single-stranded and stacks 5′ on the adjacent base pair, $r_i \cdot r_j$. When 3′stack(i) = $i + 1$ and bp($i + 1$) = 0, then r_{i+1} is single-stranded and stacks 3′ on the adjacent base pair, $r_i \cdot r_j$.

At the present time, the last two columns of ct files contain redundant information and were introduced solely so that the stacking of single-stranded bases on adjacent base pairs could be specified unambiguously. They could be used, for example, to indicate coaxial stacking or adjacent or almost adjacent helices, but UNAFold neither predicts foldings with coaxial stacking, nor does it accept ct files that specify such stacking.

3.3. Sequence File Format

The UNAFold software accepts either raw sequence files or the FASTA format. The former term denotes files that contain only nucleotides. There is no line (record) length limit and blank spaces are ignored, including tab and control characters. The letters "A," "C," "G," "U," and "T" are recognized. Upper and lowercase are accepted and case is conserved so that both can be used to distinguish, for example, one region from another. Note that "U" is treated as "T" if DNA energies are used, and "T" is recognized as "U" for RNA energies. All other characters are treated as nucleotides that cannot pair. In terms of energies, they are treated as neutral. The UNAFold software does not recognize the IUPAC ambiguous base nomenclature. So, for example, "R" and "Y" are not recognized as purine and pyrimidine, respectively, and are not permitted to form base pairs. In particular, "N" does not mean "can take on any value." Thus, in UNAFold, the base pair, G·N, cannot form. In other contexts, it could form as either G·C or G·T. The semicolon character ";" is recognized as a sequence divider. That is, multiple sequences may be stored in a single file, as long at they are separated by semicolons. They do not have to be on separate lines.

UNAFold also accepts FASTA format. The first line of a sequence file must begin with >. The remainder of the line is interpreted as the sequence name, and is read and used by the software. Subsequent lines contain pure sequence. The FASTA format may be used for files containing multiple sequences. The semicolon is no longer required to separate sequences. Only the minimum energy prediction programs use the multiple sequence file (msf) format. The partition function programs read the first sequence only.

If a sequence file name ends with .seq, which we recommend, then the "prefix" is formed by stripping the final four characters (.seq). In all other cases, the prefix is the same as the sequence file name. Output file names are standard and based on the prefix (or prefixes). Suppose then, that two sequence files are named probe.seq and target.seq. Then output files will begin with either probe or target and signify folding output. File names beginning with probe-target, probe-probe, or target-target indicate hybridization output and refer to the heterodimer and two possible homodimers. Because UNAFold automatically uses the hyphen character "-" to form output file names, sequence file names should not contain hyphens.

When raw sequence files are used, the prefix of the file name is used as the sequence name. The sequence name appears only in "ct file" output and in secondary structure or hybridization plots. Only raw sequence format should be used when melting profiles are being computed.

3.4. Folding a Sequence

We selected a Group II intron, *A.v.*I5, from the Eubacterium, *Azotobacter vinelandii (30)*. The sequence was placed in the file A_v_I5.seq using FASTA format.

The phylogenetically derived secondary structure is from the "Mobile group II introns" web site.* It is classified as type B, or "group IIB-like" *(31, 32)*. It was downloaded from this web site as a ct file named a_v_i5_r.ct, renamed here as A_v_I5phylo.ct.

---

* Zimmerly Lab Web site http://www.fp.ucalgary.ca/group2introns/

In the sequel, this folding is called the "correct" folding, or the reference folding.

A variety of programs were used to predict foldings of the intron, and results were compared to the correct folding. Our purpose is to illustrate how the software may be used rather than how to model particular ribozymes.

When a variety of UNAFold programs are run separately by a user, instead of together by a Perl script, it is easy to overwrite output previously predicted. A ct file from a stochastic sample, for example, could be overwritten by a ct file from mfe computations. In such cases, it is prudent to create one or more copies of the sequence file using different names. In this case, we used the file name A_v_I5min.seq for mfe computations. In addition, programs that write to "standard output" often require the user to direct the results into files whose names should be chosen with some care.

3.4.1. Partition Function and Stochastic Sampling

The hybrid-ss program was run using the command:

```
hybrid-ss --suffix DAT --tracebacks 100 A_v_
    15.seq
```

The --suffix flag specifies version 3.0 RNA parameters, which are recommended in situations in which melting profiles are not being computed. The --tracebacks 100 flag requests a stochastic sample of 100 secondary structures. The output comprises five new files, A_v_I5.run, A_v_I5.dG, A_v_I5.37.ct, A_v_I5.37.ext, and A_v_I5.37.plot. The contents of these text files are listed according to suffix.

run A simple description of what program was run. Some, but not all, of the flags appear.

dG The Gibbs free energy, $-RT\ln(Z)$ and the partition function value, Z. Free energy units are kcal/mol.

ct A concatenation of all 1,000 secondary structures. This is a stochastic sample, so the ordering is not relevant. The free energies are randomly distributed.

ext As indicated in the top record, this file lists, for every i between 1 and N: (the sequence size), i, the probability that r_i is single-stranded, and the probability that both r_i and r_j are single-stranded. These files are used to compute UV absorbance, as previously described.

plot This file contains a list of all possible base pairs and their probabilities. The simple format for each record is i, j and the probability of the base pair $r_i \cdot r_j$ in the Boltzmann ensemble. Unlike the ct files, these files always assume that $i < j$ for folding. (This is not the case in plot files for hybridization.) UNAFold does not output base pairs with probabilities $<10^{-6}$, and these files are easily filtered to select only higher probability base pairs.

Note that a double suffix is employed in naming output files. When appropriate, the temperature is inserted in °C. Even when the temperature or range of temperatures is not specified on the command line, it is specified in the last line of the miscloop energy file.

3.4.2. Simple mfe Computation

The command:

```
hybrid-ss-min --suffix DAT A_v_I5min.seq
```

produced a single mfe folding. The resulting ext and plot files contain zeros or ones for probabilities. A value of one means that, as the case may be, r_i is single-stranded, both r_i and r_{i+1} are single-stranded, or that the base pair $r_i \cdot r_j$ exists in the predicted folding. Zero probabilities indicate the negation of the above. The resulting dG file contains the mfe and the Boltzmann factor for that free energy alone. It is impossible to distinguish between dG files created by partition functions and those created by energy minimization. The formats are identical and the suffix dG is added to the sequence file name prefix in both cases. However, the single mfe structure is placed in a ct file named by adding only a ct suffix. The folding temperature is not added.

3.4.3. Energy Computations from ct Files

The free energies of the 100 stochastically sampled foldings were evaluated using the ct-energy program as follows:

```
ct-energy --suffix DAT A_v_I5.37.ct
```

Used in this manner, the output is a stream of free energies, one per line, containing the evaluated free energies of the successive foldings. Note that neither hybrid-ss nor hybrid computes free energies for sampled foldings, so that running ct-energy is the only way to evaluate their free energies. Using the --suffix DAT flag evaluates the free energies using the same free energies that generated them. Substituting, for example, --suffix DH, would compute the enthalpies of these structures. The output stream from ct-energy was sorted by energy and directed to a file named A_v_I5sample.dG. These random free energies have a Γ-distribution, but this file was used only to compute a mean and standard deviation of the sample free energies.

The command:

```
ct-energy --suffix DAT A_v_I5phylo.ct
```

was invoked to evaluate the free energy of the correct folding. The result was "+inf", which means $+\infty$! This situation required some "detective work," and so ct-energy was rerun with the --verbose flag. This produced a stream of output containing the energy details of all stacks and loops in the structure. The line:

Interior: 88-C 109-G, 90-U 107-G: +inf

was detected. This turns out to be a 1×1 interior loop closed by a U·G wobble pair. The version 3.0 RNA rules do not allow this. To be specific, this motif must be treated as a 2×2

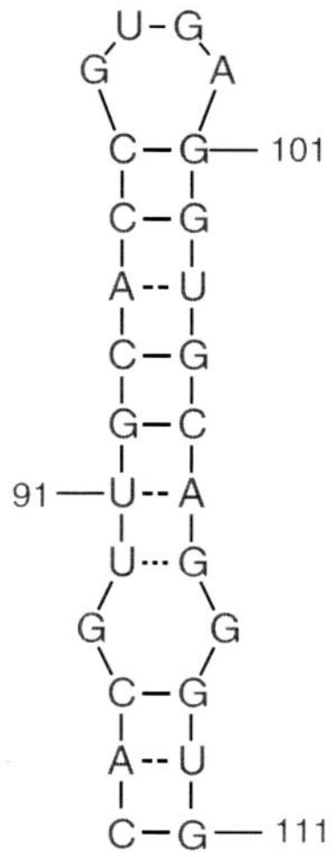

```
        U - G
      G       A
       C - G —— 101
       |   |
       C - G
       |   |
       A - U
       |   |
       C - G
       |   |
       G - C
       |   |
91 —— U - - A
       |   |
       U ··· G
      G       G
       C - G
       |   |
       A - - U
       |   |
       C - G —— 111
```

Fig. 1.1. A portion of the phylogenetically determined secondary structure for the group IIB intron, I5 from the Eubacterium, *Azotobacter vinelandii*. The UNAFold software with version 3.0 RNA energies requires that the G.G mismatch be treated as a 2×2 interior loop by deleting the $U^{90} \cdot G^{107}$ base pair.

Table 1.2
Free energy data on folding the group II intron, *A.v.*I5

ΔG_{ens}	ΔG_{min}	ΔG_{phylo}	Sample ΔG_{min}	Sample mean	Sample std
−273.2	−244.9	−235.2	−235.9	−223.6	5.6

A sample of 100 foldings was generated using stochastic traceback. From left to right, the Gibbs free energy, mfe, the free energy of the correct structure, the sample mfe, and the sample mean and standard deviation. The units are kcal/mol.

interior loop. The "offending" base pair is shown in **Fig. 1.1.** The ct-energy program does not currently flag such peculiarities and users should be aware that slight adjustments may be required when dealing with secondary structures derived in other ways. The base pair was removed from the ct file using a text editor, after which the free energy of the reference structure was evaluated with ease.

Table 1.2 gives information on the various free energies associated with these predictions. Note that the correct folding is almost 10 kcal/mol less stable than the mfe folding. Its free energy is, however, almost identical to that of the most stable folding in a sample of 100, just over two standard deviations above the mean. The message is that one should not expect the free energy of the correct folding to be close to the mfe. However, it should be within the range of energies from a stochastic sample, and in this case the result is as good as one could hope for.

3.4.4 Running Hybrid-ss-min in mfold Mode

To create an mfold-like sample of foldings the UNAFold.pl script was run in default mode:

```
UNAFold.pl A_v_I5.seq
```

It produced 18 different foldings together with structure plots and EDPs. As with mfold, the free energies are re-evaluated when version 3.0 RNA energies are used. Free energies of multi-branch loops are (re)assigned energies using a more realistic function that grows as the logarithm of the number of single-stranded bases. The same set of foldings, minus plots and energy re-evaluation, would be predicted using the more primitive command:

```
hybrid-ss-min --suffix DAT --mfold A_v_I5.seq
```

The −mfold flag indicates mfold mode. It accepts three parameters; P, W, and MAX. These are the energy increment for suboptimal foldings expressed as a Percent of the mfe, the Window parameter and the MAXimum number of foldings that may be computed, respectively. The default values are 5%, 3%, and 100%, respectively.

Note that the default is for 2.3 free energies and that the --suffix DAT flag is required to ensure the use of version 3.0 energies. The UNAFold.pl script behaves like mfold, in which the default is version 3.0 free energies.

3.4.5. Comparisons and Plots

The ct_compare program from the mfold package and the ct_boxplot program from the mfold_util package were used to compare results. The ct_compare program reads a reference structure, in this case the correct structure for the intron, and compares it with a number of other foldings on the same sequence from a second ct file. ct_compare was run on both the stochastic sample and mfold-like samples. **Table 1.3** shows the results. There are 166 base pairs in the correct structure. False-positive refers to predicted base pairs that are not correct, and false-negative refers to base pairs that are correct but not predicted.

Thus, none of the foldings in the stochastic sample do well, whereas the fourth of 18 foldings in an mfold-like sample scores very well. In fact, the free energy re-evaluation scores this folding as the most stable in the sample. This latter fact is a matter of luck. However, it is reasonable to expect that folding close to the mfe can be a fairly good approximation to the correct fold.

Secondary structure plots of several structures were generated from ct files using the interactive sir_graph program from the mfold_util package. They may be "annotated" in a variety of ways *(33–35)*. We used the command:

```
plot2ann.pl A_v_I5.37.plot A_v_I5phylo.ct
```

to create a stream of output (standard output) that was directed into a file named A_v_I5phylo.ann. This file was used to annotate the correct folding of the intron, as shown in **Fig. 1.2**.

It is not surprising that the correct folding has low probability base pairs. However, it might surprise some that mfe and close to mfe foldings may also contain low probability base pairs. The

Table 1.3
Numbers of correctly and incorrectly predicted base pairs

Structure	PF min	PF best	ΔG_{min}	mfold best
Correct	101	113	101	145
False-positive	68	51	67	23
False-negative	65	53	65	21

The reference folding contains 166 base pairs. "PF" denotes partition function, so that "PF min" is the mfe structure from the stochastic sample. "PF best" is the structure from the stochastic sample with the largest number of correctly predicted base pairs. ΔG_{min} is the overall mfe folding, always the first in an "mfold-like" sample, and mfold best refers to the structure in the "mfold-like" sample that scores best.

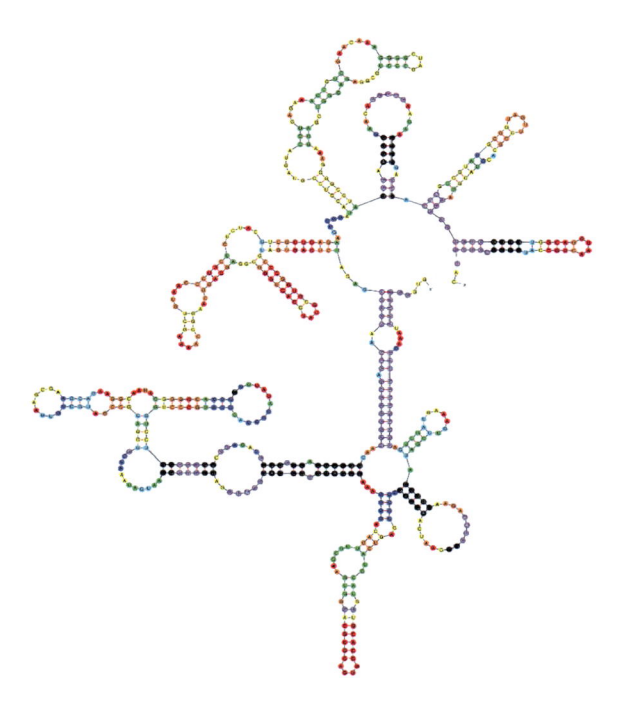

Fig. 1.2. Probability annotation of the secondary structure of the group II intron from *Azotobacter vinelandii*. Probabilities of correct base pairing correspond to colors as follows: black: <0.01, magenta: 0.01–0.10, blue: 0.10–0.35, cyan: 0.35–0.65, green: 0.65–0.90, yellow: 0.90–0.99, orange: 0.99–0.999, and red: > 0.999. Twenty-eight (17%) of the base pairs have probabilities <1%, and 31 more (19%) have probabilities <10%.

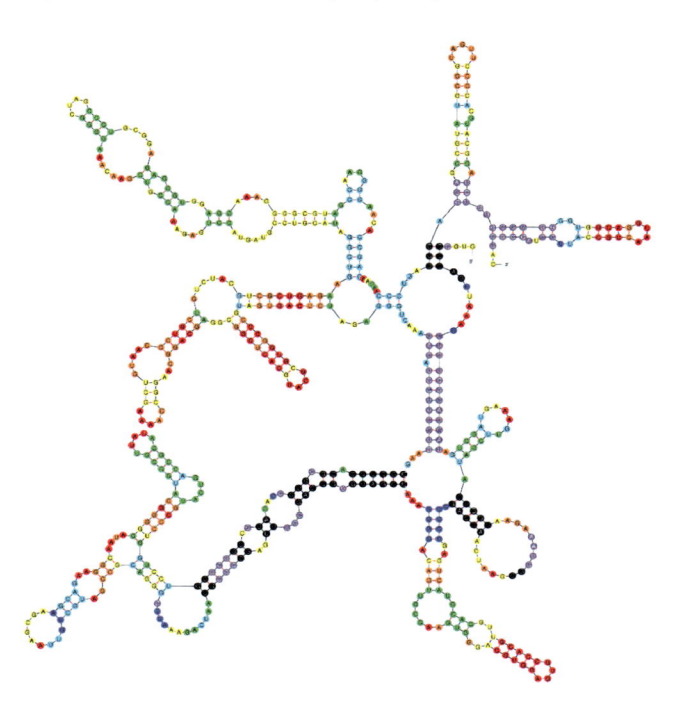

Fig. 1.3. Probability annotation of a computed folding of the group II intron from *Azotobacter vinelandii*. The color scheme is the same as in Fig. 1.2. Its free energy is only 1.9 kcal/mol above the mfe (non-revised), and yet it contains 25 base pairs (15%) with probabilities <1%. Another 24 base pairs (14%) have probabilities <10%.

structure closest to the correct folding in the mfold-like folding was the fourth folding, placed in A_v_I5_4.ct by UNAfold.pl. Proceeding as with the correct folding, an annotation file for this structure was created, resulting in **Fig. 1.3**.

We generated probability annotation files using a plot file computed by hybrid-ss. Plot files may also be computed from stochastic samples using ct-prob, as described in **Section 2.5**. However, there is no reason to restrict the use of ct-prob to stochastic samples. By concatenating the ct files for the correct folding and for the "best mfold-like" folding (A_v_I5_4.ct) and storing the result in temp.ct, the command:

```
ct-prob temp.ct
```

produces an output stream that we directed to the file phylo-mfold_best.plot. It contains a list of all the base pairs that appear in either folding. The probabilities are 1 (occur in both structures) or 0.5 (occur in just one structure). Running plot2ann.pl using this "plot" file allows us to annotate one structure with respect to another. **Figure 1.4** presents the correct intron folding and shows clearly which base pairs occur in the A_v_I5_ct structure.

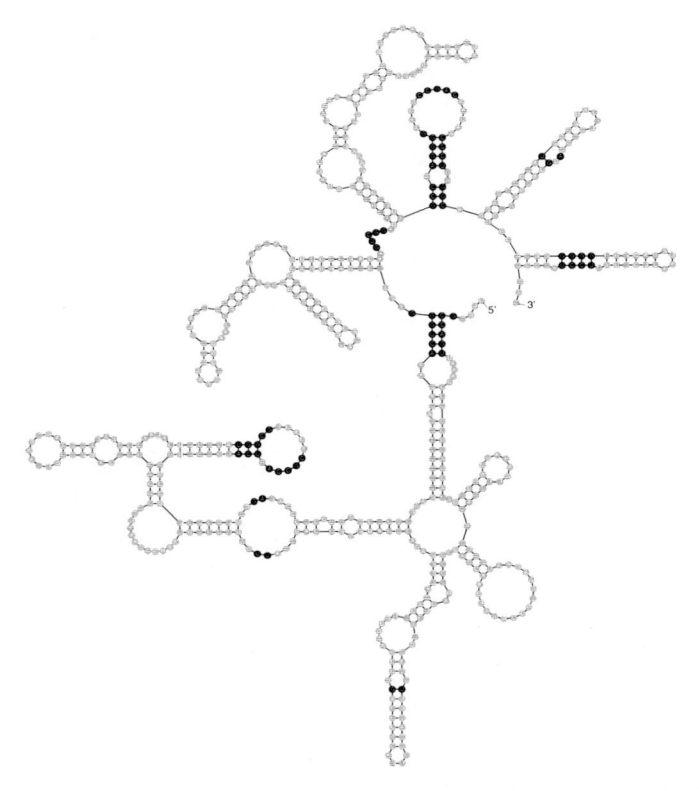

Fig. 1.4. The secondary structure of the group II intron from *Azotobacter vinelandii*. Paired gray encircled bases occur in the predicted folding, A_v_I4_4.ct, shown in Fig. 1.3. Single-stranded gray encircled bases are also single-stranded in the predicted folding. Paired black encircled bases do not occur in the predicted folding, and black encircled single-stranded bases are paired in the predicted folding.

3.5. Hybridizing Two Sequences

Both hybrid and hybrid-min predict intermolecular base pairs that link one strand with the other. They do not predict intramolecular base pairs. Software to do this will be released in a later version of UNAFold. We currently recommend that hybridization simulations be limited to "short" sequences of lengths up to at most 100. It is also appropriate to allow a small sequence (probe) to hybridize to a much larger one (target) to assess the likelihood of binding at various sites.

No assumption is made about two sequences that are hybridized. They might be perfectly complementary, there may be some mismatches and or small bulges, or they may be totally unrelated.

3.5.1. Self-Hybridization of a Short Oligo

We selected the DNA oligomer, 5′-CGCAATTCGCG-3′, and stored in A.seq, a generic name for the first of up to two sequences. The command:

```
hybrid-min --NA DNA --tmin 20 --tmax 60 A.seq
    A.seq
```

computed single, mfe self-hybridizations at temperatures from 20°C to 60°C, inclusive. Only two distinct hybridizations were computed over the entire temperature range. They are displayed in **Fig. 1.5**.

3.5.2. Hybridization PDPs (Probability Dot Plots)

The apparent sudden switch from one structure to another is artificial. In reality, many conformations would exist simultaneously at each temperature. This phenomenon is easily modeled using the hybrid program. Stochastic samples at 20°C contain the two main hybridizations depicted in **Fig. 1.5** and many more, most of which are slight variants. The same is true for a stochastic sample at 40°C, for example. What changes are the frequencies of the various structures. The ct-uniq program is a simple "standard input" "standard output" program that selects unique structures

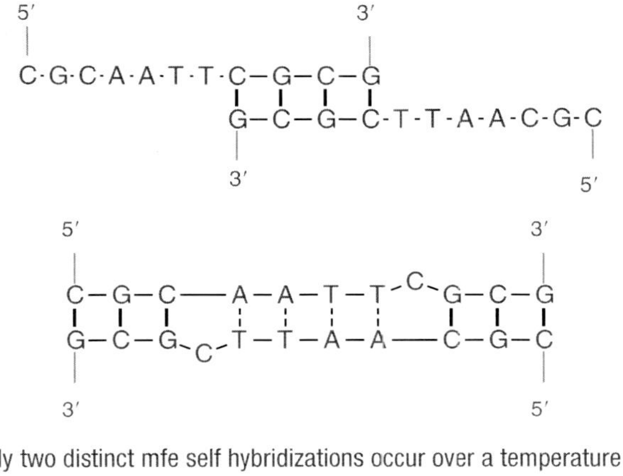

Fig. 1.5. Only two distinct mfe self hybridizations occur over a temperature range from 20°C to 60°C, inclusive. Standard DNA conditions were used (Na$^+$ = 1M, Mg^{2+} = 0). The bottom hybridization has the lowest mfe from 20°C to 28°C, inclusive. For temperatures 29°C to 60°C, the top hybridization has the lowest mfe. At 28°C, the free energies of the bottom and top hybridization are −6.43 and −6.32 kcal/mol, respectively. At 29°C, they are −6.18 and −6.22 kcal/mol, respectively.

from a multiple structure ct file. A stochastic sample of 100 structures computed at 30°C was reduced to 27 in this way, and only 24 in terms of distinct sets of base pairs, since structures with the same base pairs may differ by the addition or deletion of single-base stacking. Many of the structures are merely sub-structures of others. An effective way to view ensembles of hybridizations is to run the hybrid-plot program:

```
hybrid-plot A-A
```

and to view base pair probabilities as the temperature varies. It is easy to create PDP images at various temperatures by running the non-interactive version of the same program, hybrid-plot-ng. The command:

```
hybrid-plot-ng --temperature $T --colors linear\*
--grid 20 --cutoff 0.001 A-A
```

was run a number of times, with "$T" replaced by temperatures 15, 20, ... Because the sequence is so small, the grid display was turned off by setting the spacing to a number larger than the sequence length (--grid 20). The --colors linear flag chooses colors for the base pair dots that vary linearly with the probability. Using an "rgb" scale, the colors vary linearly with probability from $(1,0,0)$ to $(1,1,0)$ (red to yellow), from $(1,1,0)$ to $(0,1,0)$ (yellow to green), $(0,1,0)$ to $(0,1,1)$ (green to cyan), $(0,1,1)$ to $(0,0,1)$ (cyan to blue), and finally from $(0,0,1)$ to $(1,0,1)$ (blue to magenta). The --cutoff 0.001 flag is the cutoff. Base pairs with probabilities below the cutoff are not plotted. By default, UNA-Fold programs do not output probabilities $<10^{-6}$ into plot or ext files. Six of these PDP files were combined into **Fig. 1.6**.

3.6. Melting Profiles

The original motivation for creating the UNAFold software was to simulate entire melting profiles. In the most general case, two DNA or RNA molecules are interacting. In general, there may be competition between dimer formation and folding of the individual molecules. Self-hybridization may occur. Thus, competition among a total of five molecular species is computed. There is the heterodimer, two homodimers, and two folded species. When only one molecule is present, only dimer formation (a single homodimer) and folding are considered. Partition function computations are performed for each molecular species over a user-specified range of temperatures. Finally, results for individual species are combined to compute overall ensemble quantities: free energy, enthalpy, and entropy change (ΔG, ΔH, and ΔS, respectively), UV absorbance at 260 nm, heat capacity (C_p), and equilibrium molar concentrations of each of the species.

---

* Long command lines may be broken into multiple lines if a backslash, \, is the last character of all but the final line.

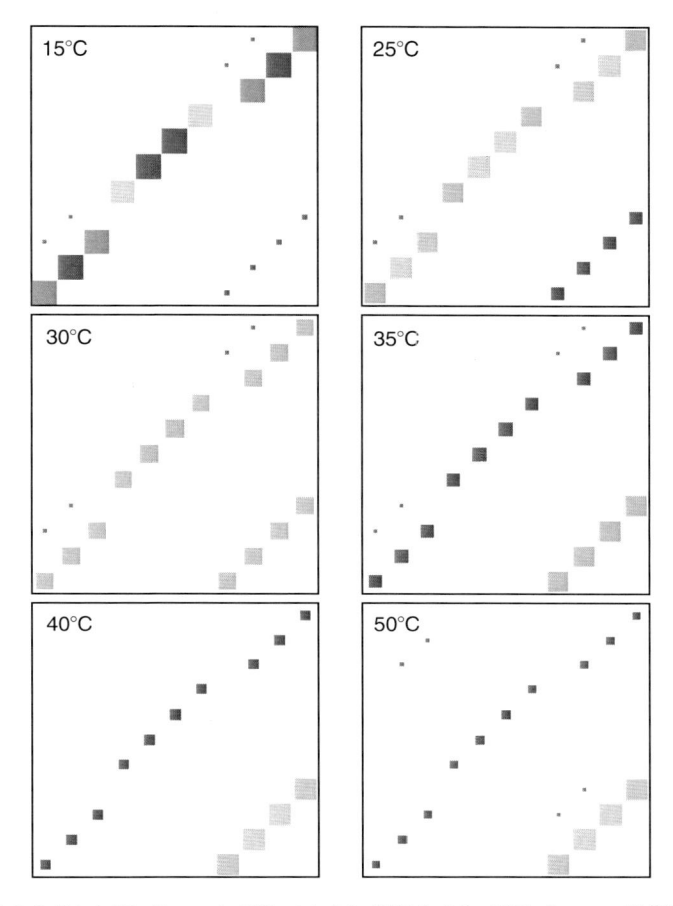

Fig. 1.6. Self-hybridization probability dot plots (PDPs) of the DNA oligomer, 5′-CGCAAT-TCGCG-3′ computed for [Na⁺] = 1M, [Mg²⁺] = 0 and for a variety of temperatures as shown in the plots. At 15°C, the large red and orange dots along the diagonal comprise the base pairs of the bottom structure from Fig. 1.5, the "low temperature" hybridization. These base pairs have probabilities that vary from 96.5% to 85%. The four small consecutive magenta dots *(lower right)* comprise the upper high temperature hybridization from Fig. 1.5. These base pairs all have probabilities of 3%. At 25°C, the base pair probabilities of the low temperature hybridization have fallen (63–73%), whereas those of the high temperature hybridization have risen to about 24%. The base pair probabilities of the two main hybridizations are roughly equal at 30°C, and the trend continues. At 50°C, the base pair probabilities of the low temperature hybridization have fallen to the 7–9% range, whereas the four base pairs of the high temperature hybridization have probabilities in the 68–73% range.

During the development of the UNAFold software, our analyses of measured melting profiles indicated that an additional, internal energy term is required for each molecule in order to avoid a systematic underestimation of overall enthalpy change. An ad hoc rule was adopted based on observations. For each strand, the enthalpy change from a perfect hypothetical hybridization to its reverse complement to total dissociation is computed using the published enthalpies. By default, 5% of this enthalpy change is assigned to that strand as an internal energy term independent of folding or hybridization. It is useful to attribute this extra energy to base stacking in unfolded, single strands. Assuming a cooperative transition (melting) from "stacked single-stranded" to "unstacked random coil" that takes place at about 50°C, an entropy change can be computed. The default behavior of the UNAFold Perl scripts that compute ensemble quantities and melting profiles is to include this additional energy term. It is also used in the DINAMelt web server *(36)*.

Melting profiles were computed for the short oligomer, 5'-CGCAATTCGCG-3' considered above. The command:

```
hybrid2.pl --NA DNA --tmin 10 --tmax 90 --sodium\
0.05 --A0 10e-6 A.seq A.seq
```

was used. The temperature range is from 10°C (--tmin 0) to 90°C (--tmax 90). The temperature increment is 1°C unless specified otherwise using the --tinc flag. --NA DNA defines the nucleic acid(s) as DNA. The default is RNA. --sodium 0.05 defines 50 mM Na$^+$ (or equivalent). The concentration of "A" (first sequence) is 10 μmol, as specified by --A0 10e-6. It is not necessary to specify another concentration in this case, but two sequences must be specified, even if they are the same. The molar concentration plot, shown in **Fig. 1.7**, reveals that no hybridization is taking place! The predicted melting profiles (not shown) are simulating the melting of a single-stranded, folded 11-mer. "What about the PDPs computed for A-A self hybridization?" The answer is that the probabilities in any of these plots are conditional. For hybridization, they are conditional on hybridization; for folding, they are conditional on

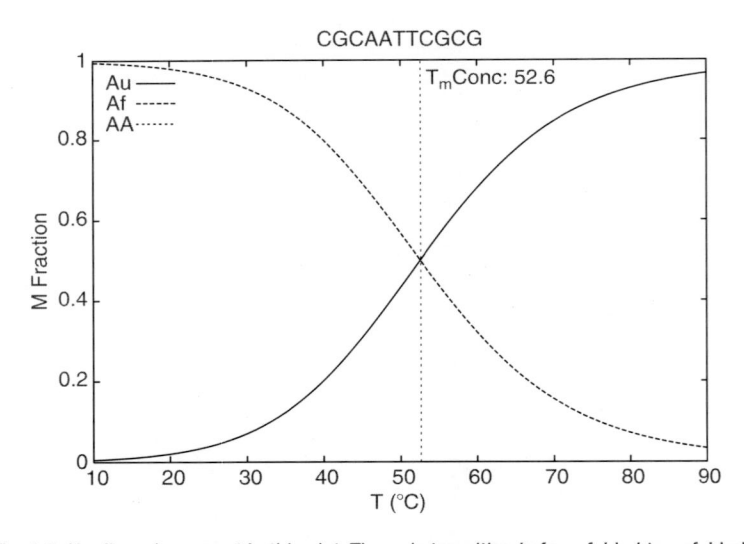

Fig. 1.7. No dimer is present in this plot. The only transition is from folded to unfolded.

folding. Thus, if a base pair has a 50% probability in an A-A PDP, but the equilibrium constant for dimer formation is 10^{-6}, the effective probability is 5×10–7. In fact, by increasing Na^+ to 1M and [A] to $500\,\mu M$, simulation predicts a mole fraction of almost 0.2 for the dimer at 10°C.

A naïve user might be tempted to suppress single-stranded folding in this simulation. Re-running the hybridization script with the folding of A excluded:

```
hybrid2.pl --NA DNA --tmin 10 --tmax 90 --sodium\
0.05 --A0 10e-6 --exclude A --reuse A.seq A.seq
```

yields results that predict a dimer mole fraction of just over 60% at 10°C. It is rapidly decreasing with heat capacity and UV absorbance melting temperatures of about 13°C and concentration melting temperature of 16.5°C. These results are nonsense. Even more dangerous would be the unthinking use of the simple melt.pl script that computes a mfe structure at a specified temperature and uses the simple formula given in Equation [1]. The command:

```
melt.pl --NA DNA --sodium 0.C5 --temperature $T\
--Ct 10e-6 A.seq A.seq
```

gives the same predicted T_m, 14.0°C, for $T = 20°C$, 50°C, and for the default value of 37°C. This, too, is nonsense. The added "danger" is that folding and homodimer formation are automatically excluded when this script is used. When hybrid2.pl is used, any excluded species must be specified by the user. In general, the melt.pl script is valid only when it may be safely assumed that the heterodimer (when two distinct sequences hybridize) or the homodimer (single sequence) is the only molecular species that forms, that it melts cooperatively, and that strand concentrations are equal when two different molecules hybridize.

3.6.1. Simulation vs. Measurements

When measured data are available, the hybrid2.pl script includes measured profiles with predicted ones. Heat capacity measurements using DSC (differential scanning calorimetry) require relatively high concentrations of nucleic acid. Total strand concentration of 100 to $200\,\mu M$ is usual. For UV absorbance measurements, total strand concentration is reduced by a factor of 10 (roughly). For this reason, if heat capacity measurements are available, then UV absorbance data will not be available, and vice versa. hybrid2.pl expects observed (measured) heat capacities to be in a .obs.Cp file. Likewise, observed absorbance data should be in a .obs.ext file. The format is simple. Each line (record) contains a temperature and a heat capacity (or absorbance) separated by spaces and/or a tab character. Additional columns are ignored. hybrid2.pl expects a .obs.Tm file containing the measured T_m and heat capacity (or absorbance) at that temperature.

Plots created by hybrid2.pl use the prefix formed from the two sequence file names, as explained above. The PostScript plot files have double suffixes. Below is a list of these suffixes and a description of the plots.

conc.ps: The concentrations of all non-excluded molecular species are plotted *versus* temperature. The concentrations are expressed as mole fractions. The values come from the .conc file, where the units are moles.

Cp.ps   A plot of ensemble heat capacity *versus* temperature. Values are from the .ens.Cp file and the plotted *Tm* is from the .ens.TmCp file.

Cp2.ps   Not for general use. The contributions from the different species are included in this additional heat capacity plot. Negative values indicate energy transfer from a species to others. The net sum must be positive.

ext.ps   A plot of ensemble absorbance of UV radiation at 260 nm vs. temperature. ("ext" stands for "extinction.") Values are from the .ens.ext file. The .ens. MaxExt contains the maximum possible absorbance assuming that no base pairs exist. The plotted T_m is from the .ens.TmExt2 file. This file contains T_m computed as the absorbance midpoint. It is the temperature at which absorbance is midway between the minimum value plotted and the theoretical maximum contained in the .ens.MaxExt file. It is the UNAFold default. The .ens.TmExt1 file contains an absorbance T_m computed as the inflection point in the absorbance curve. Multiple values might exist.

ext2.ps   Not for general use. The contributions from the different species are included in this additional absorbance plot. (The "ext2" in the suffix has nothing to do with the definition of absorbance. It is unrelated to the .ens.TmExt1 and .ens.TmExt2 files.

obs.ps   This plot contains both the computed heat capacity (left vertical axis) and the computed absorbance (right vertical axis). In addition, measured heat capacity or absorbance is plotted, together with the measured T_m.

Figure 1.8 displays the extra plot that is generated for the melting of two complementary DNA 25-mers when a measured heat capacity file is also present. The command that generated this plot is:

```
hybrid2.pl --NA DNA --tmin 25 --tinc 0.2 --tmax\
100 --A0 9e-05 --B0 9e-05 --sodium 0.069
R948.seq R949.seq
```

Total strand concentration is 180 µM; with both strands having the same molar concentration. [Na⁺] = 69 mM. The --tinc 0.2 means that computations are performed for all temperatures between 25°C and 100°C in increments of 0.2°C. It quintuples the

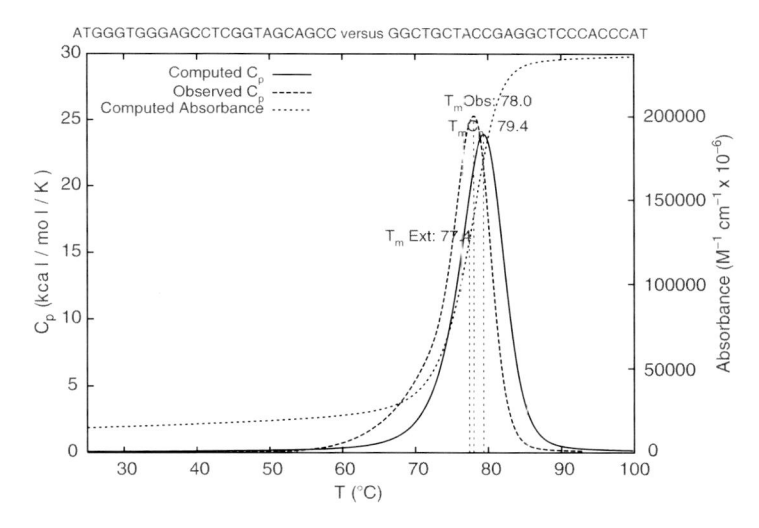

ATGGGTGGGAGCCTCGGTAGCAGCC versus GGCTGCTACCGAGGCTCCCACCCAT

Fig. 1.8. A plot of computed heat capacity, measured heat capacity and computed absorbance vs. temperature for two complementary DNA 25-mers with $[Na^+] = 69$ mM. Both strand concentrations are 90 μM. The computed $T_m (C_p)$ is 1.4°C greater than the measured one. The absorbance plot is "extra." There is no reason to expect T_m (Ext) to match T_m Obs, although the two are very close in this case.

computation time with respect to the default, but it creates a superior computed C_p curve, which is generated by taking the second derivative of the Gibbs free energy profile with respect to temperature. The default C_p plots are "fatter" and have lower peak values. This is not a serious problem, since the computed values for ΔH, ΔS, and T_m are robust with respect to the temperature increment. In these computations, the homodimers may be safely excluded in reruns, since initial computations reveal that homodimer concentrations are negligible over the entire temperature range.

Measured absorbances are available for these complementary 25-mers, but at $[Na^+] = 1$ M and with a total strand concentration of 2.25 μM, which is down by a factor of 80 from that used in the heat capacity measurements.

3.6.2. A Molecular Beacon Example

Molecular beacons *(37)* take advantage of the competition between folding and hybridization to distinguish between a gene and a mutant containing a single nucleotide polymorphism (SNP). A molecular beacon is a DNA oligomer with a fluorophore attached at the 5′ end and with "DABCYL" (a quencher) attached at the 3′ end. The first 5–7 bases are complementary to the last 5–7, so that a stem-loop structure can form that brings the fluorophore close to the quencher. In this conformation, any fluorescence generated by the beacon is immediately quenched. When the hairpin does not form, quenching does not occur and a signal may be detected. The middle portion of a beacon varies from 10 to 40 nt, and is complementary to a region of interest, called the Target. In the simulation we have chosen, the sequence of the molecular beacon, stored in Beacon.seq is:

5′-gcgagcTAGGAAACACCAAAGATGATATTTgctcgc-3′

The complementary bases that form a stem-loop structure are underlined and in lower case. The center 24 bases are complementary to a stretch of 24 nucleotides within a (much) larger DNA molecule, such as a gene. In the simulation, only the complementary region is considered. This defines the Target, stored as Target.seq. It is:

5′-AAATATCATCTtTGGTGTTTCCTA-3′

The reason for selecting this target is that a mutant gene contains a SNP within it. This defines a mutant Target, or TargetSNP, stored as TargetSNP.seq. It is:

5′-AAATATCATCTcTGGTGTTTCCTA-3′

The SNP, a T to C transition, is easy to spot, since both are lowercase. The whole point of molecular beacons is that at some temperature, ionic conditions, and strand concentrations, all beacons should hybridize to the target, whereas hybridization to the SNP should be significantly lower, and the single-stranded beacons should be folded. Preliminary runs indicated that folding and self-hybridization of the targets could be excluded. We did not bother to exclude self-hybridization of beacons. Beacon concentration was set at 50 nM and the target concentrations were twice that; 100 nM. This two-fold excess of target ensures maximum hybridization of the beacons.

The commands:

```
hybrid2.pl --tmin 10 --tmax 90 --NA DNA --sodium\
0.05 magnesium 0.0035 --exclude B --exclude BB --\
A0 5e-8 --B0 1e-7 Beacon.seq Target.seq

hybrid2.pl --tmin 10 --tmax 90 --NA DNA --sodium\
0.05 --magnesium 0.0035 --exclude B --exclude BB\
--A0 5e-8 --B0 1e-7 Beacon.seq TargetSNP.seq
```

produced the usual large number of files. Only the species concentration plots, and especially the conc files, are of interest. **Figure 1.9** shows the relevant concentrations for the two simulations. It contains parts of the two concentration plot files produced automatically by hybrid2.pl. The melting temperature of the dimer falls by 5.4°C. It is desirable that single-stranded beacons be folded. Unfolded beacons give a false signal. The optimal experimental temperature is predicted to be 58°C. At this temperature, the Beacon-Target concentration is 83% of its maximum. Single-stranded beacons are mostly folded, but the unfolded ones contribute a small false signal, bringing the total to 86% of maximum. For the Beacon-TargetSNP hybridization, the signal

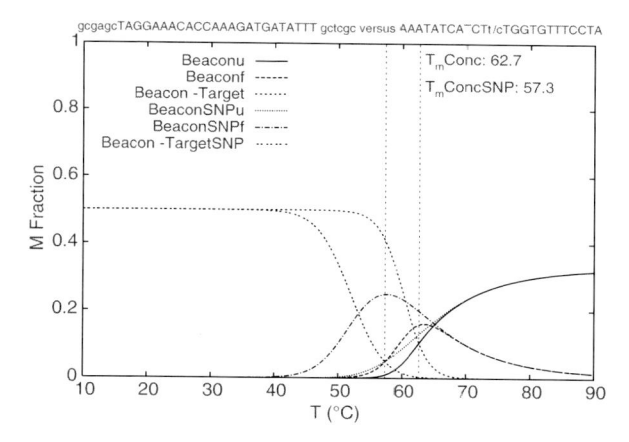

Fig. 1.9. Predicted Beacon-Target and Beacon-TargetSNP concentration plots. The suffix "u" denotes "unfolded" and "f" denotes "folded". BeaconSNP refers to Beacon concentrations in the presence of TargetSNP. As the Beacon-Target dimer melts, single-stranded Beacons are mostly folded vs. unfolded. The Beacon-TargetSNP dimer melts at a lower temperature. As it does, the released Beacons are overwhelmingly folded. However, by 58°C, the false signal from unfolded beacons, although small, has just begun to exceed the rapidly decreasing signal from the dimer.

is down to just 11% of its maximum, but enough of the released beacons unfold to bring the signal up to 28% of maximum. The bottom line is that at 58°C, the fluorescence in the SNP case is one-third of what it would be otherwise.

4. Notes

The examples presented for how some of the UNAFold programs and scripts might be used are just a small sample of possible applications and ways to combine various programs. Users are encouraged to write their own scripts or modify existing ones to better suit their needs. We believe that users without programming skills will find the package easy to use. As with mfold, feedback from users is expected to influence what new features might be added.

Some directions for further development are already clear.

- The current hybrid2.pl script is too inflexible when plots are created. Initial strand concentrations that differ by orders of magnitude can make concentration plots useless unless logarithmic scaling is used for plotting mole fractions.

- The current UNAFold.pl script behaves like the mfold script in the mfold package. It acts on a single sequence, and predicts a sample of foldings together with EDP (energy dot plot). When auxiliary software is installed (mfold_util), text files of structures and EDPs may be used to create plots. The

UNAFold.pl script will be extended to predict (stochastic) samples of structures together with PDPs.

- The internal energy option for single-stranded, unfolded sequences adds enthalpy using a simple ad hoc rule derived from a group of measured DSC profiles. This model needs to be studied and improved so that the added enthalpies can be assigned more objectively.

- The intra-molecular hybridization software requires testing to determine its usefulness when larger oligos are hybridized.

- We now realize that for individual molecular species, enthalpy and heat capacity can be computed directly from stochastic samples. How to extend these species predictions to ensemble predictions remains a bit unclear, but we are confident that it can be done in a numerically stable way. Such calculations would make it unnecessary to compute derivatives numerically.

We believe that continuing feedback from users will have a positive effect on further development.

Acknowledgments

This work was supported, in part, by grants GM54250 and GM068564 from the National Institutes of Health and by a graduate fellowship to N.R.M. from RPI.

References

1. Bellman, R. E. (1957) *Dynamic Programming*. Princeton University Press, Princeton, NJ.

2. Waterman, M. S., Smith, T. F. (1978) RNA secondary structure: a complete mathematical analysis. *Math Biosci* 42, 257–266.

3. Waterman, M. S. (1978) Secondary structure of single-stranded nucleic acids, in (Rota, G.-C., ed.), *Studies in Foundations and Combinatorics*, Academic Press, New York.

4. Nussinov, R. (1980) Some rules for ordering nucleotides in DNA. *Nucleic Acids Res* 8, 4545–4562.

5. Zuker, M., Stiegler, P. (1981) Optimal computer folding of large RNA sequences using thermodynamics and auxiliary information. *Nucleic Acids Res* 9, 133–148.

6. Sankoff, D., Kruskal, J. B., eds. (1983) *Time Warps, String Edits, and Macromolecules: The Theory and Practice of Sequence Comparison*. Addison-Wesley, Reading, MA.

7. Zuker, M. (1989) The use of dynamic programming algorithms in RNA secondary structure prediction, in (Waterman, M. S., ed.), *Mathematical Methods for DNA Sequences*. CRC Press, Boca Raton, FL.

8. Zuker, M. (1994) Prediction of RNA secondary structure by energy minimization, in (Griffin, A. M., Griffin, H. G., eds.), *Computer Analysis of Sequence Data*,. Humana Press, Totowa, NJ.

9. Zuker, M., Mathews, D. H., Turner, D. H. (1999) Algorithms and thermodynamics for RNA secondary structure prediction: a practical guide, in (Barciszewski J., Clark, B. F. C., eds.), *RNA Biochemistry and Biotechnology*. Kluwer, Dordrecht.

10. Zuker, M. (2003) Mfold web server for nucleic acid folding and hybridization prediction. *Nucleic Acids Res* 31, 3406–3415.

11. Hofacker, I. L., Fontana, W., Stadler, P. F., et al. (1994) Fast folding and compari-

son of RNA secondary structures. *Monatsh Chem* 125, 167–188.

12. Wuchty, S., Fontana, W., Hofacker, I. L., et al. (1999) Complete suboptimal folding of RNA and the stability of secondary structures. *Biopolymers* 49, 145–165.

13. Hofacker, I. L. (2003) Vienna RNA secondary structure server. *Nucleic Acids Res* 31, 3429–3431.

14. McCaskill, J. S. (1990) The equilibrium partition function and base pair binding probabilities for RNA secondary structure. *Biopolymers* 29, 1105–1119.

15. Ding, Y., Lawrence, C. E. (2001) Statistical prediction of single-stranded regions in RNA secondary structure and application to predicting effective antisense target sites and beyond. *Nucleic Acids Res* 29, 1034–1046.

16. Ding, Y., Lawrence, C. E. (2003) A statistical sampling algorithm for RNA secondary structure prediction. *Nucleic Acids Res* 31, 7280–7301.

17. Ding, Y., Lawrence, C. E. (2004) Sfold web server for statistical folding and rational design of nucleic acids. *Nucleic Acids Res* 32, W135–W141.

18. Allawi, H. A., SantaLucia, J. Jr. (1997) Thermodynamics and NMR of internal G·T mismatches in DNA. *Biochemistry* 36, 10581–10594.

19. SantaLucia, J. Jr. (1998) A unified view of polymer, dumbell, and oligonucleotide DNA nearest-neighbor thermodynamics. *Proc Natl Acad Sci U S A* 95, 1460–1465.

20. SantaLucia, J. Jr., Hicks, D. (2004) The thermodynamics of DNA structural motifs. *Annu Rev Biophys Biom* 33, 415–440.

21. Zhang, Y., Hammer, D. A., Graves, D. J. (2005) Competitive hybridization kinetics reveals unexpected behavior patterns. *Biophys J* 89, 2950–2959.

22. Kilgard, M. J. (1996) *OpenGL Programming for the X Window System*. Addison-Wesley, Boston.

23. Puglisi, J. D., Tinoco, I. Jr. (1989) Absorbance melting curves of RNA, in (Dahlberg, J. E., Abelson, J. N. eds.), *RNA Processing Part A: General Methods*. Academic Press, New York.

24. Waugh, A., Gendron, P., Altman, R., et al. (2002) RNAML: a standard syntax for exchanging RNA information. *RNA* 8, 707–717.

25. Jelesarov, I., Bosshard, H. R. (1999) Isothermal titration calorimetry and differential scanning calorimetry as complementary tools to investigate the energetics of biomolecular recognition. *J Mol Recog* 12, 3–18.

26. Walter, A. E., Turner, D. H., Kim, J., et al. (1994) Coaxial stacking of helixes enhances binding of oligoribonucleotides and improves predictions of RNA. *Proc Natl Acad Sci U S A* 91, 9218–9222.

27. Mathews, D. H., Sabina, J., Zuker, M., et al. (1999) Expanded sequence dependence of thermodynamic parameters improves prediction of RNA secondary structure. *J Mol Biol* 288:911–940.

28. De Rijk, P., De Wachter, R. (1997) RnaViz2, a program for the visualisation of RNA secondary structure. *Nucleic Acids Res* 25, 4679–4684.

29. De Rijk, P., De Wachter, R. (2003) RnaViz2: an improved representation of RNA secondary structure. *Bioinformatics* 19, 299–300.

30. Dai, L. Zimmerly, S. (2002) Compilation and analysis of group II intron insertions in bacterial genomes: evidence for retroelement behavior. *Nucleic Acids Res* 30, 1091–1102.

31. Toor, N., Hausner, G., Zimmerly, S. (2001) Coevolution of group II intron RNA structures with their intron-encoded reverse transcriptases. *RNA* 7, 1142–1152.

32. Michel, F., Umesono, K., Ozeki, H. (1989) Comparative and functional anatomy of group II catalytic introns: a review. *Gene* 82, 5–30.

33. Zuker, M., Jacobson, A. B. (1998) Using reliability information to annotate RNA secondary structures. *RNA* 4, 669–679.

34. Jacobson, A. B., Arora, R., Zuker, M., et al. (1998) Structural plasticity in RNA and its role in the regulation of protein translation in coliphage qβ. *J Mol Biol* 275, 589–600.

35. Mathews, D. H. (2004) Using an RNA secondary structure partition function to determine confidence in base pairs predicted by free energy minimization. *RNA* 10, 1174–1177.

36. Markham, N. R., Zuker, M. (2005) DINAMelt web server for nucleic acid melting prediction. *Nucleic Acids Res* 33, W577–W581.

37. Tyagi, S., Kramer, F. R. (1996) Molecular beacons: probes that fluoresce upon hybridization. *Nat Biotechnol* 4, 303–308.

Chapter 2

Protein Structure Prediction

Bissan Al-Lazikani, Emma E. Hill, and Veronica Morea

Abstract

Protein structure prediction has matured over the past few years to the point that even fully automated methods can provide reasonably accurate three-dimensional models of protein structures. However, until now it has not been possible to develop programs able to perform as well as human experts, who are still capable of systematically producing better models than automated servers. Although the precise details of protein structure prediction procedures are different for virtually every protein, this chapter describes a generic procedure to obtain a three-dimensional protein model starting from the amino acid sequence. This procedure takes advantage both of programs and servers that have been shown to perform best in blind tests and of the current knowledge about evolutionary relationships between proteins, gained from detailed analyses of protein sequence, structure, and functional data.

Key words: Protein structure prediction, homology modeling, fold recognition, fragment assembly, metaservers.

1. Introduction

In spite of many years of intense research, unravelling the algorithm by which Nature folds each amino acid (a.a.) sequence into a unique protein three-dimensional (3D) structure remains one of the great unsolved problems in molecular biology. However, analyses of the wealth of information contained in protein sequence and structural databases (DBs) have revealed the existence of a number of fundamental rules and relationships among protein sequence, structure, and function, based on which many of both the current theories about molecular evolution and protein structure prediction methods have been developed.

The first important question to ask when dealing with protein structure prediction concerns the purpose for which the

Jonathan M. Keith (ed.), *Bioinformatics, Volume II: Structure, Function and Applications, vol. 453*
© 2008 Humana Press, a part of Springer Science + Business Media, Totowa, NJ
Book doi: 10.1007/978-1-60327-429-6 Springerprotocols.com

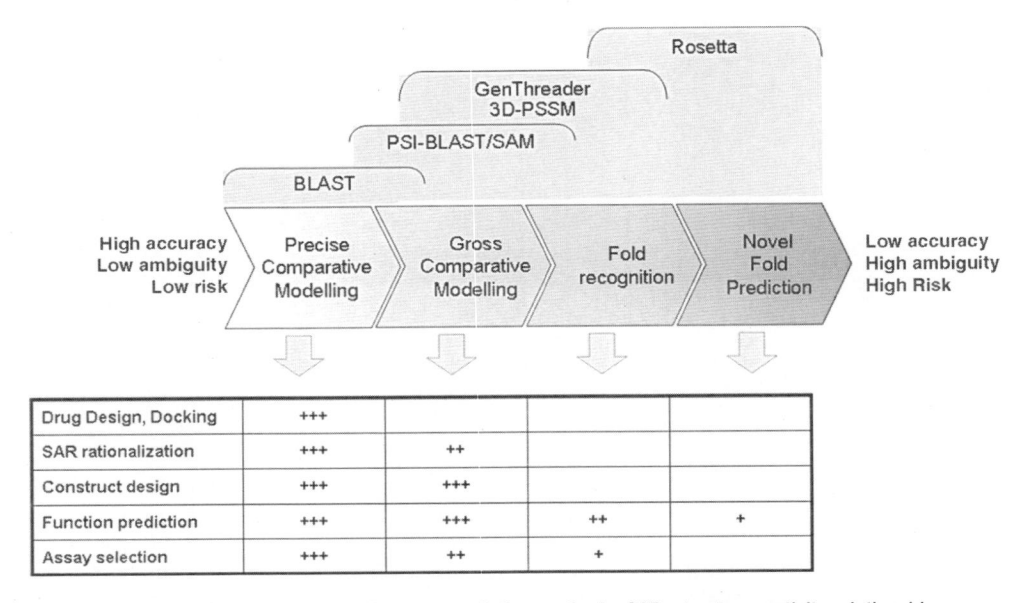

Fig. 2.1. Biological applications of protein structure prediction methods. SAR: structure–activity relationships.

model is built. This is of fundamental importance since the accuracy required of the model, that is, its similarity to the real protein structure, can be substantially different for different biological applications (**Fig. 2.1**). At one end of the spectrum, a very accurate prediction of the functional site in terms of both main- and side-chain conformation is indispensable for drug design purposes, and a correct positioning of the side-chains involved in intermolecular interactions is required for protein–protein interaction (docking) studies. At the other extreme, an approximate description of the protein topology at the level of general arrangement of secondary structure elements (SSE) or domains, or even an idea of which regions are likely to be globular, unfolded, or aggregation prone, can be valuable to those who want to cut insoluble proteins into smaller and soluble portions, which are likely to be more easily expressed and studied experimentally. In general, correct models of the overall protein fold, even with unrefined details, can be useful to rationalize experimental data at a structural level and guide the design of new experiments aimed at improving our understanding of protein function. In the era of structural genomics and fast solving of protein structures, looking for structural biologists interested in experimentally determining the structure of your protein(s) is also an option.

In choosing the procedure to follow for model building, other factors to consider are the time available and number of models to make. Production of 3D models on a large and even genomic scale is achievable with the use of automated or partially automated methods. A number of automated methods have been developed that can rapidly produce 3D models starting from the amino acid sequence of a target protein (*see* **Section 3.5.1**). However, the quality of these models can vary to a large extent. A detailed understanding of the underlying steps of the modeling procedure is required to evaluate, and often improve, the accuracy of automatically produced models.

The most reliable source of information about the accuracy of protein structure prediction methods is provided by the evaluation of their performance in blind tests. In such evaluations, 3D models of target proteins are compared with the experimentally determined structures of the same proteins using visual assessments performed by human experts and/or numerical evaluators of structural similarity (**Note 1**). Two main types of evaluations are performed on a regular basis: fully automated evaluations (devoted to fully automated methods), and human-based evaluations (examining predictions refined by human experts as well as those provided by fully automated methods). The human-based evaluation, named Critical Assessment of Structure Predictions (*CASP*), is performed every 2 years since its debut in 1994 and contributes enormously to the improvement of protein structure prediction methods as well as to the diffusion of information about their performance. (The URLs of all the Web sites, programs, servers, and databases indicated in italic in the text are reported in **Table 2.1**, along with relevant references, when available.) A full description of the experiment along with reports of the results of each of the six previous *CASP* experiments are available *(1–6)*. A seventh edition took place in 2006 and its results, published in 2007, provides up-to-date information on the most recent advances in the field of protein structure prediction (preliminary evaluation results are available from the *CASP7* Web site). In parallel with the last four *CASP* editions, the Critical Assessment of Fully Automated Structure Predictions (*CAFASP*) experiments have also been run in which the ability of automated servers to predict *CASP* targets was evaluated by other servers, without human intervention *(7–9)*. In both *CASP* and *CAFASP* the predictions are guaranteed to be "blind" by the fact that they are submitted to the evaluation before the experimental structures are released. The performance of automated servers on proteins recently released from the *PDB (10)* is also evaluated on a continuous basis by servers such as *Livebench (11)* and *EVA (12)*. Every week these servers submit the sequences of proteins newly released from the *PDB* to the prediction servers participating in these experiments, collect their results, and evaluate them using automated programs. To take part in *Livebench* and *EVA* the prediction servers must agree to delay the updating of their structural template libraries by 1 week, as the predictions evaluated in these experiments refer to targets whose structures have already been made publicly available. These predictions cannot, therefore, be considered strictly "blind" (as those evaluated in *CASP* and *CAFASP*). Nevertheless, automated assessments can provide an ongoing picture of how automated prediction methods

Table 2.1
URLs of web sites, programs, servers and DBs relevant to protein structure prediction

Web site, program, server, DB	URL	References
@TOME	bioserv.cbs.cnrs.fr/HTML_BIO/frame_meta.html	(210)
3D-JIGSAW	www.bmm.icnet.uk/servers/3djigsaw/	(164)
3D-Jury	BioInfo.PL/Meta/	(196)
3D-PSSM	www.sbg.bio.ic.ac.uk/~3dpssm/	(94)
3D-Shotgun	www.cs.bgu.ac.il/~bioinbgu/	(197–200)
ANOLEA	protein.bio.puc.cl/cardex/servers/anolea/index.html	(173, 174)
Arby	arby.bioinf.mpi-inf.mpg.de/arby/jsp/index.jsp	(80, 81)
BCM Search Launcher	searchlauncher.bcm.tmc.edu/	(34)
Belvu	www.cgb.ki.se/cgb/groups/sonnhammer/Belvu.html	(110)
BioEdit	www.mbio.ncsu.edu/BioEdit/bioedit.html	
Bioinbgu	www.cs.bgu.ac.il/~bioinbgu/	(197–200)
BLAST	www.ncbi.nlm.nih.gov/BLAST/ᵃ	(67)
CAFASP4 MQAPs	cafasp4.cse.buffalo.edu/progs/mqaps/	
CAFASP	www.cs.bgu.ac.il/~dfischer/CAFASP5/	(7–9)
CAPRI	capri.ebi.ac.uk/	(36, 37)
CASP experiments (CASP1-CASP7)	predictioncenter.org/	(1–6)
CATH	www.biochem.ucl.ac.uk/bsm/cath/cath.html	(29)
CAZy	afmb.cnrs-mrs.fr/CAZY/	(25)
CDD	www.ncbi.nlm.nih.gov/Structure/cdd/wrpsb.cgi	(41)
CE	cl.sdsc.edu/	(122)
CINEMA	umber.sbs.man.ac.uk/dbbrowser/CINEMA2.1/	(112)
ClustalW	www.ebi.ac.uk/clustalw/	(106)
ClustalX	ftp://ftp-igbmc.u-strasbg.fr/pub/ClustalX/	(108)
Cn3D	www.ncbi.nlm.nih.gov/Structure/CN3D/cn3dinstall.shtml	(147)
COACH	www.drive5.com/lobster/	(86)
COLORADO3D	asia.genesilico.pl/colorado3d/	(142)
COMPASS	prodata.swmed.edu/compass/compass.php	(82–84)
CONSURF	consurf.tau.ac.il/	(143)

(continued)

Table 2.1 (continued)

Web site, program, server, DB	URL	References
CPHmodels	www.cbs.dtu.dk/services/CPHmodels/	(168)
DALI	www.ebi.ac.uk/dali/	(31)
DaliLite	www.ebi.ac.uk/DaliLite/	(119)
DISOPRED	bioinf.cs.ucl.ac.uk/disopred/	(55)
DISpro	www.ics.uci.edu/~baldig/dispro.html	(52)
Domain Parser	compbio.ornl.gov/structure/domainparser/	(114, 115)
DomCut	www.bork.embl.de/~suyama/domcut/	(40)
DOMPLOT	www.biochem.ucl.ac.uk/bsm/domplot/index.html	(117)
DRIPPRED	www.sbc.su.se/~maccallr/disorder/	
DSSP	swift.cmbi.ru.nl/gv/dssp/	(92)
EBI	www.ebi.ac.uk	
Entrez Tutorial	www.ncbi.nlm.nih.gov/Entrez/tutor.html	
ESyPred3D	www.fundp.ac.be/sciences/biologie/urbm/bioinfo/esypred/	(169)
EVA	eva.compbio.ucsf.edu/~eva/	(12)
Expasy	www.expasy.org	(33)
FAMSBASE	daisy.nagahama-i-bio.ac.jp/Famsbase/index.html	(214)
FastA	www.ebi.ac.uk/fasta33/	(69)
Fasta format	www.ebi.ac.uk/help/formats_frame.html	
FFAS03	ffas.burnham.org	(88)
FORTE	www.cbrc.jp/forte	(89)
FRankenstein3D	genesilico.pl/frankenstein	(207)
FSSP	ekhidna.biocenter.helsinki.fi/dali/start	(30, 31)
Fugue	www-cryst.bioc.cam.ac.uk/fugue/	(99)
Genesilico	www.genesilico.pl/meta/	(202)
GenThreader, mGen-Threader	bioinf.cs.ucl.ac.uk/psipred/psiform.html	(95, 96)
Ginzu	robetta.bakerlab.org/	(39)
GROMACS	www.gromacs.org/	(179)
HBPLUS	www.biochem.ucl.ac.uk/bsm/hbplus/home.html	(137)
HHpred	toolkit.tuebingen.mpg.de/hhpred	(91)
HMAP	trantor.bioc.columbia.edu/hmap/	(101)

(continued)

Table 2.1 (continued)

Web site, program, server, DB	URL	References
HMMER	hmmer.wustl.edu/	*(72, 73)*
Homo sapiens genome	www.ensembl.org/Homo_sapiens/index.html	*(26)*
Homstrad	www-cryst.bioc.cam.ac.uk/~homstrad/	*(32)*
IMPALA	blocks.fhcrc.org/blocks/impala.html	*(77)*
InsightII / Biopolymer / Discover	www.accelrys.com/products/insight/	*(178)*
InterPro	www.ebi.ac.uk/interpro/	*(45)*
Inub	inub.cse.buffalo.edu/	*(199, 200)*
IUPred	iupred.enzim.hu/index.html	*(53)*
Jackal	trantor.bioc.columbia.edu/programs/jackal/index.html	
Joy	www-cryst.bioc.cam.ac.uk/joy/	*(135)*
Jpred	www.compbio.dundee.ac.uk/~www-jpred/	*(65, 66)*
LAMA	blocks.fhcrc.org/blocks-bin/LAMA_search.sh	*(87)*
LGA	predictioncenter.org/local/lga/lga.html	*(118)*
LIGPLOT	www.biochem.ucl.ac.uk/bsm/ligplot/ligplot.html	*(140)*
Livebench	bioinfo.pl/meta/livebench.pl	*(11)*
LOBO	protein.cribi.unipd.it/lobo/	*(156)*
LOOPP	cbsuapps.tc.cornell.edu/loopp.aspx	*(102)*
Loopy	wiki.c2b2.columbia.edu/honiglab_public/index.php/ Software:Loopy	*(155)*
Mammoth	ub.cbm.uam.es/mammoth/pair/index3.php	*(120)*
Mammoth-mult	ub.cbm.uam.es/mammoth/mult/	*(126)*
Meta-BASIC	BioInfo.PL/Meta/	*(113)*
ModBase	modbase.compbio.ucsf.edu/modbase-cgi-new/search_form.cgi	*(213)*
Modeller	salilab.org/modeller/	*(161)*
ModLoop	alto.compbio.ucsf.edu/modloop/	*(157)*
MolMol	hugin.ethz.ch/wuthrich/software/molmol/index.html	*(149)*
MQAP-Consensus	cafasp4.cse.buffalo.edu/mqap/submit.php	*(177)*
NACCESS	wolf.bms.umist.ac.uk/naccess/	*(136)*
NAMD	www.ks.uiuc.edu/Research/namd/	*(180)*
NCBI	www.ncbi.nlm.nih.gov	
NCBI NR sequence DB	ftp://ftp.ncbi.nlm.nih.gov/blast/db/FASTA/nr.gz	*(23)*

(continued)

Table 2.1 (continued)

Web site, program, server, DB	URL	References
Nest	wiki.c2b2.columbia.edu/honiglab_public/index.php/ Software:nest	(165)
ORFeus	bioinfo.pl/meta/	(90)
Pcons, Pmodeller	www.bioinfo.se/pcons/, www.bioinfo.se/pmodeller/	(201)
Pcons5	www.sbc.su.se/~bjornw/Pcons5/	(208, 209)
PDB	www.pdb.org/	(10)
PDBsum	www.ebi.ac.uk/thornton-srv/databases/pdbsum/	(141)
PDP	123d.ncifcrf.gov/pdp.html	(116)
Pfam	www.sanger.ac.uk/Software/Pfam/	(42)
Phyre	www.sbg.bio.ic.ac.uk/~phyre/	
Picasso	www.embl-ebi.ac.uk/picasso/	(85)
PMDB	a.caspur.it/PMDB/	(215)
Porter	distill.ucd.ie/porter/	(64)
POSA	fatcat.burnham.org/POSA/	(127)
PPRODO	gene.kias.re.kr/~jlee/pprodo/	(40)
PRC	supfam.mrc-lmb.cam.ac.uk/PRC/	
PRED-TMBB	Biophysics.biol.uoa.gr/PRED-TMBB/	(60)
PredictProtein	www.predictprotein.org/	(59)
PrISM	wiki.c2b2.columbia.edu/honiglab_public/index.php/ Software:PrISM	(125)
Procheck	www.biochem.ucl.ac.uk/~roman/procheck/procheck.html	(138, 139)
ProDom	protein.toulouse.inra.fr/prodom/current/html/home.php	(44)
PROF	cubic.bioc.columbia.edu/predictprotein/	(59)
ProQ, ProQres	www.sbc.su.se/~bjornw/ProQ/	(175, 176)
ProSa	www.came.sbg.ac.at/typo3/	(171)
Prosite	www.expasy.org/prosite/	(46)
Protein Explorer	proteinexplorer.org	(146)
Protinfo AB CM	protinfo.compbio.washington.edu/protinfo_abcmfr/	(159)
PSI-BLAST	www.ncbi.nlm.nih.gov/BLAST/	(67)
Psi-Pred	bioinf.cs.ucl.ac.uk/psipred/	(57, 62)
RAPTOR	ttic.uchicago.edu/~jinbo/RAPTOR_form.htm	(100)
RasMol	www.umass.edu/microbio/rasmol/getras.htm	(145)

(continued)

Table 2.1 (continued)

Web site, program, server, DB	URL	References
ReadSeq	bioweb.pasteur.fr/seqanal/interfaces/readseq-simple.html	
Robetta	robetta.bakerlab.org/	(187–189)
ROKKY	www.proteinsilico.org/rokky/	(194)
Rosetta	depts.washington.edu/ventures/UW_Technology/ Express_Licenses/Rosetta/	(19, 182–186)
Rosettadom	robetta.bakerlab.org/	(39)
SAM (download)	www.soe.ucsc.edu/research/compbio/sam2src/	
SAM-T02	www.cse.ucsc.edu/research/compbio/HMM-apps/ T02-query.html	(75, 76)
SAM-T99	www.cse.ucsc.edu/research/compbio/HMM-apps/ T99-query.html	(63)
Sanger Centre	www.sanger.ac.uk	
Scap	wiki.c2b2.columbia.edu/honiglab_public/index.php/ Software:Scap	(160)
Schistosoma mansoni genome	www.tigr.org/tdb/e2k1/sma1/	(27)
SCOP	scop.mrc-lmb.cam.ac.uk/scop/	(28)
SCRWL	www1.jcsg.org/scripts/prod/scwrl/serve.cgi	(158)
Seaview	pbil.univ-lyon1.fr/software/seaview.html	(111)
SegMod/ENCAD	csb.stanford.edu/levitt/segmod/	(162)
Sequence Manipulation Suite	bioinformatics.org/sms2/	(35)
SMART	smart.embl-heidelberg.de/	(43)
SP3	sparks.informatics.iupui.edu/hzhou/anonymous-fold-sp3. html	(104, 105)
SPARKS2	sparks.informatics.iupui.edu/hzhou/anonymous- fold-sparks2.html	(103, 104)
SPRITZ	protein.cribi.unipd.it/spritz/	(54)
SSAP	www.cathdb.info/cgi-bin/cath/GetSsapRasmol.pl	(123)
SSEARCH	pir.georgetown.edu/pirwww/search/pairwise.shtml	(70)
SSM	www.ebi.ac.uk/msd-srv/ssm/ssmstart.html	(124)
STRUCTFAST	www.eidogen-sertanty.com/products_tip_structfast.html	(195)
SUPERFAMILY	supfam.mrc-lmb.cam.ac.uk/SUPERFAMILY/	(47, 48)
Swiss-PDBViewer	www.expasy.org/spdbv/[b]	(144)

(continued)

Table 2.1 (continued)

Web site, program, server, DB	URL	References
SwissModel	swissmodel.expasy.org/	*(163)*
SwissModel Repository	swissmodel.expasy.org/repository/	*(212)*
SwissProt, TrEMBL	www.expasy.uniprot.org/database/download.shtml	
T-Coffee	igs-server.cnrs-mrs.fr/~cnotred/Projects_home_page/ t_coffee_home_page.html	*(107)*
TASSER-Lite	cssb.biology.gatech.edu/skolnick/webservice/tasserlite/ index.html	*(97)*
TBBpred	www.imtech.res.in/raghava/tbbpred/	*(61)*
Threader	bioinf.cs.ucl.ac.uk/threader/	*(16)*
Three to One	bioinformatics.org/sms2/three_to_one.html	
TIGR	www.tigr.org	
TINKER	dasher.wustl.edu/tinker/	
TMHMM	www.cbs.dtu.dk/services/TMHMM/	*(58)*
Translate	www.expasy.org/tools/dna.html	
UniProt	www.expasy.uniprot.org/	*(24)*
VAST	www.ncbi.nlm.nih.gov/Structure/VAST/vastsearch.html	*(121)*
Verify-3D	nihserver.mbi.ucla.edu/Verify_3D/	*(15, 172)*
VMD	www.ks.uiuc.edu/Research/vmd/	*(150)*
VSL2	www.ist.temple.edu/disprot/predictorVSL2.php	*(50)*
WebLogo	weblogo.berkeley.edu/logo.cgi	*(109)*
Whatcheck	www.cmbi.kun.nl/gv/whatcheck/	*(170)*
WHAT IF	swift.cmbi.kun.nl/whatif/	*(148)*
Wikipedia on Structural Alignment Software	en.wikipedia.org/wiki/Structural_alignment_software #Structural_alignment	

[a]For more details, see the BLAST tutorial (www.ncbi.nlm.nih.gov/BLAST/tutorial/) and frequently asked questions (FAQ) (www.ncbi.nlm.nih.gov/blast/blast_FAQs.shtml).
[b]Also download: Swiss-Pdb Viewer Loop Database, User Guide, and Tutorial, containing detailed information on the program commands and explanations on how to build a homology model of the target protein using this program.

perform based on a larger number of targets than those evaluated by *CASP/CAFASP* experiments.

In *CASP* and related experiments protein structure prediction methods have been traditionally grouped into three broad categories depending on the level of similarity of the target protein sequence to other proteins of known structure, which necessarily impacts the procedure that must be used to build the models: comparative or homology modeling (CM), fold recognition (FR), and new fold (NF) predictions. The separation between these categories, in particular between CM and FR and between FR and NF, has been challenged by the development of more sophisticated methods (e.g., profile-based methods and fragment-based methods, *see* **Sections 3.3.1.1 and 3.4**) able to cross the boundaries between them. The accuracy of the structures predicted in blind tests is generally highest for CM and lowest for NF methods, but there is a large overlap between the accuracy reached by neighboring categories of methods (**Fig. 2.2**).

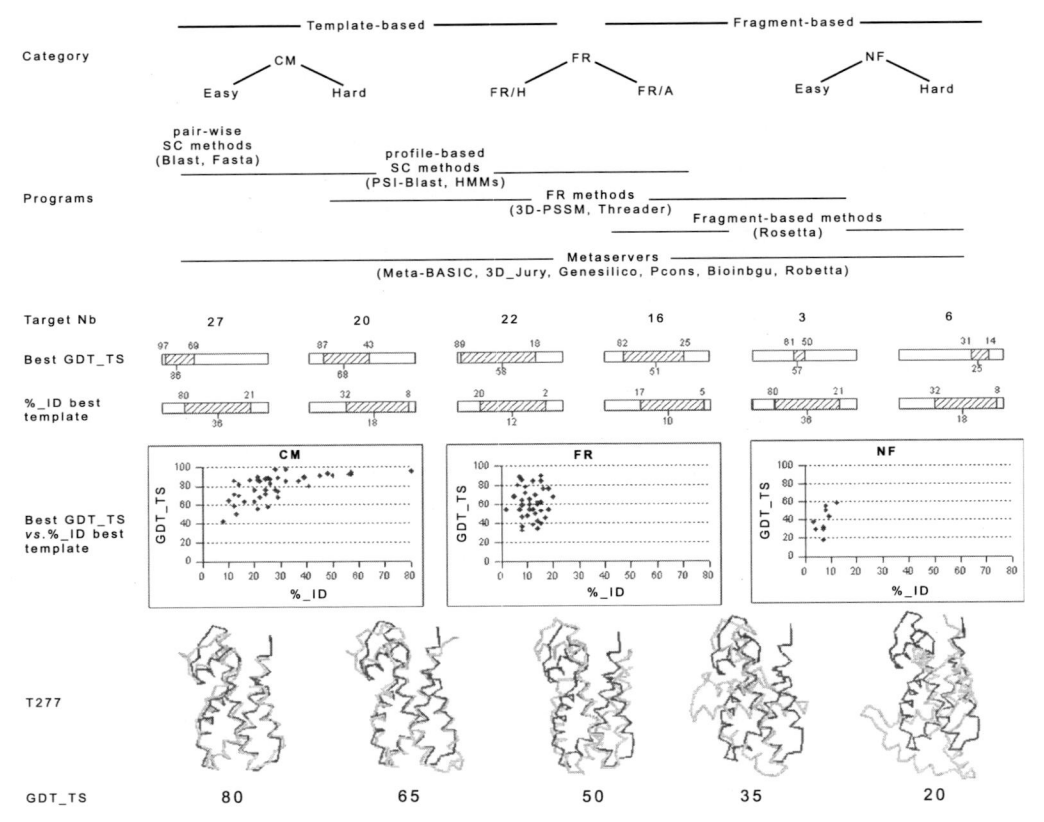

Fig. 2.2. Protein structure prediction methods used in *CASP6* and accuracy of the predictions, expressed by GDT_TS (Note 1). For a description of prediction methods and *CASP6* prediction categories see text. Target Nb, Best GDT_TS, and %_ID best template are the number of targets evaluated in each prediction category, the GDT_TS between the real structure of each target and the best model submitted for it, and the %_ID between the target sequence and the best template structure present in the *PDB*, i.e., the structure most similar to the target. The bottom panel shows the structural super-position between the experimental structure of *CASP6* target T277 (PDB ID: 1wty) and five models with varying degrees of accuracy submitted by predictors taking part in the experiment. Only Cα atoms of structure *(black)* and models *(gray)* are shown. As shown by the figure, GDT_TS values near 80 indicate that the Cα atoms of the model are well superimposed to those of the structure, except in some loop regions. For a GDT_TS value of 65 the core regions are still predicted quite accurately, although structural differences in the loop regions and protein *termini* are more pronounced. GDT_TS values around 50 correspond to an overall structural similarity, but structural variations occur even in the conserved core regions. GDT_TS values around 35 indicate lack of an overall accurate topology prediction, with similarity between only about half of the core regions of structure and model, whereas other regions differ significantly. For GDT_TS values around 20, only a small fraction of the model shows some resemblance to the target structure.

The rational basis of Comparative Modeling (CM) are the following observations: (1) proteins having highly similar a.a. sequences fold into similar 3D structures, and (2) the higher the sequence similarity in the conserved protein "core" region, the higher the similarity between their 3D structures *(13)*. Given the sequence of a protein of unknown structure (target), if another protein (template) whose 3D structure has been experimentally determined can be detected, via a.a. sequence similarity to the target, in available DBs, then the 3D structure of the target can be modeled on the basis of the template structure. Since CM is the method that results in the most detailed and accurate protein structure predictions in blind tests *(14)*, it is the elective protein structure prediction method, whenever applicable.

In the next category, FR methods exploit the observation that during evolution protein structures are more conserved than sequences; although proteins with similar sequences have similar 3D structures, similar 3D structures can also be assumed by proteins with relatively different a.a. sequences. Therefore, even if a target protein does not show recognizable sequence similarities with proteins of known 3D structure, its fold might still be similar to one of them. To identify the compatibility of the target sequence with known 3D structures, FR methods take advantage of structural information derived from statistical analyses of protein structure DBs, such as frequency of pairwise a.a. interactions and residue propensity to assume a certain type of secondary structure and/or to be solvent accessible or buried *(15–17)*. Although they have been remarkably successful at identifying suitable structural templates for target sequences without detectable sequence similarities to proteins of known structure, FR methods have two major drawbacks: (1) they are not always able to discriminate between structures truly similar to the target and those unrelated to it, the correct ranking of which remains a challenge; and (2) they are somewhat less successful in recognizing conserved regions between target and template, often producing poor-quality sequence alignments from which to produce the 3D model. As a consequence, these methods should only be used when CM methods are unable to provide an answer.

Both CM and FR methods involve the identification of a suitable structural template in the *PDB* and differ only in the way they detect it (i.e., based on target-template sequence similarity vs. target sequence-template structure compatibility). Once a suitable template has been identified, the procedure used to build the model is essentially the same for both categories. CM and FR are often grouped together in the broader category of template-based protein structure prediction methods and were evaluated in this new category in the most recent *CASP7*.

Conversely, the remaining category of NF prediction methods does not use whole structural template proteins from the *PDB*. However, the most successful of these do exploit information contained in *PDB* structures at a local level. Protein structures contain smaller sub-structures, or structural motifs, which assume the same conformation in the context of different 3D structures. Therefore, it is possible that a "new" fold, i.e., a protein whose 3D structure differs from already known structures at a global level, is formed by a combination of sub-structures that are similar to those present in known structures. Most current NF methods (called fragment-assembly or fragment-based methods) try to reconstruct the global structure of the target by combining structural fragments having a.a. sequence similar to an equivalent short segment of the target sequence, and applying a scoring function to evaluate the resulting models *(18, 19)*. Although at a global level each target sequence generally assumes only one 3D structure, short peptide sequences can assume different conformations depending on their structural context. Therefore, for each target, fragment-based prediction methods have to explore the structure-space that can be occupied by both different fragment conformations and the possible combinations of these fragments. This results in the generation of many models, frequently notably different from one another. As was a previously discussed challenge for FR methods, perhaps the biggest drawback of NF methods lies in their limited ability to discriminate between correct and incorrect models. Nevertheless, these methods represent one of the biggest innovations that have taken place in the field of protein structure prediction for at least a decade, and are the only currently available tool to obtain, however non-systematically, 3D models of small protein structures with new folds. Even when the correct or most-accurate model cannot be identified based on the score provided by the method, having a few potential models can help in experimental structure determination by x-ray crystallography using the Molecular Replacement technique (*see* **Volume I, Chapter 3**), especially in cases of proteins or protein complexes that are proving particularly tricky to solve *(20)*. Additionally, in *CASP* experiments, fragment-based methods have proved to be particularly successful in the prediction of difficult FR targets (*see* **Sections 3.3.1.2** and **3.4**).

The NF category also comprises *ab initio* methods. Contrary to CM, FR, and NF fragment-based methods, *ab initio* algorithms do not exploit information contained in the *PDB* either at a global or local level. Instead, they try to reproduce the physical laws governing protein folding starting only from the a.a. sequence of the target protein and empirical energy functions based on physicochemical principles. Although addressing an intellectually challenging and ever-stimulating problem, until now *ab initio* methods have not produced protein structure predictions competitive with those provided by the methods discussed in the preceding. The practical applications of *ab initio* methods in protein structure prediction are currently limited to short protein segments (e.g., loops), which cannot be predicted by other methods, and energy refinement of 3D models or parts of them.

One of the remarkable features of protein structure prediction, which has contributed greatly to its diffusion and progress, is that the whole process can be performed using tools that have been made freely available by their foresighted developers. The large majority of the programs and servers described in this chapter can be freely downloaded from, or used through, the Internet (*see* **Table 2.1**).

2. Systems, Software, and Databases

The variety of programs and databases used in protein structure prediction is large and ever-increasing, and a comprehensive listing goes beyond the scope of this chapter. For a full and up-date listing, the reader can refer to:

- The special database and Web server issues of *Nucleic Acids Research*, each published once a year *(21, 22)*, which are dedicated to the most established and popular, as well as recently developed, databases for the collection of biological data and software for their analysis, respectively.

- The special issue of *PROTEINS: Structure, Function, and Bioinformatics (1–6)*, published every 2 years and dedicated to the results of the previous year's *CASP* experiment, comprising both articles describing the evaluation of the performance of protein structure prediction methods in blind tests and articles about the most successful and/or innovative prediction methods.

- The *Livebench* and *EVA* Web sites, in which the performance of automated servers for protein structure prediction on newly released *PDB* targets is assessed on a continuous basis.

Many different types of sequence DBs are available from the *NCBI, EBI, Sanger Centre*, and *TIGR* Web sites. The most commonly used sequence DBs are central repositories in which sequences from many different DBs are collected, such as the comprehensive non-redundant (NR) protein sequence database at the *NCBI (23)* or *UniProt*, containing accurate annotations about protein function, subcellular localization, and/or other protein features *(24)*. For particular purposes, specialized DBs can be searched, dedicated to specific protein groups or families (e.g., carbohydrate-active enzymes (*CAZy*) *(25)*), or individual genomes (e.g., *Homo sapiens (26)* and *Schistosoma mansoni (27)*), some of which might not be sufficiently complete or refined to be included in the central repositories.

The central repository for macromolecular structures is the *PDB*. DBs derived from the *PDB* and containing classifications of proteins according to their 3D structures and, in some cases, evolutionary relationships (e.g., *SCOP (28)*, *CATH (29)*, *FSSP (30, 31)*), and structural alignment DBs (e.g., *Homstrad (32)*) are also of central importance in protein structure prediction.

The majority of the programs required for protein structure prediction can be run on remote servers accessible through the World Wide Web; therefore, any system supporting a Web browser may be used. Many of these programs are also available to be downloaded and run on most common operating systems (i.e., Linux, Unix, Mac OS X, and Windows). Using programs through the Internet is the easiest option, and the best choice for most common applications. On the other hand, downloading the programs permits a greater flexibility in the choice of parameters and amount of computer power to devote to their use. In turn, this allows for greater automation and removes any reliance on the server availability (a number of protein structure prediction servers are not accessible during the "*CASP* prediction season", which runs from June to September every even year).

The Methods section describes protein structure prediction procedures that take advantage of frequently used methods and programs, highlighting which of them have been performing best in blind protein structure prediction experiments.

3. Methods

Protein structure prediction methods are described in the order: CM, FR, and NF, in agreement with the accuracy of the predictions that they provide in blind tests (from highest to lowest).

3.1. Obtain the Protein Sequence

The starting point of any prediction procedure is the a.a. sequence of the target protein, preferably in *Fasta format*, which is accepted or required by most programs and servers. Servers that do not accept *Fasta format* usually require just a sequence of a.a. residues in one-letter code (equivalent to *Fasta format* without the first header line).

The sequence can be obtained from the *NCBI* Web site or other sequence DBs using similar procedures. (For a simple and comprehensive explanation on the use of the *NCBI* data retrieval system, see the *Entrez tutorial.*) Other programs, such as those available from *Expasy (33)*, *BCM Search Launcher (34)*, or *Sequence Manipulation Suite (35)* perform useful operations on biological sequences, such as the conversion of different sequence formats to *Fasta format* (*ReadSeq*) and of a.a. sequences from three- to one-letter code (*Three to One*), or the translation of nucleotide gene sequences to the a.a. sequence of their protein products (*Translate*).

3.2. Prediction of Domain Architecture, Transmembrane and Disordered Regions, and Secondary Structures

Most proteins, especially large ones, are comprised of two or more structural units called domains, joined to each other by linker peptides. In case of multi-domain proteins, homologous template domains of known structure may only be available for some of the target domains, they may belong to different proteins whose domain architecture (i.e., the type and linear order of domains comprised in a protein sequence) is different from the target, and/or have a different degree of evolutionary distance from the target domains, so that different techniques may be required to predict their structures. For these reasons, protein structure prediction is generally performed (and assessed, in experiments such as *CASP*) at a domain level. Predicting the spatial arrangement of interacting domains relative to one another in multi-domain proteins (docking) is still a difficult problem, which is only possible to solve in particular cases (see, for example, the results of the Critical Assessment of PRediction of Interactions (*CAPRI*) experiment) *(36, 37)*, and are not addressed in this chapter.

To perform structure prediction at a domain level it is necessary to obtain an initial picture of the domain composition of the target protein and of regions typically found outside globular domains, such as signal, linker, low-complexity, disordered, and transmembrane regions, as well as of the SSE contained in each domain.

1. Domain boundaries are predicted using a variety of approaches based on multiple sequence alignments (MSAs), frequency of amino acids found in domain and linker regions, domain size, predicted secondary structure, sequence comparison methods, neural networks, hidden Markov models, and, sometimes, even using FR and NF methods to build 3D models from which domain definitions are derived (see *(38)* and references therein). *Rosettadom* and *Ginzu (39)* were the most successful automated servers at predicting domain boundaries in *CASP6*, whereas *Phyre* and *DomCut (40)* were the programs used by the best human predictors *(38)*. Other successful groups used *PPRODO (40)* and *CDD (41)*. In *CASP6* the best methods were able to correctly assign domains to >80% of the residues of multi-domain targets. The prediction quality was shown to decrease with an increase of the number of domains in the target, from targets assigned to the CM category to those assigned to FR and NF and for domains made by segments that are not contiguous in the sequence. However, since the number of targets available for domain prediction was relatively small (63 in total, of which about half contained one domain only), the

trends observed in this experiment might not be representative of more general scenarios *(38)*. If the target domains are homologous to those already classified in specialized DBs, servers such as *Pfam (42)*, *SMART (43)*, *ProDom (44)*, *InterPro (45)*, *Prosite (46)*, and *SUPERFAMILY (47, 48)* can provide reliable domain assignments, as well as predictions about signal peptides, low complexity, and transmembrane regions.

2. Natively disordered proteins or protein regions are those that do not assume a regular secondary or tertiary structure in absence of binding partners. Proteins containing relatively long (>30 a.a.s) disordered segments are relatively common, especially in higher eukaryotes, and often have important functional roles *(49)*. The available methods to predict disordered protein regions are based on experimental data from x-ray crystallography and NMR spectroscopy and in general they use machine learning approaches such as neural networks and support vector machines *(49)*. In *CASP6* the best performing method was *VSL2 (50, 51)*, which correctly predicted 75% of disordered residues (true-positives) with an error rate of 17% (false-positives, i.e., ordered residues incorrectly predicted as disordered). Among the other best performing methods were *DISpro (52)*, *IUPred (53)*, *SPRITZ (54)*, *DISOPRED (55)*, and *DRIPPRED*, with about 50% of correctly predicted residues for an error rate <20% *(49)*. A drawback of these methods is that the probabilities associated with the disorder predictions are not always good indicators of the prediction accuracy (48). Assessment of disorder predictions in *CASP6* was confined to targets whose structures were determined mostly by x-ray crystallography, which may impose some order to regions that would be disordered in solution, and whose disordered segments were often rather short *(49)*. Therefore, although the results of this assessment are indicative of the performance of the methods on proteins similar to the *CASP6* targets, they do not necessarily reflect their performance on different types of disorder, e.g., their ability to identify entirely disordered proteins or disordered regions as measured by other experimental methods such as NMR spectroscopy *(49)*.

3. Transmembrane (TM) regions are predicted by *Psi-Pred (57)*, which also provides secondary structure predictions (see the following); *TMHMM (58)* and a number of servers accessible through the *PredictProtein* metaserver *(59)* specialize in predicting transmembrane alpha-helices, and *PRED-TMBB (60)* and *TBBpred (61)* transmembrane beta-barrels.

4. A number of programs have been developed to predict the SSE present in a target sequence at a residue level, and the

best of these have now reached a remarkable level of accuracy. In the continuous benchmarking system *EVA*, several methods are tested for their ability to correctly predict the secondary structure of target proteins showing no significant sequence identity to any protein of known structure. Currently tested methods have been subjected to the evaluation for several months on sets of tens to hundreds of proteins, and were shown to have an accuracy, defined as their ability to correctly predict each of the target residues as being in alpha-helical, beta-strand, or other (coil) conformation, of at least 70% on their overall set of target proteins. This accuracy depends on the quality of the MSA (*see* **Section 3.3.2.1**) that is automatically built for the target, ranging from about 64% for a single sequence with no homologs to nearly 90% for targets with high-quality MSAs. In addition to the high prediction accuracy they achieve, another important feature of these methods is that the confidence values associated with their residue-based predictions correlate well with the actual accuracy of the prediction. Several consistently well-performing methods, such as *PROF (59)*, *Psi-Pred (57, 62)* and *SAM-T99 (63)* predict correctly >76% of residues of the set of targets they have been tested on. *Porter (64)* achieves almost 80% correctly predicted residues and is currently the best among the methods evaluated by *EVA*, but as its evaluation started more recently than that of the other methods it has been tested on a smaller set of target proteins. Another useful server is *Jpred (65, 66)*, which provides consensus SSE predictions between several methods.

In general, searches with any of these servers are easy to run and do not take long to complete (from a few minutes to a few hours, depending on the server); therefore, it is advisable to run the target sequence through more than one server to get as accurate as possible a starting guess about the target domain boundaries and SSE composition. Ideally, since the confidence in a given prediction is higher when different methods are in agreement, it is advisable to use several methods and compare their predictions, especially when the methods performance in blind tests has not yet been assessed, the assessment has been performed on a small number of targets, or the target has significantly different features from those on which the assessment has been performed (e.g., it contains large disordered regions).

All the predictions from now on should be performed using as input single domain sequence regions as opposed to whole protein sequences. Given the uncertainty in domain boundary prediction, it is advisable to include in the target domain sequence some 10–20 a.a.s N-terminal and C-terminal to the boundaries predicted by domain prediction servers. A more precise prediction of the domain boundaries can often be obtained at a later stage, following the production of optimized target-template sequence alignments and 3D target models.

3.3. Template-Based Modeling

Template-based modeling consists of the following key steps: (1) identification of the template structure; (2) refinement of target-template(s) sequence alignments; (3) model building; (4) model evaluation; and (5) model refinement.

3.3.1. Identification of the Template Structure

3.3.1.1. Sequence Comparison Methods

Sequence comparison (SC) methods are used to retrieve proteins similar to the target from sequence DBs. Proteins that score better than certain threshold values are considered to have statistically significant sequence similarity to the target, based on which an evolutionary relationship (i.e., homology) with the target is inferred. Since similar sequences fold into similar 3D structures, proteins of known structure scoring above the threshold can be used as template(s) from which a comparative model of the target is built. Homology models are usually built using as main template the structure of the protein having the highest global sequence conservation with the target. However, regions containing insertions and deletions in the sequence alignment of the target with the best global template might be better conserved, at a local level, in other homologous proteins of known structure, which can therefore be used as templates for these regions. In principle, homologous proteins may be retrieved searching sequence DBs including only proteins of known structure (e.g., pdb at the *NCBI*). However, since protein homologs of unknown structure provide important information for model building, and the sequences of proteins of known structure are also contained in comprehensive sequence DBs such as *NR*, these are generally the DBs of choice.

SC methods can be assigned to two main categories: pairwise methods and profile-based methods. Pairwise sequence comparison methods (e.g., *BLAST*) *(67)* compare the sequence of the target with each sequence present in a sequence DB. They are able to detect proteins showing high sequence similarity to the target, based on which the target and the retrieved proteins are inferred to be close evolutionary relatives and expected to assume very similar 3D structures. Profile-based sequence comparison methods (e.g., *PSI-BLAST*) *(67)* compare each sequence in a DB with a profile created from an MSA of the target and its closest homologs, which have been previously detected using pairwise methods. Since the profile incorporates information about several family members and not just the target, these methods are able to detect proteins evolutionarily related to the target that cannot be detected by pairwise sequence comparison methods. Based on their lower sequence similarity with the target, these proteins are considered to be distant evolutionary relatives; therefore, their structural similarity with the target might be lower than that of proteins matched by pairwise methods (however, this is not always the case, *see* **Section 3.3.1.2** and **Fig. 2.2**). In *CASP6*, if a significant match with a correct template domain (as defined based on the structural similarity of the target with structures in the *PDB* measured by the *LGA* program, *see* **Section 3.3.2.4**) was found using *BLAST*, the target was assigned to the "easy" CM prediction sub-category; any targets for which structural homologs were detected by *PSI-BLAST* but not by *BLAST* were considered to be "hard" CM targets *(14, 68)*.

The most popular pairwise sequence comparison methods are *BLAST*, *FastA (69)*, and *SSEARCH (70)*. *BLAST* is the most commonly used, it is implemented by almost every sequence DB available on the Internet and can either be used interactively or downloaded, together with a number of comprehensive or specific sequence DBs, from the *NCBI* Web site. Although installing *BLAST* on a local computer can be advantageous to run multiple searches automatically and allow for greater flexibility in parameter settings, the Web versions are easier to use and provide reasonable default parameters, which might be preferable for first-time users. *BLAST* returns pairwise alignments of the target with sequences retrieved from the DB, and several parameters to help decide about the significance of each alignment, i.e., whether the matched protein is likely to a be a real homolog of the target as opposed to showing sequence similarity with it purely by chance.

1. The expectation value (E-value) of an alignment represents the number of different alignments with scores equivalent to or better than the score of that alignment (**Note 2**) that are expected to occur by chance in a database search. Therefore, the lower the E-value, the higher the probability that a matched sequence is a real homologue of the target, and vice versa. Unfortunately, there is no universal threshold value that guarantees identification of all the true homologs and rejection of all non-homologous sequences. In general, for a given threshold value, there will be both proteins with E-values better (i.e., lower) than the threshold that are not real homologs of the target but show some sequence similarity with it purely by chance (false-positives), and proteins with E-values worse (i.e., higher) than the threshold that are homologous to the target but have diverged from it to the point that the sequence similarity is not distinguishable from that occurring by chance (false-negatives). Lowering the E-value threshold results in a decrease in the number of false positives (i.e., incorrect hits) and an increase in the number of false-negatives (in that a higher number of real homologs have E-values above the threshold, and are discarded). Conversely, increasing the E-value threshold results in a lower number of false-negatives (i.e., missed hits) and a

higher number of false-positives (e.g., for E-values of 10 or higher, a considerable number of hits found by chance have E-values below the threshold, and are selected). In general, a match is considered to be significant if the E-value is around 10^{-2}–10^{-3} or lower, whereas E-values of 10^2–10^3 or higher indicate that the match is almost certainly not to be trusted. Matches with E-values between or approaching these values should be evaluated carefully taking into account other parameters.

2. The percentage of sequence identity (%_ID) represents the number of identical a.a.s found at corresponding positions in the aligned regions and it can provide indications about the homology between two proteins. For sequences aligning over about 100 a.a.s, a %_ID above 40% indicates certain homology, whereas a %_ID of 20% or less might occur purely by chance; if the %_ID is between 20 and 40% the two proteins might be homologous, but additional information is required to support this hypothesis. These threshold values vary with the length of the aligned regions. Since short segments have a higher chance of showing sequence similarity by chance, for considerably shorter and longer alignments the %_ID required to infer homology is therefore higher and lower, respectively. Similar information to the %_ID is provided by the percentage of sequence similarity, which depends on the substitution matrix used by SC methods (**Note 3**). High values of percentage of sequence similarity qualitatively support the significance of the match and the correctness of the alignment, but the relationship of this parameter with homology is even less precise than for %_ID.

3. Between closely related proteins, insertions and deletions (gaps) are usually relatively few and generally cluster in surface loop regions, rather than being spread all over the structure and interrupting regular SSE. Therefore, the lower the number of gaps and different positions in which they occur in the alignment, the higher the significance of the match and quality of the alignment.

In the absence of overall sequence similarity, i.e., in case of E-values higher and %_ID lower than the aforementioned values, the following additional sources of evidence can support the hypothesis of the existence of evolutionary relationships between sequences retrieved from the DBs and the target: the similarity of both the target and the retrieved sequence to a third "intermediate" sequence that is more closely related to each of them than they are to each other; the existence of structural relationships between the matched proteins of known structure (as shown, for example, by their classification within the same *SCOP* Family, Superfamily or Fold) (**Note 4**); a good overlap between the SSE of the templates and the SSE predicted for the target (*see* **Section 3.3.2.2**); and finally the conservation of residues that are known to play important structural and/or functional roles for the target and/or the retrieved proteins (key residues, *see* **Section 3.3.2.5**). Additionally, since DB searches are not symmetrical, confidence in the target homology with a retrieved sequence can be increased if the target sequence is matched, in turn, when searching the DB using the previously retrieved sequence as a query.

The availability of biological information about the target family and experience with alignments from many different protein families can also be of help in the evaluation of the biological significance of a match.

If proteins of known structure closely related to the target are not found by pairwise methods, it is possible to change the aforementioned parameters to tailor them to the specific problem at hand. As an example, to detect distantly related proteins it is possible to choose a lower number BLOSUM or higher number PAM matrix (*see* **Note 3**), and/or decrease the penalty associated with the insertion and elongation of gaps in the alignment (*see* **Note 2**). However, a more efficient way to detect distant homologs *(71)* consists in the use of profile-based methods such as *PSI-BLAST (67)* and HMM-based methods *(63, 72, 73)*.

In *PSI-BLAST (67)* the first iteration, coinciding with a simple *BLAST* search, is used to collect sequences similar to the target based on a pre-defined E-value threshold (a default value is provided, but it can be changed by the user). The alignments of these sequences to the target are used to build a multiple alignment from which a Position-Specific Score Matrix (PSSM), or profile, is derived that contains values related to the frequency with which each a.a. occurs at each alignment position. In the second *PSI-BLAST* iteration, the sequences in the DB are matched to this profile, rather than to the target sequence. If a third iteration is run, the profile is updated to incorporate in the alignment the new sequences found with E-values below the threshold; if no new sequences are found, the profile does not change, i.e., the program has reached convergence. *PSI-BLAST* iterations can be run until the program converges or a protein of known structure is matched below the threshold. *PSI-BLAST* results can be evaluated using the same parameters described for *BLAST*. However, from the second iteration onward *PSI-BLAST* E-values are not directly comparable to those calculated by *BLAST*. Indeed, the E-value associated with a protein retrieved by *BLAST* is different from the E-value associated with the same protein, retrieved from the same DB, by any *PSI-BLAST* iteration following the first. The reason for this is that *BLAST* scores the target sequence against each DB sequence using a matrix (e.g., BLOSUM62) containing fixed values for each a.a. pair, independent of the position where they occur in the sequence alignment, whereas *PSI-BLAST* scores the target sequence against a PSSM whose values (depending on the frequency of a.a.s observed at each position in the MSA from which the PSSM was generated) are updated after each iteration. Because it is derived from an alignment of multiple sequence homologs, the PSSM is more powerful than the fixed scoring matrices, and can give sequences homologous to the target a higher score and therefore a better E-value, thus promoting them over incorrect matches. However, while convergence is rarely reached, if sequences non-homologous to the target are matched with E-values below the defined threshold (false-positives), they will be incorporated in the *PSI-BLAST* profile leading to the matching of more non-homologous sequences in the following iteration, and ultimately to divergence from the original target. Indeed, the profile can drift away from the originating target sequence so far that eventually the target sequence itself, and not only its homologs, will score very poorly against the profile! To prevent this, the hits collected by each *PSI-BLAST* iteration should be carefully examined, adjusting the threshold for inclusion (to make it more restrictive in case of divergence and more permissive in case convergence is reached before a protein of known structure is matched) and/or selecting manually sequences to be included in or excluded from the profile.

If *PSI-BLAST* does not identify any sufficiently convincing hits, or converges before identifying any matches to proteins of known structure, hidden Markov model (HMM)-based programs can be used *(74)*. HMMs can also be run to detect additional templates and/or compare their results with those obtained by *PSI-BLAST*. Starting from an MSA, these programs build a hidden Markov model (HMM) that, similarly to a *PSI-BLAST* profile, represents the properties of all the sequences in the alignment. This HMM is then used to search the DBs for homologous proteins. The two most popular HMM-based programs are *SAM* and *HMMER (72, 73)*, both freely available for downloading. *SAM* is also accessible through a Web server interface (*SAM-T02*) *(75, 76)* that takes the target sequence as input and automatically builds both the MSA and the HMM. Although expert users might prefer to use the downloadable version to be able to modify program parameters, the Web interface is straightforward to use and provides results relatively quickly. The *SAM-T02* output provides E-values (i.e., estimates of approximately how many sequences would score equally well by chance in the database searched) and the *SCOP* classification of the matched structures to help evaluate the matches. If the E-values are higher than the suggested significance threshold (e.g., E-values $<10^{-5}$ and higher than 0.1 indicate very reliable and speculative matches, respectively) and/or proteins matched by the HMM do not belong to the same *SCOP* superfamily (**Note 4**), additional information is required to infer homology between any of the matched proteins and the target (see the preceding). To speed-up the search for homologs, several methods have been developed that allow comparison of the target sequence with pre-calculated profile libraries, such as PSSMs generated with *PSI-BLAST* (*IMPALA*) *(77)* and HMM libraries representing homologous sequences (*Pfam* and *SMART*) or proteins of known structure that are evolutionarily related at a superfamily level as defined by *SCOP* (*SUPERFAMILY*).

More recently, a number of profile–profile comparison methods have been developed capable of detecting distant relationships that are not recognized by sequence-profile matching methods *(78)*. These include: prof_sim *(79)*, *Arby (80, 81)*, *COMPASS (82–84)*, *Picasso (85)*, *COACH (86)*, *LAMA (87)*, *PRC* (the profile–profile implementation of *SUPERFAMILY*), *FFAS03 (88)*, *FORTE (89)*, *ORFeus (90)*, and *HHpred (91)*.

The most successful prediction groups in the last *CASP* editions used *SAM-T02*, *FORTE*, *ORFeus*, and *FFAS03*, either as stand-alone programs or as part of metaservers (*see* **Section 3.5.2**). The performance of all these methods, together with that of *HHpred*, *SUPERFAMILY*, and *PRC* is also subjected to continuous evaluation by the *Livebench* server.

Since several sequence–profile and profile–profile comparison methods (e.g., *SAM-T02*, *ORFeus*, and *HHpred*) exploit structural information (most often, secondary structure predictions and secondary structure assignments by *DSSP (92)*), sometimes they are classified together with FR methods or metaservers. Targets for which a correct template structure (i.e., a structure similar to the target according to the *LGA* program, *see* **Section 3.3.2.4**) was identified in the *PDB* by profile–profile sequence comparison methods were assigned to the FR/H (H: homologous) category in *CASP6*; conversely, targets for which the correct template structure could not be detected by any sequence-based methods was assigned to the FR/A (A: analogous) category *(68)*.

3.3.1.2. Fold Recognition Methods

As mentioned in the Introduction, analysis of protein sequence and structure DBs led to the observation that, as proteins diverge, overall structural similarity persists even when no significant sequence similarity can be detected. In fact, two proteins with <25–30% overall identities can have either very similar or completely different 3D structures. In order to detect an evolutionary relationship in the absence of sequence similarity, FR methods: (1) identify potential structural similarity signals within the sequence, and (2) apply confidence statistics to rank potential matches and provide confidence values for the prediction in order to distinguish "real" matches (true-positives) from spurious unrelated ones (false-positives). FR methods try to assess the likelihood of the target proteins sharing a fold with one of the proteins of known structure by comparing structural features predicted for target sequences, on the basis of statistical analysis of known protein structures, with those actually observed in each structure. Structural features commonly taken into account include 3D environmental preferences of each amino acid, such as propensity to form pairwise contacts with other residues, solvent accessible surface area, and local secondary structure. This comparison between one-dimensional (1D) sequences and 3D structures is usually done by either encoding 3D information into 1D sequences *(15)* or by threading, i.e., inscribing the 1D sequence of the target into each fold contained in a library of representative 3D structures *(93)*. The compatibility of the target sequence with each fold is evaluated using empirical or "knowledge-based" potentials that take into account the aforementioned structural properties.

The Three-Dimensional Position Specific Scoring Matrix *(3D-PSSM) (94)* server uses sequence and structure alignments, as well as secondary structure and solvent accessibility, to construct descriptive position-specific matrices for each domain in a non-redundant structural DB. These matrices can be compared with PSSMs or profiles of a query sequence and the results reported with underlying evidence to enable the user to understand the strength and confidence of the fold prediction. *3D-PSSM* and its eventual successor *Phyre*, both have simple and user-friendly Web-based interfaces. The calculations can take some time as the many-by-many profile comparisons can be intensive. Eventually the results will be displayed on a Web page which is e-mailed to the user. The page displays proteins of known structures that are predicted to fold in a similar way to the query sequence. These proteins are listed in a table in order of predicted similarity to the query, so that the proteins at the top of the table are predicted to be most similar. An E-value is provided to indicate the probability that the prediction is true, the lower the E-value the more likely the prediction is to be correct. The alignment to the target sequence and *SCOP* classification of the protein structures are provided. *3D-PSSM* also provides an automatically generated homology model based on the alignment to each of the matched structures. If the steps for using the server are simple, most attention must be paid to interpreting the results returned. With all prediction methods, a sign of a prediction being correct is whether it is persistent. For example, the confidence in the prediction can be increased if the top reported *3D-PSSM* structures matching the query belong to the same *SCOP* family or superfamily, or even fold (**Note 4**), and/or if running through the server sequences homologous to the target the same structures appear on top of the list. Conversely, if the top hits belong to different families or superfamilies, a fold similar to the target might still be present among them, even if the method cannot identify it clearly from the wrong ones. In such cases, to detect the correct fold from incorrect ones, and to further support FR predictions in general, it is possible to exploit additional information, such as those used to validate the output of SC methods (*see* **Section 3.3.1.1**). As an example, residues strongly conserved in the MSA of the target sequence can be mapped on the known structures to check whether they play key structural or functional roles (e.g., involvement in disulfide bridges, building of active sites, etc.).

There are a number of fold recognition methods available via the internet. *3D-PSSM*, *Threader (16)*, *GenThreader (95)*, and *mGenThreader (96)*, *TASSER-Lite (97)* (combining the PROSPECTOR 3.0 threading algorithm *(98)* with fragment-based NF prediction methods), *Fugue (99)*, *RAPTOR (100)*, *HMAP (101)* (using structure-based profiles in the conserved regions and sequence-based profiles in the loops), *LOOPP* (102), *SPARKS2 (103, 104)*, and *SP3 (104, 105)* have been used as stan-dalone programs or as input for metaservers (*see* **Section 3.5.2**) by the best performing group in the CM and FR categories in the last *CASP* editions, and their performance, together with that of many other servers, is continuously assessed by *Livebench*. In general, metaservers that use a consensus of SC and FR-based methods are most successful at predicting FR targets that are clearly homologous to proteins of known structure, classified in the FR/H (H: homologous) sub-category in *CASP6*. Conversely, the best predictions of FR targets assigned to the *CASP6* FR/A (A: analogous) sub-category, for which no clear evolutionary relationship with already known folds can be detected, are provided by fragment-based methods (*see* **Section 3.4**).

3.3.2. Refinement of Target-Template(s) Sequence Alignments

Together with the extent of structural similarity between target and template, the generation of structurally correct target-template sequence alignments (i.e., sequence alignments corresponding to the optimal structural superposition between the target and template structures) is one of the most important factors affecting the final quality of template-based models. Therefore, sequence alignment(s) provided by SC and/or FR methods should be critically evaluated and refined, with increasing care as the %_ID between target and template decreases. Because of the difficulty of obtaining a correct alignment, many of the best performing groups in *CASP6* generate and evaluate both a number of target-template sequence alignments obtained from different sources and/or using different alignment parameters, and a number of 3D models produced by different servers. The evaluation is based on the results of model quality assessment programs (MQAPs) (*see* **Section 3.3.4**) and/or on the assumption that consensus predictions provided by independent methods are more reliable than any single prediction; accordingly, consensus regions are taken as such while variable regions are re-aligned and subsequently re-evaluated in a cyclic procedure.

3.3.2.1. Multiple Sequence Alignments

One way to refine pairwise target-template alignments is by comparing them to the alignments produced by MSA methods (*see* **Volume I, Chapter 7**). MSA programs align all sequences in a given set to one another, producing MSAs that can be more accurate than those produced by SC or FR methods. *ClustalW (106)* and *T-Coffee (107)* are among the most widely used programs to build MSAs. Both are available for downloading (a windows interface version of *ClustalW* called *ClustalX (108)* is also available), and can be used interactively through a Web interface. However, several newer and potentially more powerful methods are now available (*see* **Volume I, Chapter 7**). Together with the specific features of the MSA program used, the set of sequences given as input to the program is one of the most important factors affecting the quality of the final MSA.

MSAs for protein structure prediction are typically generated from sequences putatively homologous to the target identified by SC methods. Sequences matching the target with E-values significantly higher than that of the selected threshold or of the most distantly related of the selected templates might be eliminated in that they should not contribute to improving the quality of the target-template(s) alignments, and might actually make it worse. However, since false-negatives might be among them, these sequences can also be kept and their relationships with the target evaluated at a later stage, on the basis of the resulting MSA itself. The pattern of conserved residues in an MSA can provide information about key structural or functional residues in a protein family and increase the confidence in the existence of evolutionary relationships between target and template (*see* **Section 3.3.2.5**). Potential templates detected by FR methods may be added to the set of target homologs retrieved by SC methods and given as input to MSA programs; however, they often have sequences too different from the target and its homologs to produce good alignments. In such cases, the sequences of the putative FR templates can be used as queries by SC methods to retrieve their homologs from sequence DBs and MSAs can be produced for the templates as well. Comparison of the MSAs produced for the target and template sequences, and in particular of the pattern of conserved and putatively essential residues, may support the hypothesis of the existence of evolutionary relationships between them. An informative way to highlight sequence conservation in MSAs is provided by the *WebLogo (109)* program, producing a picture in which residues occurring at each alignment position are shown by the one-letter code, the size of each letter being proportional to their frequency of occurrence. MSAs are most informative when they contain a relatively high number of sequences, similar enough to one another to be certain of their homology and to allow for the production of a correct alignment, and divergent enough to allow for conserved positions to be distinguished from variable ones. For this reason, redundant and outlier sequences (e.g., those that have %_ID >80% with any other sequence in the alignment or <20% with all other sequences in the alignment, respectively) are usually eliminated. Other kinds of editing include the deletion of alignment regions other than those matching the target domain, which may be present, for example, if the sequences retrieved from the DBs comprise additional domains besides that homologous to the target. Additionally, shifts in the aligned sequences can be introduced manually based on structural information such as those described in the following (**Sections 3.3.2.2, 3.3.2.4,** and **3.3.2.5**). Although guidelines such as these can be useful to start with, production of a good-quality MSA is very much a process of trial and error. In general, choices on how to edit MSAs depend on specific features of the sequences contained therein, and on a balance between the computational and human time required to analyze a high number of sequences and the accuracy requested of the final result. This, in turn, depends largely on the difficulty of the prediction: in case of "easy" targets, for which reliable templates aligning well with the target can be identified, even the pairwise *BLAST* alignments might be sufficient; conversely, when the templates are distant homologs, detectable only by profile-based or FR methods, with very low %_ID and difficult to align, all available information from sequence and structural homologs should be exploited. Often, several rounds of alignment editing and re-alignment are required to produce high-quality MSAs.

Several programs are available to visualize MSAs, allowing the user to color a.a.s according to residue type or conservation, edit the alignment to eliminate redundancies and outliers as well as alignment columns and blocks, and save them in different sequence formats. Such programs include *Belvu (110), Seaview (111), BioEdit, CINEMA (112)*, and various other tools from the *Expasy* Web site.

3.3.2.2. Comparison Between SSE Identified in the Templates and Predicted for the Target

SSE are usually among the best conserved parts of evolutionarily related proteins; therefore, they should also be found in corresponding positions in the sequence alignment. Additionally, insertions and deletions of more than one or two residues are unlikely to occur within SSE, whereas they can easily take place within the conformationally more variable and solvent exposed loop regions.

The secondary structure assignment for the template structures can be calculated by programs such as *DSSP* or obtained from the *PDB* Web site. This contains both the *DSSP* automated assignment and the manual assignment provided by the experimentalists who have solved the structure, which may be more accurate than those calculated automatically. Secondary structure predictions for the target can be obtained as described above (**Section 3.2**). Mapping the SSE calculated for the template(s) and predicted for the target on the target-template(s) pairwise alignments or MSAs might help refine the sequence alignments. If the SSE do not align and/or large insertions or deletions occur within these regions, the target-template alignment may be modified to adjust for these features, which are likely to be incorrect. However, unless it is obvious how to correct errors in the sequence alignment, modifying it by hand is only advised once one has reached a certain degree of experience. Hybrid- or meta-profiles combining sequence and secondary structure information are now used by many of the methods that have been most successful in the *CASP* experiments (e.g., *Meta-BASIC (113)*, *see* **Section 3.5.2**); additionally, several FR methods (e.g., *3D-PSSM*, *Phyre*, and *mGenThreader*) report the SSE of the templates and those predicted for the target in their output target-template alignments.

3.3.2.3. Identification of Domain Boundaries in the Template Structures

The template region aligned to the target by SC methods does not necessarily correspond to a structural domain of the template, but it can be shorter, if only a fraction of the template domain displays recognizable sequence similarity with the target, or longer, in case regions before or after the template domain have also similar sequence to the target. Indeed, the initial domain definition of the target, which has been preliminarily provided by domain prediction servers (**Section 3.2**), may be refined during the prediction procedure based on accurate target-template(s) alignments. Until this step is performed, it is advisable to use a slightly larger segment of the target sequence than that predicted to correspond to the target domain.

The boundaries of the template domains matching the target can be identified based on *SCOP*, which is widely believed to be the "gold standard" of protein structure classification, and where protein domains are classified according to sequence, structural, functional, and evolutionary criteria. In case the template structures have been made available from the *PDB* more recently than the date of the latest *SCOP* release, a domain assignment for them may be found in other structural classification DBs, such as *CATH* or *FSSP*. If no DB contains a pre-calculated domain definition for the selected templates, this can be obtained by running the template structures through one of the available programs for domain assignment, such as *Domain Parser (114, 115)*, *PDP (116)*, or *DOMPLOT (117)*. Mapping the structural domain definition of the templates on the target-template sequence alignment(s) can help to refine the initial prediction of target domain boundaries.

3.3.2.4. Structural Alignment of Selected Template Domains

The structural alignment of the template domains shows which regions are structurally conserved among them, and are therefore likely to be conserved in a target protein of unknown structure evolutionarily related to them. This alignment might be extended, by including other proteins of known structure whose evolutionary relationships with the template(s) have been ascertained on the basis of structural criteria, to get a more precise definition of the conserved "core" and variable regions among proteins homologous to the target. Proteins of known structure evolutionarily related to the template(s) can be obtained from *SCOP*, in which closely and distantly related homologs are classified within the same family or superfamily, respectively, (*see* **Note 4**), or from other structural classification DBs mentioned in **Section 2**.

Pre-compiled alignments of protein domain structures are available from specific DBs (e.g., *Homstrad*, *FSSP*, and *CE*) and may include the selected templates. Alternatively, structural alignments of the templates can be built by using one of the many available programs for protein structural alignment (a large, if incomplete, list is available from *Wikipedia*). Choosing which one(s) to use depends partially on how many proteins match the template and have to be aligned. Some programs perform only pairwise alignments (e.g., *LGA (118)*, *DaliLite (119)*, *Mammoth (120)* and *VAST (121)*) whereas others can perform multiple structure alignments (e.g., *CE (122)*, *SSAP (123)*, *SSM (124)*, *PrISM (125)*, *Mammoth-mult (126)*, and *POSA (127)*), providing a global alignment of all the input structures and, in principle, better results. Although no automated program is capable of systematically producing accurate alignments of highly divergent structures, most structural alignment programs can produce relatively good alignments of similar structures, as should be the case for templates identified based on sequence similarity to the same target. Depending on the time available and the level of accuracy required, the templates might be run through several servers to compare the results.

Many predictors successful at *CASP* exploit information deriving from the structural alignment of multiple templates to identify conserved regions, guide sequence alignments, and/or build chimeric models from fragments extracted from different templates, to be compared and evaluated in subsequent steps using consensus and/or quality assessment criteria (*see* **Section 3.3.4**).

3.3.2.5. Structural Analysis of Selected Template Domains

Regions that are structurally conserved among the templates and their homologs and, therefore, are putatively conserved in the target structure as well, in general correspond to SSE and conserved loops, and should not contain insertions or deletions in the target-template sequence alignments. Therefore, if "gaps" occur within these structurally conserved regions, they will have to be moved to regions where they might be more easily accommodated from a structural point of view.

As protein structures are more conserved than sequences during evolution, structurally conserved regions may have low sequence similarity; nevertheless, they should contain "key" structural features allowing them to assume similar conformations. Such features have been identified in the past for several protein families and found to consist, most often, of residues belonging to one of the following categories: (1) residues, in general hydrophobic in nature, having low solvent accessibility, interacting with one another within the protein core (*128*); (2) residues conferring special structural properties to the protein region where they occur: Gly, Asn, and Asp, able to assume more frequently than other residues positive φ values; or Pro, whose main-chain nitrogen atom is involved in peptide bonds found more frequently than those of other residues in *cis*, rather than *trans*, conformation and that, lacking the main-chain hydrogen bond donor capability, cannot take part in the formation of structures maintained by regular hydrogen bond patterns, e.g., α-helices or internal β-strands; (3) Cys residues, that can be involved in the formation of covalent disulfide bonds; (4) any other residues that appear to play a specific structural role in a given family (e.g., negatively charged Asp and Glu and polar Asn residues binding calcium ions in the cadherin family (*129, 130*), and polar or charged residues able to form hydrogen bonds or salt-bridges maintaining the conformation of antibody hypervariable loops (*131–134*)). Comparison of evolutionarily related structures can allow identifying conserved residues having a key structural role, which are likely to be conserved in the target structure as well, and represent useful landmarks to help refine pairwise target-template sequence alignments and MSAs. *Joy (135)* is a program that reports different types of residue-based structural information (e.g., secondary structure, solvent accessibility, positive φ values, *cis*-peptides, involvement in disulfide and hydrogen bonds) on sequence alignments. Additionally, *NACCESS (136)*, *HBPLUS (137)*, and *Procheck (138, 139)* can be used to calculate solvent accessible surface area, involvement in hydrogen bond formation, and main-chain dihedral angles (Ramachandran plots), respectively.

Residues playing a key functional role are also likely to be conserved, in either type or physicochemical properties, between the target and template structures. These may be identified from the literature and/or from template structures in complex with their ligands, in case these are available. Protein–ligand contacts can be calculated by *LIGPLOT (140)* and pre-calculated contacts can be obtained from the *PDBsum* Web site *(141)*.

In principle, residues highly conserved in MSAs might be involved in key structural and/or functional roles. However, in practice, the possibility to predict involvement in important structural or functional roles from sequence conservation is highly dependent on the MSA "quality": if the sequences comprised in the MSA are too closely related, other positions besides the essential ones will be conserved; conversely, if the sequences are distantly related, essential structural or functional roles might be played by non-identical residues sharing specific physicochemical features that might not be easy to identify from an MSA. Nevertheless, once putative key structural and/or functional residues have been identified as described, the conservation of specific features (e.g., main-chain flexibility, hydrogen bond donor ability, presence of an aromatic residue at a position involved in a cation-π interaction, etc.) at given positions can be searched in an MSA in a targeted way. Programs like *COLORADO3D* (142) and *CONSURF (143)* color each residue in a protein structure according to its conservation in the MSA given as input and, therefore, allow visualization of the structural location of conserved and variable residues.

3.3.3. Model Building

This section describes the generation of a 3D protein model based on target-template(s) sequence alignment(s).

3.3.3.1. Choice of the Best Overall Template(s)

Choosing the best template is usually done based on both structural features of the template(s) and information collected for the production of optimal target-template(s) alignments as described in **Section 3.3.1.1,** such as: E-values, and other statistical scores provided by different methods to evaluate the likelihood of the existence of structural relationships between target and template(s); %_ID; number and distribution of gaps; length of the aligned regions (i.e., coverage of the target sequence); correspondence between SSE of the templates and those predicted for the target; absence of insertions and deletions in the target-template alignment within regions corresponding to those structurally conserved among the templates; and conservation of key structural and functional residues between target and template.

If good alignments with different templates are available, structural considerations can help decide which template structure(s) is/are likely to produce most suitable models for their intended applications. In general, x-ray crystallography provides more precise pictures of protein structures than NMR spectroscopy. However, it should be kept in mind that x-ray structures and, therefore, models based on them, represent static images of proteins, which often assume multiple conformations in solution. In the case of structures determined by x-ray crystallography, the parameters to take into account are the following (*see* also **Volume I, Chapter 3**).

1. Resolution and B-factor. The lower the values of these parameters, the better the quality of the structure. In general, for resolution values <2.0 Å the quality of the structure is very high, for values >3.0 Å it is low; B-factor values <30–35, in the range 40–80, and >80 indicate well-determined, mobile, and unreliable regions, respectively.

2. Completeness. Regions corresponding to those relevant for our model in the target-template sequence alignment should not be missing from the template structure (e.g., N-terminal

regions are often cleaved out; exposed loops and N- and C-terminal regions might remain flexible in the crystal structure and not be determined; older, low-resolution structures may contain only Cα carbon atoms; etc.).

3. Protein conformation. The same protein domain can have different conformations in different *PDB* files, depending on the functional state of the domain (e.g., free vs. ligand bound), crystal packing interactions (i.e., interactions with other copies of the same molecule in the crystal), experimental conditions used (e.g., pH, ionic strength, etc.). All this information is contained in the coordinate files of the protein structures, which can be freely downloaded from the *PDB* and visualized using a number of freely available structure visualization programs such as *Swiss-PDB Viewer (144)*, *RasMol (145)*, *Protein Explorer (146)*, *Cn3D (147)*, *WHAT IF (148)*, *MolMol (149)*, and *VMD (150)*. The choice of the template(s) is a trade-off between all these different factors, and the purpose of the model may also be a useful guide (e.g., when modeling receptor–ligand interactions, templates in the ligand-bound conformation should be chosen, if available).

In experiments like *CASP*, predictors often build chimeric models assembling together fragments taken from different templates and/or build several models, based on different main templates and different target-template alignments, which are evaluated at a later stage (*see* **Section 3.5.2**).

3.3.3.2. Model Building

Template-based modeling involves taking advantage of as much as possible information from proteins of known structure putatively homologous to the target, i.e., from the selected template(s). Once refined target-template(s) sequence alignments have been obtained and one or more principal templates have been chosen, model building itself is a relatively straightforward procedure, which can be carried out interactively using structure manipulation programs such as *Swiss-PDB Viewer*, *InsightII/Biopolymer*, or *WHAT IF*.

1. Modeling of the main-chain atoms of regions conserved in the template structure(s). If a single best-template has been detected, the main-chain atoms of conserved regions in the optimized target-template alignment are imported from this template. Conversely, if different regions of the target appear to be more closely related to different structures (e.g., they contain 'gaps' in the alignment with the best template but not with other templates), the conserved main-chain regions of the templates are optimally superimposed (using the structural alignment programs mentioned in **Section 3.3.2.4**), and the regions to serve as templates for different segments

of the target are joined together and imported in the target model.

2. Modeling of the main-chain atoms of structurally variable regions. In case insertions and/or deletions are present in the sequence alignment of the target with all templates and cannot be modeled based on the coordinates of homologous structures, several techniques can be applied to model the regions containing them (usually loops). For a restricted number of loops, sequence–structure relationships have been described, based on which loop conformation can be predicted quite accurately (e.g., antibody loops *(131–134, 151)*, and β-hairpins *(152, 153)*). However, with the exception of these particular cases, and although encouraging results have been achieved in *CASP6* in the CM category, in which four groups were able to predict loops larger than five residues with RMSD <1.0 Å *(14)* (*see* **Note 1**), no method is currently available to accurately and consistently model regions of more than five residues that cannot be aligned to a template; the larger the loops, the more difficult their prediction. Therefore, these regions may either be left out of the model, especially if they are far from the sites of interest of the protein (e.g., active sites) or, if included, it should be pointed out that their reliability is much lower than that of the regions conserved in, and imported from, homologous templates.

One common way to model loops is based on structural searches of the *PDB* database for protein regions having: (1) the same length as the loop to model; (2) a similar conformation of the main-chain atoms of the residues before and after the loop; and (3) a similar pattern of "special residues" that can confer special structural properties to the protein region in which they occur (e.g., Gly, Asn, Asp, or Pro, *see* **Section 3.3.2.5**) and are often important determinants of loop conformation. Conversely, "*ab initio*" methods do not use information contained in structural DBs but try to simulate the folding process or explore the conformational space of the loop region, for example, by molecular dynamics or Monte Carlo methods, followed by energy minimization and selection of low-energy conformations *(154)*.

The interactive graphics software *Swiss-PDB Viewer* provides options to evaluate the compatibility of loops derived from structural searches of the *PDB* with the rest of the target model based on the number of unfavorable van der Waals contacts that they establish and on the results of energy calculations. The groups using *Swiss-PDB Viewer* and the *Loopy* program *(155)* of the *Jackal* package were among the most successful predictors of loops in the *CASP6* CM category. Other software for loop modeling includes *LOBO (156)* and *ModLoop (157)*.

3. Side-chains are generally modeled by "copying" the conformations of conserved residues from their structural template(s) and selecting those of mutated residues from libraries containing the most common conformations that each residue assumes in protein structures (rotamers). After all non-conserved residues have been replaced, unfavorable van der Waals contacts might be present in the model and have to be eliminated, for example, by exploring side-chain conformations different from those involved in the clashes. Alternative side-chain conformations may be examined also to try and bring potentially interacting residues next to each other, for example, in case hydrogen-bond donor or acceptors or, more rarely, positively or negatively charged groups, found in a buried region of the model do not have an interaction partner nearby.

Several programs have been developed to automatically model side-chain conformations, which take into account the above factors and explore combinatorially a number of rotamers for each residue of the model to try and find an optimal conformational ensemble. *SCRWL (158)* was used by several groups providing good rotamer predictions in the CM category in *CASP6 (14)* where, as expected, prediction of rotamers was found to be more difficult for surface than for buried residues, which are subjected to stronger constraints. However, methods providing the best rotamer predictions were not the best at predicting side-chain contacts, which are, in turn, best predicted at the expense of rotamer accuracy *(14)*. The program *Protinfo AB CM (159)* and the *Scap* program *(160)* of the *Jackal* package were used by some of the best predictors of side-chain contacts in the CM category in *CASP6*.

The ability of several programs (*Modeller (161)*, *SegMod/ENCAD (162)*, *SwissModel (163)*, *3D-JIGSAW (164)*, *Nest (165)* of the *Jackal* package, and Builder (*166*)) to build 3D models starting from target-template sequence alignments has been assessed *(167)*. In this test *Modeller*, *Nest*, and *SegMod/ENCAD* performed better than the others, although no program was better than all the others in all tests. *Modeller* is the program used by most of the successful *CASP* groups to generate 3D models from target-template alignments produced using SC- or FR-based methods. The relative performance of *SwissModel (163)*, *CPHmodels (168)*, *3D-JIGSAW*, and *ESyPred3D (169)* is continuously evaluated by *EVA*.

3.3.4. Model Evaluation

Programs like *Procheck*, *Whatcheck (170)*, and *Swiss-PDB Viewer* evaluate the quality of 3D structures based on parameters such as the number of main-chain dihedral angles lying outside the allowed regions of the Ramachandran Plot, unfavorable van der Waals contacts, and buried polar residues not involved in hydrogen bond formation. Some of these programs also evaluate the energy of bond lengths, angles, torsions, and electrostatic interactions based on empirical force fields. These evaluations can be useful to highlight errors in the model resulting from the modeling procedure, or more rarely, inherited from the template structures. However, it is worth stressing that the stereochemical correctness of the model is no guarantee of its biological accuracy, i.e., its actual similarity to the target structure.

Other model quality assessment programs (MQAPs), such as *ProSa (171)*, *Verify-3D (15, 172)*, and *ANOLEA (173, 174)* evaluate model quality based on the comparison between the 3D context of each target residue in the model and the 3D context commonly associated with each residue in known structures. Environmental features taken into account by these programs include neighboring residues or atoms, solvent accessible surface area, and secondary structure of each residue. Similar structural features are incorporated in the neural-network based *ProQ (175)* and *ProQres (176)* programs, which assign quality measures to protein models or parts of them, respectively. Other tools, such as *COLORADO3D*, help evaluate model quality by visualizing information such as sequence conservation, solvent accessibility, and potential errors, including those detected by *ProSa*, *Verify-3D*, and *ANOLEA*. *MQAP-Consensus (177)* uses a consensus of MQAP methods registered in *CAFASP* to evaluate and select models produced by different servers, all of which can be downloaded from the *CAFASP4 MQAP* Web server. One or more of these MQAPs were used by the most successful predictors in *CASP6* to evaluate their models at various stages of the model building and refinement procedures. Successful prediction strategies included collecting models from different servers or building a number of models based on different templates and/or different target-template alignments, and screening them based on quality assessments performed by MQAPs. The regions that are structurally conserved in the different models and/or are considered to be reliable by MQAPs are retained, whereas structurally variable and/or less reliable regions according to MQAPs are realigned and remodeled until they either reach an acceptable quality, as measured by MQAP methods, or their quality cannot be improved anymore in subsequent refinement cycles.

Blind predictions of both the overall and residue-based quality of protein models were analyzed by human assessors for the first time in *CASP7*. The relative assessment paper in the 2007 issue of the Journal *PROTEINS* dedicated to CASP7 provides a reliable picture of the relative performance of MQAPs in blind tests.

3.3.5. Model Refinement

Based on the results of quality assessment programs, 3D models can be refined to eliminate obvious structural mistakes (for example by selecting alternative side-chain rotamers to eliminate unfavorable residue–residue interactions) using structure visualization programs such as *Swiss-PDB Viewer* and *InsightII/Biopolymer (178)*. *Swiss-PDB Viewer*, *GROMACS (179)*, *NAMD (180)*, *TINKER*, and the *Discover* module of *InsightII* can also perform energy minimizations. Since no evidence has been collected over the various *CASP* experiments about model improvements achieved by energy minimization procedures, these should only be limited to the regions showing bad geometrical parameters (e.g., those in which fragments from different structures have been joined, for example, in loop modeling) and involved in clashes that cannot be relieved by changing side-chain conformations (e.g., those involving main-chain atoms and/or proline side-chains). However, as discussed for loop modeling, since there is no evidence that such procedures will improve the model (i.e., make it more similar to the target structure, as opposed to improving its geometry) rather than make it worse, and depending on the proximity of any problematic regions to important structural/functional sections of the model, small structural errors may also be left unrefined and indicated as potentially less reliable regions in the model.

The increase in evolutionary distance between target and template is associated with a reduction of the conserved core and an enlargement of the variable regions, which in turn makes it more and more difficult to align the target and template sequences correctly. For distantly related templates (e.g., those identified by FR methods) errors in the target-template(s) sequence alignment might result in serious mistakes that cannot be corrected by modifying the coordinates of the final model. When this occurs, it is necessary to re-evaluate the target-template alignment by making use of any 3D information contained in the model, and try to modify the alignment in such a way that the new model generated from it will not contain the aforementioned errors. Several cycles of model building and evaluation might be necessary to achieve this result.

3.4. Structure Prediction Without Template

As discussed in the preceding, evolutionarily related domains of known structure are used as templates for the whole target. When no template structure can be identified in the DBs by either sequence-based or FR methods, two scenarios are possible: either a structure similar to the target is present in the DBs, but none of the aforementioned SC- or FR-based methods is able to detect it, or no similar structure is available, i.e., the target structure is actually a NF. In both cases, different methods from those described before have to be used. The so-called *ab initio* methods, which use computationally intensive strategies attempting to recreate the physical and chemical forces involved in protein folding, have been, until now, less successful at predicting protein structures in absence of structural templates than the knowledge-based approaches. These exploit information contained in the structure databases, from which fragments potentially similar to the targets are extracted and assembled together to produce a full 3D model of the target *(18, 19)*.

Models built by fragment-assembly techniques are evaluated using knowledge-based statistical potentials and clustering procedures. Knowledge-based potentials contain terms derived from statistical analyses of protein structures as well as physicochemical terms (e.g., terms for pairwise interactions between residues or atoms, residue hydrophobicity, hydrogen bonds, and main-chain and side-chain dihedral angles). Since large numbers of conformations are generated for each target, a clustering analysis is performed based on structure similarity and the cluster centroid model is usually chosen. Models representative of highly populated clusters are assumed to be more likely to be correct than models from less populated clusters. Several NF methods take advantage of homologous sequence information, either for secondary structure predictions or as input for model building, and some use 3D models produced by automated CM, FR, and/or NF prediction servers and consensus SSE predictions to derive structural restraints that are used to guide or constrain a subsequent fragment assembly procedure or folding simulation. Manual intervention can occur at different stages, for example, to choose templates or fragments, or inspect models.

Although originally developed for NF predictions, fragment-based methods have been used by several successful predictors in the FR category in both *CASP6* and *CASP5*, and the group that developed the program *Rosetta (19, 182–186)* and the server *Robetta (187–189)* has been the most successful at predicting the structure of difficult FR targets, i.e., targets for which no evolutionary relationship with known structures is apparent and are classified in the *CASP* FR/A (A: analogous) sub-category *(190, 191)*.

Unfortunately, the performance of these methods on real NF targets is somewhat less successful. In *CASP6*, nine targets whose structures did not show overall similarities with already known folds based on the results of the *LGA* program *(68)* were evaluated in the NF category *(190)*. Three of them, however, turned out to be variants of known folds, in that they contain sub-structures that match sub-structures in known proteins, and were therefore defined as NF "easy" *(190)*. In the NF category, models bearing an overall structural similarity with the target structures were submitted only for these three "easy" targets. For the remaining six NF "hard" targets, which did not show any similarity to known folds (and were, therefore, the only truly "novel" folds), no globally correct prediction was submitted. It should be mentioned, however, that all of the NF "hard" targets were relatively large proteins (*115–213* a.a.s vs. *73–90* a.a.s of NF "easy" targets), which are particularly difficult to predict by NF methods. The best structure predictions for these targets were limited to partial regions, although often larger than standard supersecondary structures, which could not be assembled into correct global topologies. Since it is not obvious how to compare predictions for different target fragments in the context of globally incorrect models, it is particularly difficult to evaluate these predictions and decide which of the methods were the most successful. Further, the difference between the best and average models submitted for these targets is not very large, and since most models are poor even large differences in ranking are unlikely to be significant. Another general drawback of methods used to predict NF targets is that predictors were not particularly good at recognizing their best models, which highlights a problem with ranking. Finally, even considering all the nine NF targets, the sample size is too small to draw statistically significant conclusions *(190)*. Taking all these caveats into account, FRAGFOLD *(192)*, CABS *(193)*, and *Rosetta* were used by the best performing groups in the NF category, although several other groups submitted each at least one best model for different targets. Another freely available program that was used by a relatively well-performing group is *ROKKY* (194).

The *Rosetta* folding simulation program by the Baker group *(19, 182–184)* is not only one of the best available methods for NF and difficult FR targets predictions (the Baker group was the best or among the best performing at predictions without template in the last *CASP* editions) but, being freely available, it is also one of the most popular and one that has been incorporated in a large number of metaservers (*see* **Section 3.5.2**). The output of *Rosetta* contains a number of 3D models of the target protein ordered according to an energy score. As for FR methods, this score cannot be solely relied upon, and all additional sequence, structural, functional, and evolutionary information that can be gathered about the target should be exploited. Since a sequence for which no prediction can be provided by either SC- or FR-methods might be a difficult FR target (FR/A) rather than an NF, top scoring models and other models that might be correct based on all available information about the target should be used to search structural DBs with structural alignment programs (e.g., *DALI (31)*, *VAST, CE, SSM, Mammoth-mult*). If proteins of known structure showing significant structural similarity with these models can be found (i.e., the sequence is likely to be a difficult FR target), these should be examined to identify sequence, structural, and/or functional evidence (e.g., key residues, *see* **Section 3.3.2.5**) that might support the hypothesis of the correctness of the model(s) produced for the target. Conversely, if no protein showing global structural similarity with the target is detected, the target might indeed have a novel fold. Nevertheless, searches of structural DBs might detect proteins containing sub-structures similar to those of the target, and the analysis of these structurally similar regions might provide important clues supporting some models over others.

Given their independence from structural templates, NF techniques can be useful to try to predict not only complete 3D structures for targets that elude CM and FR methods, but also loops and other regions of models generated by template-based methods that are not conserved with respect to the available templates.

3.5. Automated Methods, Metaservers, and 3D Model DBs

3.5.1. Automated Methods

A number of automated CM, FR, and NF methods and metaservers for protein structure prediction are described in the relevant sections of this chapter (e.g., *see* **Sections 3.3.1.1, 3.3.1.2, 3.3.3.2, 3.4, and 3.5.2**), together with their performance in the hands of expert users. As far as the performance of automated methods without user intervention is concerned, *SPARKS2* and *SP3* have been the first and third best performing servers in the CM category in *CASP6*, and the 14th and 22nd overall best predictors in ranking including human predictors *(14)*. The second best server was *STRUCTFAST (195)*. In the FR category *Robetta* was the best server and the overall ninth best predictor in ranking including human predictors. In the NF category, *Robetta* was the best server. The best performing servers in *Livebench* and *CAFASP* are all metaservers (*see* **Section 3.5.2**), in particular *3D-Jury* (196), *3D-Shotgun (197–200)*, *Pcons*, and *Pmodeller (201)*.

In spite of the good performance of the best automated methods in blind tests, intervention by expert users (often by the same authors of the automated procedures) consistently provides significantly improved results with respect to fully automated methods, indicating that, in spite of much effort, the prediction community has not yet managed to encode all expert knowledge on protein structure prediction into automated programs. Therefore, depending on the number of models required, the purpose for which they are built, and the user's level of expertise, automated programs might be best used as a support for, rather than a replacement of, human predictions.

3.5.2. Metaservers

Metaservers collect a variety of predictions and structural information from different automated servers, evaluate their quality, and combine them to generate consensus predictions that might consist either in the result provided by a single structure prediction server and reputed to be the best among those examined, or in the combination of results provided by different servers. Putative *PDB* templates, target-template sequence alignments, and 3D structural models are collected from fully automated SC, FR, or NF method-based servers and in some cases other metaservers that perform regularly well in blind test experiments such as *CASP, CAFASP*, or *Livebench*. This information is often integrated with domain, secondary structure, and function predictions, as well as with information derived from analyses of protein structures, e.g., protein structure classification by *SCOP* and/or *FSSP* and secondary structure assignment by *DSSP*. Depending on the metaserver, different types of predictions are returned, consisting of 3D model coordinates *(202)* or target-template sequence alignments from which 3D models can then be generated *(113)*. Preliminary models may be fragmented and spliced to produce new hybrid models and/or subjected to refinements to eliminate errors such as overlapping regions or broken chains. Predictions of domain boundaries, secondary structure, transmembrane helices, and disordered regions are often provided as well.

Metaservers exploit a variety of criteria and methods to evaluate the quality of alignments and models that they collect and produce, including: the scores associated with the predictions by contributing servers; the extent to which different alignments agree with one another; contact order, i.e., the average sequence separation between all pairs of contacting residues normalized by the total sequence length *(203, 204)*; structural comparisons of independent models and detection of large clusters of structures with similar conformations *(205)*; and model/structure evaluation programs mentioned before, such as *ProSa*, *Verify-3D*, *ProQ*, *ProQres*, and *COLORADO3D*. Well-scoring model fragments (e.g., consensus predictions provided by unrelated methods) are considered to be reliable and are included in the model, whereas for poorly scoring regions models based on alternative alignments are generated and assessed using MQAPs until they reach acceptable quality or their quality cannot be further improved.

Many of the available metaservers have implemented completely automated procedures to cover the full spectrum of structure predictions, from CM to NF, by combining template-based and *de novo* structure prediction methods. First, several well-performing CM and FR servers are queried with the sequence of the target protein. If homologs with experimentally characterized 3D structure are detected, the conserved regions of the target are predicted by template-based modeling and the variable regions (e.g., loops, N- and C-terminal extensions) by NF methods such as the *Rosetta* fragment-insertion protocol. If no reliable structural homolog is found, the target sequence is sent to NF prediction servers (most often *Robetta*).

Predictors that used metaservers performed among the best in the more recent *CASP* editions, in both CM *(14)* and FR *(191)* categories. Although metaservers owe much of their predictive power to the quality of the results provided by remote servers, their ability to choose, evaluate, and combine different predictions has often resulted in better performances than those achieved by the remote prediction servers alone.

Meta-BASIC and *3D-Jury* were used by the top ranking predictor in both *CASP6* CM and FR/H categories *(14, 191)*, and

3D-Jury was also used by another of the top four predictors in the CM category *(206)*.

The group that developed the *Robetta* server ranked best in FR/A, among the best in FR and NF, and well in the CM category in *CASP6*. The *Robetta* server itself performed well in both the CM and FR categories in *CASP6*.

The *Genesilico* metaserver *(202)* was used by one of the most successful groups in the CM and FR categories in *CASP5* and was one of the best in all categories in *CASP6*, where *Genesilico* was used in combination with the *FRankenstein3D* program for splicing of FR models *(207)* and fragment assembly techniques for NF targets *(193)*.

Pcons, *Pmodeller* and *Pcons5 (208, 209)*, and *Bioinbgu*, *3D-Shotgun* and *Inub (197–200)* were among the best performing servers in *CASP5* and groups using them performed well, although not among the best, in both the CM and FR categories in *CASP6*.

The groups using *@TOME (210)* and a combination of CHIMERA and FAMS *(211)* performed well, although they were not among the best groups, in *CASP6*.

3.5.3. 3D Model DBs

Before embarking on the modeling procedure, however automatically, it might be worth checking whether 3D models for the target of interest are available among protein model DBs. The *SwissModel Repository (212)* and *ModBase (213)* contain 3D models of all sequences in the *SwissProt* and *TrEMBL* databases that have detectable sequence similarity with proteins of known structure, generated by the *SwissModel* and *Modeller* programs, respectively. *FAMSBASE (214)*, built using the fully automated modeling system FAMS, contains comparative models for all the proteins assigned to many sequenced genomes. Of course, models produced automatically, even those for "easy" CM targets, may contain significant errors; to detect them, models should be subjected to all the sequence, structure, function, and evolutionary analyses described before. The Protein Model Database (*PMDB*) *(215)* stores manually built 3D models together with any supporting information provided. It contains, among others, the models submitted to the past *CASP* editions, and it allows users to submit, as well as to retrieve, 3D protein models.

3.6. Expected Accuracy of the Model

A general indication of the accuracy of template-based models is provided by sequence–structure relationships derived from the analysis of homologous proteins *(13)*. Based on these analyses, if the sequence identity between target and template in the conserved core regions is about 50% or higher, we can expect the main-chain atoms of the core regions of the target structure to be superimposable to the best template and, consequently, to the model built on this template, with an RMSD value (*see* **Note 1**) within 1.0 Å. However, for sequence identities as low as 20%, the structural similarity between two proteins can vary to a large extent *(13)*.

Similar results are provided by the structural comparison of the target domain structures and the best models produced for them in *CASP6* (**Fig. 2.2**). For %_ID values between the sequences of target and best-template of 50% or higher, the GDT_TS (*see* **Note 1**) between the real target structure and the best model submitted for it was 90 or higher. For %_ID values between 20% and 30% the GDT_TS varied between about 100 and 50. For %_ID values <20%, the GDT_TS was between 90 and 40 for CM targets, between 90 and 30 for FR targets, and from 60 to <20 for NF targets. In other words, for sequence identities of 20% or lower, the model built can vary from either having a globally very similar structure to the target (GDT_TS >80) to only having a few small fragments showing some structural similarity to the target (GDT_TS <20). A visual example of structural variations corresponding to representative GDT_TS values is shown in **Fig. 2.2**.

In *CASP6*, for "easy" CM targets template detection and production of accurate target-template sequence alignments were carried out successfully and the accuracy of the produced models was quite high, at least in the conserved regions. Conversely, model refinement and improvement over the best template were still open issues: In only a few cases were the best predictors able to produce 3D models more similar to their target structures than the structures of the best templates. In most cases, improvements over the best templates were obtained only for the easiest targets, whereas for harder targets the best templates were generally much more similar to the target structures than any model. In *CASP7*, for the first time, models of such "easy" CM targets were assessed in a separate category to establish whether there has been any improvement in these areas. For "hard" *CASP6* CM targets choosing the right template and producing the correct target-template alignment proved to be challenging tasks *(14)*, and both became more and more difficult with an increase in the evolutionary distance between targets and best templates. For most FR targets the top predictions contained all or most of the SSE in the right place or with small shifts with respect to the target structure. However, for many targets most predictions had little resemblance to the target structure *(191)*. For "hard" NF targets, only fragments of the best models resembled the real target structures *(190)*.

Structural comparisons of the templates selected for a specific target (*see* **Section 3.3.2.4**) permit identification of relationships between %_ID and structural similarity for the target family, based on which a more accurate prediction about the expected model accuracy for targets belonging to that family can be obtained. In any case, information about model accuracy provided by sequence–structure relationships is limited to the conserved regions between target and template(s). Unless loop regions belong to one of the restricted number of categories for which sequence–structure relationships have been defined (*see* **Section 3.3.3.2**), or side-chain conformations are strongly conserved for structural and/or functional reasons, in general it is not possible to know how accurate the predictions of loops or side-chain conformations are before comparing them with the real structures.

Automated programs can generally produce models of "easy" CM targets (e.g., sequence identity with the templates >40%) that are good enough for many practical applications. However, these models may contain serious mistakes even in regions close to the functional sites, the more so the higher the evolutionary distance between target and template; therefore, it is wise to carefully compare their output with the template structures, especially in the key structural and functional regions. This is a point of utmost importance: In principle, it is always possible to build homology-based 3D models automatically; all that is needed is a template structure and an alignment between the target and template sequences. However, if the chosen template is not a real target homologue, or if the target-template sequence alignment is wrong, the model is bound to be incorrect ("garbage *in*, garbage *out*!"). In *CASP*, expert human predictors using semi-automated procedures consistently produce better predictions than fully automated methods. However, in choosing between the use of a fully automated method and manual intervention it is important to remember that in both CM and FR categories there was a big difference between the best predictors and many of the others, and that the best automated programs (*see* **Section 3.5.1**) performed remarkably well compared with many human predictors.

4. Notes

1. The two commonly used measures to evaluate similarity between protein structures used in this chapter are root-mean-square deviation (RMSD) and global distance test total score (GDT_TS). The RMSD between two structures A and B is the average distance between a specific set of atoms in structure A (e.g., Cα or main-chain atoms belonging to a given set of residues) and the corresponding atoms in structure B, after optimal superposition of their coordinates. The formula to calculate the RMSD is:

$$RMSD = \sqrt{\frac{\sum_{i=1}^{N} d_i^2}{N}}$$

where d_i is the distance between the pair of atoms i and N is the total number of superimposed atom pairs in the dataset. The GDT_TS between two structures A and B represents the percentage of residues of protein A whose Cα atoms are found within specific cut-off distances from the corresponding Cα atoms in structure B after optimal superimposition of the two structures. In *CASP* the GDT_TS is calculated according to the formula:

$$GDT\_TS = (GDT\_P1 + GDT\_P2 + GDT\_P4 + GDT\_P8)/4,$$

where GDT_P1, GDT_P2, GDT_P4, and GDT_P8 represent the number of Cα atoms of a target model whose distance from the corresponding atoms of the real structure, after optimal model-to-structure superposition, is ≤1, 2, 4, and 8 Å, respectively, divided by the number of residues in the target structure.

2. The score of the *BLAST* alignment is calculated as the sum of substitution and gap scores. Substitution scores are given by the chosen substitution matrix (e.g., PAM, BLOSUM, *see* **Note 3**), whereas gap scores are calculated as the sum of the gap opening and gap extension penalties used in the search. The choice of these penalties is empirical, but in general a high value is chosen for gap opening and a low value for gap extensions (e.g., *BLAST* default values for gap insertions and extensions are 7–12 and 1–2, respectively). The rational bases for these choices are that: (1) only a few regions of protein structures (e.g., loops) can accommodate insertions and deletions (hence, a high value for gap openings); and (2) most of these regions can usually accept insertions and deletions of more than a single residue (hence, the lower value used for gap extension).

3. Substitution matrices are 20 × 20 matrices containing a value for each a.a. residue pair that is proportional to the probability that the two a.a.s substitute for each other in alignments of homologous proteins. The two most commonly used matrix series, PAM (percent accepted mutation) *(216)* and BLOSUM (blocks substitution matrix) *(217)*, comprise several matrices that have been designed to align proteins with varying extent of evolutionary distance. PAM matrices with high numbers (e.g., PAM 250) and BLOSUM matrices with low numbers (e.g., BLOSUM 45) are suitable for aligning distantly related sequences; conversely, low PAM (e.g., PAM 1) and high BLOSUM (e.g., BLOSUM 80) number matrices are appropriate to align closely related sequences. *BLAST* default matrix is BLOSUM 62.

4. In *SCOP*, protein domains are assigned to the same family if they either have %_ID ≥ 30% or, in case their %_ID is lower,

if they have very similar structures and functions. Different families are assigned to the same superfamily if they comprise proteins presenting structural and, often, functional features suggesting a common evolutionary origin. Families and superfamilies having the same major SSE in the same spatial arrangement and with the same topological connections are assigned to the same fold. Proteins assigned to the same families and superfamilies are thought to be close and remote evolutionary relatives, respectively, whereas proteins having the same fold might still have a common origin but no evidences strong enough to support this hypothesis are available yet.

Acknowledgments

The authors gratefully acknowledge Claudia Bertonati, Gianni Colotti, Andrea Ilari, Romina Oliva, and Christine Vogel for manuscript reading and suggestions, and Julian Gough and Martin Madera for discussions.

References

1. Moult, J., Pedersen, J. T., Judson, R., et al. (1995) A large-scale experiment to assess protein structure prediction methods. *Proteins* 23, ii–v.

2. Moult, J., Hubbard, T., Bryant, S. H., et al. (1997) Critical assessment of methods of protein structure prediction (CASP): round II. *Proteins* Suppl. 1, 2–6.

3. Moult, J., Hubbard, T., Fidelis, K., et al. (1999) Critical assessment of methods of protein structure prediction (CASP): round III. *Proteins* Suppl. 3, 2–6.

4. Moult, J., Fidelis, K., Zemla, A., et al. (2001) Critical assessment of methods of protein structure prediction (CASP): round IV. *Proteins* Suppl. 5, 2–7.

5. Moult, J., Fidelis, K., Zemla, A., et al. (2003) Critical assessment of methods of protein structure prediction (CASP): round V. *Proteins* 53, Suppl. 6, 334–339.

6. Moult, J., Fidelis, K., Rost, B., et al. (2005) Critical assessment of methods of protein structure prediction (CASP): round 6. *Proteins* 61, Suppl. 7, 3–7.

7. Fischer, D., Barret, C., Bryson, K., et al. (1999) CAFASP-1: critical assessment of fully automated structure prediction methods. *Proteins* Suppl. 3, 209–217.

8. Fischer, D., Elofsson, A., Rychlewski, L., et al. (2001) CAFASP2: the second critical assessment of fully automated structure prediction methods. *Proteins* Suppl. 5, 171–183.

9. Fischer, D., Rychlewski, L., Dunbrack, R. L., Jr., et al. (2003) CAFASP3: the third critical assessment of fully automated structure prediction methods. *Proteins* 53, Suppl. 6, 503–516.

10. Berman, H. M., Westbrook, J., Feng, Z., et al. (2000) The Protein Data Bank. *Nucleic Acids Res* 28, 235–242.

11. Rychlewski, L., Fischer, D. (2005) Live Bench-8: the large-scale, continuous assessment of automated protein structure prediction. *Protein Sci* 14, 240–245.

12. Koh, I. Y., Eyrich, V. A., Marti-Renom, M. A., et al. (2003) EVA: Evaluation of protein structure prediction servers. *Nucleic Acids Res* 31, 3311–3315.

13. Chothia, C., Lesk, A. M. (1986) The relation between the divergence of sequence and structure in proteins. *Embo J* 5, 823–826.

14. Tress, M., Ezkurdia, I., Grana, O., et al. (2005) Assessment of predictions submitted for the CASP6 comparative modeling category. *Proteins* 61, Suppl. 7, 27–45.

15. Bowie, J. U., Luthy, R., Eisenberg, D. (1991) A method to identify protein sequences that fold into a known three-dimensional structure. *Science* 253, 164–170.

16. Jones, D. T., Taylor, W. R., Thornton, J. M. (1992) A new approach to protein fold recognition. *Nature* 358, 86–89.

17. Sippl, M. J., Weitckus, S. (1992) Detection of native-like models for amino acid sequences of unknown three-dimensional structure in a data base of known protein conformations. *Proteins* 13, 258–271.

18. Jones, D. T. (1997) Successful ab initio prediction of the tertiary structure of NK-lysin using multiple sequences and recognized supersecondary structural motifs. *Proteins* Suppl. 1, 185–191.

19. Simons, K. T., Kooperberg, C., Huang, E., et al. (1997) Assembly of protein tertiary structures from fragments with similar local sequences using simulated annealing and Bayesian scoring functions. *J Mol Biol* 268, 209–225.

20. Sprague, E. R., Wang, C., Baker, D., et al. (2006) Crystal structure of the HSV-1 Fc receptor bound to Fc reveals a mechanism for antibody bipolar bridging. *PLoS Biol* 4, e148.

21. Galperin, M. Y. (2006) The Molecular Biology Database Collection: 2006 update. *Nucleic Acids Res* 34, D3–5.

22. Fox, J. A., McMillan, S., Ouellette, B. F. (2006) A compilation of molecular biology web servers: 2006 update on the Bioinformatics Links Directory. *Nucleic Acids Res* 34, W3–5.

23. Benson, D. A., Boguski, M. S., Lipman, D. J., et al. (1997) GenBank. *Nucleic Acids Res* 25, 1–6.

24. Wu, C. H., Apweiler, R., Bairoch, A., et al. (2006) The Universal Protein Resource (UniProt): an expanding universe of protein information. *Nucleic Acids Res* 34, D187–191.

25. Coutinho, P. M., Henrissat, B. (1999) Carbohydrate-active enzymes: an integrated database approach. In *Recent Advances in Carbohydrate Bioengineering*. H.J. Gilbert, G. Davies, B. Henrissat and B. Svensson eds., The Royal Society of Chemistry, Cambridge, UK, pp. 3–12.

26. Lander, E. S., Linton, L. M., Birren, B., et al. (2001) Initial sequencing and analysis of the human genome. *Nature* 409, 860–921.

27. LoVerde, P. T., Hirai, H., Merrick, J. M., et al. (2004) Schistosoma mansoni genome project: an update. *Parasitol Int* 53, 183–192.

28. Andreeva, A., Howorth, D., Brenner, S. E., et al. (2004) SCOP database in 2004: refinements integrate structure and sequence family data. *Nucleic Acids Res* 32, D226–229.

29. Pearl, F., Todd, A., Sillitoe, I., et al. (2005) The CATH Domain Structure Database and related resources Gene3D and DHS provide comprehensive domain family information for genome analysis. *Nucleic Acids Res* 33, D247–251.

30. Holm, L., Ouzounis, C., Sander, C., et al. (1992) A database of protein structure families with common folding motifs. *Protein Sci* 1, 1691–1698.

31. Holm, L., Sander, C. (1997) Dali/FSSP classification of three-dimensional protein folds. *Nucleic Acids Res* 25, 231–234.

32. Mizuguchi, K., Deane, C. M., Blundell, T. L., et al. (1998) HOMSTRAD: a database of protein structure alignments for homologous families. *Protein Sci* 7, 2469–2471.

33. Gasteiger, E., Gattiker, A., Hoogland, C., et al. (2003) ExPASy: the proteomics server for in-depth protein knowledge and analysis. *Nucleic Acids Res* 31, 3784–3788.

34. Smith, R. F., Wiese, B. A., Wojzynski, M. K., et al. (1996) BCM Search Launcher—an integrated interface to molecular biology data base search and analysis services available on the World Wide Web. *Genome Res* 6, 454–462.

35. Stothard, P. (2000) The sequence manipulation suite: JavaScript programs for analyzing and formatting protein and DNA sequences. *Biotechniques* 28, 1102, 1104.

36. Janin, J. (2005) Assessing predictions of protein-protein interaction: the CAPRI experiment. *Protein Sci* 14, 278–283.

37. Janin, J., Henrick, K., Moult, J., et al. (2003) CAPRI: a Critical Assessment of PRedicted Interactions. *Proteins* 52, 2–9.

38. Tai, C. H., Lee, W. J., Vincent, J. J., et al. (2005) Evaluation of domain prediction in CASP6. *Proteins* 61, Suppl. 7, 183–192.

39. Kim, D. E., Chivian, D., Malmstrom, L., et al. (2005) Automated prediction of domain boundaries in CASP6 targets using Ginzu and RosettaDOM. *Proteins* 61, Suppl. 7, 193–200.

40. Suyama, M., Ohara, O. (2003) DomCut: prediction of inter-domain linker regions in amino acid sequences. *Bioinformatics* 19, 673–674.

41. Marchler-Bauer, A., Anderson, J. B., Cherukuri, P. F., et al. (2005) CDD: a Conserved Domain Database for protein classification. *Nucleic Acids Res* 33, D192–196.

42. Finn, R. D., Mistry, J., Schuster-Bockler, B., et al. (2006) Pfam: clans, web tools and services. *Nucleic Acids Res* 34, D247–251.

43. Letunic, I., Copley, R. R., Pils, B., et al. (2006) SMART 5: domains in the context of genomes and networks. *Nucleic Acids Res* 34, D257–260.

44. Bru, C., Courcelle, E., Carrere, S., et al. (2005) The ProDom database of protein domain families: more emphasis on 3D. *Nucleic Acids Res* 33, D212–215.

45. Mulder, N. J., Apweiler, R., Attwood, T. K., et al. (2005) InterPro, progress and status in 2005. *Nucleic Acids Res* 33, D201–205.

46. Hulo, N., Bairoch, A., Bulliard, V., et al. (2006) The PROSITE database. *Nucleic Acids Res* 34, D227–230.

47. Gough, J., Chothia, C. (2002) SUPER-FAMILY: HMMs representing all proteins of known structure. SCOP sequence searches, alignments and genome assignments. *Nucleic Acids Res* 30, 268–272.

48. Madera, M., Vogel, C., Kummerfeld, S. K., et al. (2004) The SUPERFAMILY database in 2004: additions and improvements. *Nucleic Acids Res* 32, D235–239.

49. Jin, Y., Dunbrack, R. L., Jr. (2005) Assessment of disorder predictions in CASP6. *Proteins* 61, Suppl. 7, 167–175.

50. Obradovic, Z., Peng, K., Vucetic, S., et al. (2005) Exploiting heterogeneous sequence properties improves prediction of protein disorder. *Proteins* 61, Suppl. 7, 176–182.

51. Peng, K., Radivojac, P., Vucetic, S., et al. (2006) Length-dependent prediction of protein intrinsic disorder. *BMC Bioinformatics* 7, 208.

52. Cheng, J., Sweredoski, M., Baldi, P. (2005) Accurate prediction of protein disordered regions by mining protein structure data. *Data Mining Knowl Disc* 11, 213–222.

53. Dosztanyi, Z., Csizmok, V., Tompa, P., et al. (2005) IUPred: web server for the prediction of intrinsically unstructured regions of proteins based on estimated energy content. *Bioinformatics* 21, 3433–3434.

54. Vullo, A., Bortolami, O., Pollastri, G., et al. (2006) Spritz: a server for the prediction of intrinsically disordered regions in protein sequences using kernel machines. *Nucleic Acids Res* 34, W164–168.

55. Ward, J. J., Sodhi, J. S., McGuffin, L. J., et al. (2004) Prediction and functional analysis of native disorder in proteins from the three kingdoms of life. *J Mol Biol* 337, 635–645.

56. Bryson, K., McGuffin, L. J., Marsden, R. L., et al. (2005) Protein structure prediction servers at University College London. *Nucleic Acids Res* 33, W36–38.

57. Krogh, A., Larsson, B., von Heijne, G., et al. (2001) Predicting transmembrane protein topology with a hidden Markov model: application to complete genomes. *J Mol Biol* 305, 567–580.

58. Rost, B., Yachdav, G., Liu, J. (2004) The PredictProtein server. *Nucleic Acids Res* 32, W321–326.

59. Bagos, P. G., Liakopoulos, T. D., Spyropoulos, I. C., et al. (2004) PRED-TMBB: a web server for predicting the topology of beta-barrel outer membrane proteins. *Nucleic Acids Res.* 32, W400–404.

60. Natt, N. K., Kaur, H., Raghava, G. P. (2004) Prediction of transmembrane regions of beta-barrel proteins using ANN- and SVM-based methods. *Proteins* 56, 11–18.

61. Jones, D. T. (1999) Protein secondary structure prediction based on position-specific scoring matrices. *J Mol Biol* 292, 195–202.

62. Karplus, K., Barrett, C., Hughey, R. (1998) Hidden Markov models for detecting remote protein homologies. *Bioinformatics* 14, 846–856.

63. Pollastri, G., McLysaght, A. (2005) Porter: a new, accurate server for protein secondary structure prediction. *Bioinformatics* 21, 1719–1720.

64. Cuff, J. A., Barton, G. J. (2000) Application of multiple sequence alignment profiles to improve protein secondary structure prediction. *Proteins* 40, 502–511.

65. Cuff, J. A., Clamp, M. E., Siddiqui, A. S., et al. (1998) JPred: a consensus secondary structure prediction server. *Bioinformatics* 14, 892–893.

66. Altschul, S. F., Madden, T. L., Schaffer, A. A., et al. (1997) Gapped BLAST and PSI-BLAST: a new generation of protein database search programs. *Nucleic Acids Res* 25, 3389–3402.

67. Tress, M., Tai, C. H., Wang, G., et al. (2005) Domain definition and target classification for CASP6. *Proteins* 61, Suppl. 7, 8–18.

68. Pearson, W. R. (1990) Rapid and sensitive sequence comparison with FASTP and FASTA. *Methods Enzymol* 183, 63–98.

69. Pearson, W. R. (1995) Comparison of methods for searching protein sequence databases. *Protein Sci* 4, 1145–1160.

70. Park, J., Karplus, K., Barrett, C., et al. (1998) Sequence comparisons using

multiple sequences detect three times as many remote homologues as pairwise methods. *J Mol Biol* 284, 1201–1210.

71. Eddy, S. R. (1996) Hidden Markov models. *Curr Opin Struct Biol* 6, 361–365.

72. Eddy, S. R. (1998) Profile hidden Markov models. *Bioinformatics* 14, 755–763.

73. Madera, M., Gough, J. (2002) A comparison of profile hidden Markov model procedures for remote homology detection. *Nucleic Acids Res* 30, 4321–4328.

74. Karplus, K., Karchin, R., Draper, J., et al. (2003) Combining local-structure, fold-recognition, and new fold methods for protein structure prediction. *Proteins* 53, Suppl. 6, 491–496.

75. Karplus, K., Katzman, S., Shackleford, G., et al. (2005) SAM-T04: what is new in protein-structure prediction for CASP6. *Proteins* 61, Suppl. 7, 135–142.

76. Schaffer, A. A., Wolf, Y. I., Ponting, C. P., et al. (1999) IMPALA: matching a protein sequence against a collection of PSI-BLAST-constructed position-specific score matrices. *Bioinformatics* 15, 1000–1011.

77. Ohlson, T., Wallner, B., Elofsson, A. (2004) Profile-profile methods provide improved fold-recognition: a study of different profile-profile alignment methods. *Proteins* 57, 188–197.

78. Yona, G., Levitt, M. (2002) Within the twilight zone: a sensitive profile-profile comparison tool based on information theory. *J Mol Biol* 315, 1257–1275.

79. von Ohsen, N., Sommer, I., Zimmer, R. (2003) Profile-profile alignment: a powerful tool for protein structure prediction. *Pac Symp Biocomput* 252–263.

80. von Ohsen, N., Sommer, I., Zimmer, R., et al. (2004) Arby: automatic protein structure prediction using profile-profile alignment and confidence measures. *Bioinformatics* 20, 2228–2235.

81. Sadreyev, R., Grishin, N. (2003) COMPASS: a tool for comparison of multiple protein alignments with assessment of statistical significance. *J Mol Biol* 326, 317–336.

82. Mittelman, D., Sadreyev, R., Grishin, N. (2003) Probabilistic scoring measures for profile-profile comparison yield more accurate short seed alignments. *Bioinformatics* 19, 1531–1539.

83. Sadreyev, R. I., Baker, D., Grishin, N. V. (2003) Profile-profile comparisons by COMPASS predict intricate homologies between protein families. *Protein Sci* 12, 2262–2272.

84. Heger, A., Holm, L. (2001) Picasso: generating a covering set of protein family profiles. *Bioinformatics* 17, 272–279.

85. Edgar, R. C., Sjolander, K. (2004) COACH: profile-profile alignment of protein families using hidden Markov models. *Bioinformatics* 20, 1309–1318.

86. Pietrokovski, S. (1996) Searching databases of conserved sequence regions by aligning protein multiple-alignments. *Nucleic Acids Res* 24, 3836–3845.

87. Jaroszewski, L., Rychlewski, L., Li, Z., et al. (2005) FFAS03: a server for profile–profile sequence alignments. *Nucleic Acids Res* 33, W284–288.

88. Tomii, K., Akiyama, Y. (2004) FORTE: a profile-profile comparison tool for protein fold recognition. *Bioinformatics* 20, 594–595.

89. Ginalski, K., Pas, J., Wyrwicz, L. S., et al. (2003) ORFeus: detection of distant homology using sequence profiles and predicted secondary structure. *Nucleic Acids Res* 31, 3804–3807.

90. Soding, J., Biegert, A., Lupas, A. N. (2005) The HHpred interactive server for protein homology detection and structure prediction. *Nucleic Acids Res* 33, W244–248.

91. Kabsch, W., Sander, C. (1983) Dictionary of protein secondary structure: pattern recognition of hydrogen-bonded and geometrical features. *Biopolymers* 22, 2577–2637.

92. Sippl, M. J. (1995) Knowledge-based potentials for proteins. *Curr Opin Struct Biol* 5, 229–235.

93. Kelley, L. A., MacCallum, R. M., Sternberg, M. J. (2000) Enhanced genome annotation using structural profiles in the program 3D-PSSM. *J Mol Biol* 299, 499–520.

94. Jones, D. T. (1999) GenTHREADER: an efficient and reliable protein fold recognition method for genomic sequences. *J Mol Biol* 287, 797–815.

95. McGuffin, L. J., Bryson, K., Jones, D. T. (2000) The PSIPRED protein structure prediction server. *Bioinformatics* 16, 404–405.

96. Zhang, Y., Arakaki, A. K., Skolnick, J. (2005) TASSER: an automated method for the prediction of protein tertiary structures in CASP6. *Proteins* 61, Suppl. 7, 91–98.

97. Skolnick, J., Kihara, D., Zhang, Y. (2004) Development and large scale benchmark testing of the PROSPECTOR_3 threading algorithm. *Proteins* 56, 502–518.

98. Shi, J., Blundell, T. L., Mizuguchi, K. (2001) FUGUE: sequence-structure homology

recognition using environment-specific substitution tables and structure-dependent gap penalties. *J Mol Biol* 310, 243–257.

99. Xu, J., Li, M., Kim, D., et al. (2003) RAP-TOR: optimal protein threading by linear programming. *J Bioinform Comput Biol* 1, 95–117.

100. Tang, C. L., Xie, L., Koh, I. Y., et al. (2003) On the role of structural information in remote homology detection and sequence alignment: new methods using hybrid sequence profiles. *J Mol Biol* 334, 1043–1062.

101. Teodorescu, O., Galor, T., Pillardy, J., et al. (2004) Enriching the sequence substitution matrix by structural information. *Proteins* 54, 41–48.

102. Zhou, H., Zhou, Y. (2004) Single-body residue-level knowledge-based energy score combined with sequence-profile and secondary structure information for fold recognition. *Proteins* 55, 1005–1013.

103. Zhou, H., Zhou, Y. (2005) SPARKS 2 and SP3 servers in CASP6. *Proteins* 61, Suppl. 7, 152–156.

104. Zhou, H., Zhou, Y. (2005) Fold recognition by combining sequence profiles derived from evolution and from depth-dependent structural alignment of fragments. *Proteins* 58, 321–328.

105. Thompson, J. D., Higgins, D. G., Gibson, T. J. (1994) CLUSTAL W: improving the sensitivity of progressive multiple sequence alignment through sequence weighting, position-specific gap penalties and weight matrix choice. *Nucleic Acids Res* 22, 4673–4680.

106. Notredame, C., Higgins, D. G., Heringa, J. (2000) T-Coffee: a novel method for fast and accurate multiple sequence alignment. *J Mol Biol* 302, 205–217.

107. Thompson, J. D., Gibson, T. J., Plewniak, F., et al. (1997) The CLUSTAL_X windows interface: flexible strategies for multiple sequence alignment aided by quality analysis tools. *Nucleic Acids Res* 25, 4876–4882.

108. Crooks, G. E., Hon, G., Chandonia, J. M., et al. (2004) WebLogo: a sequence logo generator. *Genome Res* 14, 1188–1190.

109. Sonnhammer, E. L., Hollich, V. (2005) Scoredist: a simple and robust protein sequence distance estimator. *BMC Bioinformatics* 6, 108.

110. Galtier, N., Gouy, M., Gautier, C. (1996) SEAVIEW and PHYLO_WIN: two graphic tools for sequence alignment and molecular phylogeny. *Comput Appl Biosci* 12, 543–548.

111. Parry-Smith, D. J., Payne, A. W., Michie, A. D., et al. (1998) CINEMA—a novel colour INteractive editor for multiple alignments. *Gene* 221, GC57–63.

112. Ginalski, K., von Grotthuss, M., Grishin, N. V., et al. (2004) Detecting distant homology with Meta-BASIC. *Nucleic Acids Res* 32, W576–581.

113. Xu, Y., Xu, D., Gabow, H. N. (2000) Protein domain decomposition using a graph-theoretic approach. *Bioinformatics* 16, 1091–1104.

114. Guo, J. T., Xu, D., Kim, D., et al. (2003) Improving the performance of Domain-Parser for structural domain partition using neural network. *Nucleic Acids Res* 31, 944–952.

115. Alexandrov, N., Shindyalov, I. (2003) PDP: protein domain parser. *Bioinformatics* 19, 429–430.

116. Todd, A. E., Orengo, C. A., Thornton, J. M. (1999) DOMPLOT: a program to generate schematic diagrams of the structural domain organization within proteins, annotated by ligand contacts. *Protein Eng* 12, 375–379.

117. Zemla, A. (2003) LGA: a method for finding 3D similarities in protein structures. *Nucleic Acids Res* 31, 3370–3374.

118. Holm, L., Park, J. (2000) DaliLite workbench for protein structure comparison. *Bioinformatics* 16, 566–567.

119. Ortiz, A. R., Strauss, C. E., Olmea, O. (2002) MAMMOTH (matching molecular models obtained from theory): an automated method for model comparison. *Protein Sci* 11, 2606–2621.

120. Gibrat, J. F., Madej, T., Bryant, S. H. (1996) Surprising similarities in structure comparison. *Curr Opin Struct Biol* 6, 377–385.

121. Shindyalov, I. N., Bourne, P. E. (1998) Protein structure alignment by incremental combinatorial extension (CE) of the optimal path. *Protein Eng* 11, 739–747.

122. Orengo, C. A., Taylor, W. R. (1996) SSAP: sequential structure alignment program for protein structure comparison. *Methods Enzymol* 266, 617–635.

123. Krissinel, E., Henrick, K. (2004) Secondary-structure matching (SSM), a new tool for fast protein structure alignment in three dimensions. *Acta Crystallogr D Biol Crystallogr* 60, 2256–2268.

124. Yang, A. S., Honig, B. (1999) Sequence to structure alignment in comparative modeling using PrISM. *Proteins* Suppl. 3, 66–72.

125. Lupyan, D., Leo-Macias, A., Ortiz, A. R. (2005) A new progressive-iterative algorithm for multiple structure alignment. *Bioinformatics* 21, 3255–3263.

126. Ye, Y., Godzik, A. (2005) Multiple flexible structure alignment using partial order graphs. *Bioinformatics* 21, 2362–2369.

127. Hill, E. E., Morea, V., Chothia, C. (2002) Sequence conservation in families whose members have little or no sequence similarity: the four-helical cytokines and cytochromes. *J Mol Biol* 322, 205–233.

128. Chothia, C., Jones, E. Y. (1997) The molecular structure of cell adhesion molecules. *Annu Rev Biochem* 66, 823–862.

129. Hill, E., Broadbent, I. D., Chothia, C., et al. (2001) Cadherin superfamily proteins in *Caenorhabditis elegans* and *Drosophila melanogaster. J Mol Biol* 305, 1011–1024.

130. Chothia, C., Lesk, A. M. (1987) Canonical structures for the hypervariable regions of immunoglobulins. *J Mol Biol* 196, 901–917.

131. Chothia, C., Lesk, A. M., Tramontano, A., et al. (1989) Conformations of immunoglobulin hypervariable regions. *Nature* 342, 877–883.

132. Al-Lazikani, B., Lesk, A. M., Chothia, C. (1997) Standard conformations for the canonical structures of immunoglobulins. *J Mol Biol* 273, 927–948.

133. Morea, V., Tramontano, A., Rustici, M., et al. (1998) Conformations of the third hypervariable region in the VH domain of immunoglobulins. *J Mol Biol* 275, 269–294.

134. Mizuguchi, K., Deane, C. M., Blundell, T. L., et al. (1998) JOY: protein sequence-structure representation and analysis. *Bioinformatics* 14, 617–623.

135. Hubbard, S. J., Thornton, J. M., (1993) NACCESS. Department of Biochemistry and Molecular Biology, University College London.

136. McDonald, I. K., Thornton, J. M. (1994) Satisfying hydrogen bonding potential in proteins. *J Mol Biol* 238, 777–793.

137. Morris, A. L., MacArthur, M. W., Hutchinson, E. G., et al. (1992) Stereochemical quality of protein structure coordinates. *Proteins* 12, 345–364.

138. Laskowski, R. A., MacArthur, M. W., Moss, D. S., et al. (1993) PROCHECK: a program to check the stereochemical quality of protein structures *J Appl Cryst* 26, 283–291.

139. Wallace, A. C., Laskowski, R. A., Thornton, J. M. (1995) LIGPLOT: a program to generate schematic diagrams of protein-ligand interactions. *Protein Eng* 8, 127–134.

140. Laskowski, R. A., Hutchinson, E. G., Michie, A. D., et al. (1997) PDBsum: a Web-based database of summaries and analyses of all PDB structures. *Trends Biochem Sci* 22, 488–490.

141. Sasin, J. M., Bujnicki, J. M. (2004) COLORADO3D, a web server for the visual analysis of protein structures. *Nucleic Acids Res* 32, W586–589.

142. Landau, M., Mayrose, I., Rosenberg, Y., et al. (2005) ConSurf 2005: the projection of evolutionary conservation scores of residues on protein structures. *Nucleic Acids Res* 33, W299–302.

143. Guex, N., Peitsch, M. C. (1997) SWISS-MODEL and the Swiss-PdbViewer: an environment for comparative protein modeling. *Electrophoresis* 18, 2714–2723.

144. Sayle, R. A., Milner-White, E. J. (1995) RASMOL: biomolecular graphics for all. *Trends Biochem Sci* 20, 374.

145. Martz, E. (2002) Protein Explorer: easy yet powerful macromolecular visualization. *Trends Biochem Sci* 27, 107–109.

146. Wang, Y., Geer, L. Y., Chappey, C., et al. (2000) Cn3D: sequence and structure views for Entrez. *Trends Biochem Sci* 25, 300–302.

147. Vriend, G. (1990) WHAT IF: a molecular modeling and drug design program. *J Mol Graph* 8, 52–56.

148. Koradi, R., Billeter, M., Wuthrich, K. (1996) MOLMOL: a program for display and analysis of macromolecular structures. *J Mol Graph* 14, 51–55, 29–32.

149. Humphrey, W., Dalke, A., Schulten, K. (1996) VMD: visual molecular dynamics. *J Mol Graph* 14, 33–38, 27–38.

150. Tramontano, A., Chothia, C., Lesk, A. M. (1990) Framework residue 71 is a major determinant of the position and conformation of the second hypervariable region in the VH domains of immunoglobulins. *J Mol Biol* 215, 175–182.

151. Sibanda, B. L., Thornton, J. M. (1985) Beta-hairpin families in globular proteins. *Nature* 316, 170–174.

152. Sibanda, B. L., Blundell, T. L., Thornton, J. M. (1989) Conformation of beta-hairpins in protein structures. A systematic classification with applications to modelling by homology, electron density fitting and protein engineering. *J Mol Biol* 206, 759–777.

153. Bruccoleri, R. E. (2000) Ab initio loop modeling and its application to homology modeling. *Methods Mol Biol* 143, 247–264.

154. Xiang, Z., Soto, C. S., Honig, B. (2002) Evaluating conformational free energies: the colony energy and its application to the problem of loop prediction. *Proc Natl Acad Sci U S A* 99, 7432–7437.

155. Tosatto, S. C., Bindewald, E., Hesser, J., et al. (2002) A divide and conquer approach to fast loop modeling. *Protein Eng* 15, 279–286.

156. Fiser, A., Sali, A. (2003) ModLoop: automated modeling of loops in protein structures. *Bioinformatics* 19, 2500–2501.

157. Canutescu, A. A., Shelenkov, A. A., Dunbrack, R. L., Jr. (2003) A graph-theory algorithm for rapid protein side-chain prediction. *Protein Sci* 12, 2001–2014.

158. Hung, L. H., Ngan, S. C., Liu, T., et al. (2005) PROTINFO: new algorithms for enhanced protein structure predictions. *Nucleic Acids Res* 33, W77–80.

159. Xiang, Z., Honig, B. (2001) Extending the accuracy limits of prediction for side-chain conformations. *J Mol Biol* 311, 421–430.

160. Marti-Renom, M. A., Stuart, A. C., Fiser, A., et al. (2000) Comparative protein structure modeling of genes and genomes. *Annu Rev Biophys Biomol Struct* 29, 291–325.

161. Levitt, M. (1992) Accurate modeling of protein conformation by automatic segment matching. *J Mol Biol* 226, 507–533.

162. Schwede, T., Kopp, J., Guex, N., et al. (2003) SWISS-MODEL: an automated protein homology-modeling server. *Nucleic Acids Res* 31, 3381–3385.

163. Bates, P. A., Kelley, L. A., MacCallum, R. M., et al. (2001) Enhancement of protein modeling by human intervention in applying the automatic programs 3D-JIGSAW and 3D-PSSM. *Proteins* Suppl. 5, 39–46.

164. Petrey, D., Xiang, Z., Tang, C. L., et al. (2003) Using multiple structure alignments, fast model building, and energetic analysis in fold recognition and homology modeling. *Proteins* 53, Suppl. 6, 430–435.

165. Koehl, P., Delarue, M. (1994) Application of a self-consistent mean field theory to predict protein side-chains conformation and estimate their conformational entropy. *J Mol Biol* 239, 249–275.

166. Wallner, B., Elofsson, A. (2005) All are not equal: a benchmark of different homology modeling programs. *Protein Sci* 14, 1315–1327.

167. Lund, O., Frimand, K., Gorodkin, J., et al. (1997) Protein distance constraints predicted by neural networks and probability density functions. *Protein Eng* 10, 1241–1248.

168. Lambert, C., Leonard, N., De Bolle, X., et al. (2002) ESyPred3D: prediction of proteins 3D structures. *Bioinformatics* 18, 1250–1256.

169. Hooft, R. W., Vriend, G., Sander, C., et al. (1996) Errors in protein structures. *Nature* 381, 272.

170. Sippl, M. J. (1993) Recognition of errors in three-dimensional structures of proteins. *Proteins* 17, 355–362.

171. Luthy, R., Bowie, J. U., Eisenberg, D. (1992) Assessment of protein models with three-dimensional profiles. *Nature* 356, 83–85.

172. Melo, F., Devos, D., Depiereux, E., et al. (1997) ANOLEA: a www server to assess protein structures. *Proc Int Conf Intell Syst Mol Biol* 5, 187–190.

173. Melo, F., Feytmans, E. (1998) Assessing protein structures with a non-local atomic interaction energy. *J Mol Biol* 277, 1141–1152.

174. Wallner, B., Elofsson, A. (2003) Can correct protein models be identified? *Protein Sci* 12, 1073–1086.

175. Wallner, B., Elofsson, A. (2006) Identification of correct regions in protein models using structural, alignment, and consensus information. *Protein Sci* 15, 900–913.

176. Fischer, D. (2006) Servers for protein structure prediction. *Current Opin Struct Biol* 16, 178–182.

177. Dayringer, H. E., Tramontano, A., Sprang, S. R., et al. (1986) Interactive program for visualization and modeling of protein, nucleic acid and small molecules. *J Mol Graph* 4, 82–87.

178. Spoel, D. v. d., Lindahl, E., Hess, B., et al. (2005) GROMACS: fast, flexible and free. *J Comp Chem* 26, 1701–1718.

179. Phillips, J. C., Braun, R., Wang, W., et al. (2005) Scalable molecular dynamics with NAMD. *J Comput Chem* 26, 1781–1802.

180. Simons, K. T., Ruczinski, I., Kooperberg, C., et al. (1999) Improved recognition of native-like protein structures using a combination of sequence-dependent and sequence-independent features of proteins. *Proteins*. 34, 82–95.

181. Bonneau, R., Tsai, J., Ruczinski, I., et al. (2001) Rosetta in CASP4: progress in ab initio protein structure prediction. *Proteins* Suppl. 5, 119–126.

182. Bonneau, R., Strauss, C. E., Rohl, C. A., et al. (2002) De novo prediction of

three-dimensional structures for major protein families. *J Mol Biol* 322, 65–78.

183. Rohl, C. A., Strauss, C. E., Chivian, D., et al. (2004) Modeling structurally variable regions in homologous proteins with rosetta. *Proteins* 55, 656–677.

184. Bradley, P., Malmstrom, L., Qian, B., et al. (2005) Free modeling with Rosetta in CASP6. *Proteins* 61, Suppl. 7, 128–134.

185. Chivian, D., Kim, D. E., Malmstrom, L., et al. (2003) Automated prediction of CASP-5 structures using the Robetta server. *Proteins* 53, Suppl. 6, 524–533.

186. Chivian, D., Kim, D. E., Malmstrom, L., et al. (2005) Prediction of CASP6 structures using automated Robetta protocols. *Proteins* 61, Suppl. 6, 157–166.

187. Kim, D. E., Chivian, D., Baker, D. (2004) Protein structure prediction and analysis using the Robetta server. *Nucleic Acids Res* 32, W526–531.

188. Vincent, J. J., Tai, C. H., Sathyanarayana, B. K., et al. (2005) Assessment of CASP6 predictions for new and nearly new fold targets. *Proteins* 61, Suppl. 7, 67–83.

189. Wang, G., Jin, Y., Dunbrack, R. L., Jr. (2005) Assessment of fold recognition predictions in CASP6. *Proteins* 61, Suppl. 7, 46–66.

190. Jones, D. T., Bryson, K., Coleman, A., et al. (2005) Prediction of novel and analogous folds using fragment assembly and fold recognition. *Proteins* 61, Suppl. 7, 143–151.

191. Kolinski, A., Bujnicki, J. M. (2005) Generalized protein structure prediction based on combination of fold-recognition with de novo folding and evaluation of models. *Proteins* 61, Suppl. 7, 84–90.

192. Fujikawa, K., Jin, W., Park, S. J., et al. (2005) Applying a grid technology to protein structure predictor "ROKKY". *Stud Health Technol Inform* 112, 27–36.

193. Debe, D. A., Danzer, J. F., Goddard, W. A., et al. (2006) STRUCTFAST: protein sequence remote homology detection and alignment using novel dynamic programming and profile-profile scoring. *Proteins* 64, 960–967.

194. Ginalski, K., Elofsson, A., Fischer, D., et al. (2003) 3D-Jury: a simple approach to improve protein structure predictions. *Bioinformatics* 19, 1015–1018.

195. Fischer, D. (2003) 3DS3 and 3DS5 3D-SHOTGUN meta-predictors in CAFASP3. *Proteins* 53, Suppl. 6, 517–523.

196. Sasson, I., Fischer, D. (2003) Modeling three-dimensional protein structures for CASP5 using the 3D-SHOTGUN meta-predictors. *Proteins* 53, Suppl. 6, 389–394.

197. Fischer, D. (2003) 3D-SHOTGUN: a novel, cooperative, fold-recognition meta-predictor. *Proteins* 51, 434–441.

198. Fischer, D. (2000) Hybrid fold recognition: combining sequence derived properties with evolutionary information. *Pac Symp Biocomput* 119–130.

199. Lundstrom, J., Rychlewski, L., Bujnicki, J., et al. (2001) Pcons: a neural-network-based consensus predictor that improves fold recognition. *Protein Sci* 10, 2354–2362.

200. Kurowski, M. A., Bujnicki, J. M. (2003) GeneSilico protein structure prediction metaserver. *Nucleic Acids Res* 31, 3305–3307.

201. Plaxco, K. W., Simons, K. T., Baker, D. (1998) Contact order, transition state placement and the refolding rates of single domain proteins. *J Mol Biol* 277, 985–994.

202. Bonneau, R., Ruczinski, I., Tsai, J., et al. (2002) Contact order and ab initio protein structure prediction. *Protein Sci* 11, 1937–1944.

203. Shortle, D., Simons, K. T., Baker, D. (1998) Clustering of low-energy conformations near the native structures of small proteins. *Proc Natl Acad Sci U S A* 95, 11158–11162.

204. Venclovas, C., Margelevicius, M. (2005) Comparative modeling in CASP6 using consensus approach to template selection, sequence-structure alignment, and structure assessment. *Proteins* 61, Suppl. 7, 99–105.

205. Kosinski, J., Gajda, M. J., Cymerman, I. A., et al. (2005) FRankenstein becomes a cyborg: the automatic recombination and realignment of fold recognition models in CASP6. *Proteins* 61, Suppl. 7, 106–113.

206. Wallner, B., Fang, H., Elofsson, A. (2003) Automatic consensus-based fold recognition using Pcons, ProQ, and Pmodeller. *Proteins* 53, Suppl. 6, 534–541.

207. Wallner, B., Elofsson, A. (2005) Pcons5: combining consensus, structural evaluation and fold recognition scores. *Bioinformatics* 21, 4248–4254.

208. Douguet, D., Labesse, G. (2001) Easier threading through web-based comparisons and cross-validations. *Bioinformatics* 17, 752–753.

209. Takeda-Shitaka, M., Terashi, G., Takaya, D., et al. (2005) Protein structure prediction in CASP6 using CHIMERA and FAMS. *Proteins* 61, Suppl. 7, 122–127.

210. Kopp, J., Schwede, T. (2004) The SWISS-MODEL Repository of annotated three-dimensional protein structure homology models. *Nucleic Acids Res* 32, D230–234.

211. Pieper, U., Eswar, N., Braberg, H., et al. (2004) MODBASE, a database of annotated comparative protein structure models, and associated resources. *Nucleic Acids Res* 32, D217–222.

212. Yamaguchi, A., Iwadate, M., Suzuki, E., et al. (2003) Enlarged FAMSBASE: protein 3D structure models of genome sequences for 41 species. *Nucleic Acids Res* 31, 463–468.

213. Castrignano, T., De Meo, P. D., Cozzetto, D., et al. (2006) The PMDB Protein Model Database. *Nucleic Acids Res* 34, D306–309.

214. Dayhoff, M. O., Schwartz, R. M., Orcutt, B. C., (1978) A model of evolutionary change in proteins. In *Atlas of Protein Sequence and Structure*. M.O. Dayhoff, ed. National Biomedical Research Foundation, Washington, DC.

215. Henikoff, S., Henikoff, J. G. (1992) Amino acid substitution matrices from protein blocks. *Proc Natl Acad Sci U S A* 89, 10915–10919.

Chapter 3

An Introduction to Protein Contact Prediction

Nicholas Hamilton and Thomas Huber

Abstract

A fundamental problem in molecular biology is the prediction of the three-dimensional structure of a protein from its amino acid sequence. However, molecular modeling to find the structure is at present intractable and is likely to remain so for some time, hence intermediate steps such as predicting which residues pairs are in contact have been developed. Predicted contact pairs have been used for fold prediction, as an initial condition or constraint for molecular modeling, and as a filter to rank multiple models arising from homology modeling. As contact prediction has advanced it is becoming more common for 3D structure predictors to integrate contact prediction into structure building, as this often gives information that is orthogonal to that produced by other methods. This chapter shows how evolutionary information contained in protein sequences and multiple sequence alignments can be used to predict protein structure, and the state-of-the-art predictors and their methodologies are reviewed.

Key words: Protein structure prediction, contact prediction, contact map, multiple sequence alignments, CASP.

1. Introduction

Analysis of a protein's evolutionary history may seem irrelevant to protein structure prediction. Indeed, a protein's folded structure depends entirely on the laws of physics and is dictated by the protein's amino acid sequence and its environment. It will not make any difference whether the protein sequence was designed *de novo*, resulted from random shuffling experiments, or evolved naturally. However, for the purpose of protein structure prediction, the evolutionary context of a protein provides important information that can be used to increase success. The

Jonathan M. Keith (ed.), *Bioinformatics, Volume II: Structure, Function and Applications, vol. 453*
© 2008 Humana Press, a part of Springer Science + Business Media, Totowa, NJ
Book doi: 10.1007/978-1-60327-429-6 Springerprotocols.com

most simplistic way to use such information is to base structure predictions on average (or consensus) properties of a family of related proteins instead of on properties unique to a single member. Noise due to variations in individual proteins is thus reduced and the accuracy of the prediction is increased. Most, if not all, prediction methods today take advantage of this fact and use so-called *profiles* from homologous sequences. Some methods go even further and average over similar but not necessarily homologous protein structures.

This chapter reviews some of the more sophisticated evolutionary information that can be obtained from protein sequences that are relevant to protein structure. It also describes how such information can be used for protein contact prediction. The next section discusses methods for using multiple sequence alignments to make contact predictions. In general, it is found that predictors, although performing well above chance levels, make predictions that are not in fact physically realizable. Hence, filtering methods are often used to improve predictions; these are described in **Section 3**. It is fair to say that the contact prediction literature has suffered from predictors being tested on different datasets using different measures of predictive quality, making comparison difficult. For this reason, blind tests of methodologies on a common dataset, such as the CASP experiments, are invaluable. **Section 4** describes the more common measures of predictive quality and presents results of the state-of-the-art predictors participating in the most recent round of CASP experiments.

2. Sequence-Based Approaches to Contact Prediction

A variety of approaches have been developed for contact prediction using multiple sequence alignments (MSAs) and other information. At the most fine-grained level, an MSA is constructed for a given protein and the information in pairs of columns is used to predict which residues are in contact. The fundamental assumption is that for residues that are in contact, the corresponding columns will be in some way more highly related to each other than for residues that are not.

Several measures of the relatedness of MSA columns have been developed. One approach is to look for *correlated mutations* in the MSA *(1)*. If residue substitutions in one column of an MSA are correlated to those in another, then one reason for this may be that the residues are physically close. For instance, substitutions might occur in pairs so as to preserve the total charge in a region and so maintain the structural integrity of the protein. To calculate a mutational correlation score for a pair of columns in an MSA, a measure of the physicochemical similarity of any pair of residues is required, and for this the McLachlan matrix *(2)* is often used. For each column of the MSA of *n* sequences, an $n \times n$ matrix is constructed with entry *(i,j)* being the McLachlan interchange score for the *i*th and *j*th residues of the column. The correlated mutation score between two columns is then calculated as the standard Pearson correlation between the entries of their matrices. A weighting scheme is also used in the correlation calculation to favor those row pairs that are least similar. Column pairs are then ranked according to their correlated mutation score, those with high scores being deemed more likely to be in contact. Formally, the correlation score between columns i and j is given as:

$$r_{ij} = \frac{1}{N^2} \sum_{k,l} \frac{w_{kl}\left(s_{ijk} - \overline{s}_i\right)\left(s_{jkl} - \overline{s}_i\right)}{\sigma_i \sigma_j}$$

where N is the length of the sequence alignment, w_{kl} is a weighting function that measures the similarity between rows k and l of the alignment, s_{ijk} is the McLachlan interchange score for the *j*th and *k*th residues in column *i*, and $\overline{s}_i$ and σ_i are the mean and standard deviation of the interchange scores in column *i*.

Rather than using a general "physical similarity" score, researchers have attempted to determine which physical factors affect compensatory mutations, and look for *biophysical complementarity principles*. For a given physical quantity, such as side-chain charge, the correlation between the values of this quantity for pairs of residues in two columns may be calculated *(3, 4)*. In this way, it has been found that for residues in contact, compensatory mutations are highly correlated with side-chain charge. In other words, for residues in contact, the sum of their side-chain charges tends to be preserved if the pair mutate. In contrast, it has been found that there is little correlation between side-chain volume and compensatory mutations.

The likelihood of a given residue pair being in contact may be predicted based on empirical observations of how often such a pair has been observed to be in contact in other proteins of known structure. *Contact likelihood* tables for all residue pairs have been constructed based on a set of 672 proteins of known structure *(5)*, and these have been used to construct a contact likelihood score for columns of an MSA by summing the likelihoods of all residue pairs in the columns. Similarly, Markov models for mutations of residues in contact have been built *(6)*, and these may also be used as a (crude) predictor of when residues are in contact.

On a more abstract level, *information theory* has been applied to contact prediction from MSAs *(7–9)*. Here the *mutual information* between two columns of an MSA is used as an estimate of contact likelihood and is defined as:

$$\sum_{(a_i, a_j)} P^2_{a_i, a_j} \log(P_{a_i, a_j} / (P_{a_i} P_{a_j}))$$

where a_i and a_j are the amino acids found in columns i and j, P_{ai} is the observed relative frequency of amino acid a_i in column i, similarly for a_j in column j, and $P_{ai,aj}$ is the observed relative frequency of the pair (a_i, a_j). This is based on the definition of entropy from information theory, and gives an estimate of the degree to which the identity of one amino acid in one column allows us to predict an amino acid in another. The column pairs with the highest degree of mutual information are then predicted to be in contact.

Many of these measures have also been useful in predicting the function of groups of residues *(7, 10)*. Although all of these approaches have been moderately successful in that they usually predict contacts at a level significantly above random chance, the accuracy of the predictions is generally low and certainly not high enough to enable accurate reconstruction of the 3D structure of a protein from its sequence. One essential problem is that phylogenetic relationships are disregarded, and these can lead to strong artifactual correlations between physically distant residues due simply to the phylogenetic relatedness of the sequences. Several attempts have been made to separate out phylogenetic covariation from mutational covariation *(4, 9, 11–13)*, but predictive accuracy remains in need of improvement.

Interest has hence turned to combining contact prediction methods with other information to improve predictions. Other sources of information include predicted secondary structure *(14)*, residue sequence separation, sequence length, residue conservation, hydrophobicity of residues, predicted disulfide bridges, and conservation weight. Although each of these is not a good predictor of protein contacts considered alone, by combining several relatively poor predictors a better predictor can be obtained. (In the theory of boosting *(15)* it has been proved that weak predictors may be combined to provide a predictor that is significantly more accurate than any of the weak predictors, though the boosting method appears not to have been exploited in contact prediction yet.)

Another technique that has been shown to improve the accuracy of contact predictions is to use information about the residues in windows around the pair being scored. Information such as correlation scores, residue frequencies in columns of the MSA, and predicted secondary structure for these neighboring residues can be taken into consideration. This improves prediction accuracy because if two residues are in contact, then it is likely that the residues around them will also be in contact and correlate in some way.

Once a set of measures has been chosen, the problem becomes how to combine the information into a single prediction score for a given protein and residue pair. Virtually every type of learning algorithm has been applied to this problem, including neural networks *(16–20)*, self-organizing maps *(21)*, support vector machines (SVMs) *(22–24)*, and hidden Markov models (HMMs) *(25–27)*. Typically, an algorithm will develop a predictor using a training set of data, and validate it on an independent test set. Each of these learning methods has its own advantages and disadvantages, and it might be said that training a learning algorithm is something of a black art. Some claim that neural networks have a tendency to over-train (that is, to fit the training data too closely and hence lose their ability to generalize to unseen datasets), although this can be to some extent avoided by the use of a validation set to halt the learning during training. Support vector machines are less prone to suffer from this problem and are thought to be more tolerant of noisy data *(28)*. Balanced training, which uses equal numbers of each category of data in training *(18, 20)*, is often favored, but some researchers have obtained better results training on data in which the categories are in the proportions in which they naturally occur *(19)*. Encoding of inputs is certainly important. In one study of predicting disulfide connectivity it was found that taking the log of the sequence separation of the residues (together with other inputs), improved the predictive accuracy of the SVM by 4% compared with simply using the linear sequence separation *(29)*, and choice of window size varied the predictive accuracy by up to 10% (larger window sizes were also found to increase accuracy in *(21)*). In general, training a good predictor involves much testing of encoding schemes and experimenting with data as well as a little good luck. **Figure. 3.1** summarizes the main components of sequence based contact prediction.

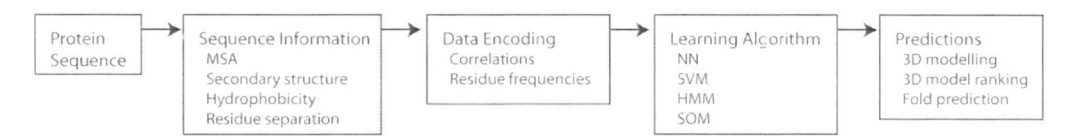

Fig. 3.1. From protein sequence to contact prediction. A typical workflow taking a protein's sequence, extracting sequence/amino acid properties, encoding the information, applying a learning algorithm and finally making contact pair predictions to be used for structure prediction.

3. Contact Filtering

In many cases a contact predictor will incorrectly predict residue pairs to be in contact. For this reason, contact filtering is often applied to a given set of predicted contacts, the aim being to remove those predicted contacts that are in some way physically unrealizable.

The simplest and perhaps most effective method of filtering is *contact occupancy*. For a given residue in a protein sequence there is a limit to the number of residues with which it may be in contact. This number varies according to the type of residue, the secondary structure around the residue, whether the residue is exposed on the surface of the protein, and other factors. By examining proteins of known structure, tables of the average number of contacts a given residue of a particular class has can be created. Linear regression *(30)*, support vector machine *(31)*, and neural network *(32)* approaches to predicting contact occupancy have also been developed. A list of predicted contacts can then be filtered by removing those pairs for which one or both of the residues is predicted to be in contact with more than the average for its class. Significant improvements of up to 4% in predictive accuracy have been found using this methodology *(20, 33)*. Similarly, for any pair of residues that are in contact there is a limit to the number of residues that may be in contact with both of the residues, and this can also be used as a filter *(27)*.

Bootstrapping can also be applied to assess the stability of a prediction of contact pairs. Olmea and Valencia performed bootstrapping experiments by excluding 10% of sequences in the alignments given to their prediction method *(33)*. By excluding predicted contact pairs that occurred in <80% of bootstraps, a 20% improvement in accuracy was obtained.

More detailed consideration of the (predicted) secondary structure can also be applied. For instance, within a helix, the ith residue should only be in contact with residues $i+4$ and $i-4$; a residue cannot be in contact with residues on opposite sides of a helix; and within a single strand of a β-sheet only adjacent residues should be in contact *(27)*. A more direct method of checking the physical possibility of contact maps is to align fragments of a predicted contact map to template contact maps, in which the templates are fragments of contact maps of proteins of known structure. The predicted contact map fragment then becomes the contact map of the most closely aligned template *(27)*. However, care needs to be taken with the definition of contact and the application of these rules and maps since, say, a 4Å cutoff for contact will give a very different pattern of contact than an 8Å cutoff.

4. Evaluating Contact Predictors and the CASP6 Experiment

There are many definitions of residue contact used in the literature. Some use the C-α distance, that is, the distance between the α carbon atoms of the residue pair *(34)*, whereas others prefer the C-β distance *(20, 35)* or even the minimal distance between the heavy atoms of the side chain or backbone of the two residues *(36)*. The most common minimum separation used to define a contact pair is 8Å. It is also usual to exclude residue pairs that are separated along the amino acid sequence by less than some fixed number of residues, since short-range contacts are less interesting and easier to predict than long-range ones.

For a given target protein, the *prediction accuracy* A_N on N predicted contacts is defined to be $A_N = N_c/N$, where N_c is the number of the predicted contacts that are indeed contacts for a given minimum sequence separation. Typically N is taken be one of L, $L/2$, $L/5$, or $L/10$, where L is the length of the sequence. For most proteins, the actual number of contacts (using the 8Å definition) is in the range L to $2L$. It has become relatively standard to report results on the best $L/2$ predictions with a maximum distance of 8Å between C-β atoms (C-α for glycine), with a minimum sequence separation of six. This standardization is in large part due to the CASP *(37)* and EVA *(38)* protein structure prediction blind tests, and has been invaluable in enabling comparison between predictors.

The *prediction coverage* is defined to be N_c/T_c, where T_c is the total number of contacts pairs for the protein. The *random accuracy* is given by the fraction of all residue pairs that are in contact (for a given sequence separation), and gives a measure of the probability of picking a pair to be in contact by chance. The *improvement over random* is then prediction accuracy divided by the random accuracy, and gives a measure of how much better than chance the predictor performs. This can be a useful measure since the number of contacts can vary widely between proteins, and prediction accuracy may be artificially high due to an unusually large number of contacts in a given protein.

Another measure that is sometimes used is the *weighted harmonic average distance (39)*:

$$X_d = \sum_{i=1}^{15} \frac{P_{ip} - P_{ia}}{15d_i}$$

Where the sum runs over 15 distance bins in the range 0 to 60Å, d_i is the upper bound of each bin, normalized to 60, P_{ip} is the percentage of predicted pairs whose distance is included in bin i, and P_{ia} is the same percentage for all pairs. The harmonic average is designed to reflect the difference between the real and predicted distances of residue pairs: when the average distance between predicted residue pairs is less than the average distance between all pairs in the structure then $X_d > 0$, although interpreting the meaning of a particular value of X_d can be difficult.

Prediction accuracy and coverage are the most commonly reported measures in the literature. However, the choice of sequence separation can greatly affect the prediction accuracy since residues that are close on the sequence are more likely to be in contact. Choosing a minimum sequence separation of twelve instead of six may reduce the accuracy by 50% or more depending on the characteristics of the predictor. Similarly, the accuracy is usually strongly dependent on the number of predictions made. A predictor that has an accuracy of 0.3 on its best $L/5$ predictions may well drop to 0.15 on its best L predictions. Also, a contact predictor that does relatively well on one dataset may predict poorly on another, perhaps due to there being many proteins in the first dataset for which several homologous structures are known. For these reasons it can be difficult to evaluate the relative performance of contact predictors in the literature.

To overcome these problems, standardized blind tests of protein structure and protein contact predictors have been introduced. The Critical Assessment of Techniques for Protein Structure Prediction (CASP) experiments are run biannually and involve releasing over several months the sequences of a set of proteins for which the structure has been solved, but which are not yet publicly available *(32, 37)*. Groups from around the world submit their predictions, and these are independently evaluated by a team of experts. The results are then published. The experiment also includes an automated section in which sequences are submitted to prediction servers and predictions are returned without human intervention (for additional details, *see* **Chapter 2**). Similarly, the EVA project provides a continuous, fully automatic analysis of structure prediction servers *(38)*. Both EVA and CASP include sections for comparative 3D modeling, fold recognition, contact, and secondary structure prediction.

In the sixth round of CASP in 2004 there were 87 target proteins released, and 16 groups competed in the contact prediction category, five of which were registered as automated servers *(37)*. Unfortunately, the only published evaluation performed was for the eleven hard new fold (NF) targets for which additional and structural information was not available *(37)*. These targets are not representative of the wide range of possible protein folds, and with such a small sample set it is difficult to evaluate the effectiveness of each contact predictor accurately. Fortunately, the raw prediction scores for each contact predictor are available from the CASP6 web site (http://predictioncenter.org/casp6/Casp6.html), so we can present results for the full set of targets here.

In **Table 3.1**, average accuracy and coverage results are shown for the contact predictors submitted to CASP6. The data shown are for the best $L/2$ predictions with a minimum sequence separation of six residues. The tables are separated according to the target type. Not all groups submitted predictions for all targets, and so the averages presented are over those proteins for which a prediction was submitted. The number of targets predicted by each group is also shown. For most purposes, accuracy is more

Table 3.1
Performance results from all contact predictors submitted to the CASP6 experiment for L/2 predicted contacts and a minimum separation of six residues along the sequence

Group		#Submitted	All targets Accuracy		Coverage		Xd		#Submitted	Comparative modelling targets Accuracy		Coverage		Xd	
			av	stddev	av	stddev	av	stddev		av	stddev	av	stddev	av	stddev
RR011	MacCallum	83	15.5	8.4	4.4	2.5	7.4	3.3	41	15.0	8.0	4.2	2.3	8.1	3.3
RR012	GPCPred	82	19.3	11.5	6.9	4.6	9.3	3.8	40	21.0	12.1	6.2	3.2	10.5	3.4
RR018	Baldi	31	33.6	15.3	10.7	9.2	14.1	4.7	14	32.0	16.0	11.7	12.6	13.7	5.1
RR019	Baldi-server	85	36.8	17.3	10.0	9.0	15.4	5.0	42	36.2	15.5	9.9	8.8	15.9	4.6
RR088	Bystroff	34	13.4	11.7	6.4	8.2	5.6	4.8	11	17.0	13.9	9.7	12.3	7.0	3.8
RR089	KIAS	85	15.3	13.3	3.2	2.5	7.3	4.2	42	15.0	10.1	3.3	2.3	7.8	3.8
RR100	Baker	83	40.1	22.2	19.9	14.4	15.8	8.0	41	52.1	20.8	24.0	15.2	20.0	7.2
RR166	SAMT04-hand	43	20.8	11.8	6.3	3.4	9.5	4.3	30	23.3	12.5	7.0	3.5	10.7	4.0
RR185	Huber-Torda	75	38.3	30.3	17.0	18.9	13.8	10.2	41	57.5	25.2	24.5	19.1	20.3	8.3
RR301	rostPROF-con	85	28.3	13.5	13.5	9.4	12.7	4.8	42	31.3	14.4	13.3	9.9	14.2	5.0

(continued)

Table 3.1 (continued)

		#Submitted	Accuracy av	stddev	Coverage av	stddev	Xd av	stddev	#Submitted	Accuracy av	stddev	Coverage av	stddev	Xd av	stddev
RR327	Hamilton-Huber-	65	24.0	16.2	5.6	3.8	10.7	5.9	34	26.3	14.4	6.6	3.7	12.4	5.1
RR348	Distill	81	9.1	7.8	5.9	9.4	5.0	3.8	41	8.8	7.9	6.5	11.9	5.3	3.3
RR361	karypis	74	14.7	11.9	4.6	4.5	10.4	3.8	34	16.6	14.5	4.7	4.4	11.6	3.6
RR491	comet	72	4.7	6.3	1.0	1.4	7.1	4.1	36	5.2	7.2	1.0	1.3	7.2	4.6
RR545	cracow.pl	19	8.8	6.4	2.4	1.8	3.4	3.7	8	9.3	6.0	2.2	1.8	5.1	2.8

Group		Fold recognition targets							New fold targets						
		#Submitted	Accuracy av	stddev	Coverage av	stddev	Xd av	stddev	#Submitted	Accuracy av	stddev	Coverage av	stddev	Xd av	stddev
RR011	MacCallum	31	16.4	9.1	4.9	2.9	6.6	3.4	11	14.4	7.3	3.9	1.8	6.7	2.4
RR012	GPCPred	31	16.6	9.9	7.8	5.9	7.6	3.9	11	20.6	11.5	7.0	4.7	9.6	3.3
RR018	Baldi	14	31.6	12.9	9.6	4.7	13.9	4.4	3	50.2	12.3	10.8	4.5	16.6	1.8
RR019	Baldi-server	32	35.8	19.6	11.2	10.3	14.4	5.3	11	41.7	16.2	6.6	3.7	16.5	5.2
RR088	Bystroff	18	12.8	10.8	4.8	4.4	5.5	5.3	5	8.0	5.7	4.7	4.3	2.8	2.9
RR089	KIAS	32	16.3	17.8	3.4	3.1	6.8	5.0	11	13.7	7.0	2.2	1.2	6.9	2.8
RR100	Baker	31	32.0	16.9	18.1	12.7	12.7	6.8	11	18.3	10.7	9.7	8.0	8.7	4.3
RR166	SAMT04-hand	9	12.2	6.7	4.0	2.6	5.5	3.6	4	21.6	2.2	6.1	1.4	9.9	1.9
RR185	Huber-Torda	29	15.3	17.8	8.7	15.3	5.9	6.2	5	14.5	7.3	4.3	1.4	6.9	3.4

RR301	rostPROF-con	32	25.8	13.1	14.6	9.8	11.1	4.7	11	24.2	6.5	10.8	4.3	11.4 2.1
RR327	Hamilton-Huber-	22	21.7	18.1	5.1	4.2	8.7	6.5	9	20.5	16.1	3.5	2.1	9.6 5.6
RR348	Distill	29	9.7	8.5	5.8	6.5	5.0	4.6	11	8.5	4.1	4.0	2.2	4.1 3.0
RR361	karypis	30	13.6	9.3	4.7	4.7	9.5	3.6	10	11.7	7.4	4.0	3.8	9.3 3.7
RR491	comet	26	4.1	5.6	1.1	1.5	7.3	3.8	10	4.9	3.8	0.9	0.8	6.1 2.0
RR545	cracow.pl	9	8.4	7.0	2.6	2.0	1.7	3.9	2	8.2	3.9	2.3	1.0	3.6 2.5

important than coverage, since the aim is to get a number of high-quality contact predictions.

Several groups attained accuracy of 20% or better on most classes of protein. Here we emphasize those that do not involve 3D modeling, or in which 3D modeling incorporates a contact predictor. For more information on the predictors, see also the CASP6 methods abstracts, available from http://predictioncenter. org/casp6/abstracts/abstract.html. Links to contact prediction services and methodology references for the best performing predictors are given in **Table 3.2**.

4.1. RR100 Baker

The Baker predictor is particularly interesting in that while it takes a whole-structure 3D modeling approach, a contact predictor is integrated into the structure building for fold recognition targets. The approach to contact prediction is to train several neural networks on the predictions made by a set of 24 protein (3D) structure predictors that participated in recent LIVEBENCH experiments *(40)*. For a given residue pair, the principal input to a neural network is the ratio of the number of servers that predict the residues to be in contact (contact meaning closer than 11Å), along with other inputs, such as secondary structure prediction and amino acid property profiles. Ten neural networks were trained and validated on different subsets of the same training

Table 3.2
Contact prediction services, software and methods references for the top predictors in CASP6

Predictor Group	URL	Reference
Baker	http://www.jens-meiler.de/ contact.html	*(37, 40)*
Huber-Torda	http://www.zbh. uni-hamburg.de/wurst	*(42)*
Baldi-server	http://www.igb.uci.edu/ servers/pass.html	*(43, 44)*
rost_PROFCon	http://www.predictprotein.org/ submit_profcon.html	*(18)*
Hamilton-Huber-Torda	http://foo.maths.uq.edu. au/~nick/Protein/ contact.html	*(19)*
GPCPred	http://www.sbc.su.se/~maccallr/ contactmaps	*(21)*
Bystroff	http://www.bioinfo.rpi.edu/ ~bystrc/downloads.html	*(27)*

set. For prediction, the average score of the 10 trained networks is taken, and the highest scoring residue pairs are taken as predicted contacts. The consensus contact predictor is then used as an indictor of distant contacts that should be present in the *de novo* predicted models.

4.2. RR185 Huber-Torda

The Huber-Torda predictor is not a dedicated contact predictor but builds 3D models by threading, which combines structure- and sequence-based terms for scoring alignments and models. Protein contacts are extracted from the models in a post-processing step. It is interesting to observe the performance of a threading method that is based on a fundamentally different philosophy than protein contact predictors, since it shows limitations of the methods and may suggest new ways to improve overall performance.

4.3. RR019 and RR018 Baldi-Server and Baldi

Similarly to the Baker group, the Baldi group predictors are whole structure 3D modelers that incorporate contact predictions. The energy function used in constructing the 3D coordinates incorporates a contact map energy term that "encourages" the models to follow the predicted contact structure. The contact predictor is a 2D recursive neural network in which outputs feed back as inputs *(41)*. The recursive approach allows local information to be combined with more distant contextual information to provide better prediction. The inputs include the residue type of the pair being predicted, the residue frequencies in an MSA for the corresponding columns, the frequencies of residue pairs in the columns of the MSA, the correlated mutation score for the column pair, secondary structure classification, and solvent accessibility. To make a contact map from the residue contact probabilities given by the neural network, two approaches are taken. One method is to use a fixed threshold that maximizes precision and recall on a test set, the other is a variable, band-dependent threshold determined by estimating the number of contacts in a band from the sum of all the predicted contact probabilities in that band.

4.4. RR301 rost_PROFCon

The rost_PROFCon server takes a neural network approach to contact prediction. For each residue pair, information in two windows of length nine centered on each residue is encoded. For each residue position in the windows there are 29 inputs to the network, including frequency counts of the residue types in the corresponding MSA column, predicted secondary structure and the reliability of that prediction, predicted solvent accessibility and conservation weight. Inputs are also included to give a biophysical classification of the central residues, as well as whether or not the residues are in a low complexity region. Unusually for a contact predictor, inputs describing a window of length five half way between the pair of residues being considered are also included. In this window, the same 29 input encoding scheme for each position is used as for the windows of length nine. A binary encoding scheme is used to describe the separation of the residues of interest. Finally, there are inputs describing global information such as the length (via a coarse-grained binary encoding), the composition of amino acids and secondary structure for the protein.

4.5. RR327 Hamilton-Huber-Torda

The Hamilton-Huber-Torda server (recently named PoCM "possum" for Patterns of Correlated Mutations) is also a neural network predictor. The approach is to train the network on patterns of correlation. For a given residue pair, there are two windows of length five centered on the residues. The correlated mutation score for all 25 pairs of residues between the windows are then calculated, the idea being that if the central residues are in contact, then adjacent residues are also likely to be in contact and so correlated. Inputs are also included for predicted secondary structure, biophysical classification of residues, a residue affinity score based on observed contacts in proteins of known structure, sequence length, and residue separation.

4.6. RR166 SAMT04-Hand

This is a whole structure 3D modeler based on homology and templates.

4.7. RR012 GPCPred

Perhaps the most unusual approach to contact prediction in CASP6 is via "stripped sheets." For a given protein, a PSI-BLAST sequence profile is constructed, that is, a $21 \times L$ matrix that records the frequencies of amino acids in each of the positions of an MSA, where L is the length of the protein. From this matrix, windows of length w are extracted, with $w = 1, 5, 9, 25$. During training, to reduce the number of dimensions in the data in the windows, a self-organizing map (SOM) (42) was created for each w, with output three integers in the range 0 to 5. Any profile window, or indeed central residue, could then be mapped to three integers by the trained SOMs. Genetic programming techniques were used to classify whether a pair of residues were in contact from the SOM outputs for the windows around them.

4.8. RR088 Bystroff

The Bystroff predictor uses a threading method to predict contact maps. The target sequence is aligned to a set of template sequences with template contact maps, and target contact maps are generated. Predicted contact maps are then scored using a "contact free energy" function, and physicality rules similar to those outlined in **Section 3** are applied.

In the CASP6 experiment the contact predictors that performed best were those that took a whole structure 3D modeling approach, although several "pure" contact predictors also performed well.

It is interesting to note that it is becoming more common for 3D model builders to rely on pure contact predictors to refine and select among models. No doubt as the pure predictors improve and the newer ones are incorporated into the 3D predictors, this will lead to both better 3D structure and contact prediction.

For the pure contact predictors there are a number of general trends in accuracy that have been observed in the literature. Since most contact prediction methods rely on multiple sequence alignments, they tend to have lower accuracy on proteins for which there are few homologs. Many predictors also report a decrease in accuracy for longer sequence proteins *(1, 5, 21, 35)*, although there are exceptions *(18, 19)*. In some cases the average predictive accuracy may be reduced by up to a factor of two for long proteins. This may be due to the fact that for shorter proteins a randomly chosen residue pair is more likely to be in contact than for a longer one. Similarly, residues that are close on a sequence are more likely to be in contact and so are usually easier to predict than distant residues. Also, most predictors will significantly improve their accuracy if allowed to make fewer predictions. For instance, on a test set of 1033 proteins the PoCM predictor gave average accuracies of 0.174, 0.217, 0.27, 0.307, on the best L, $L/2$, $L/5$, and $L/10$ predictions, respectively *(19)*. This can be useful if only a few higher-quality predictions are required.

Predictive accuracies also tend to vary widely between secondary structure classes such as those of the SCOP classification *(43)*. Proteins classified as "all α" are almost always poorly predicted in comparison to other classes. For example, the rost-PROFCon server obtained an average accuracy of 0.24 on all α proteins, but 0.35 and 0.36 on the other classes, on a test set of 522 proteins with minimum sequence separation of six residues and the best $L/2$ predictions taken *(18)*. This order of decrease in accuracy is typical of contact predictors and may be due to a number of factors. It may be that to maintain the structure of the α-helices the kinds of substitutions possible are restricted, and so there is less information within the areas of the multiple sequence alignments corresponding to helices. Another problem may be the "windows" approach that some of the predictors take. Since, on average, a single turn of a regular α-helix is 3.6 residues long, if two sequence-distant residues contained in alpha helices are in contact, the residues adjacent to these residues are unlikely to be in contact. Hence one approach that might improve prediction on residues contained in alpha helices would be to use non-standard windows for these residues. For instance, for a window of size five around a given residue, the window would be taken as the fourth and seventh residues before and after the central residue. In this way it would be ensured that the residues in the window were on the same side of the helix.

All of these factors in combination can lead to a wide variation in predictive accuracy. On a dataset of 1033 proteins, the PoCM predictor had an average accuracy of 0.174 on the best L predictions, with sequence separation of at least five *(19)*. Taking the subset of 64 proteins of class $\alpha + \beta$ for which there were at least 100 sequences in the alignment, the average accuracy rises to 0.457 for the best $L/10$ predictions.

5. Conclusions

Fariselli et al. *(20)* state that their goal is to obtain an accuracy of 50%, for then the folding of a protein of less than 300 residues length could be reconstructed with good accuracy (within 0.4-nm RMSD). Although current contact predictors are still well short of this aim, predictive accuracy has significantly improved in recent years and has provided a valuable source of additional information in protein structure prediction.

Interestingly, contact predictions are not yet widely used in combination with 3D structure prediction and only a few approaches use them routinely. However, 3D modeling approaches that do use evolutionary analyses to predict contacts in protein structures seem to also be the better performing ones. One reason why contact prediction is generally omitted in fold recognition is simply algorithmic difficulties. Dynamic programming, as it is used in sequence(s)-sequence(s) alignment and sequence(s)-structure(s) alignment approaches, is not easy to reconcile with these residue-pair distance constraints. The problem does not exist with 3D prediction methods that use heuristic optimization methods instead. Well performing programs of this kind include Skolnick's TASSER protein-folding approach *(44)* and Baker's fragment assembly approach Robetta *(45)*. Even when it is not possible to integrate contact predictions into a structure predictor it may still be useful to use the predictions as a way of selecting the "best" structure from a number of models. Our own experiments have shown that if a set of 3D models is ranked according to how many predicted contacts it is in agreement with, then the (known) real structure is ranked most highly against other predicted structures in almost all cases.

As we have seen, a number of different methodologies for protein contact prediction have been developed in recent years. The question is: How can contact prediction be improved? One approach would be to attempt to construct a predictor that combines the best and most novel aspects of each. Most predictors have a similar core of inputs to a training algorithm, such as predicted secondary structure, but each has some unique feature such as using the predicted solvent accessibility, a stringent contact filtering algorithm, or a totally novel encoding as in the stripped sheets approach of McCallum. Also, within 3D structure prediction, meta-servers that make predictions based on the predictions of other servers have proved highly successful in the CASP experiments, often out-performing all other methods. As more contact predictors come online it will be interesting to see if meta-contact predictors will enjoy similar success.

Acknowledgments

The authors gratefully acknowledge financial support from the University of Queensland, the ARC Australian Centre for Bioinformatics and the Institute for Molecular Bioscience. The first author would also like to acknowledge the support of Prof. Kevin Burrage's Australian Federation Fellowship.

References

1. Gobel, U., Sander, C., Scheider, R., et al. (1994) Correlated mutations and residue contacts in proteins. *Proteins* **18**, 309–317.

2. McLachlan, A.D. (1971) Tests for comparing related amino acid sequences. *J Mol Biol* **61**, 409–424.

3. Neher, E. (1994) How frequent are correlated changes in families of protein sequences? *Proc Natl Acad Sci USA* **91**(1), 98–102.

4. Vicatos, S., Reddy, B.V.B., and Kaznessis, Y. (2005) Prediction of distant residue contacts with the use of evolutionary information. *Proteins: Structure, Function, and Bioinformatics* **58**, 935–949.

5. Singer, M.S., Vriend, G., and Bywater, R.P. (2002) Prediction of protein residue contacts with a PDB-derived likelihood matrix. *Protein Eng* **15**(9), 721–725.

6. Lin, K., Kleinjung, J., Taylor, W., et al. (2003) Testing homology with CAO: A contact-based Markov model of protein evolution. *Comp Biol Chem* **27**, 93–102.

7. Clarke, N.D. (1995) Covariation of residues in the homeodomain sequence family. *Protein Sci.* **7**(11), 2269–78.

8. Korber, B.T.M., Farber, R.M., Wolpert, D.H., et al. (1993) Covariation of Mutations in the V3 Loop of Human Immunodeficiency Virus Type 1 Envelope Protein: An Information Theoretic Analysis. *Proc Natl Acad Sci* **90**, 7176–7180.

9. Martin, L.C., Gloor, G.B., Dunn, S.D., et al. (2005) Using information theory to search for co-evolving residues in proteins. *Bioinformatics* **21**(22), 4116–4124.

10. Oliveira, L., Paiva, A.C.M., and Vriend, G. (2002) Correlated Mutation Analyses on Very Large Sequence Families. *Chem Bio Chem* **3**(10), 1010–1017.

11. Akmaev, V.R., Kelley, S.T., and Stormo, G.D. (2000) Phylogenetically enhanced statistical tools for RNA structure prediction. *Bioinformatics* **16**(6), 501–512.

12. Tillier, E.R.M. and Lui, T.W.H. (2003) Using multiple interdependency to separate functional from phylogenetic correlations in protein alignments. *Bioinformatics* **19**(6), 750–755.

13. Wollenberg, K.R., and Atchley, W.R. (2000) Separation of phylogenetic and functional associations in biological sequences by using the parametric bootstrap. *Proc Natl Acad Sci USA* **97**, 3288–3291.

14. McGuffin, L.J., Bryson, K., and Jones, D.T. (2000) The PSIPRED protein structure prediction server. *Bioinformatics* **16**, 404–405.

15. Shapire, R.E., *The boosting approach to machine learning: An overview.* MSRI Workshop on Nonlinear Estimation and Classification. 2002: Springer.

16. Haykin, S., *Neural Networks.* 2nd ed. 1999: Prentice Hall.

17. Zell, A., Marnier, M., Vogt, N., et al, *Stuttgart Neural Network Simulator User Manual Version 4.2*. 1998: University of Stuttgart.

18. Punta, M., and Rost, B. (2005) PROFcon: novel prediction of long range contacts. *Bioinformatics* **21**(13), 2960–2968.

19. Hamilton, N., Burrage, K., Ragan, M.A., et al. (2004) Protein contact prediction using patterns of correlation. *Proteins: Structure, Function, and Bioinformatics* **56**, 679–684.

20. Fariselli, P., Olmea, O., Valencia, A., et al. (2001) Prediction of contact maps with neural networks and correlated mutations. *Protein Eng* **14**, 835–843.

21. MacCallum, R.M. (2004) Stripped sheets and protein contact prediction. *Bioinformatics* **20**(1), i224–i231.

22. Cortes, C., and Vapnik, V. (1995) Support vector network. *Machine and learning* **20**, 273–297.

23. Boser, B., Guyon, I., and Vapnik, V. *A training algorithm for optimal margin classifiers.* in *Proceedings of the fifth annual workshop on computational learning theory*. 1992.

24. Chang, C-C, and Lin, C-J, *LIBSVM: a library for support vector machines. Software available at http://www.csie.ntu.edu.tw/ cjlin/libsvm*. 2001.

25. Koski, T., *Hidden Markov Models for Bioinformatics*. 2002: Springer.

26. Karplus, K., Karchin, R., Draper, J., et al. (2003) Combining local-structure, fold-recognition, and new-fold methods for protein structure prediction. *Proteins: Structure, Function, and Genetics* **53**(S6), 491–496.

27. Shao, Y. and Bystroff, C. (2003) Predicting Interresidue contacts using templates and pathways. *Proteins* **53**, 497–502.

28. Conrad, C., Erfle, H., Warnat, P., et al. (2004) Automatic Identification of Subcellular Phenotypes on Human Cell Arrays. *Genome Research* **14**, 1130–1136.

29. Tsai, C-H, Chen, B-J, Chan, C-h, et al. (2005) Improving disulphide connectivity prediction with sequential distance between oxidized cysteines. *Bioinformatics* **21**(4), 4416–4419.

30. Hu, J., Shen, X., Shao, Y., et al., eds. *Mining protein contact maps*. In 2nd BIOKDD Workshop on Data Mining in Bioinformatics. 2002.

31. Yuan, Z. (2005) Better prediction of protein contact number using a support vector regression analysis if amino acid sequence. *BMC Bioinformatics* **6**, 248–257.

32. Aloy, P., Stark, A., Hadley, C., et al. (2003) Predictions without templates: new folds, secondary structure, and contacts in CASP5. *Proteins* **Suppl. 6**, 436–456.

33. Olmea, O., and Valencia, A. (1997) Improving contact predictions by the combination of correlated mutations and other sources of sequence information. *Fold Design* **2**, S25–S32.

34. Mirny, L. and Domany, E. (1996) Protein Fold Recognition and Dynamics in The Space of Contact Maps. *Proteins* **26**, 319–410.

35. Fariselli, P., Olmea, O., Valencia, A., et al. (2001) Progress in predicting inter-residue contacts of proteins with neural networks and correlated mutations. *Proteins* **Suppl 5**, 157–162.

36. Fariselli, P. and Casadio, R. (1999) Neural network based prediction of residue contacts in protein. *Protein Eng* **12**, 15–21.

37. Graña, O., Baker, D., Maccallum, R.M., et al. (2005) CASP6 assessment of contact prediction. *Proteins: Structure, Function, and Bioinformatics* **61 Suppl 7**, 214–24.

38. Koh, I.Y.Y., Eyrich, V.A., Marti-Renom, M.A., et al. (2003) EVA: evaluation of protein structure prediction servers. *Nucleic Acids Research* **31**, 3311–3315.

39. Pazos, F., Helmer-Citterich, M., and Ausiello, G. (1997) Correlated mutations contain information about protein-protein interaction. *J Mol Biol* **271**, 511–523.

40. Rychlewski, L., and Fischer, D. (2005) LiveBench-8: The large-scale, continuous assessment of automated protein structure prediction. *Protein Science* **14**, 240–245.

41. Pollastri, G. and Baldi, P. (2002) Prediction of contact maps by GIOHMMs and recurrent neural networks using lateral propagation from all four cardinal corners. *Bioinformatics* **18**(Suppl. 1), S62–S70.

42. Kohonen, T., and Makisari, K. (1989) The self-organizing feature maps. *Phys Scripta* **39**, 168–172.

43. Andreeva, A., Howorth, D., Brenner, S.E., et al. (2004) SCOP database in 2004: refinements integrate structure and sequence family data. *Nucleic Acids Research* **32**(Database issue), D226–9.

44. Zhang, Y., Arakaki, A.K., and Skolnick, J. (2005) TASSER: An automated method for the prediction of protein tertiary structures. *Proteins: Structure, Function, and Bioinformatics* **Suppl. 7**, 91–98.

45. Kim, D.E., Chivian, D., and Baker, D. (2004) Protein structure prediction and analysis using the Robetta server. *Nucleic Acids Research* **32**, W526–W531.

Chapter 4

Analysis of Mass Spectrometry Data in Proteomics

Rune Matthiesen and Ole N. Jensen

Abstract

The systematic study of proteins and protein networks, that is, proteomics, calls for qualitative and quantitative analysis of proteins and peptides. Mass spectrometry (MS) is a key analytical technology in current proteomics and modern mass spectrometers generate large amounts of high-quality data that in turn allow protein identification, annotation of secondary modifications, and determination of the absolute or relative abundance of individual proteins. Advances in mass spectrometry–driven proteomics rely on robust bioinformatics tools that enable large-scale data analysis. This chapter describes some of the basic concepts and current approaches to the analysis of MS and MS/MS data in proteomics.

Key words: Database searching, *de novo* sequencing, peptide mass fingerprinting, peptide fragmentation fingerprinting, quantitation.

1. Introduction

1.1. Protein Identification, Annotation of PTMs, and Quantitation

DNA sequencing and microarray technology have provided large-scale, high throughput methods to quantitate cellular mRNA levels and determine distributions of single nucleotide polymorphisms (SNPs) at a genome-wide scale. However, DNA-based technologies provide little, if any, information about dynamic biomolecular events, such as protein interactions, protein-based regulatory networks, and post-translational modifications of proteins. For example, regulatory events governed at the level of mRNA translation have been reported (1), and the activity of proteins is often controlled by post-translational modifications (2, 3). Proteomics, the systematic study of proteins, grew out of protein chemistry during the 1980s and 1990s. Initially based mainly on two-dimensional gel electrophoresis, proteomics has now embraced a range of

Jonathan M. Keith (ed.), *Bioinformatics, Volume II: Structure, Function and Applications, vol. 453*
© 2008 Humana Press, a part of Springer Science + Business Media, Totowa, NJ
Book doi: 10.1007/978-1-60327-429-6 Springerprotocols.com

biochemical, immunological, and computational fields as well as a series of sensitive analytical technologies, including mass spectrometry (3–5). Currently, most large-scale protein identification work in proteomics is based on mass spectrometry. The mass spectrometer is used to determine the accurate molecular mass of proteins and the derived peptides, and tandem mass spectrometry (MS/MS) facilitates amino acid sequencing and mapping of post-translational modifications (2, 4). Modern mass spectrometers generate a wealth of proteomics data in a short time and the computational processing of this data remains a bottleneck in many proteomics projects (5, 6). The aim of this chapter is to introduce the basic concepts for analysis and processing of mass spectrometry data obtained in proteomics experiments.

1.2. Mass Spectrometry and Proteomics Workflows

Protein analysis by mass spectrometry is typically performed in a "bottom-up" fashion, meaning that proteins are digested into peptides, which are in turn analyzed by mass spectrometry. The protein sequence and secondary modifications are subsequently assembled based on peptide MS and MS/MS data (8). Recently, the concept of "top-down" analysis of intact proteins was introduced (9), but it is beyond the scope of this chapter to describe this approach (see **Note 1**). Peptides are normally generated by trypsin cleavage of protein. Trypsin specifically and efficiently cleaves the amide bond C-terminal to arginine and lysine residues, unless a proline is the next residue (see **Notes 2** and **3**). The peptide fragments are then analyzed using mass spectrometry–based strategies, including peptide mass mapping by MS and peptide sequencing by MS/MS.

Peptide mass mapping strategies are usually used for characterization of simple protein samples containing only one or a few protein species and it is often used in combination with protein separation by 2D gel electrophoresis (**Fig. 4.1**). In the peptide mass mapping approach, the molecular masses of a set of tryptic peptides derived from a protein sample is measured. Since each individual protein has a distinct amino acid sequence, the pattern of tryptic peptides for each protein will be unique and it can be used to identify the protein by database searching combined with scoring algorithms to retrieve the best matches (see **Note 4** and **Fig. 4.2**). Protein identification by peptide mass mapping relies on accurate molecular weight determination by mass spectrometry and the assumption that trypsin faithfully cleaves at Arg and Lys residues (10). More detailed analysis of individual peptides is achieved by MS/MS, which allows amino acid sequencing of selected peptides (see **Fig. 4.1**). Peptide separation by liquid chromatography (LC) is advantageous when analyzing complex peptide samples as LC equipment is readily interfaced to electrospray ionization tandem mass spectrometers, so called LC-MS/MS systems.

LC-MS/MS strategies are used for detailed analysis of complex protein mixtures and for mapping of post-translational modifications. First, the MS/MS instrument records a mass spectrum

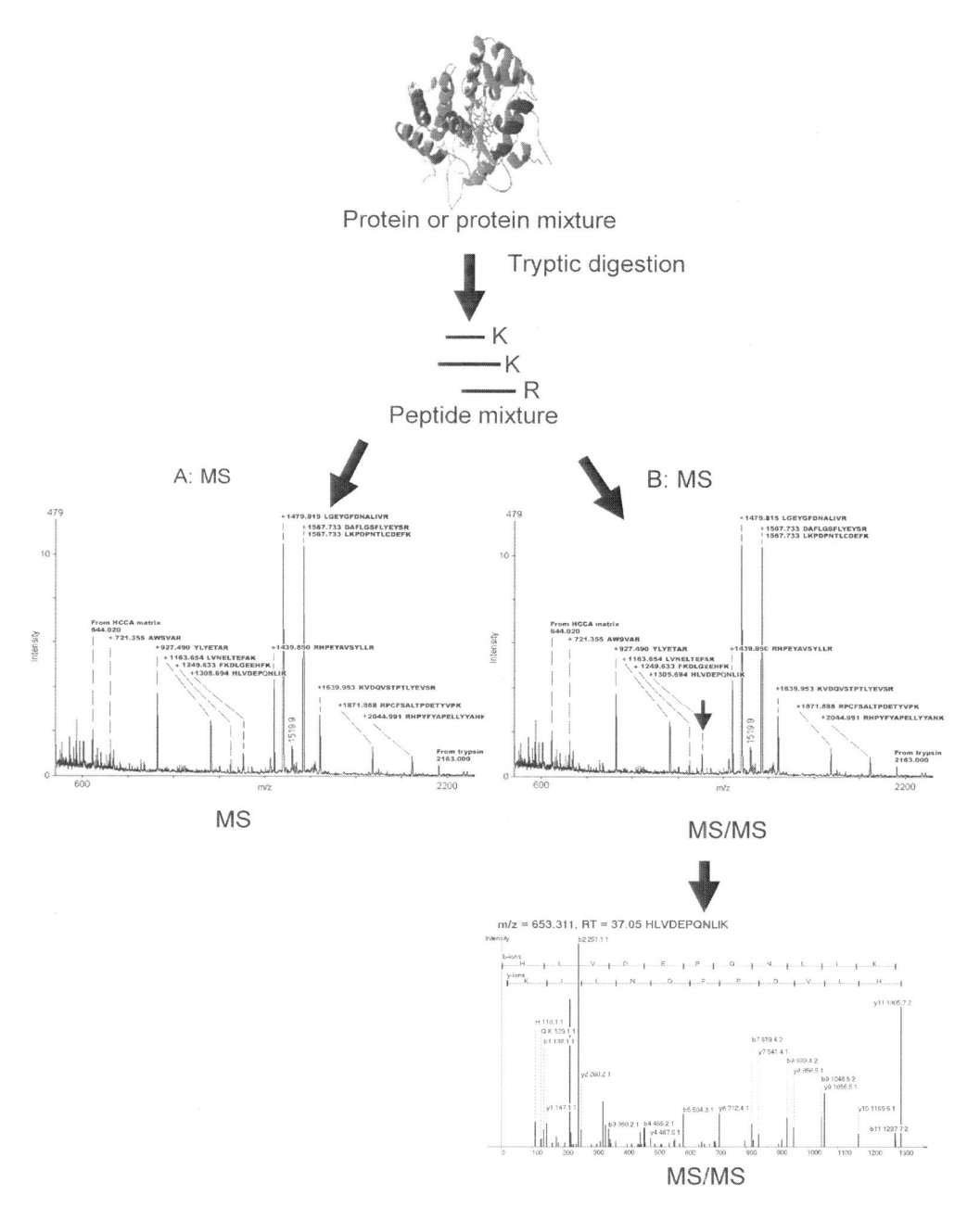

Fig. 4.1. The workflow for the MS based peptide mass mapping (**A**) and MS/MS-based peptide sequencing (**B**) methods. The latter method goes one step further by recording the MS/MS spectrum/spectra of chosen peptides.

to determine the masses of all peptides that are eluting from the LC-column at a given time. Next, the mass spectrometer control software determines the masses (m/z values) of the two to five most intense peptide signals for further analysis by MS/MS, i.e., for sequencing (*see* **Note 5**). Each of these gas-phase peptide ions is, in turn, isolated and fragmented inside the mass spectrometer.

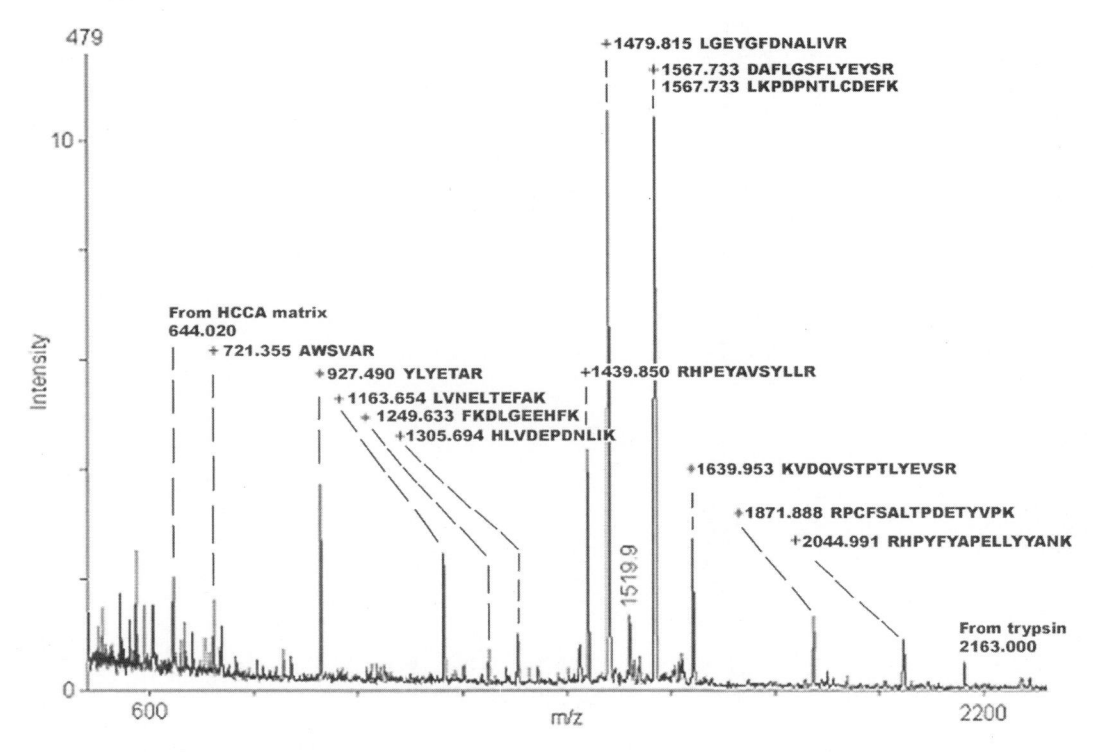

Fig. 4.2. MALDI-TOF MS spectrum (PMF/PMM strategy). (+) indicates mass peaks that matched the masses of theoretical tryptic peptides from BSA.

The set of peptide fragments is then recorded by a second mass analyzer to generate the MS/MS spectrum for that peptide. Notably, the most frequent type of fragmentation occurs by cleavage of the amide bond between amino acid residues to generate y-ions containing the C-terminal parts and b-ions containing the N-terminal parts of the peptide (**Fig. 4.3A**). The mass differences between the peptide fragment ion signals in the MS/MS spectrum can be correlated with amino acid residue masses (*see* **Fig. 4.3B** and **Table 4.1**): In this way amino acid sequence information can be obtained by tandem mass spectrometry (*8, 11*). The precise fragmentation chemistry depends on the fragmentation method and mass analyzer used, and has been reviewed elsewhere (*12, 13*).

The analysis of peptide mass spectra requires several steps that are described in the following (**Fig. 4.4**): *(1)* conversion of the continuous mass spectral data to a list of peptide masses (MS) or peptide fragment masses (MS/MS) (**Fig. 4.5**); *(2)* spectral interpretation and sequence annotation to generate a list of peptides and proteins; *(3)* quantification of peptides and proteins and comparative analysis; *(4)* bibliographic and computational sequence analysis; *(5)* data storage in a relational database.

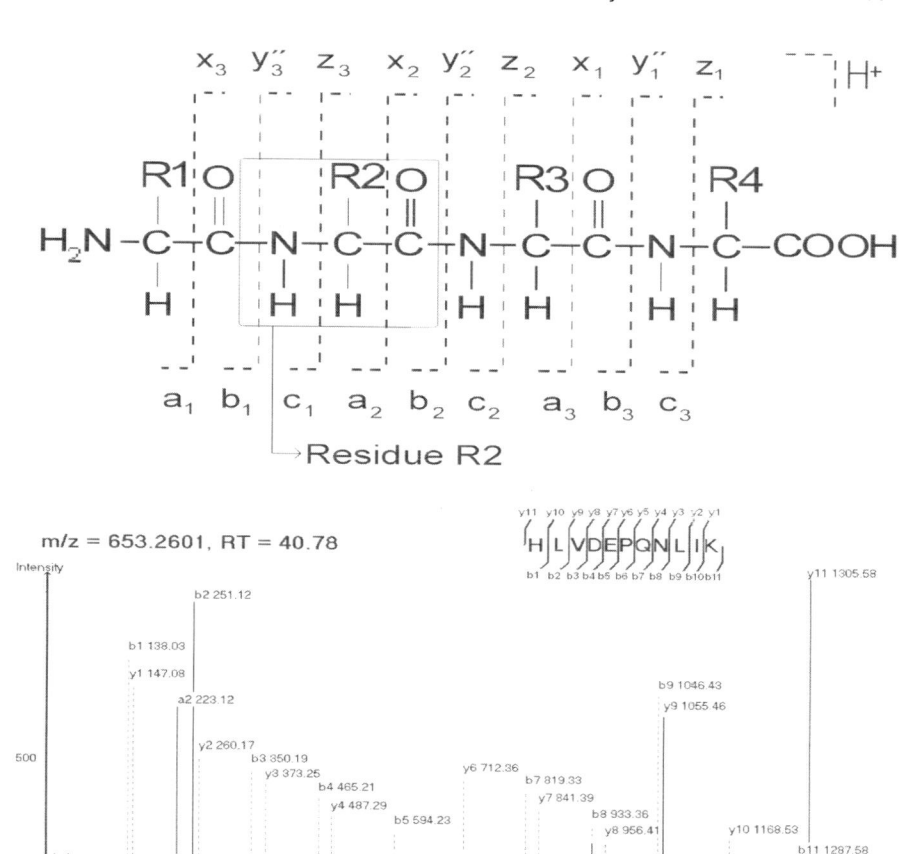

Fig. 4.3. (**A**) The nomenclature used for observed MS/MS fragments of peptides. (**B**) MS/MS spectrum obtained from the double charged peptide HLVDEPQNLIK. For simplicity only the intense y-, b-, and a-ions are annotated. Notice that the mass difference between the peaks from ions in the b-ion series gives the amino acid sequence directly. The mass difference between the peaks from ions in the y-ion series gives the reverse amino acid sequence.

2. Data Analysis

2.1. Data Preprocessing: The Need for Standardization of Data Formats

Proteomics is still a rather young scientific field, and it is only recently that proposals for common data standards and data exchange formats have emerged: mzXML *(14)* and mzDATA *(15)*. There is now an ongoing effort to merge these two standard formats *(16)*.

The mzXML and mzDATA converters either keep the continuous data obtained from the mass spectrometers or convert them into a peak list by using the centroid method (*see* **Note 6**). These converters do not use any advanced spectra processing techniques. More advanced spectra processing algorithms can reduce the "background signals" in MS and MS/MS data

Table 4.1
Delta masses that can be observed between peaks in a MS/MS spectrum and their corresponding interpretations

Identity	Delta mass (Da)
G	57.0215
A	71.0371
S	87.0320
P	97.0528
V	99.0684
T	101.0477
L	113.0841
I	113.0841
N	114.0429
D	115.0269
Q	128.0586
K	128.0950
E	129.0426
M	131.0405
H	137.0589
F	147.0684
R	156.1011
Y	163.0633
W	186.0793
C unmodified	103.0092
C carbamidomethylated	160.0307
M oxidation	147.0354

(*see* **Note 7**) (*see* **Fig. 4.5A,B**), isotope-deconvolute the peptide ion signals, and charge-deconvolute the protonated peptide ion signals (*see* **Fig. 4.5**). The final result from the advanced processing algorithms is a "peak list" of peptide masses (MS) or peptide fragment masses (MS/MS) that correspond to singly charged

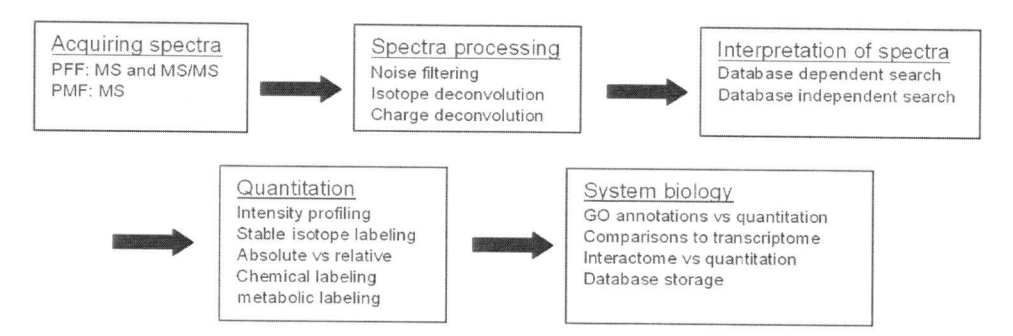

Fig. 4.4. Steps involved in interpretation of mass spectrometry data.

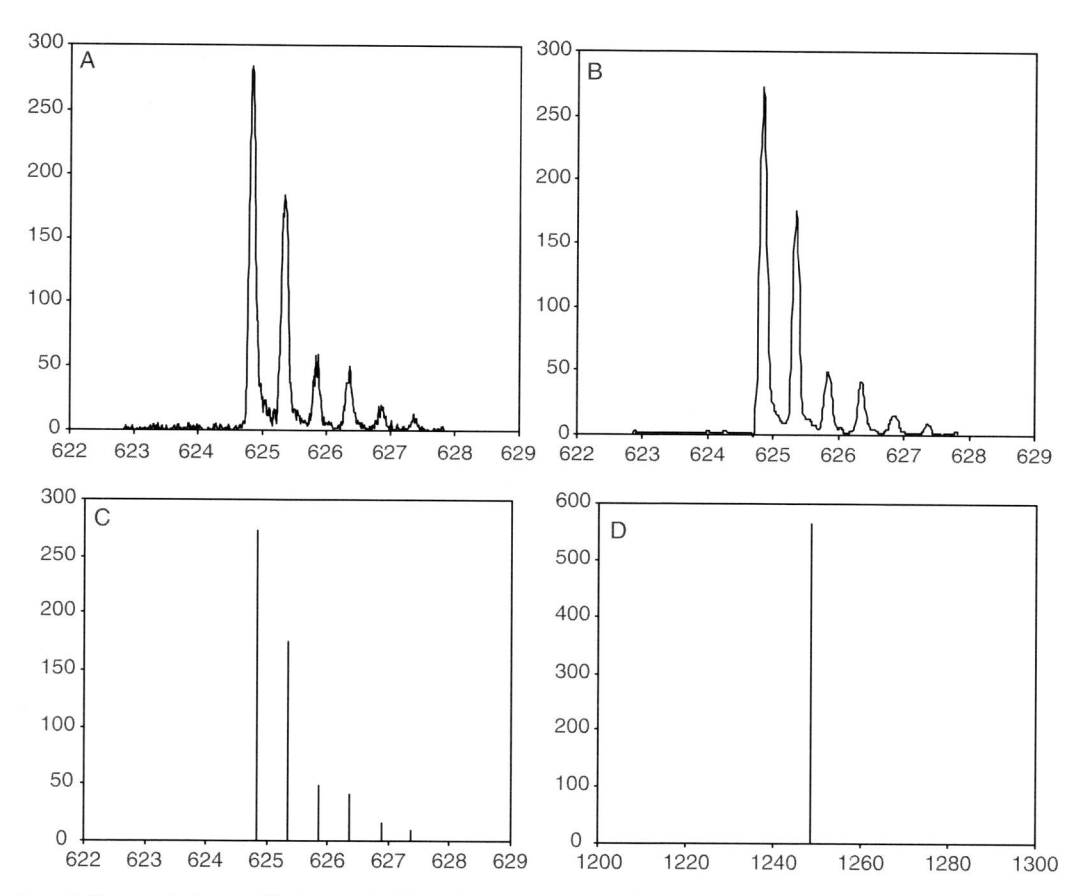

Fig. 4.5. The result of noise filtering, centroiding, charge, and isotope deconvolution. Mass over charge is plotted on the x-axis and intensity on the y-axis. (**A**) The raw data MS spectrum of the double charged peptide LGGQTYNVALGR. (**B**) The data after three iterations of a nine-point Savitsky-Golay filtering algorithm. (**C**) The peak list obtained by centroiding the peaks in (**B**). (**D**) Charge and isotope deconvolution of the peak list in (**C**).

monoisotopic peaks. Such algorithms and data reduction make the subsequent data analysis much simpler. To our knowledge, there is no publicly available application that can process raw data from any of the standard MS data formats into singly charged monoisotopic peak lists. However, this can be done by the Max-Ent algorithm, which is available in the commercial programs MassLynx v4.0 and PLGS v2.05 from Waters.

In the VEMS *(17)* platform, the program ExRaw interfaces to the mzXML converters *(14)* and the Maxent algorithm from PLGS v2.05 for convenient automated data processing of several LC-MS/MS runs. The spectra mass and intensity values are stored by Base64 encoding in both the mzXML and mzDATA formats.

2.2. Database Searching

Peak lists of peptide masses (MS) and/or peptide fragment masses (MS/MS) are searched against protein sequence databases *(18)*. It is also possible to search DNA and EST databases *(19, 20)*, if needed. The techniques of comparing mass spectrometry data against a sequence database are also called database-dependent search algorithms *(21)* since these algorithms are dependent on a sequence database for interpreting the data. A large number of algorithms and programs have been proposed; many of them are publicly available and others are commercial (**Table 4.2**). It is often useful to compare search results from different search algorithms since current software, e.g., VEMS, Mascot *(22)*, and X!Tandem *(23)* generate slightly different results.

Table 4.2
Useful programs for interpreting MS and MS/MS spectra of peptides

Name	Input Data	Interfaced from VEMS	Public
VEMS v3.0	MS, MS/MS	No	Yes
Mascot	MS, MS/MS	Yes	Semi
X!Tandem	MS/MS	Yes	Yes
P3	MS/MS	No	Yes
Inspect	MS/MS	No	Yes
Phenyx	MS/MS	No	Semi
PepNovo	MS/MS	No	Yes
Lutefisk	MS/MS	Yes	Yes
OpenSea	MS/MS	No	Yes
De Novo peaks	MS/MS	No	Yes
PepHMM	MS/MS	No	Yes
ProteinProphet	MS/MS	No	Yes

The user-defined search parameter settings of the different search engines are rather similar and are briefly discussed in the following. The preprocessing of raw data into peak lists, which can be done with the mass spectrometry instrument vendor software, is necessary before MS and MS/MS data can be submitted to database search engines (*see* **Section 2.1**).

1. Database. The FASTA format used for the sequence databases can be obtained from resources such as NCBI *(24)*, EBI *(25)*, or Swiss-Prot *(26)*. The choice of database is an important issue. It is essential to understand that the search engines work by finding the optimum match in the database. This means that if an incomplete database is searched then some spectra might be matched to a wrong peptide sequence. Another issue is that data sets often have cross-contamination from organisms other than the one used in the study (e.g., keratin from *Homo sapiens* and peptides from porcine trypsin may contaminate the search results obtained from an *A. thaliana* [plant] protein sample). The generally accepted procedure in the mass spectrometry field is to search the international protein indexed database IPI and then include the most likely contaminants from other organisms. The IPI database is the top-level database for a number of other databases (**Fig. 4.6**). IPI is a non-redundant database composed of sequences from Uniprot knowledgebase, Ensembl, and RefSeq. Ensembl and RefSeq are composed of translated predicted genes. Uniprot knowledgebase is composed of sequences from Swiss-Prot, PIR (protein information resource), and TrEMBL. Swiss-Prot and PIR are manually curated databases with many cross-references to other resources, including literature references. TrEMBL contains translated nucleotide sequences from EMBL. A great effort has been made to make IPI non-redundant without removing isoforms. Uniprot knowledge base is further divided into UniREF 100, 90, and 50. In the UniREF databases, all sequences with a length longer than 11 amino acids and sequence identity of 100, 90 or 50 percent are represented as one sequence entry.

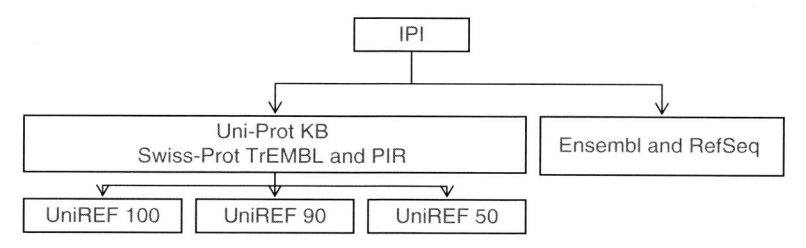

Fig. 4.6. The organization of a selected set of databases frequently used for searching mass spectrometry data.

2. Cleavage method. There are both enzymatic and chemical cleavage methods that can be used to generate peptide fragments. However, trypsin digestion is used most often (*see* **Notes 2** and **3**). Some database search programs allow specification of enzyme specificity. In such cases it is useful to know that trypsin does not cleave all sites followed by Arg and Lys equally well. If there are some charged amino acids in the neighborhood of the cleavage site or if the Lys or Arg is followed by a Pro, then the cleavage is generally less efficient. If the sample contains proteases other than trypsin, then one can expect non-tryptic cleavages as well.

3. Fixed and variable modifications. Most search algorithms allow specification of both fixed and variable modifications. Fixed modifications are considered to be present on all of the specified amino acids. Specifying a fixed modification is equivalent to changing the mass of an amino acid and will not have an effect on the computational search time. A frequently used modification is alkylation of Cys residues by, e.g., iodoacetamide which should be specified as a fixed modification. A variable modification may or may not occur on the specified amino acid, for example partial phosphorylation of Ser, Thr, or Tyr residues. Introducing variable modifications has an exponential effect on the computational search time. In practice this means that it is not always possible to test for all the possible variable modifications in one search. It is therefore essential to mine the literature to narrow down the most relevant variable modification for the sample in question. Mascot has a threshold of nine variable modifications. VEMS has searched the human IPI database with 16 variable modifications *(17)*.

4. Mass accuracy. The mass accuracy should be set to include the largest possible mass deviation in the data set. This will depend on the mass spectrometer used. The mass accuracy can be specified as deviation in Da (absolute mass error) or in ppm (parts per million, relative mass error). In mass spectrometry there is often a linear systematic error in the data which is largest for high mass values. Specifying the mass accuracy in ppm has the advantage that it allows higher mass deviation for larger peptides, which is often observed. If the maximum allowed deviation in ppm is set to 10 ppm, then it would correspond to 0.01 Da for a 1 kDa peptide and 0.02 Da for 2 kDa peptide. In Mascot the scores and the significance values are dependent on the specified mass accuracy. The VEMS program gives the same score and significance for the same data as long as the mass deviation is set appropriately high to include the worst mass deviation.

2.3. Search Specificity

An important issue when performing database dependent searches is an evaluation of the rate of false-positives versus the rate of true positives. These rates can be visualized using receiver operating characteristics curves (ROC curves) *(27)*. The ROC curves can be made by searching a data set of known proteins against various databases using different search algorithms. Each search algorithm will give one ROC curve.

The search specificity can also be studied by evaluating the false discovery rate. This is often done by searching the protein sequence database first in the forward direction and then in the reverse direction *(28)*. The search in the reverse direction can then be used to evaluate the false discovery rate at different score thresholds. A more realistic approach would be to reverse all the tryptic peptide sequences, leaving all the arginines and lysines at the C-terminal position. The false discovery rate can also be evaluated by searching a large set of random peptides with the same masses as the identified peptides.

2.4. Database Independent Interpretation

Database independent sequencing a.k.a. *de novo* sequencing, is often used if no complete protein sequence database is available for the organism under study. It is also useful to validate results obtained from database-dependent searches. Database-independent interpretation only uses the molecular mass information available in the peptide MS/MS spectra. There are a large number of algorithms available for database independent sequencing (*see* **Table 4.1**). Some database independent algorithms work by generating all possible peptide sequences with a theoretical peptide mass corresponding to the measured mass of the intact peptide, allowing for an instrument-specific mass inaccuracy. Most algorithms use graph theory to find the best amino acid sequence path through the MS/MS spectrum, by "jumping" from peak to peak and finding a matching amino acid residue mass (*see* **Fig. 4.3B**) *(29)*. There are some variations in the methods used to trace the amino acid sequence through the spectrum. For example, some algorithms continuously calculate and exclude fragment masses from other ion series every time a new fragment ion is included in an ion series.

The main differences between various algorithms are that they use different scoring functions, which are often optimized for a specific instrument. PepNovo *(30)*, Lutefisk *(31)*, and VEMS are examples of programs that offer database independent (*de novo*) sequencing algorithms.

When one or several peptide sequences are obtained from a protein by database-independent assignment, they can be merged to search for homology or similarity by using the very useful online tool "MS-BLAST" (http://dove.embl-heidelberg.de/Blast2/msblast.html) *(32)*.

2.5. Quantitation by Mass Spectrometry

Biological systems are always studied by comparing different states, for example, a control state vs. a perturbed state, or temporal protein expression profiles. Quantitative methods for proteomic analysis are required in these situations. Quantitative proteomic analysis by MS is achieved by peptide intensity profiling or by stable isotope labeling as described in more detail below. The publicly available programs VEMS *(17)*, MSquant *(33)*, RelEx *(34)*, and ASAPratio *(35)* are useful programs in quantitative proteomics.

2.5.1. Quantitation by Intensity Profiling

Quantitative proteomics by intensity profiling is done by comparing the measured intensity signals in the different samples *(36)*. Since there is some degree of instrument-introduced variation in the intensity of the signals, it is important to have replicate measurements. In a recent publication, the dose-response of a histone deacetylase inhibitor (PXD101) on human cell cultures was studied by peptide intensity profiling *(37)*. Histones were purified from six different states, corresponding to human cell cultures exposed to 6 different concentrations of histone deacetylase inhibitor. Triplicate LC-MS/MS analysis was performed on the tryptic digest obtained from each of the six states. Quantitation of the peptides that were covalently modified by acetylation on lysine residues required several analytical and computational steps (**Fig. 4.7**). Protein identifications and PTM assignments were performed using the Mascot and VEMS

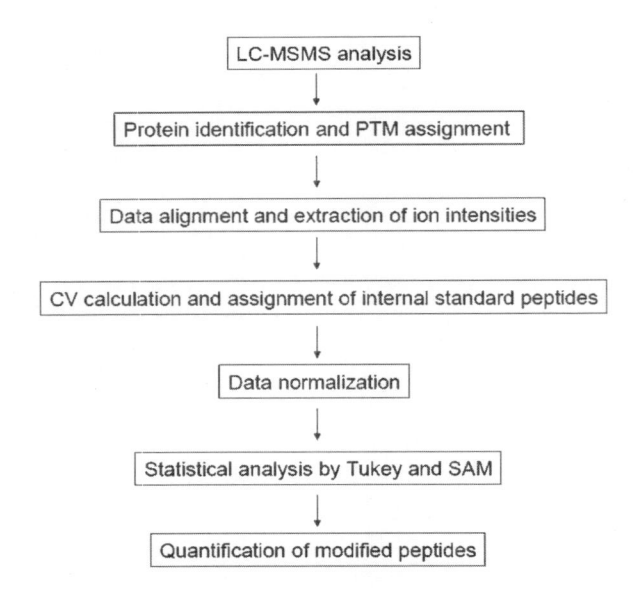

Fig. 4.7. Flow diagram showing the steps in quantitation of post-translational modification by intensity profiling.

search programs. Peptide retention times, peptide masses, and intensities were extracted from the LC-MS raw data. The peptide mass and retention times from the different runs were aligned using the VEMS program. The coefficient of variance (CV) of the ion intensities of all detected and sequenced peptides was calculated and used to identify unmodified peptides with CVs <30% that could be used to normalize the intensity values in the 18 LC-MS/MS runs. Finally, the significance of difference between the intensity values obtained for the different peptide from the six samples was evaluated by a Tukey Q-test (38) and SAM analysis (39). This method helped us identify and quantify a variety of acetylated and methylated peptides derived from human histones, thereby revealing the molecular action of the histone deacetylase inhibitor PDX101 (37).

2.5.2. Stable Isotope Labeling

Stable isotope labeling can be used to perform absolute or relative quantitation. In absolute quantitation, the measured peptide mass intensity is compared with the intensity for that peptide labeled with stable isotopes (and added to the sample in known concentrations) (40). In relative quantitation, one or more stable isotope labels are used to label peptides from different samples (**Fig. 4.8**) (35). The different intensities for the peptides with the same sequence but labeled with different stable isotopes is used to calculate the relative quantitations for the samples. The labels can either be introduced chemically (41) or by metabolic labeling (42, 43). A broad variety of amino acids labeled with stable isotopes can be used for quantitation. An example of peaks used for relative quantitation for two samples is shown in **Fig. 4.9**. Stable isotope labeling is a very accurate way to make relative comparisons between biological samples. The literature gives many examples of its use, such as studying mitogen activations (44, 45), comparing different cell lines (46), studying the differentiation of cells (47), and differentiating between unspecific and specific binding (48).

2.6. Data Storage

A number of database systems are available for storing proteomics mass spectrometry data. The most difficult part is often to get the data parsed into a database system. Unfortunately, there are currently no good publicly available tools that are able to parse search results and all the experimental settings of importance from the different search engines into one of the publicly available databases such as YASSdb (49), CPAS (16), GPMdb (50), PRIDE (51), Trans-Proteome Pipeline (6), and Proteios (52). However, we are aware of groups that have initiated the development of such parsers. Such functionality is useful to compare search results obtained from different instruments using different instrument settings and/or search engines.

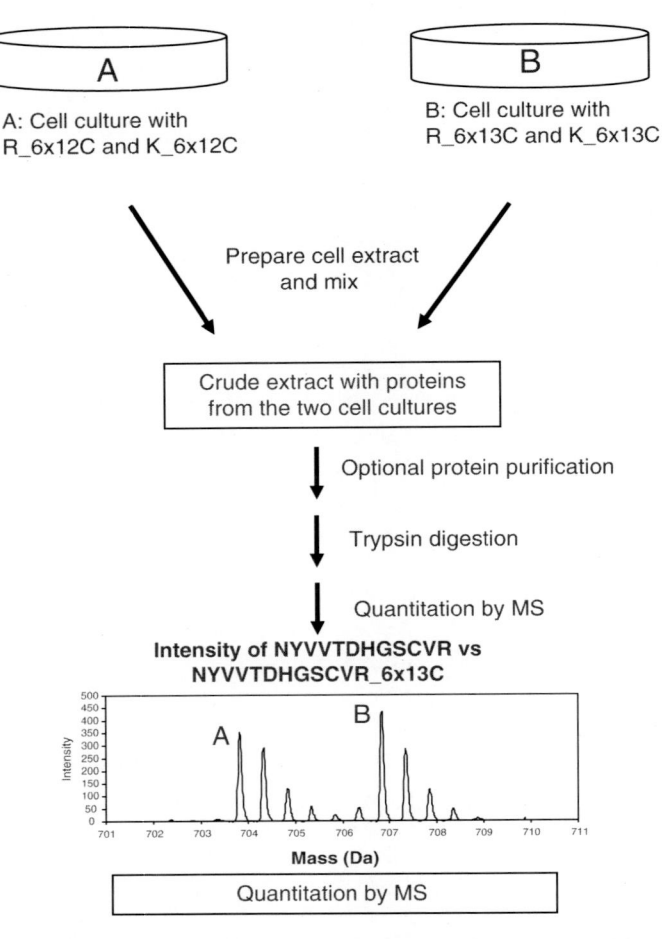

Fig. 4.8. Flow schema of the SILAC method. The result is MS spectra in which peptides from cell culture (**A**) and (**B**) are separated by n*~6 Da where n is the number of stable isotope labeled arginines and lysines in the corresponding peptide. R_6x^{13}C is abbreviation for arginine where six ^{12}C are substituted with six ^{13}C and similar symbols (K instead of R) are used for lysine.

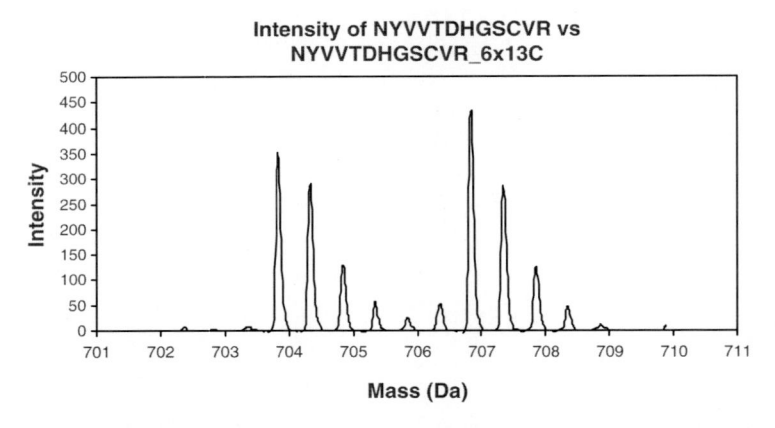

Fig. 4.9. Relative quantification between the light and heavy version of the peptide NYV-VTDHGSCVR originating from two different samples. R_6x^{13}C is abbreviation for arginine where six 12C are substituted with six ^{13}C.

3. Conclusion

Mass spectrometry is a very effective proteomics tool for identification and quantitation of proteins and for mapping and quantitation of their post-translational modifications. The diversity of experiments in the proteomics field makes the data analysis challenging and exciting. Computational tools for interpretation of mass spectra, sequence database searching, protein quantitation, and data storage are continuously developed. From the computational point of view, there are still many challenges and problems to solve. With the continuous emergence of novel experimental strategies and analytical instruments, the demand for advanced computational algorithms for data analysis and data integration in proteomics will grow in the coming years.

4. Notes

1. In top-down sequencing, intact proteins are fragmented directly in the mass spectrometer. This is a technique that is under development; therefore, it is not a technique that is widespread in different laboratories.

2. The cleavage method used should be compatible with the mass spectrometer employed. For example, the peptide fragments should be in a mass range that can be measured by the mass spectrometer. Trypsin generates suitable peptide masses in the m/z range covered by most mass spectrometers, i.e., m/z 400 to 3,500. The cleavage method should also generate peptides with a low number of basic amino acids, so that peptides of low charge states are formed in the gas phase. Low charge states simplify the MS and MS/MS spectra and ensure that the m/z values for peptides are within the range of the mass spectrometer. Most tryptic peptides have one basic amino acid as the C-terminal residue.

3. Good efficiency and specificity of the cleavage methods makes it easier to identify peptides in a database since it can be used as an extra constraint during the search. Trypsin has a high specificity; however, the cleavage efficiency is known to be lower for cleavage sites that have neighboring charged residues. For example, the peptide ..ESTVKKT.. or ..ESTVDKT.. will not be cleaved with the same efficiency by trypsin as the peptide ..ESTVAKT..

4. Leucine and isoleucine have identical mass. Therefore, standard MS and MS/MS strategies cannot distinguish proteins that

only differ by having leucine substituted for isoleucine. Further complications arise when two distinct, but near-identical peptides (e.g., EVESTK and VEESTK) have the same mass.

5. The mass spectrometry software normally selects the most intense peaks with a charge state of +2 or higher. The reason for this is that peptides with a charge state above +2 are more likely to produce good fragmentation spectra.

6. The centroid mass m_c and the corresponding intensity I_c can be calculated by the following expressions:

$$m_c = \frac{\sum\limits_{y_i > y_{i,max}^x} m_i I_i}{I_c}.$$

$$I_c = \sum\limits_{y_i > y_{i,max}^x} I_i$$

where m_i is the mass at a certain mass bin and I_i is the corresponding intensity. x is a specified percentage of the maximum intensity.

7. Convolution is a process in which a function g convolves (transforms) another function:

$$f * g = h$$

f can, for example, be a function that describes the physical quantity of interest. The function g has convolved the function f to the measured function h. Deconvolution is the reverse of convolution; therefore, it can be used to obtain the function f, which describes the physical quantity of interest.

Acknowledgments

R.M was supported by the EU TEMBLOR (IntAct) project and by a Carlsberg Foundation Fellowship. O.N.J. is a Lundbeck Foundation Research Professor and the recipient of a Young Investigator Award from the Danish Natural Science Research Council.

References

1. Kozak, M. (2006) Rethinking some mechanisms invoked to explain translational regulation in eukaryotes. *Gene* Available online 22 June.

2. Seet, B. T., Dikic, I., Zhou, M. M., et al. (2006) Reading protein modifications with interaction domains. *Nat Rev Mol Cell Biol* 7, 473–483.

3. Jensen, O. N. (2006) Interpreting the protein language using proteomics. *Nat Rev Mol Cell Biol* 7, 391–403.

4. Aebersold, R., Mann, M. (2003) Mass spectrometry-based proteomics. *Nature* 422, 198–207.

5. Patterson, S. D., Aebersold, R. (1995) Mass spectrometric approaches for the

identification of gel-separated proteins. *Electrophoresis* 16, 1791–1814.

6. Domon, B, Aebersold, R. (2006) Challenges and opportunities in proteomic data analysis. *Mol Cell Proteomics*. Available online 8 August.

7. Patterson S. D. (2003) Data analysis: the Achilles heel of proteomics. *Nat Biotechnol* 21, 221–222.

8. Steen, H., Mann, M. (2004) The ABC's (and XYZ's) of peptide sequencing. *Nat Rev Mol Cell Biol* 5, 699–711.

9. Fridriksson, E. K., Beavil, A., Holowka, D., et al. (2000) Heterogeneous glycosylation of immunoglobulin E constructs characterized by top-down high-resolution 2-D mass spectrometry. *Biochemistry* 39, 3369–3376.

10. Jensen, O. N., Larsen, M. R., Roepstorff, P. (1998) Mass spectrometric identification and microcharacterization of proteins from electrophoretic gels: strategies and applications. *Proteins* 2, 74–89.

11. Roepstorff, P., Fohlman, J. (1984) Proposal for a common nomenclature for sequence ions in mass spectra of peptides. *Biomed Mass Spectrom* 11, 601.

12. Wysocki, V. H., Tsaprailis, G., Smith, L. L., et al. (2000) Mobile and localized protons: a framework for understanding peptide dissociation. *J Mass Spectrom* 35, 1399–1406.

13. Laskin, J., Futrell, J. H. (2003) Collisional activation of peptide ions in FT-ICR. *Mass Spectrom Rev* 22, 158–181.

14. Pedrioli, P. G., Eng, J. K., Hubley, R., et al. (2004) A common open representation of mass spectrometry data and its application to proteomics research. *Nat Biotechnol* 22, 1459–1466.

15. Orchard, S., Kersey, P., Hermjakob, H., et al. (2003) The HUPO Proteomics Standards Initiative meeting: towards common standards for exchanging proteomics data. *Comp Funct Genom* 4, 16–19.

16. Cottingham, K. (2006) CPAS: a proteomics data management system for the masses. *J Proteome Res* 5, 14.

17. Matthiesen, R., Trelle, M. B., Højrup, P., et al. (2005) VEMS 3.0: Algorithms and computational tools for tandem mass spectrometry based identification of post-translational modifications in proteins. *J Proteome Res* 4, 2338–2347.

18. Fenyo, D., Qin, J., Chait, B.T. (1998) Protein identification using mass spectrometric information. *Electrophoresis* 19, 998–1005.

19. Matthiesen, R., Bunkenborg, J., Stensballe, A., et al. (2004) Database-independent, data-base-dependent, and extended interpretation of peptide mass spectra in VEMS V2.0. *Proteomics* 4, 2583–2593.

20. Fermin, D., Allen, B. B., Blackwell, T. W., et al. (2006) Novel gene and gene model detection using a whole genome open reading frame analysis in proteomics. *Genome Biol* 7, R35.

21. Fenyö, D., Beavis, R. C. (2003) A method for assessing the statistical significance of mass spectrometry-based protein identifications using general scoring schemes. *Anal Chem* 75, 768–774.

22. Creasy, D. M., Cottrell, J. S. (2002) Error tolerant searching of tandem mass spectrometry data not yet interpreted. *Proteomics* 2, 1426–1434.

23. Craig, R., Beavis, R. C. (2004) TANDEM: matching proteins with tandem mass spectra, *Bioinformatics* 20, 1466–1467.

24. Woodsmall, R. M., Benson, D. A., (1993) Information resources at the National Center for Biotechnology Information. *Bull Med Libr Assoc* 81, 282–284.

25. LinksKersey, P. J., Duarte, J., Williams, A., Karavidopoulou, Y., Birney, E., Apweiler, R. (2004) The International Protein Index: an integrated database for proteomics experiment. *Proteomics* 4, 1985–1988.

26. LinksBairoach, A., Apweiler, R. (1998) The SWISS-PROT protein sequence data bank and its supplement TrEMBL in 1998. *Nucleic Acids Res* 26, 38–42.

27. Colinge, J., Masselot, A., Cusin, I., et al. (2004) High-performance peptide identification by tandem mass spectrometry allows reliable automatic data processing in proteomics. *Proteomics* 4, 1977–1984.

28. López-Ferrer, D., Martínez-Bartolomé, S., Villar, M., et al. (2004) Statistical model for large-scale peptide identification in databases from tandem mass spectra using SEQUEST. *Anal Chem* 76, 6853–6860.

29. Dancik, V., Addona, T., Clauser, K., et al. (1999) De novo peptide sequencing via tandem mass spectrometry. *J Comput Biol* 6, 327–342.

30. Frank, A., Pevzner, P. (2005) PepNovo: de novo peptide sequencing via probabilistic network modeling. *Anal Chem* 77, 964–973.

31. Johnson, R. S., Taylor, J. A. (2002) Searching sequence databases via de novo peptide sequencing by tandem mass spectrometry. *Mol Biotechnol* 22, 301–315.

32. Shevchenko, A., Sunyaev, S., Loboba, A., et al. (2001) Charting the proteomes of organisms with unsequenced genomes by

MALDI-Quadrupole time-of flight mass spectrometry and BLAST homologuey searching. *Anal Chem* 73, 1917–1926.

33. Andersen, J. S., Wilkinson, C. J., Mayor, T., Mortensen, P., Nigg, E. A., Mann, M. (2003) Proteomic characterization of the human centrosome by protein correlation profiling. *Nature* 426, 570–574.

34. MacCoss, M. J., Wu, C. C., Liu, H., et al. (2003) A correlation algorithm for the automated quantitative analysis of shotgun proteomics data. *Anal Chem* 75, 6912–6921.

35. Venable, J. D., Dong, M. Q., Wohlschlegel, J., et al. (2004) Automated approach for quantitative analysis of complex peptide mixtures from tandem mass spectra. *Nat Methods* 1, 39–45.

36. Listgarten, J., Emili, A. (2005) Statistical and computational methods for comparative proteomic profiling using liquid chromatography-tandem mass spectrometry. *Mol Cell Proteomics* 4, 419–434.

37. Beck, H. C., Nielsen, E. C., Matthiesen, R., et al. (2006) Quantitative proteomic analysis of post-translational modifications of human histones. *Mol Cell Proteomics* 5, 1314–1325.

38. Zar, J. H. (1999) *Biostatistical Analysis.* Prentice-Hall, Upper Saddle River, NJ.

39. Tusher, V. G., Tibshirani, R., Chu, G., et al. (2001) Significance analysis of microarrays applied to the ionizing radiation response. *PNAS* 98, 5116–5121.

40. Gerber, S. A., Rush, J., Stemman, O., et al. (2003) Absolute quantification of proteins and phosphoproteins from cell lysates by tandem MS. *Proc Natl Acad Sci U S A* 100, 6940–6945.

41. Turecek, F. (2002) Mass spectrometry in coupling with affinity capture-release and isotope-coded affinity tags for quantitative protein analysis. *J Mass Spectrom* 37, 1–14.

42. Ong, S. E., Blagoev, B., Kratchmarova, I., et al. (2002) Stable istotope labeling by amino acids in cell culture, SILAC, as a simple and accurate approach to expression proteomics. *Mol Cell Proteom* 1, 376–386.

43. Yang, W. C., Mirzaei, H., Liu, X., et al. (2006) Enhancement of amino Acid detection and quantification by electrospray ionization mass spectrometry. *Anal Chem* 78, 4702–4708.

44. Gruhler, A., Schulze, W. X., Matthiesen, R., et al. (2005) Stable isotope labeling of *Arabidopsis thaliana* cells and quantitative proteomics by mass spectrometry. *Mol Cell Proteom* 4, 1697–709.

45. Ballif, B. A., Roux, P. P., Gerber, S. A., et al. (2005) Quantitative phosphorylation profiling of the ERK/p90 ribosomal S6 kinase-signaling cassette and its targets, the tuberous sclerosis tumor suppressors. *Proc Natl Acad Sci U S A* 102, 667–672.

46. Fierro-Monti, I., Mohammed, S., Matthiesen, R., et al. (2005) Quantitative proteomics identifies Gemin5, a scaffolding protein involved in ribonucleoprotein assembly, as a novel partner for eukaryotic initiation factor 4. *J Proteome Res* 5, 1367–1378.

47. Romijn, E. P., Christis, C., Wieffer, M., et al. (2006) Expression clustering reveals detailed co-expression patterns of functionally related proteins during B cell differentiation. *Molecular & Cellular Proteomics* 4, 1297–1310.

48. Blagoev, B., Kratchmarova, I., Ong, S. E., et al. (2003) A proteomics strategy to elucidate functional protein-protein interactions applied to EGF signaling. *Nat Biotechnol* 21, 315–318.

49. http://www.yass.sdu.dk/yassdb/

50. Craig, R.., Cortens, J. P., Beavis, R. C. (2004) Open source system for analyzing, validating, and storing protein identification data. *J Proteome Res* 3, 1234–1242.

51. Jones, P., Cote, R. G., Martens, L., et al. (2006) PRIDE: a public repository of protein and peptide identifications for the proteomics community. *Nucleic Acids Res* 34, D659–663.

52. Gärdén, P., Alm, R., Häkkinen, J. (2005) Proteios: an open source proteomics initiative. *Bioinformatics* 21, 2085–2087.

## Chapter 5

The Classification of Protein Domains

Russell L. Marsden and Christine A. Orengo

Abstract

The significant expansion in protein sequence and structure data that we are now witnessing brings with it a pressing need to bring order to the protein world. Such order enables us to gain insights into the evolution of proteins, their function, and the extent to which the functional repertoire can vary across the three kingdoms of life. This has led to the creation of a wide range of protein family classifications that aim to group proteins based on their evolutionary relationships.

This chapter discusses the approaches and methods that are frequently used in the classification of proteins, with a specific emphasis on the classification of protein domains. The construction of both domain sequence and domain structure databases is considered and the chapter shows how the use of domain family annotations to assign structural and functional information is enhancing our understanding of genomes.

Key words: Protein domain, sequence, structure, clustering, classification, annotation.

1. Introduction

The arrival of whole-genome sequencing, heralded by the release of the yeast genome in 1995, promised significant new advances in our understanding of protein evolution and function. Over a decade on we are now beginning to understand the full potential of the genome sequencing projects, with over 90 completed genomes released to the biological community in 2005 alone (**Fig. 5.1**). This rapid increase in genomic sequence data challenges us to unravel the patterns and relationships that underlie the pathways and functions that form genome landscapes.

Detailed comparisons among genomes require the annotation of gene names and functions. Such annotations can be derived from biochemical analysis—a limited approach given the

Jonathan M. Keith (ed.), *Bioinformatics, Volume II: Structure, Function and Applications, vol. 453*
© 2008 Humana Press, a part of Springer Science+Business Media, Totowa, NJ
Book doi: 10.1007/978-1-60327-429-6 Springerprotocols.com

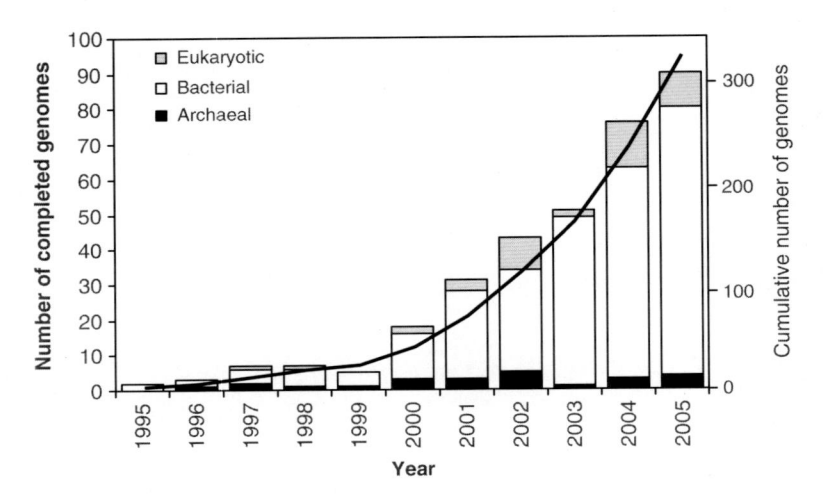

Fig. 5.1. The number of completed genomes released each year since 1995 has grown rapidly, although the number of published eukaryotic genomes is still small in comparison to the prokaryotic genomes. The black line shows the cumulative growth in completed genomes—>330 between 1995 and 2005.

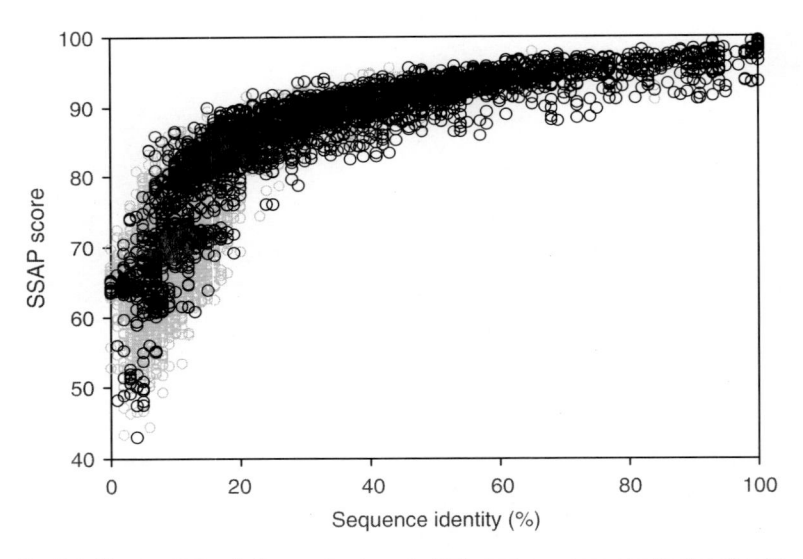

Fig. 5.2. The correlation between structure similarity and sequence identity for all pairs of homologous domain structures in the CATH domain database. Structural similarity was measured by the SSAP structure comparison algorithm, which returns a score in the range of 0 to 100 for identical protein structures. (Dark gray circles represent pairs of domains with the same function; light gray those with different functions).

number of sequences we need to functionally classify—or through computational analysis to identify evolutionary relationships. The exploitation of sequence homology for genome annotation is of considerable use because sequence homologues generally share the same protein fold and often share similarities in their function, depending on the degree of relatedness between them (**Fig. 5.2**). The diversity of such relationships can be seen within families of proteins that group together protein sequences that are thought to share a common ancestor. Protein families often exhibit considerable divergence in sequence, structure, and sometimes function. Over the past decade a varied range of protein family classifications have been developed with a view to transferring functional information from experimentally characterized genes to families of sequence relatives.

This chapter discusses the different approaches to classifying proteins, with a particular emphasis on the classification of protein domains into families. We will see how the use of domain-centric annotations is pivotal to maximizing our understanding of genomes, and enables experimentally obtained annotation to be imputed to a vast number of protein sequences.

2. What is a Protein Domain?

The protein domain has become a predominant concept in many areas of the biological sciences. Despite this, a single universally accepted definition does not exist, an ambiguity that is partly attributable to the subjective nature of domain assignment. At the sequence level, protein domains are considered to be homologous portions of sequences encoded in different gene contexts that have remained intact through the pathways of evolution. From this perspective, domains are the "evolutionary unit" of protein sequences, and increasing evidence is emerging through the analysis of completed genome data that a large proportion of genes (up to 80% in eukaryotes) comprise more than one domain *(1, 2)*. In structural terms, domains are generally observed as local, compact units of structure, with a hydrophobic interior and a hydrophilic exterior forming a globular-like state that cannot be further subdivided. As such, domains may be considered as semi-independent globular folding-units. This property has enabled them to successfully combine with other domains and evolve new functions.

The duplication and recombination of domains has been a staple of evolution: Genes have been divided and fused using a repertoire of pre-existing components. During the course of evolution domains derived from a common ancestor have diverged at the sequence level through residue substitution or mutation and by the insertion and deletion of residues, giving rise to families of homologous domain sequences.

Domain assignment provides an approximate view of protein function in terms of a "molecular parts bin": The overall function is a combination of the constituent parts. This view of protein function is sometimes too simplistic, particularly when compared with experimentally derived evidence. Nevertheless, it does provide accurate functional information. Moreover, the classification of domains facilitates study of the evolution of proteins, in particular through the analysis and comparison of domain sequence and structure families.

3. Classification of Domains from Sequence

As sequence databases became more populated in the early 1970s, an increasing amount of research focused on the development of methods to compare and align protein sequences. Needleman and Wunsch *(3)* devised an elegant solution using dynamic programming algorithms to identify optimal regions of local or global similarity between the sequences of related proteins. More recently, algorithms such as FASTA *(4)* and BLAST *(5)* have approximated these approaches in order to make sequence searching several orders of magnitude faster. This has been a significant development, enabling us to work with the ever-increasing collections of sequence data by performing millions of sequence comparisons within reasonable time frames. Such efforts are essential if we are to identify a significant proportion of the evolutionary relationships within the sequence databases and provide order to these databases through protein clustering and classification.

When sequence comparisons are performed among proteins containing more than one domain, complications can arise when homologous sequence regions are encountered in otherwise non-homologous contexts. This can lead to the assignment of a single function to a multi-functional multi-domain protein or to the "chaining problem," which results in unrelated proteins being grouped together by sequence clustering methods *(6)*. For example, the pairwise comparison of two protein sequences may reveal significant sequence similarity and consequently they may be considered homologous. However, closer inspection of the alignment might reveal that the sequence similarity is only shared over partial regions of each sequence (**Fig. 5.3**), suggesting the identification of homologous domains that belong to the same family. Homologous domains like these are frequently found embedded within different proteins with different domain partners and it is this domain recurrence that many domain classification resources exploit when identifying and clustering domain sequences.

3.1. Automatic Domain Sequence Clustering

Automated domain clustering algorithms are used in the construction of many domain sequence classifications to generate an initial clustering of domain-like sequences. This method is often followed by varying levels of expert-driven validation of the proposed domain families.

a   Align whole - sequences

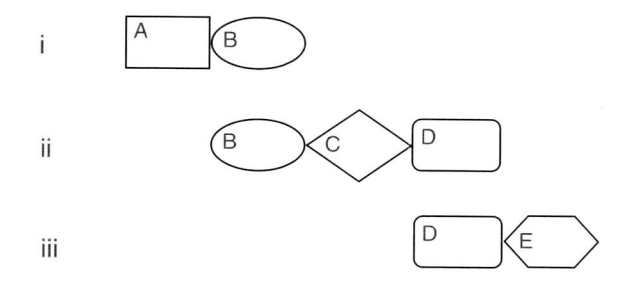

b   Cluster domain - sequences

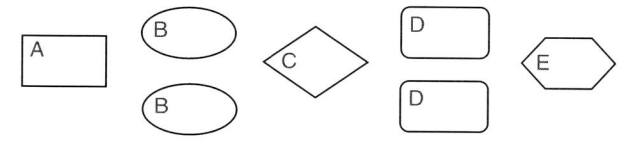

Fig. 5.3. The domain problem in protein sequence comparison. (**A**) The three sequences (i, ii, and iii) are incorrectly clustered due to sequence similarity with shared domains (**B** and **D**) in sequence ii. (**B**) The identification and excision of domains (**A–E**) enables subsequent clustering of the domains into five domain families.

To implicitly predict domain boundaries from sequence data alone is an immensely challenging problem, although a number of methods have been devised using predicted secondary structure, protein folding, or domain-linker patterns recognized through neural networks. Despite extensive research, none of these methods is reliable enough or well enough established to use in large-scale sequence analysis. Instead, most automatic domain clustering methods rely on pairwise alignment information to identify domain sequences and classify them in two main steps. These are, first, to divide protein sequences into domain-like regions corresponding to local regions of similarity (i.e., identify recurring domain sequences); and second, to cluster the sequence fragments. Although conceptually simple, in practice the output of different approaches can vary widely, for example, in the number and size of domain families that are generated (*see* **Note 1**). Nevertheless, the clustering of homologous domains remains a rational approach to organizing protein sequence data and many domain clustering approaches have been developed, including ProDom *(7)*, EVEREST *(8)*, and ADDA *(9)*.

ProDom is an automated sequence clustering and assignment method that clusters all non-fragment sequences (partial gene sequences that may not correspond to whole domains are excluded) in the SWISS-PROT and TrEMBL *(10)* sequence databases. First the database sequences are sorted by size order into a list. The smallest sequence is identified and related sequence regions in the list are identified using PSI-BLAST *(11)* similarity scores and removed from the list into the current domain cluster. The remaining sequences are once again sorted by size, and the next smallest sequence used as a query to build a new domain cluster. This process is repeated until the supply of sequences in the database is exhausted.

The EVEREST (EVolutionary Ensembles of REcurrent SegmenTs) program uses a combination of machine learning and statistical modeling to interpret the data produced from large-scale all-against-all sequence comparisons. Domain sequences are first identified using the alignment data and clustered into putative domain families. Machine learning methods are then applied to identify the best families, upon which a statistical model is built. An iterative process is then used to scan the models against the original sequence database to recreate a database of domain-like segments, which are subsequently re-clustered.

The ADDA algorithm (Automatic Domain Decomposition Algorithm) is another entirely automated approach to domain delineation and clustering with a particular emphasis on the clustering of very large sets of protein sequences. Alignment results from all-against-all BLAST sequence comparisons are used to define a minimal set of domains with as many possible overlapping aligned regions. Pairwise profile–profile comparisons are then used to form links between assigned domains and domain families are generated for sets of domains with links above a given threshold.

3.2. Whole-Chain Sequence Clustering

A number of methods attempt to circumvent the problem of assigning domains by clustering whole-chain protein sequences (e.g., Tribe-MCL *(12)*, SYSTERS *(13)*, ProtoNet *(14)*, HAMAP *(15)*). Such conservative approaches tend to generate many more sequence families compared with clustering at the domain level, in which recurrence of domains leads to a greater reduction in redundancy. However, this cautious approach to clustering sequence space can provide insights into the evolution of domain architectures (the types and orderings of domains along a chain) and associated functions across protein families.

3.3. Families Represented by Multiple Sequence Alignments

The earliest domain family classifications were only able to capture a limited number of evolutionary relationships because the reliability of pairwise sequence comparison quickly descends in the so-called twilight zone *(16)* of sequence similarity (<30% sequence identity). Comparison of protein structures has shown that evolutionarily related domains can diverge to a point of insignificant sequence similarity: Pairwise sequence comparisons do not recognize these relationships even with the use of substitution matrices that allow variation in amino acid residues caused by mutation. The expansion of sequence databases over the past decade, however, has enlarged the populations of protein families and this has improved homologue recognition in the twilight zone through the derivation of family-specific multiple sequence alignments. Such multiple alignments are an authoritative analytical tool, as they create a consensus summary of the domain family, encapsulating the constraints of evolutionary change, and can reveal highly conserved residues that may correspond to function, such as an active site. The unique features that define each domain family can then be captured in the form of patterns, or as profiles represented by position specific scoring matrices (PSSMs) or hidden Markov models (HMMs).

3.3.1. Patterns

Family specific patterns are built by automated analysis of multiple sequence alignments in order to characterize highly conserved structural and functional sequence motifs that can be used as a core "fingerprint" to recognize distant homologues. These are usually represented in the form of regular expressions that search for conserved amino acids that show little or no variability across constituent sequences. Patterns are relatively simple to build and tend to represent small contiguous regions of conserved amino acids, such as active sites or binding sites, rather than providing information across a domain family. They are rigid in their application of pattern matching, which can mean that even small divergences in amino acid type can result in matches to related sequence being missed.

3.3.2. Profiles

Like patterns, profiles can also be automatically built from a multiple sequence alignment, but unlike patterns, they tend to be used to form a consensus view across a domain family. A profile is built to describe the probability of finding a given amino acid at a given location in the sequence relatives of a domain family using position-specific amino acid weightings and gap penalties. These values are stored in a table or matrix and are used to calculate similarity scores between a profile and query sequence. Profiles can be used to represent larger regions of sequence and allow a greater residue divergence in matched sequences in order to identify more divergent family members. A threshold can be calculated for a given set of family sequences to enable the reliable identification and inclusion of new family members into the original set of sequences. Profile methods such as HMMs have been shown to be highly discriminatory in identifying distant homologues when searching the sequence databases *(17)*. It is also clear that the expansion of diversity within the sequence and domain databases will bring an increase in the sensitivity of profile methods, allowing us to go even further in tracing the evolutionary roots of domain families.

3.4. Domain Sequence Classifications

A variety of domain sequence databases are now available, most of which, as has been discussed, use patterns or profiles (or in some cases both) to build a library of domain families. Such libraries can often be browsed via the Internet or used to annotate sequences over the Internet using comparison servers. Some classifications, such as Pfam, provide libraries of HMMs and comparison software to enable the user to generate automated up-to-date genome annotations on their local computer systems. A list of the most popular databases is shown in **Table 5.1**; a selection of these (PROSITE *(18)*, PFAM *(19)*, SMART *(20)*) are described in greater detail in the following.

3.4.1. Prosite

PROSITE began as a database of sequence patterns, its underlying principle being that domain families could be characterized by the single most conserved motif observed in a multiple sequence alignment. More recently it has also provided an increasing number of sequence profiles. PROSITE multiple sequence alignments are derived from a number of sources, such as a well-characterized protein family, the literature, sequence searching against SWISS-PROT and TrEMBL, and from sequence clustering. PROSITE motifs or patterns are subsequently built to represent highly conserved stretches of contiguous sequence in these alignments. These typically correspond to specific biological regions, such as enzyme active sites or substrate binding sites. Accordingly, PROSITE patterns tend to embody short conserved biologically active regions within domain families, rather than representing the domain family over its entire length. Each pattern is manually tested and refined by compiling statistics that reflect how often a certain motif matches sequences in SWISS-PROT. The use of patterns provides a rapid method for database searching, and in cases in which global sequence comparison becomes unreliable, these recurring "fingerprints" are often able to identify very remote sequence homologues.

PROSITE also provides a number of profiles to enable the detection of more divergent domain families. Such profiles include a comparison table of position-specific weights representing the frequency distribution of residues across the initial PROSITE multiple sequence alignment. The table of weights is then used to generate a similarity score that describes the alignment between the whole or partial PROSITE profile and a whole or partial sequence. Each pair of amino acids in the alignment is scored based on the probability of a particular type of residue substitution at each position, with scores above a given threshold constituting a true match.

3.4.2. Pfam

Pfam is a comprehensive domain family resource providing a collection of HMMs for the identification of domain families, repeats, and motifs. The database is built upon two distinct

Table 5.1
Protein domain sequence classifications

Database	Description	URL
ADDA	Automatic domain decomposition and clustering based on a global maximum likelihood model	http:/echidna.biocenter.helsinki.fi:9801/sqgraph/pairsdb
Blocks	Database of domain families represented by multiple-motif	http://blocks.fhcrc.org
CDD	Conserved domain database	http://www.ncbi.nlm.nih.gov/Structure/cdd/cdd.shtml
CluSTr	Clusters of SWISS-PROT and TrEMBL sequence databases	http://www.ebi.ac.uk/clustr/
DOMO	Protein domain database built from sequence alignments	http://www.infobiogen.fr/services/domo/
EVEREST	Large-scale automatic determination of domain families	http://www.everest.cs.huji.ac.il
InterPro	Integrated domain family resource	http://www.ebi.ac.uk/interpro/
IProClass	Integrated protein classification database	http://pir.georgetown.edu/iproclass/
MetaFam	Protein family information database	http://metafam.umn.edu
Pfam	Database of multiple sequence alignments and hidden Markov models	http://www.sanger.ac.uk/Software/Pfam/
PIR	Protein Information Resource	http://pir.georgetown.edu/
PIR-ALN	Curated database of protein sequence alignments	http://pir.georgetown.edu/pirwww/dbinfo/piraln.html
PRINTS	Database of protein fingerprints based on protein motifs	http://umber.sbs.man.ac.uk/dbbrowser/PRINTS/
ProClass	Non-redundant sequence database organized into protein families	http://pir.georgetown.edu/gfserver/proclass.html
ProDom	Automatic classification of protein domains	http://prodes.toulouse.inra.fr/prodom/doc/prodom.html
PROSITE	Database of domain families described by patterns and profiles	http://www.expasy.org/prosite/
ProtoMap	Automated hierarchical classification of SWISS-PROT database	http://www.protomap.cornell.edu
SBASE	Curated domain library based on sequence clustering	http://hydra.icgeb.trieste.it/sbase/
SMART	Collection of protein domain families focusing on those that are difficult to detect	http://smart.embl-heidelberg.de
SYSTERS	Systematic sequence alignment and clustering	http://systers.molgen.mpg.de
TIGRFAMs	Protein families represented by hidden Markov models	http://www.tigr.org/TIGRFAMs/

classes of domain-like "seed" alignments that are initially generated using the ProDom algorithm (described in the preceding). These seed families, initially labeled as Pfam-B domains, are then verified using the literature and where possible extended by expert-driven HMM profile searches, to generate Pfam-A families. Consequently, the database is organized into two parts: Pfam-A domains are considered to be more reliable, as they are represented by the manually curated sequence alignments and HMMs that cover most of the common protein domain families. Pfam-B domains, although having a higher coverage of sequence space, are less reliable, representing the automatically generated supplement of HMMs derived from the automated clustering of sequences.

The high-quality seed alignments are used to build HMMs to which sequences are automatically aligned to generate the final full alignments. If the initial alignments are deemed to be diagnostically unsound, then the seed is manually checked, and the process repeated until a sound alignment is generated. The parameters that produce the best alignment are saved for each family so that the result can be reproduced.

3.4.3. SMART

The SMART database (simple modular architecture research tool) represents a collection of domain families with an emphasis on those domains that are widespread and difficult to detect. As with Pfam, the SMART database holds manually currated HMMs but unlike Pfam, the "seed" alignments are derived from sequence searching using PSI-BLAST or, where possible, are based on tertiary structure information. An iterative process is used to search for additional relatives for inclusion into each sequence alignment, until no further sequence relatives can be identified. SMART domain families are selected with a particular emphasis on mobile eukaryotic domains and as such are widely found among nuclear, signaling, and extracellular proteins. Annotations detailing function, sub-cellular localization, phyletic distribution, and tertiary structure are also included within the classification. In general, SMART domains tend to be shorter than structural domains.

4. Classification of Domains from Structure

As we have seen so far, the classification of domains at the sequence level is most commonly based on the use of sequence comparison methods to identify homologous domains belonging to the same domain family. However, despite advances in sequence comparison algorithms, such as the use of mutation matrices and development of profile-based methods, very remote homologues often remain undetected. Consequently, most sequence-based domain family classifications tend to group closely related sequences sharing significant sequence similarity and possessing similar or identical biological functions.

It is well recognized that many distant relationships can only be identified through structure comparison. This is because protein structure is much more highly conserved through evolution than is its corresponding amino acid sequence. The comparison of proteins at the structure level often allows the recognition of distant ancestral relationships hidden within and beyond the twilight zone of sequence comparison, in which sequences share <30% sequence identity (*see* **Fig. 5.2**). Under this principle, when sufficient structural data was available, a number of protein structure classifications were developed in the 1990s, each of which aimed to classify the evolutionary relationships between proteins based on their three-dimensional structures. Using structure-based comparisons in a manner comparable to sequence-based searching, it becomes possible to traverse the barriers that prevent sequence searches from identifying distant homologues. The ability to classify newly structurally characterized proteins into pre-existing and novel structural families has allowed far reaching insights to be gained into the structural evolution of proteins. Although the structural core is often highly conserved across most protein families, revealing constraints on secondary structure packing and topology, analysis can also identify considerable structural embellishments, which often correspond to changes in domain partners and function.

Like many of the sequence classifications, most structure classifications have also been established at the domain level (**Table 5.2**). Each has been constructed using a variety of algorithms and protocols to recognize similarities between proteins. Some groups use all-against-all comparisons to calculate structural relationships with less emphasis on the construction of a formal classification (e.g., the Dali Dictionary) whilst other resources have developed hierarchies based upon these sequence and structure relationships (e.g., the CATH domain database).

4.1. Identification of Domain Boundaries at the Structural Level

Over 40% of known structures in the Protein Data Bank *(21)* are multi-domain proteins, a percentage that is likely to increase as structure determination methods, such as x-ray crystallography, become better able to characterize large proteins. It is very difficult to reliably assign domain boundaries to distant homologues by sequence based methods; however, in cases in which the three-dimensional structure has been characterized, putative

Table 5.2
Protein domain sequence classifications

Database	Description	URL
3DEE	Multi-hierarchical classification of protein domains	http://jura.ebi.ac.uk:8080/3Dee/help/help_into.html
CAMPASS	CAMbridge database of Protein Alignments organized as Structural Superfamilies. Includes a library of sequence alignments for each superfamily	http://www-cryst.bioc.cam.ac.uk/~campass/
CATH	CATH is a hierarchical classification of protein domain structures that clusters proteins at four major levels, Class, Architecture, Topology, and Homologous superfamily	http://cathdb.info
CE	Combinatorial extension of the optimal path. Provides pairwise alignments of structures in the PDB.	http://cl.sdsc.edu/ce.html
Dali Database	The Dali Database represents a structural classification of recurring protein domains with automated assignment of domain boundaries and clustering	http://ekhidna.biocenter.helsinki.fi/dali/start
ENTREZ / MMDB	MMDB contains pre-calculated pairwise comparison for each PDB structure. Results are integrated into ENTREZ	http://www.ncbi.nlm.nih.gov/Structure/
HOMSTRAD	HOMologous STRucture alignment database. Includes annotated structure alignments for homologous families	http://www-cryst.bioc.cam.ac.uk/~homstrad/
SCOP	A Structural Classification Of Proteins. Hierarchical classification of protein structure that is manually curated. The major levels are family, superfamily, fold and class	http://scop.mrc-lmb.cam.ac.uk/scop/

domain boundaries can be delineated by manual inspection through the use of graphical representations of protein structure. Nonetheless, the delineation of domain boundaries by eye is a time-consuming process and is not always straightforward, especially for large proteins containing many domains or discontinuous domains in which one domain is interrupted by the insertion of another domain.

The concept of a structural domain was first introduced by Richardson, defining it as a semi-independent globular folding unit *(22)*. The following criteria, based on this premise, are often used to characterize domains:

- A compact globular core

- More intra-domain residue contacts than inter-domain contacts

- Secondary structure elements are not shared between domains, most significantly beta-strands

- Evidence of domain as an evolutionary unit, such as recurrence in different structural contexts

The growth in structure data has led to the development of a variety of computer algorithms that automatically recognize domain boundaries from structural data, each with varying levels of success. Such methods often use a measure of geometric compactness, exploiting the fact that there are more contacts between residues within a domain than between neighboring domains, or searching for hydrophobic clusters that may represent the core of a structural domain. Many of these algorithms perform well on simple multi-domain structures in which few residue contacts are found between neighboring domains, although the performance levels tend to be disappointing for more complex structures in which domains are intricately connected. Despite this, their speed (often less than 1 second per structure) does allow a consensus approach to be used for domain assignment, where the results from a series of predictions from different methods are combined, and in cases of disagreement, can be manually validated. To some extent the conflicting results produced by different automated methods is expected if we consider the fact that there is no clear quantitative definition of a domain and the high levels of structural variability in many domain families.

The advent of structure comparison enabled the principle of recurrence to again play a central role in domain definition. The observation of a similar structure in a different context is a powerful indicator of a protein domain, forming the rationale for methods that match domains in newly determined structures against libraries of classified domains. This concept is rigorously applied in the SCOP *(23)* database, in which domains are manually assigned by visual inspection of structures. In other domain classifications such as CATH *(24)*, automated methods are also used to identify putative domains that may not yet have been observed in other contexts. The Dali Database *(25)* includes an automated algorithm that identifies putative domains using domain recurrence in other structures.

4.2. Methods for Structural Comparison

The use of automated methods for structural comparison of protein domains is essential in the construction of domain structure classifications. Structural comparison and alignment algorithms were first introduced in the early 1970s and methods such as rigid body superposition are still used today for superimposing structures and calculating a similarity measure (root mean square deviation). This is achieved by translation and rotation of structures in space relative to one another in order to minimize the number of non-equivalent residues. Such approaches use dynamic programming, secondary structure alignment, and fragment comparison to enable comparison of more distantly related structures in which extensive residue insertions and deletions or shifts in secondary structure orientations have occurred. More recently, some domain structure classifications have employed rapid comparison methods, based on secondary structure, to approximate these approaches (e.g., SEA *(26)*, VAST *(27)*, CATHEDRAL *(28)*). This enables a large number of comparisons to be performed that are used to assess the significance of any match via a rigorous statistical analysis. Where necessary, potential relatives can then be subjected to more reliable, albeit more computationally intensive, residue-based comparisons (e.g., SSAP *(29)*, Dali *(30)*).

Differences between methods, and consequently classifications, can arise due to differences in how structural relationships are represented. However, recent attempts to combine sequence and structure based domain family resources in the InterPro database *(31)* should maximize the number of distant relationships detected (*see* **Note 2**).

4.3. Domain Structure Classification Hierarchies

The largest domain structure classification databases are organized on a hierarchical basis corresponding to differing levels of sequence and structure similarity. The terms used in these hierarchies are summarized in **Table 5.3**.

At the top level of the hierarchy is domain class, a term that refers to the proportion of residues in a given domain adopting an alpha-helical or beta-strand conformation. This level is usually divided into four classes: mainly-alpha, mainly-beta, alternating alpha-beta (in which the different secondary structures alternate along the polypeptide chain), and alpha plus beta (in which mainly-alpha and mainly-beta regions appear more segregated). In the CATH database, these last two classes are merged into a single alpha-beta class as a consequence of the automated assignment of class. CATH also uses a level beneath class classifying the architecture of a given domain according to the arrangement of secondary structures regardless of their connectivity (e.g., barrel-like or layered sandwich). Such a description is also used in the SCOP classification, but it is less formalized, often appearing for a given structural family rather than as a completely separate level in the hierarchy.

Within each class, structures can then be further clustered at the fold (also known as topology) level according to equivalences in the orientation *and* connectivity of their secondary structures. Cases in which domains adopt highly similar folds are often indicative of an evolutionary relationship. However,

Table 5.3
Overview of hierarchical construction of domain structure classifications

Level of hierarchy	Description
Class	The class of a protein domain reflects the proportion of residues adopting an alpha-helical or beta-strand conformation within the three-dimensional structure. The major classes are mainly-alpha, mainly-beta, alternating alpha/beta and alpha+beta. In CATH the alpha/beta and alpha+beta classes are merged.
Architecture	This is the description of the gross arrangement of secondary structures in three-dimensional space independent of their connectivity.
Fold/topology	The gross arrangement of secondary structures in three-dimensional space and the orientation and connectivity between them.
Superfamily	A group of proteins whose similarity in structure and function suggests a common evolutionary origin.
Family	Proteins clustered into families have clear evolutionary relationships. This generally means that pairwise residue identities between the proteins are 30% and greater. However in some cases, similar functions and structure provides sufficient evidence of common descent in the absence of high sequence identity.

care must be taken: Fold similarity can be observed between two domains that share no significant sequence similarity or features that indicate common functional properties (i.e., are evolutionarily unrelated). As structure databases have grown, an increasing number of domain pairs that adopt similar folds but possess no other characteristics, implying homology have been found. Such pairs may consist of two extremely distant relatives from the same evolutionary origins that have diverged far beyond sequence and functional equivalence, or alternatively two domains that have evolved from different ancestors but have converged on the same fold structure (fold analogues).

For this reason, structure relationships must be verified by further evidence such as similar sequence motifs or shared functional characteristics. Such evidence allows fold groups to be further subdivided into broader evolutionary families or superfamilies: A term first coined by Margaret Dayhoff to describe related proteins that have extensively diverged from their common ancestors *(32)*. Analysis of domain structure classifications shows that some fold groups are particularly highly populated and occur in a diverse range of superfamilies. In fact, approximately 20% of the superfamilies currently assigned in CATH belong to fewer than 10 fold groups. These folds, frequently occupied by many domains with no apparent evolutionary relationship, were defined as superfolds by Orengo and co-workers *(33)*. The occurrence of superfolds suggests that the diversity of possible protein folds in nature is constrained by the need to maintain an optimal packing of secondary structures to preserve a hydrophobic core. Currently, the number of reported folds varies from about 950 (SCOP database) to 1100 (CATH database) due to the different clustering criteria used and the difficulty in distinguishing folds in more continuous regions of fold space.

Within superfamilies, most classifications sub-cluster members into families of close relatives based on similar functional properties. Family clusters are often identified by sequence comparison; for example, in the CATH and Dali databases, close homologues are clustered at 35% sequence identity. In the SCOP classification, similarity in functional properties is also manually identified and used to generate families.

4.4. Structural Domain Classifications

In this section, three of the most comprehensive structural domain classifications are briefly discussed, SCOP, CATH and the Dali Database, representing manual, semi-automated, and automated approaches to classification respectively. A more comprehensive list, together with internet links, is also shown in **Table 5.2**.

The SCOP database uses an almost entirely manual approach for the assignment of domain boundaries and recognition of structural and functional similarities between proteins to generate superfamilies. This has resulted in an extremely high-quality resource, even though it requires a significant level of input from the curators.

Unlike SCOP, the CATH approach to domain classification aims to automate as many steps as possible, alongside the use of expert manual intervention to differentiate between homologous proteins and those merely sharing a common fold. Domain boundaries in CATH are automatically assigned through the identification of recurrent domains using CATH domain family HMMs, and structure comparison using the CATHEDRAL algorithm. In addition, a consensus method (DBS) is used to predict novel domains (i.e., those that have not been previously observed) directly from structure, with manual validation being applied for particularly difficult domain assignments. Sequence comparison and the CATHEDRAL and SSAP structural comparison algorithms are then used to identify structural and functional relatives within the existing library of CATH domains. Again, manual validation is often used at this stage in order to verify fold and superfamily assignments.

In contrast the Dali Database established by Holm and co-workers uses a completely automated protocol that attempts to provide the user with lists of putative structural neighbors rather than explicitly assigning domains into a hierarchy of fold and superfamilies. The PUU algorithm *(34)* identifies domain boundaries and assigned domains are clustered using the Dali structure comparison method. The thresholds used for clustering structures are based on Dali Z-scores calculated by scanning new domains against all representative domains in the database. Domains are further grouped according to similarities in functional annotations that are identified automatically using data-mining methods that search text for matching keywords.

The Entrez database *(35)* at the NCBI also provides a resource that, like Dali, uses a "neighborhood" approach for domain classification. It uses the VAST algorithm to identify structural matches. More recently, the Protein Data Bank has also developed a comparison facility that uses the CE algorithm to detect structural relatives *(36)*.

In an analogous manner to sequence database searching, some domain structure classifications make it possible to compare a chosen structure against all structures held in the Protein Data Bank. For example, the Dali Database provides an Internet-based comparison server (www.ebi.ac.uk/dali) that enables users to scan a given structure against the Dali Database to identify putative relatives. The CATH database also provides a server that produces a list of structural neighbors identified by one of its automated comparison algorithms, CATHEDRAL, which uses E-values to quantify the degree of similarity (www.cathdb.info).

4.5. Multiple Structural Alignments of Protein Domains

Multiple structural alignments enable detection of conserved structural features across a family that can be encoded in a template. The HOMSTRAD and CAMPASS databases developed by Blundell and colleagues *(37)* use various domain databases, including SCOP and Pfam, to cluster families of evolutionary relatives that are known to have significant levels of sequence similarity. These clusters are then represented by validated multiple structural alignments for families and superfamilies that can be used to derive substitution matrices or encode conserved structural features in a template in order to identify additional relatives. The CAMPASS database groups more distant structural homologues than HOMSTRAD by using the structural comparison algorithms COMPARER and SEA to generate multiple structure alignments of SCOP superfamilies.

In CATH, multiple structure alignments have been generated for each CATH superfamily using the CORA algorithm *(38)*. These have been made available through the Dictionary of Homologous Superfamilies (DHS; www.biochem.ucl.ac.uk/bsm/dhs) in which the different levels of structural variability can be observed across each domain family.

To date, multiple structure alignments have not performed as well as those alignments derived from sequence-based families. However, with the diversity of structural data increasing, and especially with the advent of structural genomics initiatives, these approaches should become increasingly useful.

4.6. Consistency of Structure Domain Databases

A comprehensive analysis by Hadley and Jones of the CATH, SCOP, and Dali Dictionary classifications revealed that a considerable level of agreement existed between the resources, with 75% of fold and 80% of superfamily levels having comparable assignments (39). Discrepancies tended to be attributable to the different protocols applied to each classification, each of which has a different threshold for the identification of structural relatives and the assignment of fold and superfamily groupings. For example, the manual assessment of structural similarity in SCOP allows a considerable degree of flexibility in the recognition of very distant homologues—a level that is difficult to recapitulate in automatic assignments methods that must maintain a low assignment error rate.

The most significant recent development in structure comparison has been the improvement to the protocols used both to measure similarities and assess statistical significance. Although many resources have long exploited Z-scores to highlight the most significant matches to a non-redundant database of structures, recent methods have provided better statistical models by representing the results returned by database scans as extreme value distributions. Match statistics can then be given as an expectation value (E-value) that captures the probability of an unrelated structure returning a particular score within a database of a given size.

The assignment of distinct fold groups is hindered by the recurrence of structural motifs that can be found in non-homologous structures. Such motifs are often described as supersecondary motifs since they comprise two or more secondary structures, such as an alpha-beta-motif or alpha-hairpins. The occurrence of such motifs supports the proposal of a fold continuum in which some regions of fold space are densely populated, to the degree that many folds can be linked by common motifs (**Fig. 5.4**). In such cases the granularity of fold grouping is dependent on the degree of structural overlap required for a significant match and in difficult cases manual (and therefore subjective) validation must be applied. In fact, the structural similarity score between overlapping fold groups can sometimes be as high as structural similarity measures recorded within very diverse superfamilies.

These recurrent structure motifs may well give clues to the origin of protein domains. It is possible that they represent ancient

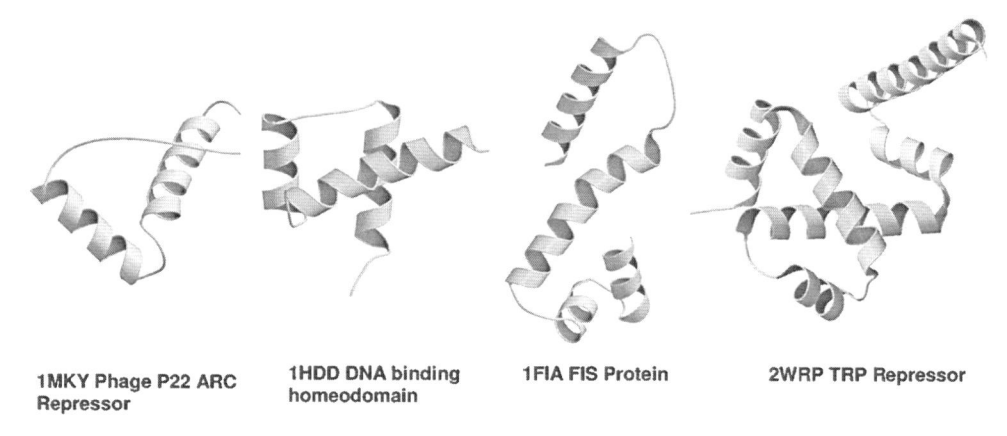

1MKY Phage P22 ARC Repressor   **1HDD DNA binding homeodomain**   **1FIA FIS Protein**   **2WRP TRP Repressor**

Fig. 5.4. Some areas of fold space are densely populated. Although the four all-alpha folds shown in this figure are different enough to be assigned to different fold groups by most classifications, other neighboring folds have intermediate degrees of similarity. This continuum in fold structure can lead to links being formed between discrete fold groups.

conserved structural cores that have been retained because they are crucial to maintaining both structure and function, whereas the surrounding structure has been modified beyond recognition. Another theory suggests that they represent small gene fragments that have been duplicated and incorporated into non-homologous contexts *(40)*. It is therefore a possibility that in the past protein domains were built up from the fusion of short polypeptides, now observed as internal repeats and supersecondary motifs.

5. Domain Family Annotation of Genomes

This section considers how the domain classifications can be used to annotate genomes and what these domain-level annotations are beginning to reveal.

The assignment of domain families to individual proteins or complete genomes is typically an automatic procedure, with little or no manual intervention involved. Providing up-to-date and automatic domain-level annotation of genomic data will be essential as genome releases continue to grow. How can we accurately assign domain families to genome sequences on such a large scale? Work pioneered by several groups, including Chothia and co-workers *(41)*, used datasets of structurally validated remote homologues to develop benchmarking protocols in order to identify reliable thresholds for accurate homologue detection. Several groups (e.g., SUPERFAMILY *(42)*, Gene3D *(43)*) have recently used automated HMM assignments to build databases of sequence and structure domain family assignments to the completed genomes.

In comparison to the growth in sequence data, relatively few protein structures have been determined. Despite this, current mappings of structural domains to completed genomes assign >50% of genes or partial genes to known structural domain families (e.g., CATH), suggesting that we currently have structural representation for many of the most highly populated domain families in nature *(2)*. An additional 20% of remaining sequences can also be assigned to Pfam domain families, revealing that a significant proportion of genome sequences can now be classified into fewer than 2,500 of the largest domain families (**Fig. 5.5**).

A significant use of domain-level genome annotations has been the identification of structurally uncharacterized super-families that are likely to possess a novel fold or function, with the aim that they might be targeted for structural determination. Structural genomics aims to characterise a large number of novel structures on a genome scale through the implementation of high throughput methods. Choosing the largest structurally uncharacterized families will help to provide structural represent-atives for the majority of genome sequences. It is interesting to note that, despite structural genomics initiatives targeting such families, only ~15% of the structures are found to contain novel folds upon structure characterization *(44)*.

Many of the sequences (20–30%) that cannot be assigned to domain families belong to very small or even single member

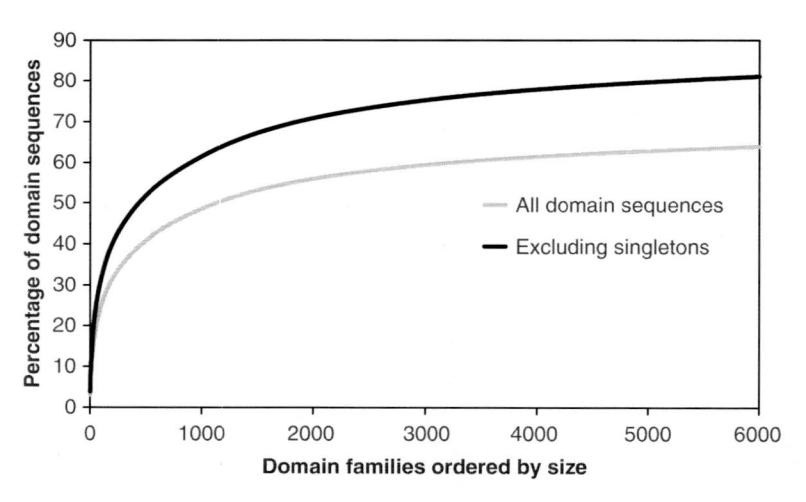

Fig. 5.5. Proportion of domain sequences in the completed genomes that can be assigned to domain families in the CATH and Pfam domain databases. The domain families (from the CATH database and Pfam) are ordered by size (large to small). Over 70% of the non-singleton (i.e., sequences that can be assigned to a family) domain sequences in the genomes can be assigned to the largest 2,500 families.

(singleton) families. Often found in a single species, they may contribute in some way to the functional repertoire of the organism. Each newly sequenced genome generated brings with it new domain families and currently it seems difficult to estimate an upper limit for the number of domain families in nature. It is quite possible that the majority will be small, organism-specific families, whereas the bulk of domain sequences within an organism will be assigned to a few thousand characterized domain families.

Another trend that has been revealed from the domain annotation of genomes is the extreme bias in the distribution of families. A very small proportion of families recur extensively in the genomes, with the remaining families occurring only once or a few times within a specific organism or sub-kingdom. Such a distribution suggests that some domains that duplicated early in evolution were retained because they performed important biochemical activities. Alternatively, some folds have particularly stable structural frameworks that support many diverse sequences, allowing paralogous relatives to tolerate changes in structure that modulate function.

Finally, the analysis of domain architectures and their distribution across genomes *(1)* has shown that some domain families are partnered with a small number of domains, limited to a kingdom or species, whereas others appear to be much more versatile, combining with a wide range of different partners.

6. Conclusions

A large number of high-quality protein domain classifications can be used to provide large-scale automated protein domain family annotations. However, a challenge remains in the transfer of useful biological knowledge to protein sequences. Domain classification resources annotate the proteome according to a parts list of component domains rather than protein or gene names. Accordingly, proteome annotation by domain content implies that gene function can be viewed as a generalized function of its component parts. Although such predictions cannot be expected to compete with experimentally derived characteristics, they provide the best approach we have for annotating the majority of functionally uncharacterized genome sequences. The comparison of entire genomes is still in its infancy and it is hoped that through these studies we can progress beyond the simple cataloguing of homologous genes and domains toward an understanding of the biology that underlies the variability within families of protein domains.

7. Notes

1. a. There are limitations in clustering domains by single or multi-linkage methods.

   b. One problem for automated domain classification methods is the choice of hierarchical clustering method used to group related domains. The aim is to compare every domain with all other domains, based on some biologically meaningful similarity score represented as a distance. Domains are then grouped into related clusters based on a distance threshold. In general, two main clustering methods tend to be used; single linkage clustering and multi-linkage clustering.

   c. Single linkage clustering defines the distance between any two clusters as the minimum distance between them. Consequently single-linkage clustering can result in the chaining problem, in which new sequences are assigned to a cluster on the basis of a single member being related (as defined by the similarity threshold), regardless of the positions of the other (unrelated) domains in that cluster.

   d. In multi-linkage clustering all members of the cluster must be linked to one another with some minimum similarity. Multi-linkage clustering, in comparison with single linkage, tends to be conservative, producing compact clusters which may have an excessive overlap between other domain clusters.

   e. Several other clustering methods are possible, including average linkage clustering, and complete linkage clustering. The difficulty in assigning meaningful domain families is that there is no unambiguous measure to evaluate the efficacy of a given method.

2. a. The consolidation of domain classifications is a logical progression toward a comprehensive classification of all protein domains.

   b. With so many domain sequence and structure domain classifications, each with unique formats and outputs, it can be difficult to choose which one to use or how to meaningfully combine the results from separate sources.

   c. One solution is the manually curated InterPro database (Integration Resource of Protein Families) at the EBI in the United Kingdom. This resource integrates all the major protein family classifications and provides regular mappings from these family resources onto primary sequences in Swiss-Prot and Trembl. Contributing databases include the Prosite, Pfam, ProDom, PRINTS & SMART, SCOP, and CATH databases. This integration of domain family resources into homologous groups not only builds on the individual strengths of the component databases, but also provides a measure of objectivity for the database curators, highlighting domain assignment errors or areas in which individual classifications lack representation.

d. Each InterPro entry includes one or more descriptions from the individual member databases that describe an overlapping domain family and an abstract providing annotation to a list of precompiled SWISS-PROT and TrEMBL sequence matches. This match list is also represented in a tabulated form, detailing protein accession numbers and the corresponding match boundaries within the amino acid sequences for each matching InterPro entry. This is complemented by a graphical representation in which each unique InterPro sequence match is split into several lines, allowing the user to view profile matches from the same and other InterPro entries. Proteins can also be viewed in the form of a consensus of domain boundaries from all matches within each entry, thus allowing proteins sharing a common architecture to be grouped.

References

1. Vogel, C., Bashton, M., Kerrison, N. D., et al. (2004) Structure, function and evolution of multidomain proteins. *Curr Opin Struct* 14, 208–216.

2. Marsden, R. L., Lee, D., Maibaum, M., et al. (2006) Comprehensive genome analysis of 203 genomes provides structural genomics with new insights into protein family space. *Nucleic Acids Res* 34, 1066–1080.

3. Needleman, S., Wunsch, C. (1970) A general method applicable to the search for similarities in the amino acid sequence of two proteins. *J Mol Biol* 48, 443–453.

4. Pearson, W. R., Lipman, D. J. (1998) Improved tools for biological sequence comparison. *Proc Natl Acad Sci U S A* 85, 2444–2448.

5. Altschul, S. F., Gish, W., Miller, W., et al. (1990) Basic local alignment search tool. *J Mol Biol* 215, 403–410.

6. Ponting, C. P. (2001) Issues in predicting protein function from sequence. *Brief Bioinformat* 2, 19–29.

7. Bru, C., Courcelle, E., Carrere, S., et al. (2005) The ProDom database of protein domain families: more emphasis on 3D. *Nucleic Acids Res* 33, D212–215.

8. Portugaly, E., Linial, N., Linial, M. (2007) EVEREST: a collection of evolutionary conserved protein domains. *Nucleic Acids Res* 35, D241–D246.

9. Heger, A., Wilton, C. A., Sivakumar, A., et al. (2005) ADDA: a domain database with global coverage of the protein universe. *Nucleic Acids Res* 33, D188–191.

10. Leinonen, R., Nardone, F., Zhu, W., et al. (2006). UniSave: the UniProtKB sequence/annotation version database. *Bioinformatics* 22, 1284–1285.

11. Altschul, S. F., Madden, T. L., Schaffer, A. A., et al. (1997) Gapped BLAST and PSI-BLAST: a new generation of protein database search programs. *Nucleic Acids Res* 25, 3389–3402.

12. Enright, A. J., Kunin, V., Ouzounis, C. A. (2003) Protein families and TRIBES in genome sequence space. *Nucleic Acids Res* 31, 4632–4638.

13. Krause, A., Stoye, J., Vingron, M. (2000) The SYSTERS protein sequence cluster set. *Nucleic Acids Res* 28, 270–272.

14. Kaplan, N., Friedlich, M., Fromer, M., et al. (2004) A functional hierarchical organization

of the protein sequence space. *BMC Bioinformatics* 5, 190–196.

15. Gattiker, A., Michoud, K., Rivoire, C., et al. (2003) Automated annotation of microbial proteomes in SWISS-PROT. *Comput Biol Chem* 27, 49–58.

16. Feng, D. F., Doolittle, R. F. (1996) Progressive alignment of amino acid sequences and construction of phylogenetic trees from them. *Methods Enzymol* 266, 368–382.

17. Eddy, S. R. (1996) Hidden Markov models. *Curr Opin Struct Biol* 6, 361–365.

18. Hulo, N., Bairoch, A., Bulliard, V., et al. (2006) The PROSITE database. *Nucleic Acids Res* 34, D227–230.

19. Finn, R. D., Mistry, J., Schuster-Bockler, B., et al. (2006) Pfam: clans, web tools and services. *Nucleic Acids Res* 34, D247–251.

20. Letunic, I., Copley, R.R., Pils, B., Pinkert, S., Schultz, J. and Bork, P. (2006) SMART 5: domains in the context of genomes and networks. *Nucleic Acids Res* 34, D257–260.

21. Bourne, P. E., Westbrook, J., Berman, H. M. (2004) The Protein Data Bank and lessons in data management. *Brief Bioinform* 5, 23–30.

22. Richardson, J. S. (1981) The anatomy and taxonomy of protein structure. *Adv Prot Chem* 34, 167–339.

23. Murzin, A. G., Brenner, S. E., Hubbard, T., et al. (2000) SCOP: a structural classification of proteins for the investigation of sequences and structures. *J Mol Biol* 247, 536–540.

24. Orengo, C. A., Mitchie, A. D., Jones, S., et al. (1997) CATH—a hierarchical classification of protein domain structures. *Structure* 5, 1093–1108.

25. Holm, L., Sander, C. (1998) Dictionary of recurrent domains in protein structures. *Proteins* 33, 88–96.

26. Sowdhamini, R., Rufino, S. D., Blundell, T. L. (1996) A database of globular protein structural domains: clustering of representative family members into similar folds. *Fold Des* 1, 209–220.

27. Gibrat, J. F., Madej, T., Bryant, S. H. (1996) Surprising similarities in structure comparison. *Curr Opin Struct Biol* 6, 377–385.

28. Pearl, F. M., Bennett, C. F., Bray, J. E., et al. (2003) The CATH database: an extended protein family resource for structural and functional genomics. *Acids Res* 31, 452–455.

29. Taylor, W. R., Flores, T. P., Orengo, C. A. (1994) Multiple protein structure alignment. *Protein Sci* 3, 1858–1870.

30. Holm, L., Sander, C. (1993) Protein structure comparison by alignment of distance matrices. *J Mol Biol* 233, 123–128.

31. Quevillon, E., Silventoinen, V., Pillai, S., et al. (2005) InterProScan: protein domains identifier. *Nucleic Acids Res* 33, W116–120.

32. Dayhoff, M. O., ed. (1965) *Atlas of Protein Sequence and Structure*. National Biomedical Research Foundation, Washington, DC.

33. Orengo, C. A., Jones, D. T., Thornton. J. M. (1994) Protein superfamilies and domain superfolds. *Nature* 372, 631–634.

34. Wernisch, L., Hunting, M., Wodak, S. J. (1999) Identification of structural domains in proteins by a graph heuristic. *Proteins* 35, 338–352.

35. Marchler-Bauer, A., Panchenko, A. R., Shoemaker, B. A., et al. (2002) SH CDD: a database of conserved domain alignments with links to domain three-dimensional structure. *Nucleic Acids Res* 30, 281–283.

36. Guda, C., Lu, S., Scheeff, E. D., et al. (2004) CE-MC: a multiple protein structure alignment server. *Nucleic Acids Res* 32, W100–103.

37. Sowdhamini, R., Burke, D. F., Deane, C., et al. (1998) Protein three-dimensional structural databases: domains, structurally aligned homologues and superfamilies. *Acta Crystallogr D Biol Crystallogr* 54, 1168–1177.

38. Orengo, C. A. (1999) CORA—topological fingerprints for protein structural families. *Protein Sci* 8, 699–715.

39. Hadley C., Jones, D. T. (1999) A systematic comparison of protein structure classifications: SCOP, CATH and FSSP. *Struct Fold Des* 7, 1099–1112.

40. Lupas, A. N., Ponting, C. P., Russell, R. B. (2001) On the evolution of protein folds. Are similar motifs in different protein folds the result of convergence, insertion or relics of an ancient peptide world? *J Struct Biol* 134, 191–203.

41. Park. J., Karplus, K., Barrett, C., et al. (1998) Sequence comparisons using multiple sequences detect three times as many remote homologues as pairwise methods. *J Mol Biol* 284, 1201–1210.

42. Gough, J., Chothia, C. (2002) SUPERFAMILY: HMMs representing all proteins of known structure. SCOP sequence searches, alignments and genome assignments. *Nucleic Acids Res* 30, 268–272.

43. Yeats, C., Maibaum, M., Marsden, R., et al. (2006) Gene3D: modeling protein structure, function and evolution. *Nucleic Acids Res* 34, D281–284.

44. Todd, A. E., Marsden, R. L., Thornton, J. M., et al. (2005) Progress of structural genomics initiatives: an analysis of solved target structures. *J Mol Biol* 348, 1235–1260.

Section II

Inferring Function

Chapter 6

Inferring Function from Homology

Richard D. Emes

Abstract

Modern molecular biology approaches often result in the accumulation of abundant biological sequence data. Ideally, the function of individual proteins predicted using such data would be determined experimentally. However, if a gene of interest has no predictable function or if the amount of data is too large to experimentally assess individual genes, bioinformatics techniques may provide additional information to allow the inference of function.

This chapter proposes a pipeline of freely available Web-based tools to analyze protein-coding DNA sequences of unknown function. Accumulated information obtained during each step of the pipeline is used to build a testable hypothesis of function.

The basis and use of sequence similarity methods of homologue detection are described, with emphasis on BLAST and PSI-BLAST. Annotation of gene function through protein domain detection using SMART and Pfam, and the potential for comparison to whole genome data are discussed.

Key words: Comparative genomics, homology, orthology, paralogy, BLAST, protein domain, Pfam, SMART, Ensembl, UCSC genome browser.

1. Introduction

Continued refinement of molecular biology methods and reduced cost of sequencing technology allows comparatively large-scale DNA cloning and sequencing within even modest research laboratories, resulting in a proliferation of sequence data. With the availability of these data an obvious question a researcher may ask is, "What does this protein do?" Experimental validation of a defined hypothesis will answer this question but what can be done if no hypothesis is known *a priori*, or if the number of candidate genes is greater than can be tackled comprehensively. In these situations bioinformatic prediction techniques are invaluable. Bioinformatics

Jonathan M. Keith (ed.), *Bioinformatics, Volume II: Structure, Function and Applications, vol. 453*
© 2008 Humana Press, a part of Springer Science + Business Media, Totowa, NJ
Book doi: 10.1007/978-1-60327-429-6 Springerprotocols.com

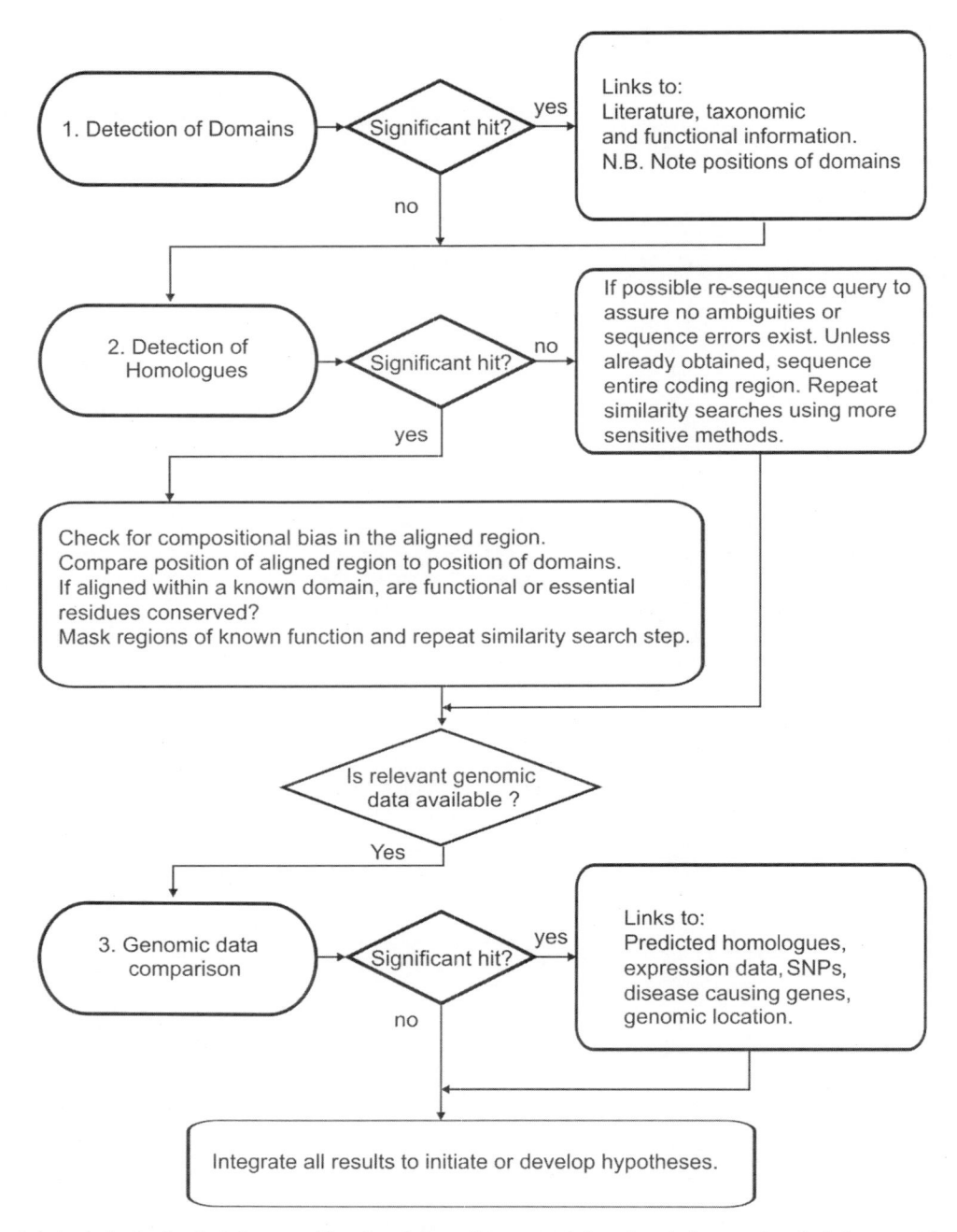

Fig. 6.1. Analysis pipeline for inference of function. Schematic representation of analysis procedure for inference of function by similarity methods.

tools may simply be used to confirm the outcome of an experimental procedure, but more powerfully these tools can be used to formulate hypotheses and influence the direction of future experiments. This chapter aims to describe an analysis pipeline of freely available sequence comparison methods that can be used to infer a potential function of protein-coding DNA sequences (**Fig. 6.1**).

The central thesis of similarity-based sequence comparison is that if two or more genes have evolved slowly enough to allow the detection of statistically significant similarity we can infer common ancestry (homology) between the genes. This follows from the assumption that the most parsimonious explanation of greater sequence similarity than we would expect to see by chance is shared rather than independent ancestry *(1, 2)*. The possession of sequence similarity is indicative of underlying structural similarity, which may impose similar function; however, it does not follow directly that all similar structures share a common function *(2)*. Often, no single method can be used to confidently infer a function. Inherent problems are associated with *in silico* predictions of protein function *(3, 4)*. Therefore, additional tools must be used in conjunction with similarity searches to assign a putative function to the protein sequence. Thus information regarding gene history, protein domain architecture and other properties such as spatial and temporal expression patterns should be collected. In this way a "guilt by multiple associations" approach can be used to infer with greater confidence a testable predicted function for an unknown protein.

1.1. What Is Homology?

It was stated that through the detection of statistically significant sequence similarity one infers sequence homology. The term *homology* means that the sequences in question share a common ancestor and diverged from this ancestral sequence. Sequence comparison methods seek to distinguish such relationships by detection of statistically significant similarity between sequences. The history of the homologues gives further information and also additional terms to explain their relationship. Homologous genes which are most recently related by a speciation event are termed orthologues, whereas genes related by gene duplication are termed paralogues (**Fig. 6.2**) *(5–8)*. Additionally, the terms *in-paralogues* (defined as duplications that occurred following a speciation event leading to lineage specific duplications) and *out-paralogues* (defined as gene duplications that occurred prior to a speciation event) have been coined to address the complications of comparing gene families with more complex life histories *(7)*. Orthologous genes are likely to share a common function; and so relation of an unknown gene to an orthologue with a well characterized role provides the best indicator of putative functionality *(9)*. The fate of paralogues following duplication is less clear. Genes may either retain different but related roles—a process known as *sub-functionalization*, or rapidly diverge and undertake new roles—a process known as *neofunctionalization (10)*. Therefore the detection of homology may suggest a potential function but additional evidence of shared domains and genealogy of the homologues will aid in refinement of such hypotheses (**Note 1**).

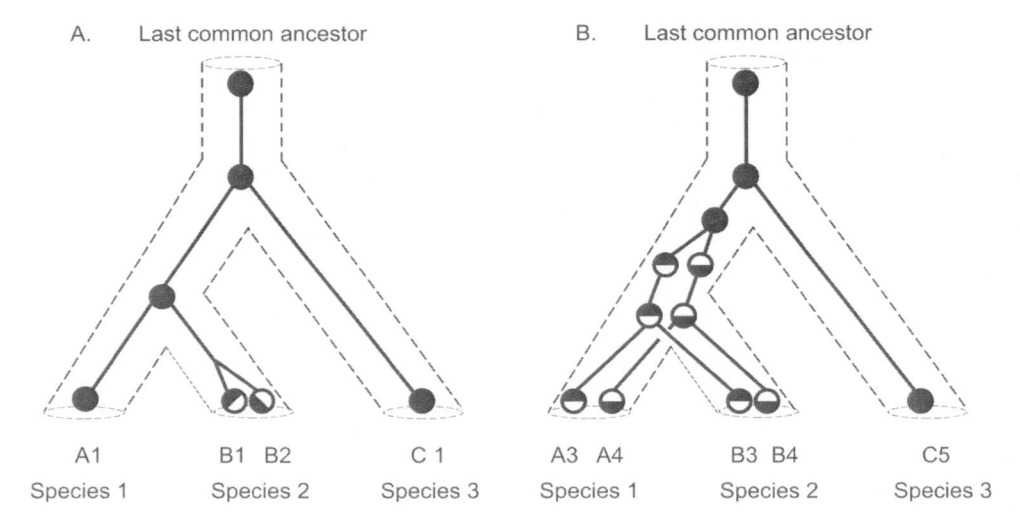

Fig. 6.2. Homology and paralogy. Dotted lines represent the relationship of species 1, species 2, and species 3, separated by two speciation events. Genes are represented by filled circles. All genes represented are homologous because they have descended from a single gene in the last common ancestor. Definition of genes as orthologues or paralogues depends on their shared ancestry. If the genes in question are most recently related by a gene duplication event they are termed paralogues, whereas if the genes are most recently related by a speciation event, they are termed orthologues. If the gene duplication is an intra-genome event; occurring following speciation, the genes are further defined as in-paralogues. If the duplication is prior to a speciation event they are termed as out-paralogues *(7, 8)*. (**A**). The intra-genome duplication within species 2 has resulted in a pair of in-paralogous sequences B1 and B2. Both B1 and B2 are orthologous to A1 and all genes are orthologous to C1. (**B**). For a different set of genes a gene duplication prior to the speciation of species 1 and 2 results in a single copy of each duplicated gene being retained in both species. As a result genes A3 and B4 are termed as out-paralogues, as are genes A4 and B3. Genes A3 and B3 share an orthologous relationship as do A4 and B4.

2. Materials

Tools used in the analysis pipeline (*see* **Section 3.2.2.2** for additional methods of interest).

General

EBI Toolbox	www.ebi.ac.uk/ Tools	Links to tools and analysis packages.
ExPASy Server	www.expasy.ch	Links to tools and analysis packages.
Genome Annotation Browsers		
COGs	www.ncbi.nlm.nih. gov/COG	Clusters of orthologous genes from multiple archaeal, bacterial, and eukaryotic genomes (*9, 11*)

Ensembl	www.ensembl.org	Whole genome annotation (12, 13).
UCSC Genome Browser	genome.cse.ucsc. edu	Whole genome annotation (14, 15).
Domain Identification Tools		
CDD (16–18)	www.ncbi.nlm.nih. gov/Structure/ cdd/	Conserved domain database. Options to search Pfam SMART and COG databases.
Interpro	www.ebi.ac.uk/ InterProScan	Multiple databases of protein families and domains (19, 20) (Includes Pfam, SMART, PRINTS, Prosite…).
Pfam	www.sanger.ac.uk/ Software/Pfam	Library of protein domain HMMs (21–23).
SMART	smart.embl-heidelberg.de	Library of protein domain HMMs (24–26).
Similarity Tools		
FASTA	www.ebi.ac.uk/ fasta33	Local alignment search tool (30).
NCBI-BLAST	www.ncbi.nlm.nih. gov/BLAST	Local alignment search tool at the NCBI (27, 28).
WU-BLAST	blast.wustl.edu	Local alignment search tool at Washington University (Gish, W. 1996–2004), also available at the EBI (29).

3. Methods

Users may assume that the obligatory first step in sequence analysis is to rush to a BLAST or FASTA search page to identify a likely homologue of a query sequence. If a single hit to a closely related species is detected with high significance and the alignment extends to the full length of both query and subject sequence then it may be safe to assume that a true orthologue has been detected and function can be assigned. However, in a less than utopian world the results are often a little messier. The query may not be full length, the subject may be one of a group of paralogues with related but distinct functions, or a number of highly significant hits to different proteins may be obtained. Therefore, although the proposed pipeline is not strictly linear it is recommended that a domain search is conducted first in the analysis, as the information gained will help in the assessment of subsequent steps.

3.1. Analysis Pipeline Step 1: Domain Identification

In comparative analysis a domain constitutes a region of conserved sequence between different proteins, which may equate to a functional unit of a protein, and often encompass a hydrophobic core *(31)*. Domains can be thought of as the building blocks of protein functionality, and hence the possession or indeed lack of a protein domain will not only provide annotation and a necessarily simplistic approach to the formulation of a testable hypothesis of function, but can later aid our assessment of homology predictions. When developing hypotheses, one should keep in mind that this simplification does not take into account biologically relevant caveats such as requisite co-occurrence of domains, domain and protein interactions and cellular location, all of which may influence protein function.

By definition, protein domains are conserved and although they can appear in a gene in different combinations, they are rarely fragmented *(31)*. This conservation allows alignment of members of a domain family and subsequent generation of a hidden Markov model (HMM), a statistical model representing the alignment *(32–34)*. HMMs incorporate information reflecting regions of high or low conservation and the probability of residue occurrence at each position in the alignment. A collection of HMMs can be queried to test for presence of domains. This approach is widely used in the annotation of whole genomes *(35–38)* and single gene families *(39–42)* and is a powerful tool to infer common function between proteins sharing domains and domain combinations.

Two widely used tools that allow comparison of single or multiple query sequences to libraries of HMMs are Pfam *(21, 22)* and SMART *(25, 26)* (*see* also Chapter 5). Pfam release 18.0 contains profile-HMMs of 7973 protein domains *(23)* and links to extensive data relating to any detected domains including taxonomic abundance, analysis of potential evolutionary origins, and relationships of domains via Pfam clans *(23)*. Version 5.0 of the SMART database contains fewer HMMs (699 domains) but offers the option to include the Pfam HMM library in a search *(24)*. Like Pfam, the SMART database gives extensive details of function, evolution, and links to structural information and relevant literature. Recent developments have included protein interaction networks for a subset of proteins, searching of a reduced dataset of proteins from 170 completely sequenced genomes (choose "SMART MODE: Genomic" at home page) and highlighting of inactive domains if key functional residues differ between the query sequence and domain HMM *(24)*. Both Pfam and SMART databases can be searched independently or via meta-search tools such as CDD *(16–18)* or Interpro *(19, 20)*, which allow searching of multiple domain and motif libraries from a single site.

A query of Pfam (global and fragment search and SEG low-complexity filter on) and SMART (Pfam domains, signal peptides, and internal repeats selected), with an ADAM 2 precursor from *Cavia porcellus* (swissprot accession Q60411) detects multiple domains with significant *E*-value (expected number of alignments that would match a given score by chance; *see* **Section 3.2.1** for description of *E*-value statistics). The output (**Fig. 6.3**) provides relative and absolute positions of domains within the query. The set of domains and their positions is often referred to as the domain architecture *(4, 26, 31)*. Annotation associated with the domains provides the user with information regarding function, co-occurrence of domains, domain evolution, and residue conservation. For example, the ACR (ADAM Cysteine-Rich Domain) SMART annotation links to an Interpro abstract that informs the user of domain architecture in ADAM proteins and potential function. The evolution section displays the abundance of ACR domains (535 domains in 533 proteins) all found within the metazoa. Similarly, the Pfam annotation of the Reprolysin domain provides information regarding function. **Figure 6.3** highlights the overall overlap using both these methods. However, slight discrepancies are evident (for example, the position of the ACR/ADAM_CR domain predictions, reflecting different length HMMs contained in the two databases). One is therefore encouraged to use multiple applications and consolidate results to achieve a consensus. Armed with this information, one can begin to build a hypothetical function of the query sequence. Additionally, both Pfam and SMART allow alignment of the query sequence to the HMM so that conserved residues in the alignment can be determined. These residues and the positions of domains will be used in the assessment of later similarity searches (**Note 2, Section 3.2**).

3.2. Analysis Pipeline Step 2: Detection of Homologues

Sequence comparison relies on tools to accurately align homologous positions in truly related sequences. When comparing sequences to identify homologues, protein, or translated nucleotide sequences should always be used. This allows the information of the class of amino acid and the relative abundance of amino acid types to be used in the scoring of alignments. Therefore, a conservative substitution; such as an isoleucine for a valine (both possess aliphatic R groups) in an alignment position or the alignment of rare amino acids such as tryptophan or cysteine

A

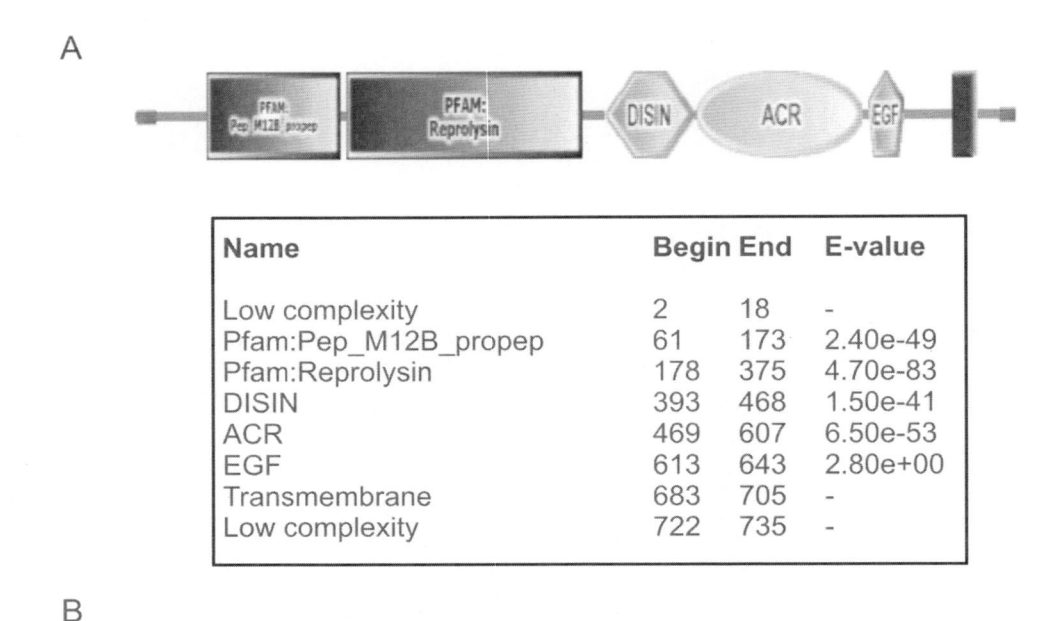

Name	Begin	End	E-value
Low complexity	2	18	–
Pfam:Pep_M12B_propep	61	173	2.40e-49
Pfam:Reprolysin	178	375	4.70e-83
DISIN	393	468	1.50e-41
ACR	469	607	6.50e-53
EGF	613	643	2.80e+00
Transmembrane	683	705	–
Low complexity	722	735	–

B

Domain	Start	End	Evalue
Pep_M12B_propep	61	173	1.8e-50
Reprolysin	178	375	3.6e-84
Disintegrin	393	468	1.8e-37
ADAM_CR	470	587	2.3e-48

Fig. 6.3. Domain detection by Pfam and SMART. Graphical and textual representations of domains detected in an ADAM 2 precursor from *Cavia porcellus*, (swissprot accession Q60411). (**A**) Domains detected by SMART version 5.0 *(24, 26)*. With additional parameters, Pfam domains, signal peptides and internal repeats selected. (**B**) Domains detected by Pfam version 19.0 *(21–23)*. A global and fragment search was conducted with SEG low complexity filter on and *E*-value cutoff = 1.0. Significant Pfam-B domains are not shown.

are weighted more heavily than a radical substitution or alignment of common amino acids. A number of schemes have been developed to weight all possible amino acid substitutions. PAM (percent accepted mutation) *(43)* and BLOSUM (Blocks Substitution matrix) *(44)* substitution matrices are the most common. The PAM series of matrices are based on an evolutionary model of point acceptable mutations per million years. The BLOSUM series of matrices are based on empirical data sets of aligned sequences, and are more common in current comparison tools. The suffix of the BLOSUM matrices denotes the maximum percentage similarity of the alignments used to generate the matrix. Thus the scores in BLOSUM45 are generated from sequences of 45% similarity or less and BLOSUM80 of 80% similar sequences or less (**Note 3**). Equipped with these substitution matrices, various algorithms are available to align sequences in such a way so as to maximise the overall alignment score. Algorithms that produce a guaranteed optimal local alignment, such as the Smith-Waterman algorithm *(45)* are often impractical for large datasets. To accelerate identification of the most likely alignments, heuristic algorithms such as BLAST *(27, 28)*, and FASTA *(30)* have been developed and are extensively used.

3.2.1. Detection of Homologues by BLAST

The most widely used sequence comparison tool is BLAST, either NCBI-BLAST *(27, 28)* or WU-BLAST (http://blast.wustl.edu/). Readers are encouraged to experiment with both tools and discover their personal preferences. However, the remainder of this section concentrates on the basic use of NCBI-BLAST for sequence comparisons. Many of the nuances of BLAST and detailed descriptions of the statistics will not be discussed here but are covered in detail elsewhere (27, 28, 46–50). A particularly thorough explanation is given in *(50)*. Also refer to BLAST tutorials, and *see* **Fig. 6.4** and **Note 4**.

The NCBI version of BLAST can be used as a Web-based or stand-alone tool, both of which are available from the NCBI. Each encompasses a variety of BLAST algorithms which have their own uses, depending on the hypothesis being tested (*see* **Fig. 6.4**).

(A) NCBI-BLAST search types

Program	Query	Database	Search type	Common Uses
blastn	DNA	DNA	DNA-DNA	Search for near identical DNA sequences. Confirmation of DNA sequencing experiment. Compare query to genomic DNA to identify splicing patterns.
blastp	Protein	Protein	Protein-Protein	Search for homologous protein sequences. Annotation of genes of unknown function.
blastx	DNA	Protein	Translated DNA-Protein	Search for homologous sequences. Particularly useful if no open reading frame can be readily predicted as compares in six reading frames.
tblastn	Protein	DNA	Protein-Translated DNA	Gene finding within DNA sequences.
tblastx	DNA	DNA	Translated DNA- Translated DNA	Identify protein coding structure in DNA sequences.

(B) NCBI-BLAST Help and Tutorials.

Blast Help: http://www.ncbi.nlm.nih.gov/BLAST/blastcgihelp.shtml

Blast Tutorials: http://www.ncbi.nlm.nih.gov/Education/BLASTinfo/information3.html

PSI-BLAST Tutorial: http://www.ncbi.nlm.nih.gov/Education/BLASTinfo/psi1.html

The NCBI handbook (Part 3, chapter 16 "The BLAST Sequence Analysis Tool"): http://www.ncbi.nlm.nih.gov/books/bv.fcgi?rid=handbook

Descriptions of Protein databases: http://www.ncbi.nlm.nih.gov/BLAST/blastcgihelp.shtml#protein_databases

Descriptions of Nucleotide databases: http://www.ncbi.nlm.nih.gov/blast/blastcgihelp.shtml#nucleotide_databases

Fig. 6.4. NCBI-BLAST. (**A**) Search parameters and common uses of NCBI-BLAST variants. (**B**) Where to get help for the NCBI-BLAST search tool.

The basic options required for a BLAST search are the query sequence, database to search, the type of search to conduct and the search parameters. The query sequence can be entered either as plain text, as a fasta formatted sequence where the first line containing identifier information is demarked by an initial greater than sign (>) followed by the sequence on subsequent lines, or as a valid NCBI sequence identifier number. It is good practice to create a fasta formatted sequence as the identifier is reported in the BLAST output and will help to keep track of multiple search results. The database searched will relate to the hypothesis of the user's experiment and may have implications for the test statistics (**Note 5**). NCBI-blast has access to 6 protein and 17 nucleotide databases to search, *see* **Fig. 6.4**. For an initial search to identify potential homologues it is best practice to search the nr database, which contains non-redundant (non-identical) entries from Gen-Bank translations, RefSeq Proteins, PDB, SwissProt, PIR, and PRF databases (*see* **Fig. 6.4**).

As discussed, we wish to search in protein space either by a blastp, if protein sequence is available, or via a translated search; blastx, tblastn, or tblastx (*see* **Fig. 6.4**). The search settings are shown in the "options for advanced blasting" section of the search page. The low complexity filter masks regions of sequences with low complexity, as these regions are likely to result in alignments of statistical but not biological significance. Unless the user is confident in the analysis of sequence alignments the low complexity filter should always remain on. The default filters are SEG masking for protein searches *(51)* and DUST (Tatusov and Lipman, unpublished) for nucleotide sequences. The expect score is the score threshold for inclusion in the BLAST results. The score known as the *E*-value is the expected number of alignments that would match a given score by chance in a search of a random database of the same size. Therefore, an *E*-value of 10 is not significant as 10 sequences of the same or more significant score would be expected by chance, whereas a score of 1×10^{-3} or less would occur infrequently by chance and therefore are good candidates as homologues of the query sequence *(27, 28, 46, 50)*. The word size is the number of matching residues required to seed an alignment extension algorithm. For our search we will keep these parameters at their default settings. Clicking the "BLAST!" button will submit your search and generate a format page. On the format page change the number of alignments to match the number of descriptions. This allows the assessment of all alignments directly. The format page also allows filtering of results by taxonomic group using a taxon id code (http://www.ncbi.nlm. nih.gov/Taxonomy/) or from predefined taxonomic groups, for example primate, eukaryote (click drop down box "or limit by"). Note that this only formats your BLAST search by taxonomy, it does not search a subset of the database corresponding to required taxonomy and so the BLAST statistics will be those calculated for a search of the complete database. Hit "Format!" and your blast search is now underway! A new status window will open and will be updated until the search is complete and a results page appears.

The blast results page consists of a header containing information of query and database searched; a graphical representation of the results; a single line summary of each significant hit, the alignments; and a footer containing details of the search statistics. The graphical alignment view shows the regions of the query sequence (which is depicted as a colored bar at the top of the graphic) with significant hits to database sequences. The colors represent the degree of similarity as defined by the color key for alignment scores. This view gives the user ready information regarding the region(s) of the query sequence that produce significant hits. This information should be related to the position of putative domains detected in Step 1. The one-line output summary ranks each hit with an E-value below the cut-off from most to least significant. Each hit has an identifier hyperlinked to the entrez database entry, with links to associated genes, structures, taxonomy, and publications. Scrolling down or clicking on an individual score will show the alignments. Each aligned region (known as high scoring segment pair or hsp) has a header with gene identifier and score summary. The bit score "Score = x bits" is determined from the raw scores of the alignment as defined in the substitution matrix. From this the E-value "Expect = x" is calculated. The alignment section shows each significant hsp between the query and subject sequences. Sandwiched between these sequences, identical matches are highlighted by the corresponding amino acid and conserved matches by a plus sign.

3.2.1.1. Assessing the Results of a BLAST Search

Confidence can be placed in an homology assignment if all relevant domains are shared and form the same architecture and key residues known to be important for function are shared in the correct spatial context by both sequences in the hsp. Thus, hsps should be assessed critically in the context of any predicted protein domains and utilising information determined from domain searches (**Section 3.1**). When viewing the alignments the user should pose the questions, "do the start and end positions of the hsp correspond to a predicted domain?" If so, "do aligned residues correspond to critical residues described in the domain annotation?" If conserved functional residues are not aligned or the hsp results from alignment of compositionally biased residues then caution should be exercised in reflection of function onto the query sequence. Alignments should also be checked for residue bias which has escaped the low-complexity filters. Certain proteins have inherent compositional biases such as myosins, which are lysine and glutamic acid rich. If using such a protein as a query, the user should assess if corresponding residues and hence the significance of the alignment is due to both protein types sharing a common bias rather than common function.

3.2.2. Detection of More Distant Homologs

The BLAST method is useful in identifying homologues of a query sequence, but relies on identification of homologues by direct sequence similarity with scoring determined by a similarity matrix. BLAST will undoubtedly miss some true ancestral relationships that are more divergent and can not be detected in this way; in this case more powerful methods are required, most often by utilizing a model of the query sequence and close homologues to search a database. When viewing an alignment of homologues it can be seen that some regions of the alignment are more conserved and that some positions are invariable, whereas others accept a greater number of substitutions. A sequence model, or *profile*, attempts to incorporate all this information in a representation of a sequence alignment. This can then be used to search for new sequences with similar characteristics, using the assumption that these are likely to represent additional homologues. Alignment to a model is therefore weighted toward high scoring of matches and high penalization of mismatches in conserved areas, coupled with greater acceptance of non-matches in non-conserved positions. This approach to database searching utilizes the increased information content of an alignment model to detect more distant homologues. A very common tool that employs a profile:sequence comparison technique is PSI-BLAST described in the following. Other methods for comparison of distant sequences are briefly introduced in **Section 3.2.2.2**.

3.2.2.1. PSI-BLAST

The PSI-BLAST tool (position-specific iterated BLAST) *(28)* utilizes a profile called a position specific score matrix (PSSM) as a model of sequence alignment *(52, 53)*. PSSMs may also be referred to as PWMs. (*See* **Chapter 11** for a discussion of the generation and visualization of PWMs.)

Briefly, PSI-BLAST first compares the query sequence to a defined database using the standard gapped BLAST algorithm *(27, 28)*. From this initial search, significantly matching sequences scoring below an E-value threshold (NCBI default 0.005) are aligned using the query sequence as a template, producing local alignments of varying depths of coverage from query alone to multiple matches. Matches identical to the query sequence or >98% identical to another match are purged to avoid redundancy in the alignment. From this multiple sequence alignment the PSSM is produced. The PSSM is an $L \times 20$ amino acid matrix of scores where L is the length of the query sequence used to initiate the PSSM and the 20 amino acid positions are weighted according to conservation within the multiple sequence alignment. High positive scores are assigned to conserved positions in the multiple sequence alignment, whereas regions of high variability are assigned scores close to zero *(47, 54)*. The PSSM is then used to search the same database. Results for sequences which match the PSSM with E-value lower than the cutoff are returned (**Note 6**). In later iterations, any significant hits are added to the multiple sequence alignment and the PSSM is regenerated. The steps of PSSM generation, searching and alignment are iterated for a specific number of runs or until convergence, that is, when no new sequences with significance below the E-value cutoff are detected. The iterative nature of the procedure further improves sensitivity by incorporation of increasingly distant homologues in the profile resulting in increasingly wider searches of the sequence space. The basic assumption of the PSI-BLAST approach is outlined in **Fig. 6.5**. If we search a comprehensive database with a protein of unknown function, "A," and BLAST identifies proteins "B" and "C" in the first search. In a second iteration protein "D"

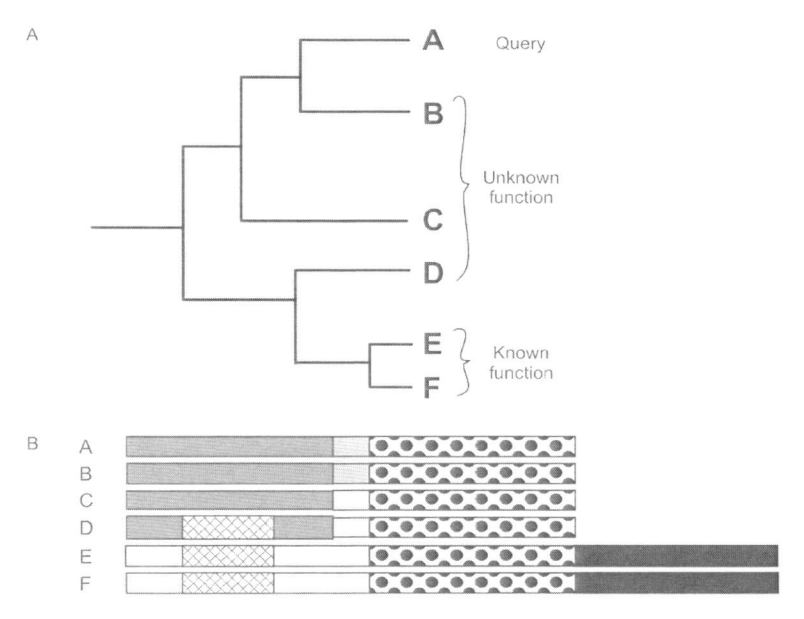

Fig. 6.5. Detection of homologues by PSI-BLAST. (**A**) Hypothetical sequences *A–F* are distantly related homologues. Their unknown relationship and similarity are represented by distance on the stylized phylogenetic tree. Initiating a standard gapped BLAST *(27, 28)* search with sequence of unknown function *A* would result in identification of similar sequences *B* and *C*. If no other sequences were identified we have no functional information with which to annotate sequence *A*. However, if PSI-BLAST is run, the additive information of sequences *A*, *B*, and *C* allows detection of sequence *D* and subsequently functionally annotated sequences *E* and *F* in later iterations of the algorithm. The BLAST methodology means that if sequences *A* and *D* are homologous as are sequences *D* and *E*, it follows that *A* and *E* must also be homologous allowing annotation of the initial query sequence. (**B**) Schematic representation of an alignment of sequences *A–F*. Aligned domains share a color, whereas non-aligned regions are represented by open boxes. To correctly annotate a gene with functional information, the alignments described in the preceding must occur in the same region of the alignment. However, if the functional annotation of sequences *E* and *F* is related to the presence of the solid black domain, reflection of function onto sequences *A–D* may not be valid as these sequences do not contain the domain.

is identified. Inclusion of protein "D" in the next iteration allows identification of proteins "E" and "F," which have a known function. If the alignments pass our check criteria we can consider protein "A," "E," and "F" to all be homologous even if not significant by pairwise comparison. This approach assumes that the hsps correspond to the same regions of each sequence. However, if the hsps shared with a known protein only occur in part of the alignment, it can only be assumed that the proteins share a common element, possibly a domain and hence, share some functionality. However, if the annotation of the known protein is largely based on a domain not shared with the query we are in danger of annotating the query sequence incorrectly. Therefore, as with standard BLAST searches, the user should exhibit caution when interpreting the results *(28, 47, 49, 50, 55)*. Briefly, the user should be aware that the primary concern for false prediction of homology is inclusion of a non-homologous sequence into the PSSM, which can be particularly problematic if the profile is compositionally biased. Incorporation of a non-homologous sequence will lead to identification and subsequent profile-inclusion of sequences with high similarity to the erroneous sequence rather than to the query sequence. As with any BLAST search the alignment should be inspected carefully. Due to the iterative nature of the PSI-BLAST search, any sequences included when they should not be, usually leads to an amplification of problems and this may go unnoticed if the user is not vigilant. The user should look for a similar conservation pattern in each of the alignments, corresponding to conserved likely functional residues. If a sequence seems to be included erroneously, it can be excluded from the PSSM and subsequent search by unchecking the relevant radio button in the BLAST output. If the sequence returns in later iterations seek corroboration of the finding by other means such as reciprocal searching (**Note 7**).

3.2.2.2. Additional Methods of Homologue Detection

As described in **Section 3.2.2**, the use of a gene model to search sequence space can result in detection of distant homologs. However, additional methods that employ profile or HMM sequence comparison techniques are available. These tools should be investigated by the interested user as they can potentially offer greater sensitivity for detection of distant homologues *(2, 56)*.

- Blocks: (http://blocks.fhcrc.org/) *(57, 58)*. Blocks are ungapped multiple alignments that correspond to the highly conserved regions of proteins. Query sequences can be compared to the block database via the block searcher tool, IMPALA (comparison of query to database of PSSMs) *(59)* and LAMA (comparison of multiple alignment to Blocks using profile: profile method) *(60)*.

- COMPASS: ftp://iole.swmed.edu/pub/compass/ *(61–63)*. Statistical comparison of multiple protein alignments via profile generation.

- HH-pred: (http://toolkit.tuebingen.mpg.de/hhpred) *(64, 65)*. Uses HHsearch algorithm to search protein and domain databases by pairwise alignment of profile-HMMs. Alignment incorporates predicted structural information and can generate an HMM from a single submitted query sequence by automated PSI-BLAST search.

- Hmmer: (http://hmmer.janelia.org/) (Eddy S.R Unpublished) *(33, 34)*. Profile-HMM *(32)* method of sequence comparison. Tools include: hmmbuild (HMM construction based on a multiple sequence alignment), hmmalign (align sequence(s) to an existing HMM) and hmmsearch (search a database of sequences for a significant match to an HMM).

3.3. Analysis Pipeline Step 3: Genomic Sequence Comparison

The recent and ongoing emphasis on whole-genome sequencing has resulted in an abundance of sequence and comparative genome data. The availability of these data and of complimentary annotation projects provides the user with vast amounts of pre-computed and catalogued information. Linking of a query sequence to a gene in these databases allows rapid access to functional annotation including predicted orthologues and paralogues, gene structure, gene expression, splice variation, association with disease, chromosomal location, and gene polymorphism data.

Thus, inferring homology (either using keyword or similarity searches as described previously) to a gene from a multi-genome resource as those described in this section should be a final step in the analysis pipeline. The annotation of a gene in this way may corroborate findings determined during previous steps and may offer additional data to reinforce a working hypothesis. These databases also have an advantage in that they are regularly updated with improved gene prediction and annotation by research groups with specialist expertise.

The resource used will depend on the organism from which the query sequence was obtained (**Note 8**). Although some data overlap is inevitable, users are encouraged to try each tool to survey the breadth of information available (**Note 9**).

For many eukaryotic organisms the UCSC genome browser *(14, 15)* and Ensembl *(12, 13)* genome servers are an ideal source of information. Both include a wealth of annotation data for the genomes of multiple organisms, and direct links between the two tools are provided. As of January 2006, UCSC allows searching of the genomic sequence of 29 species, and Ensembl of 19 species. The contents of these databases are regularly updated and reflect the current trend for whole-genome sequencing of biologically relevant model organisms and increasingly organisms of interest for comparative evolutionary analysis *(66)*. Searching of these databases can be via a gene identifier or by a similarity query (BLAST at Ensembl and BLAT *(67)* at UCSC genome browser).

The COGs database housed at the NCBI is a database of clusters of orthologous genes (COGs) which are typically associated with a specific and defined function (9, 11). Although primarily a tool for comparison of prokaryotes and unicellular eukaryotes, the COG database currently includes seven eukaryotic genomes *(11)*. Of particular use is the interlinking of the COG database with the other databases at the NCBI *(68)*, allowing direct links from a BLAST hit to a predicted COG via the Blast Link (BLink) tool.

3.4. Conclusion

The prediction of function from sequence data alone is a complex procedure. Appropriate prior information regarding data such as tissue or developmental stage collected from should be added into a working hypothesis as the analysis is conducted. It should also be remembered that predictive tools, although based on robust algorithms, can produce inconsistent or incorrect results. Therefore, the experimenter should look for the convergence of multiple analyses to allow greater confidence in any prediction, and seek experimental verification where possible.

4. Notes

1. In describing any predicted relationship it must be remembered that the terms similarity and homology are not interchangeable. Often sequences are described as containing *n* percent similarity. It is not, however, correct to use the term percent homologous; genes either are homologous (implying an ancestral relationship) or they are not *(8)*.

2. When conducting domain analysis, note positions of domains detected and conduct some background research of the essential residues and predicted function of these domains. Record the probability associated with any domain prediction and the database version searched.

3. To specifically identify recent homologues or in-paralogues, search an appropriate database with a shallow scoring matrix (BLOSUM80, PAM20) that will have a shorter look-back time, thus biasing towards more recent homologues (see Blast Protocols *(50)*).

4. When using sequence similarity to detect homology, always conduct a search with a protein sequence or a translated DNA sequence. Only use PSI-BLAST if you are attempting

to identify distant homologues; if looking for closer relationships use BLAST. If an abundant domain known to be present in many different protein types (e.g., zf-C2H2 Zinc finger domain, which has 8,133 domains in the human genome; Pfam January 2006) is present in your query sequence, consider masking this region before running a BLAST search to avoid detection of an excess of hits which provide little additional predictive information. If a representative structural sequence is available, comparison of a query sequence to the protein data bank PDB (http://www.rcsb.org/pdb/) can help in identifying structural and functional conserved residues. Increasing the gap penalty may decrease the significance of unrelated sequences, improving the signal to noise ratio for a true hit but at a cost of missing true homologues.

5. E-value scores are correlated to database size. Therefore, choosing which database to search will affect the significance or interpretation of results obtained. For example, to identify an orthologue in bacterial genomes, searching a database of only bacterial sequences will reduce the search space and improve the significance of an E-value for a given alignment. In relation to search database size, searching the large numbers of near identical sequences held in the nr database could potentially result in missing a true homologue with threshold significance. As a response to this problem databases such as nrdb95 (http://dove.embl-heidelberg.de/Blast2), which retain a single member of gene clusters with over 95% identity, can be searched (69). A corollary of this is that significant hits close to the threshold resulting from a search of a small database should be checked carefully and corroborated by other methods to avoid false-positives.

6. The relevant E-value for a hit sequence is the value when it is first identified, not at convergence or at completion of a set number of iterations. This is because inclusion of a sequence refines the PSSM for subsequent searches and will lead to greater significance of that sequence in subsequent iterations.

7. Confirm an erroneous PSI-BLAST hit by reciprocal searching. If a PSI-BLAST search using query sequence "A_{seq}" identifies a doubtful hit "B_{seq}." Initiate a PSI-BLAST search using B_{seq} as the query sequence and search the same database as before. If the PSI-BLAST identifies the original query A_{seq}, with significance and a conserved residue pattern as other homologues of B_{seq} we have greater confidence in the original PSI-BLAST identification of B_{seq} being a homologue of A_{seq}.

8. If the organism from which the query sequence was obtained is not currently available, compare to taxonomically related

organisms to build a working hypothesis of a similar function for the query gene.

9. Why is there no significant hit when I BLAST the genome of the same or closely related organism? Current methods of whole genome sequencing utilize a process of fragmentation, sequencing, and computational re-assembly of genomic DNA (Whole Genome Shotgun sequencing). Depending on the depth of coverage and the heterozygosity of the genomic DNA, this approach will result in varying degrees of incomplete non-contiguous sequences. Genes apparently missing from the genome may be located in these gaps or in repetitive hard-to-sequence regions of the genome. An alternative possibility is that the gene prediction tools used to annotate the genome and predict genes may have not predicted the query gene correctly.

Acknowledgments

Thanks to Pauline Maden and Caleb Webber for reading and commenting on the manuscript. This work was supported by a Medical Research Council UK Bioinformatics Fellowship (G90/112) to R.D.E.

References

1. Doolittle, R. F. (1981) Similar amino acid sequences: chance or common ancestry? *Science* 214, 149–159.

2. Pearson, W. R., Sierk, M. L. (2005) The limits of protein sequence comparison? *Curr Opin Struct Biol* 15, 254–260.

3. Ponting, C. P. (2001) Issues in predicting protein function from sequence. *Brief Bioinform* 2, 19–29.

4. Ponting, C. P., Dickens, N. J. (2001) Genome cartography through domain annotation. *Genome Biol* 2, Comment 2006.

5. Fitch, W. M. (2000) Homology a personal view on some of the problems. *Trends Genet* 16, 227–231.

6. Henikoff, S., Greene, E. A., Pietrokovski, S., et al. (1997) Gene families: the taxonomy of protein paralogs and chimeras. *Science* 278, 609–614.

7. Sonnhammer, E. L., Koonin, E. V. (2002) Orthology, paralogy and proposed classification for paralog subtypes. *Trends Genet* 18, 619–620.

8. Webber, C., Ponting, C. P. (2004). Genes and homology. *Curr Biol* 14, R332–333.

9. Tatusov, R. L., Galperin, M. Y., Natale, D. A., et al. (2000) The COG database: a tool for genome-scale analysis of protein functions and evolution. *Nucleic Acids Res* 28, 33–36.

10. Hurles, M. (2004) Gene duplication: the genomic trade in spare parts. *PLoS Biol* 2, E206.

11. Tatusov, R. L., Fedorova, N. D., Jackson, J. D., et al. (2003) The COG database: an updated version includes eukaryotes. *BMC Bioinformatics* 4, 41.

12. Hubbard, T., Andrews, D., Caccamo, M., et al. (2005) Ensembl 2005. *Nucleic Acids Res* 33, D447–453.

13. Hubbard, T., Barker, D., Birney, E., et al. (2002) The Ensembl genome database project. *Nucleic Acids Res* 30, 38–41.

14. Hinrichs, A. S., Karolchik, D., Baertsch, R., et al. (2006) The UCSC Genome Browser Database: update 2006. *Nucleic Acids Res* 34, D590–598.

15. Karolchik, D., Baertsch, R., Diekhans, M., et al. (2003) The UCSC genome browser database. *Nucleic Acids Res* 31, 51–54.

16. Marchler-Bauer, A., Anderson, J. B., Cherukuri, P. F., et al. (2005) CDD: a Conserved Domain Database for protein classification. *Nucleic Acids Res* 33, D192–196.

17. Marchler-Bauer, A., Anderson, J. B., DeWeese-Scott, C., et al. (2003) CDD: a curated Entrez database of conserved domain alignments. *Nucleic Acids Res* 31, 383–387.

18. Marchler-Bauer, A., Panchenko, A. R., Shoemaker, B. A., et al. (2002) CDD: a database of conserved domain alignments with links to domain three-dimensional structure. *Nucleic Acids Res* 30, 281–283.

19. Apweiler, R., Attwood, T. K., Bairoch, A., et al. (2001) The InterPro database, an integrated documentation resource for protein families, domains and functional sites. *Nucleic Acids Res* 29, 37–40.

20. Zdobnov, E. M., Apweiler, R. (2001) InterProScan—an integration platform for the signature-recognition methods in InterPro. *Bioinformatics* 17, 847–848.

21. Bateman, A., Birney, E., Durbin, R., et al. (2000) The Pfam protein families database. *Nucleic Acids Res* 28, 263–266.

22. Bateman, A., Coin, L., Durbin, R., et al. (2004) The Pfam protein families database. *Nucleic Acids Res* 32, D138–141.

23. Finn, R. D., Mistry, J., Schuster-Bockler, B., et al. (2006) Pfam: clans, web tools and services. *Nucleic Acids Res* 34, D247–251.

24. Letunic, I., Copley, R. R., Pils, B., et al. (2006) SMART 5: domains in the context of genomes and networks. *Nucleic Acids Res* 34, D257–260.

25. Letunic, I., Goodstadt, L., Dickens, N. J., et al. (2002) Recent improvements to the SMART domain-based sequence annotation resource. *Nucleic Acids Res* 30, 242–244.

26. Schultz, J., Copley, R. R., Doerks, T., et al. (2000) SMART: a web-based tool for the study of genetically mobile domains. *Nucleic Acids Res* 28, 231–234.

27. Altschul, S. F., Gish, W., Miller, W., et al. (1990) Basic local alignment search tool. *J Mol Biol* 215, 403–410.

28. Altschul, S. F., Madden, T. L., Schaffer, A. A., et al. (1997) Gapped BLAST and PSI-BLAST: a new generation of protein database search programs. *Nucleic Acids Res* 25, 3389–3402.

29. Lopez, R., Silventoinen, V., Robinson, S., et al. (2003) WU-Blast2 server at the European Bioinformatics Institute. *Nucleic Acids Res* 31, 3795–3798.

30. Pearson, W. R., Lipman, D. J. (1988) Improved tools for biological sequence comparison. *Proc Natl Acad Sci U S A* 85, 2444–2448.

31. Ponting, C. P., Russell, R. R. (2002) The natural history of protein domains. *Annu Rev Biophys Biomol Struct* 31, 45–71.

32. Durbin, R., Eddy, S. R., Krogh, A., et al. (1998) *Biological Sequence Analysis: Probabilistic Models of Proteins and Nucleic Acids.* Cambridge University Press, Cambridge, UK.

33. Eddy, S. R. (1998) Profile hidden Markov models. *Bioinformatics* 14, 755–763.

34. Eddy, S. R. (2004) What is a hidden Markov model? *Nat Biotechnol* 22, 1315–1316.

35. Gibbs, R. A., Weinstock, G. M., Metzker, M. L., et al. (2004) Genome sequence of the Brown Norway rat yields insights into mammalian evolution. *Nature* 428, 493–521.

36. Hillier, L. W., Miller, W., Birney, E., M., K., et al. (2004) Sequence and comparative analysis of the chicken genome provide unique perspectives on vertebrate evolution. *Nature* 432, 695–716.

37. Lander, E. S., Linton, L. M., Birren, B., D., E., et al. (2001) Initial sequencing and analysis of the human genome. *Nature* 409, 860–921.

38. Waterston, R. H., Lindblad-Toh, K., Birney, E., et al. (2002) Initial sequencing and comparative analysis of the mouse genome. *Nature* 420, 520–562.

39. Bateman, A. (1997) The structure of a domain common to archaebacteria and the homocystinuria disease protein. *Trends Biochem Sci* 22, 12–13.

40. Emes, R. D., Ponting, C. P. (2001) A new sequence motif linking lissencephaly, Treacher Collins and oral-facial-digital type 1 syndromes, microtubule dynamics and cell migration. *Hum Mol Genet* 10, 2813–2820.

41. Goodstadt, L., Ponting, C. P. (2004) Vitamin K epoxide reductase: homology, active site and catalytic mechanism. *Trends Biochem Sci* 29, 289–292.

42. Morett, E., Bork, P. (1999) A novel transactivation domain in parkin. *Trends Biochem Sci* 24, 229–231.

43. Dayhoff, M. O., Schwartz, R. M., Orcutt, B. C. (1978) A model for evolutionary change, in (Dayhoff, M. O., ed.), *Atlas of Protein Sequence and Structure*, vol. 5. National Biomedical Research Foundation, Washington, DC.

44. Henikoff, S., Henikoff, J. G. (1992) Amino acid substitution matrices from protein blocks. *Proc Natl Acad Sci U S A* 89, 10915–10919.

45. Smith, T. F., Waterman, M. S. (1981) Identification of common molecular subsequences. *J Mol Biol* 147, 195–197.

46. Altschul, S. F., Gish, W. (1996) Local alignment statistics. *Methods Enzymol* 266, 460–480.

47. Altschul, S. F., Koonin, E. V. (1998) Iterated profile searches with PSI-BLAST—a tool for discovery in protein databases. *Trends Biochem Sci* 23, 444–447.

48. Altschul, S. F., Wootton, J. C., Gertz, E. M., et al. (2005) Protein database searches using compositionally adjusted substitution matrices. *Febs J* 272, 5101–5109.

49. Jones, D. T., Swindells, M. B. (2002) Getting the most from PSI-BLAST. *Trends Biochem Sci* 27, 161–164.

50. Korf, I., Yandell, M., Bedell, J. (2003) *BLAST*. O'Reilly, Sebastopol CA.

51. Wootton, J. C., Federhen, S. (1996) Analysis of compositionally biased regions in sequence databases. *Methods Enzymol* 266, 554–571.

52. Gribskov, M., Luthy, R., Eisenberg, D. (1990) Profile analysis. *Methods Enzymol* 183, 146–159.

53. Gribskov, M., McLachlan, A. D., Eisenberg, D. (1987) Profile analysis: detection of distantly related proteins. *Proc Natl Acad Sci U S A* 84, 4355–4358.

54. Henikoff, S. (1996) Scores for sequence searches and alignments. *Curr Opin Struct Biol* 6, 353–360.

55. Schaffer, A. A., Aravind, L., Madden, T. L., et al. (2001) Improving the accuracy of PSI-BLAST protein database searches with composition-based statistics and other refinements. *Nucleic Acids Res* 29, 2994–3005.

56. Sierk, M. L., Pearson, W. R. (2004) Sensitivity and selectivity in protein structure comparison. *Protein Sci* 13, 773–785.

57. Henikoff, J. G., Pietrokovski, S., McCallum, C. M., et al. (2000) Blocks-based methods for detecting protein homology. *Electrophoresis* 21, 1700–1706.

58. Henikoff, S., Pietrokovski, S., Henikoff, J. G. (1998) Superior performance in protein homology detection with the Blocks Database servers. *Nucleic Acids Res* 26, 309–312.

59. Schaffer, A. A., Wolf, Y. I., Ponting, C. P., et al. (1999) IMPALA: matching a protein sequence against a collection of PSI-BLAST-constructed position-specific score matrices. *Bioinformatics* 15, 1000–1011.

60. Pietrokovski, S. (1996) Searching databases of conserved sequence regions by aligning protein multiple-alignments. *Nucleic Acids Res* 24, 3836–3845.

61. Sadreyev, R., Grishin, N. (2003) COMPASS: a tool for comparison of multiple protein alignments with assessment of statistical significance. *J Mol Biol* 326, 317–336.

62. Sadreyev, R. I., Baker, D., Grishin, N. V. (2003) Profile-profile comparisons by COMPASS predict intricate homologies between protein families. *Protein Sci* 12, 2262–2272.

63. Sadreyev, R. I., Grishin, N. V. (2004) Quality of alignment comparison by COMPASS improves with inclusion of diverse confident homologs. *Bioinformatics* 20, 818–828.

64. Soding, J. (2005) Protein homology detection by HMM-HMM comparison. *Bioinformatics* 21, 951–960.

65. Soding, J., Biegert, A., Lupas, A. N. (2005) The HHpred interactive server for protein homology detection and structure prediction. *Nucleic Acids Res* 33, W244–248.

66. Emes, R. D., Goodstadt, L., Winter, E. E., et al. (2003) Comparison of the genomes of human and mouse lays the foundation of genome zoology. *Hum Mol Genet* 12, 701–709.

67. Kent, W. J. (2002) BLAT–the BLAST-like alignment tool. *Genome Res* 12, 656–664.

68. Wheeler, D. L., Barrett, T., Benson, D. A., et al. (2006) Database resources of the National Center for Biotechnology Information. *Nucleic Acids Res* 34, D173–180.

69. Holm, L., Sander, C. (1998) Removing near-neighbour redundancy from large protein sequence collections. *Bioinformatics* 14, 423–429.

Chapter 7

The Rosetta Stone Method

Shailesh V. Date

Abstract

Analysis of amino acid sequences from different organisms often reveals cases in which two or more proteins encoded for separately in a genome also appear as fusions, either in the same genome or that of some other organism. Such fusion proteins, termed Rosetta stone sequences, help link disparate proteins together, and suggest the likelihood of functional interactions between the linked entities, describing local and global relationships within the proteome. These relationships help us understand the role of proteins within the context of their associations, and facilitate assignment of putative functions to uncharacterized proteins based on their linkages with proteins of known function.

Key words: Rosetta stone sequences, fusion proteins, functional linkages, BLAST.

1. Introduction

The processes of gene fusion and fission are well-known contributors to genome evolution, and their effects are readily visible when sequences of genes or their products are analyzed *(1, 2)*. One consequence of these processes operating on a genome is the occurrence of gene products that represent a fusion of two or more independently transcribed and translated genes, either in the same genome, or the genome of another organism. A well-known example of a fusion protein is the bifunctional dihydrofolate reductase-thymidylate synthase (DHFR-TS) enzyme found in organisms such as the malarial parasite *Plasmodium falciparum (3)*. The DHFR-TS protein shows distinct activities at each end— the N-terminal portion of the protein displays dihydrofolate reductase (DHFR) activity, whereas the C-terminal portion displays thymidylate synthase (TS) activity. In contrast, DHFR and

Jonathan M. Keith (ed.), *Bioinformatics, Volume II: Structure, Function and Applications, vol. 453*
© 2008 Humana Press, a part of Springer Science+Business Media, Totowa, NJ
Book doi: 10.1007/978-1-60327-429-6 Springerprotocols.com

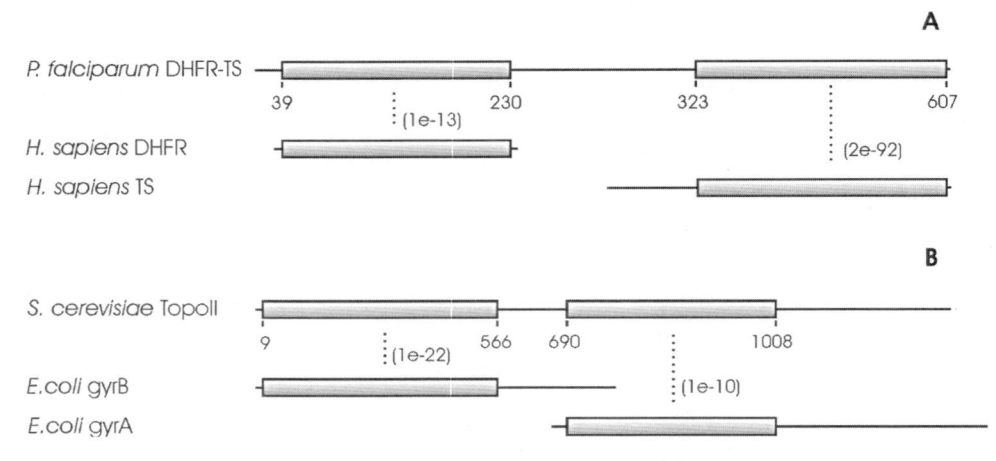

Fig. 7.1. Examples of known fusions between independently transcribed and translated genes. (**A**) Dihydrofolate reductase (DHFR) and thymidylate synthase (TS), which are products of two separate genes in the human genome, appear as a single fused protein (DHFR-TS) in the *Plasmodium falciparum* genome. (**B**) The yeast topoisomerase (TopoII) enzyme is another example of a fusion protein, representing the gyrA and gyrB proteins that are individually expressed in *Escherichia coli*.

TS are encoded as two separate proteins in the human genome (**Fig. 7.1A**). Yeast DNA topoisomerase II is yet another example, representing a fusion of the *Escherichia coli* proteins gyrase A and gyrase B (**Fig. 7.1B**).

Such fusion gene products, although yielding important evolutionary information, also suggest the existence of a strong functional connection between the linked members. Disparate proteins linked through a fusion are highly likely to functionally interact with each other, by means of a direct physical interaction or by sharing membership in the same pathway or cellular system. Therefore, fusion proteins, termed Rosetta stone sequences after Marcotte and co-workers *(4)*, can be used to establish local and genome-wide relationships between proteins in a genome *(5)*.

A computational technique that enables the prediction of functional linkages between proteins based on detection of fusions in a set of reference genomes is described in this chapter. This method, referred to as the Rosetta stone method, can be applied to investigate the existence of functional linkages between a specific group of proteins, or can be extended to the entire genome, allowing the reconstruction of a genome-wide network of functional associations. Examples of results from a genome-wide screen for functional linkages in the *Plasmodium falciparum* genome, using the Rosetta stone method, are described in **Table 7.1**.

Although many aspects of the protocol described here are simplified for ease of understanding and implementation by first time

Table 7.1
Functional linkages in the *Plasmodium falciparum* genome identified using the Rosetta stone method

Query	Linked proteins		Conf. Score	Fusions detected		
				Total	In query genome	Unique organisms involved
PF14_0401 Methionine — tRNA ligase, putative	PF10_0053	tRNA ligase, putative	0.00	47	0	47
	PF10_0340	methionine—tRNA ligase, putative	0.00	50	0	50
	PF11_0051	phenylalanine— tRNA ligase, putative	0.00	1	0	1
	PF08_0011	leucine—tRNA ligase	0.00	3	0	3
	MAL8P1.125	tyrosyl—tRNA synthetase, putative	0.00	2	0	2
	PF13_0179	isoleucine—tRNA ligase, putative	0.00	3	0	3
MAL13P1.152 hypothetical protein	PFA0155c	hypothetical protein	0.00	31	30	2
	PF11_0507	antigen 332, putative	0.00	10	9	2
	PF14_0637	rhoptry protein, putative	0.00	19	18	2

A genome-wide screen reveals high-confidence links between PF14_0401, a methionine-tRNA ligase, and five other tRNA ligases, indicating that the method captures valid biological associations. Hypothetical proteins that cannot be readily characterized by homology-based techniques alone can also be assigned putative functions using this method, based on their links with known proteins. For instance, a BLAST search for sequences matching the uncharacterized protein MAL13P1.152 against the GenBank non-redundant database (NRDB) proves inconclusive, as more than 400 of the top 500 matches involve other proteins with no known function. Matches to characterized proteins are mixed, and include kinases, carboxylases, cell-binding proteins, and malaria antigens, with E-values ranging from 4e-05 to 3e-13. Further, no conserved domains can be detected with the NCBI conserved-domain search tool (see CD-Search; http://www.ncbi.nlm.nih.gov/Structure/cdd/wrpsb.cgi) (6). However, when functional links generated by the Rosetta stone method are examined, the protein is seen to associate with antigen 332 and a rhoptry protein, suggesting a possible role for MAL13P1.152 in processes of pathogenesis. Confidence scores for the links are the highest possible (confidence in the predictions increases as the scores tend to zero, as described in **Section 3.5.2**), and fusion proteins are seen in two different species. Given the number of fusions seen in the query genome, it is likely that MAL13P1.152 is a part of a large family of proteins that share domains and domain architecture. Rosetta stone links for *Plasmodium falciparum* proteins PF14_0401 and MAL13P1.152 were downloaded from the plasmoMAP web site (http://cbil.upenn.edu/plasmoMAP) (7).

users, its practical application requires advanced computational skills and good familiarity with programming concepts. In a number of places, users are expected to create programs with variable inputs and outputs, implement algorithmic structures, and encapsulate programs within programmatic wrappers. Users lacking these skills need not be discouraged, but are rather advised to collaborate with an experienced computer programmer. Working knowledge of the BLAST package for sequence comparison is assumed.

2. Materials

2.1. Hardware Requirements

A computer with a preferably modern CPU (e.g., Pentium IV class) is required. If available, the use of computer clusters will reduce analysis time.

2.2. Software Requirements

Primary software requirements include an operating system that allows a user to write and execute custom-made computer programs or scripts. This chapter assumes that a UNIX or UNIX-like operating system (e.g., Linux) is being employed. This operating system supports a variety of programming languages (*see* **Note 1**).

1. BLAST: The Basic Local Alignment Search Tool (BLAST) software *(8)* is required, which is used for finding amino acid sequences that match the sequence of the query protein. BLAST software is available for license-free download from the National Center for Biotechnology Information (NCBI) web site (precompiled versions are available for many operating systems; http://www.ncbi.nlm.nih.gov/BLAST/download.shtml). A modified version of BLAST, which requires a license for use, is also available from the Washington University in St. Louis (WU-BLAST, http://blast.wustl.edu). Output obtained from these two programs, although similar in many cases, is not directly comparable. The protocol described in this chapter assumes the use of NCBI BLAST.

2. BLAST results parser: A computer program capable of extracting relevant information such as bit scores, expectation (E)-values, and start and stop positions from BLAST results is required. Many such parser programs are freely available over the web (*see* **Note 2**), and users are certainly free to write their own. The BLAST results parser should be modified based on the distribution of BLAST used (NCBI BLAST or WU-BLAST).

3. BLAST wrapper: A program that allows BLAST to be run sequentially for every protein in the query set is a necessity. The program should be able to read or parse the input file for query sequences, run BLAST, and store the output separately for each protein in a retrievable fashion. Using a wrapper program also facilitates application of the method to the entire genome, in case genome-wide relationships are to be investigated. Computational efficiency is increased if the parser program for BLAST results is also included in this wrapper, allowing concurrent processing of the BLAST results.

4. Fusion finder program: A computer program, capable of implementing an algorithm for identifying fusion proteins that match two or more query sequences based on certain rules, is a key requirement of this protocol (*see* **Section 3.5** for details). This program has to be created and implemented by the user. First time users/programmers are advised to seek help from an experienced programmer if this task appears daunting.

3. Methods

3.1. Creating an Input File

The input for this process consists of a text file containing sequences of the query proteins in FASTA format (*see* **Note 3** on FASTA format). For purposes of illustration, let us call this file *myInputFile* and assume that it contains sequences of proteins *A* and *B*:

```
>proteinA
MMDHIKKIHNSAIDTMKKMKEELNTSLDSDRVEFIIEEIGHMVEKFNLHLSKMRYGADYI
KNIDSQKIESYVYQVELRTLFYVAAKHYADFKFSLEHLKMFENLSKSKEKMLYSTFEKLE
GDLLNKINTLMGSEQSTSDLTSIIADSEKIIKSAESLINSSSEEIAKYALDSNEKINEIK

>proteinB
MIKNAEKNREKFNTLVQTLEAHTGEKDPNVHDSLEKFKTNLENLNLSKLETEFKSLIDSA
STTNKQIENIIKNIDTIKSLNFTKNSSDSSKLSLEKIKENKADLIKKLEQHTQEIEKYTF
IEKEETLPLLSDLREEKNRVQRDMSEELISQLNTKINAILEYYDKSKDSFNGDDETKLEQ
LDEFKKECQYVQQEIEKLTTNYKVLENKINDIINEQHEKVITLSENHITGKDKKINEKIQ
```

3.2. Creating a BLAST Searchable Database

For the sake of simplicity, this protocol restricts the search space for gene fusions to organisms with fully sequenced genomes (*see* **Note 4**). Further, instead of querying each genome separately, we will combine data from all the genomes together to create a single database (*see* **Note 5**).

1. The protein complement (all predicted and translated ORFs) of each genome is downloaded and assembled into a single file in FASTA format (*see* **Note 5**). Let us call this file, *myDatabaseFile*. Contents of a mock database file are described as follows.

```
>genome1|protein 20093441
MSLIEIDGSYGEGGGQILRTAVGMSALTGEPVRIYNIRAN-
RPRPGLSHQHLHAVKAVAEICDAE>genome1|protein
20093442
MGVIEDMMKVGMRSAKAGLEATEELIKLFREDGRLVGSILK-
EMEPEEITELLEGASSQLIRMIR>genome1|protein
20093443
MSGNPFRKMPEVPDPEELIDVAFRRAERAAEGTRKSFY-
GTRTPPEVRARSIEIARVNTACQLVQ>genome2|protein
1453778
MEYIYAALLLHAAGQEINEDNLRKVLEAAGVDVDDARLK-
ATVAALEEVDIDEAIEEAAVPAAAP>genome2|protein
1453779
MVPWVEKYRPRSLKELVNQDEAKKELAAWANEWARGSIPE-
PRAVLLHGPPGTGKTSAAYALAHD
...
>genome100|protein 6799765
MAEHELRVLEIPWVEKYRPKRLDDIVDQEHVVERLKAYVN-
RGDMPNLLFAGPPGTGKTTAALCL
```

2. This database file is formatted with *formatdb*, a utility available as a part of the BLAST software package. This results in the creation of a set of files (with extensions such as *.psq* and *.pin*) that together represent a database searchable by BLAST. A number of different options are available for use with the *formatdb* command. Since the database contains protein sequences, *formatdb* can be used in its simplest form as:

```
shell$> /path/to/BLAST/dir/formatdb -i myData-
baseFile
```

Users are free to explore other flags and options available with *formatdb*. The database is now ready for use.

3.3. Running BLAST

For this protocol, running BLAST for each protein in the query set is handled by the wrapper program. Let us refer to this program as *myBlastWrapper.pl* (assuming that the PERL programming language is used for scripting). There are a number of different ways in which the program can be configured, including making the input/options available via the command line, such as:

```
shell$> perl /path/to/wrapper/myBlastWrapper.
pl –jointoutput –i myInputFile –d myDatabase File –
b /path/to/BLAST/executable –p myBlastParser.
pl –l myLogFile
```

Here:

–jointoutput Flag indicating that the output for all proteins in the query set be stored in a combined fashion (*–jointoutput | separateoutput*)

–i Name of the input file containing query protein sequences

–d Name of the reference genome database to use

–b The directory where the BLAST programs are located

–p The BLAST results parser program to use

–l Name of the log file to use for recording progress and errors

Note the use of a log file for capturing messages and errors produced during the BLAST run. It is important to log these, especially since BLAST is run multiple times with different inputs. Calls to the BLAST executable from inside the wrapper program can also be made a variety of different ways. One easy way to start BLAST when using a PERL wrapper is to use PERL "system" calls:

```
foreach $aminoAcidSeqFile (from query set) {
system "/path/to/BLAST/dir/blastall -p blastp
-i $aminoAcidSeqFile -d myDatabaseFile >
$aminoAcidSeqFile.blast.out";
system    "/path/to/parser/myBlastParser    <
$aminoAcidSeqFile.blast.out > $aminoAcidSeq-
File.blast.out.parsed";
}
system "cat (all.blast.out.parsed.files) >>
myOutputFile if $jointOutput eq 1";
```

In this simple example, a system call invokes BLAST and captures the output of an individual run in a separate file. The raw BLAST output is parsed and stored, and finally, all files containing parsed output are merged together into a single file, based on the flag *–jointOutput*. It is certainly not necessary to use the PERL programming language; PERL commands are used here only for illustration purposes.

3.4. Parsing BLAST Results

The results of sequence comparison using BLAST include a variety of details associated with each hit to the database (or matching protein in the database). These details include statistical information such as the E-value of the high scoring pair (HSP), the percentage of positions that are either similar or identical, as well as the start and the stop positions of the matching amino acids for both the query sequence (referred to by BLAST as the "query") and the matching protein from the database (referred to by BLAST as the "subject"). Using a suitable parser program, either written by the user or obtained from another source (*see* **Note 2**), the subject protein identifier, genome name, the E-value of the HSP, and the subject start/stop positions need to be extracted from the BLAST results and stored separately. Additional information such as the percentage of identities and similarities can also be collected and used as a filtering step (discussed later). Following is an example of a mock parsed BLAST output:

```
>yeast|protein_A|123450  >celegans|protein_
P|678912  evalue|1e-180|subject_start|45|
subject_end|188
>yeast|protein_A|123450 >celegans|protein_
Q|345678 evalue|1e-120|subject_start|124|subject_
end|180
>yeast|protein_A|123450  >celegans|protein_
P|098765evalue|1e-73|subject_start|125|subject_
end|199
```

Parsed BLAST results serve at the input for the next step in the protocol.

3.5. Identifying Rosetta Stone Sequences and Functional Links with Confidence

The search for fusions involves identifying proteins that are similar to two or more proteins in diverse, non-overlapping regions. The core of the algorithm can thus be stated as follows: For any two proteins A and B in a genome, identify all proteins from a set of completely sequenced genomes (N), sharing similarities with both A and B in distinctly different regions, where:

3.5.1. Finding Rosetta Stone Fusion Proteins

1. $A_p \neq B_p \neq C_{ij}$;
2. $S(C_{ij}, A_p)^{BEGIN} > S(C_{ij}, B_p)^{END}$, or
   $S(C_{ij}, B_p)^{BEGIN} > S(C_{ij}, A_p)^{END}$; *and*
3. $p \in N$

Here, S represents the maximum *similarity span* identifiable by BLAST (*see* **Note 6**) between any protein C_i from any genome j contained in N, and proteins A and B in genome p (**Fig. 7.2**). *BEGIN* and *END* positions of the similarity span are associated with the sequence of the protein C_{ij}, and not A or B. If genome p is not a completely sequenced genome and not included in the set N, identification of fusions is restricted to all genomes other

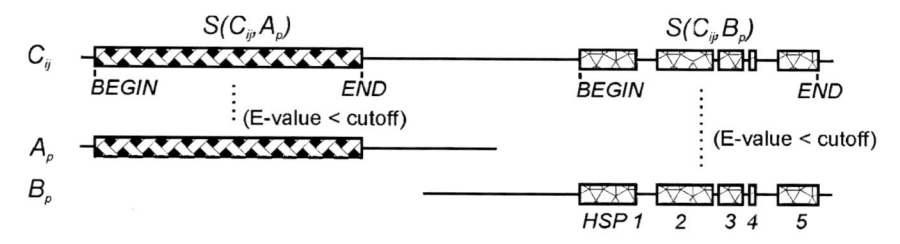

Fig. 7.2. A schematic representation of the concepts encapsulated in the fusion finder algorithm. Importantly, the protocol described in this chapter allows the existence of only one similarity span between the query protein and potential Rosetta stone sequence. If a number of distinct high scoring pairs are found when comparing the two proteins, all pairs are considered together as a single region of similarity (*see* **Section 3.5.1**).

than p, and the statistical test described in **Section 3.5.2** cannot be applied. (*See* **Note 7** for other important observations related to the identification of Rosetta stone sequences.)

This algorithm can be coded into a program, and used for identifying linkages between the query proteins. It is important to check the output of the program for accuracy using known examples (*see* **Note 8**). Although it is difficult to imagine a quantitative test for testing the reliability of the predictions, one statistical test designed to measure the accuracy of functional linkages compared with predictions made by random chance is described below.

3.5.2. Assigning Confidence Values to Predicted Linkages

One way of assessing prediction accuracy is to employ a test described by Verjovsky Marcotte and Marcotte (*9*), which attempts to calculate the probability of encountering fusions between domains based on the hypergeometric distribution, and includes an additional term to correct for within-genome ambiguity introduced by paralogs. The final probability associated with a predicted functional linkage, according to this test, is a product of the probability of finding k number of fusions by random chance, and the probability of two proteins being linked together by random chance given the existence of paralogs of each protein in the query genome. The test is simple to implement, and has been shown to be adequate against a benchmark set created from the KEGG database. Rather than include all the details of the test here, along with the logic, users are directed to the original paper for more information, and tips on its computational implementation.

3.5.3. Additional Confirmation of Putative Functions Assigned Using Rosetta Stone Linkages

It is certainly important to ascertain whether functional information obtained using this protocol is accurate and biologically valid. The gold standard for confirming predictions involves designing experiments and assays; however, corroborative data from other sources also serves to enhance the value of the functional assignments. Evidence of direct physical interaction via yeast two-hybrid assays, or indirect physical interaction via tandem affinity purification (TAP) tag experiments or other methods of protein complex purification provide strong proof of interaction, and therefore of prediction accuracy. Results of functional genomics studies, such as microarray measurements of gene expression, also help substantiate results obtained using the Rosetta stone method, for cases in which proteins found linked by the Rosetta stone sequence also happen to have strongly correlated expression profiles. If a set of curated protein–protein interactions is available, users can also ascertain accuracy and coverage of the results based on the number of true- and false-positives observed in the predicted set.

3.6. Enhancing the Scope of the Rosetta Stone Method

The described protocol predicts functional linkages between two or more gene products from an input set, based on the identification of fused proteins from a database of reference genomes. The method can be extended and applied to the entire genome, by including all coding ORFs from a particular genome in the input file.

In addition to finding functional links between the input proteins, it is often useful to apply the protocol in reverse, i.e., identify proteins in the query genome that represent a fusion of independent proteins found in other genomes. This is especially useful if the method is being applied to determine functional linkages between proteins on a genome-wide scale. Although this objective is partly achieved by including the sequence of the query genome (the genome in which the query proteins are included) in the database of reference genomes, proteins that represent a fusion of independent entities in other genomes cannot be identified. Thus, ideally, the method achieves its full potential in an *all vs. all* search scenario, where both functional linkages and fusion proteins are identified for every organism in the database.

4. Notes

1. If any other operating system is preferred, the user has to ensure that a suitable programming environment is supported, along with the ability to compile and run a distribution of the BLAST software. Commonly used platforms such as Microsoft Windows and Mackintosh OS support many popular programming languages, such as PERL, C, C++, and Java.

2. The BioPerl package provides BLAST parsing modules.

3. The FASTA format allows inclusion of a descriptor line along with the sequence lines. For more information on FASTA formats, see http://www.ebi.ac.uk/help/formats_frame.html. (*See* **Section 3.1** for an example.)

4. It is certainly possible to include all available protein sequences (e.g., from the NCBI non-redundant database) when searching for fusions between the input genes. However, if incompletely sequenced genomes are included, the search space for fusions is indeterminate, making it difficult to apply a statistical test (e.g., the one described in **Section 3.5.2**) for measuring confidence in the predicted linkages.

5. Protein complements of fully sequenced genomes are available from the respective sequencing center web sites or via FTP from NCBI (either ftp://ftp.ncbi.nih.gov/genbank/genomes/ or ftp://ftp.ncbi.nih.gov/genomes/, based on degree of sequence curation). Files containing amino acid sequences from individual genomes are combined together to create one large FASTA file. Combining data from all genomes into one file results in more accurate BLAST E-values when a BLAST searchable database is created, and greatly decreases the computational time required for sequence comparison. As always, it is better to note and follow any restrictions on use of genome sequences that are not yet published.

6. This protocol is restricted to using non-overlapping domain information when identifying fusion proteins. Identification of fusions based on the presence of fragmented, interspersed domains from different proteins is also possible, but is exceedingly complex, and likely to produce false-positives. If HSPs are separated by gaps, the gaps and HSPs are merged to produce a single, contiguous region of similarity and the ends of this span are noted.

7. The search process is greatly affected by the presence of "promiscuous" domains, such as the ATP-binding cassette (ABC) or the SH2 domains in proteins *(2)*. Every effort should be made to compile a list of proteins with promiscuous domains, which should then be excluded from the process. To enhance chances of finding true fusions, the algorithm can also be modified to take into account identities and similarities associated with the HSPs, along with the E-values. If an E-value cutoff was not applied during the BLAST search, the fusion program should be configured to ignore $C\sim A$ and $C\sim B$ matches above a certain cutoff. Empirically, E-values $>10^{-5}$ cannot be trusted, and should not be used in the analyses. This 10^{-5} cutoff for using E-values in sequence analysis is generally accepted within the scientific community and its use can be frequently observed in published studies. It is always advisable to use a low E-value cutoff to avoid including possible false-positive sequence matches. Users are free to explore and use other filtering rules as well. For instance, in some cases, the presence of fusions in more than

one genome might be desirable, and can be used to filter out possible low-confidence matches.

8. As the process for identifying fusion proteins and the corresponding functional linkages involves a number of computational steps, the chances of generating erroneous output are high. Any small glitch, in the FASTA input file format, parsing of the BLAST output, or an error in programming logic, can result in output that often proves practically useless. In this author's experience, incorrect parsing of gene/protein identifiers is common, and often leads to misidentification of fusion products. Therefore, it is important to check the input data and program code, as well as the results obtained. One easy test involves checking the results for identification of known fusions, such as the DHFR protein, and for known functional links between products, such as the *Escherichia coli* gyrA-gryB or parC-parE proteins.

References

1. Snel, B., Bork, P., Huynen, M. (2000) Genome evolution. Gene fusion versus gene fission. *Trends Genet* 1, 9–11.

2. Kummerfeld, S. K., Teichmann, S. A. (2005) Relative rates of gene fusion and fission in multidomain proteins. *Trends Genet* 1, 25–30.

3. Bzik, D. J., Li, W. B., Horii, T., et al. (1987) Molecular cloning and sequence analysis of the *Plasmodium falciparum* dihydrofolate reductase-thymidylate synthase gene. *Proc Natl Acad Sci USA* 84, 8360–8364.

4. Marcotte, E. M., Pellegrini, M., Ng, H-L., et al. (1999) Detecting protein function & protein-protein interactions from genome sequences. *Science* 285, 751–753.

5. Enright, A. J., Iliopoulos, I., Kyrpides, N. C., et al. (1999) Protein interaction maps for complete genomes based on gene fusion events. *Nature* 402, 86–90.

6. Marchler-Bauer, A., Bryant, S. H. (2004) CD-Search: protein domain annotations on the fly. *Nucleic Acids Res* 32, W327–331.

7. Date, S. V., Stoeckert, C. J. (2006) Computational modeling of the *Plasmodium falciparum* interactome reveals protein function on a genome-wide scale. *Genome Res* 4, 542–549.

8. Altschul, S. F., Gish, W., Miller, W., et al. (1990). Basic local alignment search tool. *J Mol Biol* 215, 403–410.

9. Verjovsky Marcotte, C. J., Marcotte, E. M. (2002) Predicting functional linkages from gene fusions with confidence. *App Bioinform* 1, 1–8.

Chapter 8

Inferring Functional Relationships from Conservation of Gene Order

Gabriel Moreno-Hagelsieb

Abstract

The idea behind the gene neighbor method is that conservation of gene order in evolutionarily distant prokaryotes indicates functional association. The procedure presented here starts with the organization of all the genomes into pairs of adjacent genes. Then, pairs of genes in a genome of interest are mapped to their corresponding orthologs in other, informative, genomes. The final step is to determine whether the orthologs of each original pair of genes are also adjacent in the informative genome.

Key words: Conservation of gene order, operon, genomic context, functional inference, gene neighbor method.

1. Introduction

Although two independent works first presented data supporting the idea *(1, 2)*, probably the first thoroughly described method to infer functional relationships using conservation of gene order in evolutionarily distant genomes was published by Overbeek et al. *(3)*. The method was inspired by the knowledge that genes in operons, stretches of adjacent genes in the same DNA strand transcribed into a single mRNA *(4, 5)* (**Fig. 8.1**), are functionally related, by the expectation that operons should be conserved throughout evolution, and by the finding that gene order in general is not conserved *(6)* and is lost much faster than protein sequence identity *(7)*. Thus, conservation of gene order at long evolutionary distances should indicate a functional relationship. Some divergently transcribed genes (**Fig. 8.1**) might also be

Jonathan M. Keith (ed.), *Bioinformatics, Volume II: Structure, Function and Applications, vol. 453*
© 2008 Humana Press, a part of Springer Science + Business Media, Totowa, NJ
Book doi: 10.1007/978-1-60327-429-6 Springerprotocols.com

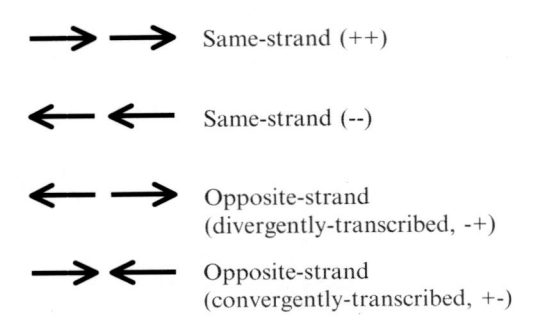

Fig. 8.1. Pairs of genes used in the study of conservation of gene order. Genes are represented by arrows indicating the direction of transcription. Same-strand genes would be the ones that might be in operons, and thus functionally related. Genes in opposite strands can be either divergently or convergently transcribed. Comparing the conservation of gene order of genes in the same strand against that of genes in opposite strands helps calculate a confidence value for predictions of functional interactions.

functionally related *(8)*. However, Overbeek et al. *(3)* found that the conservation of divergently transcribed genes in evolutionarily distant genomes was minimal when compared with the conservation of genes in the same strand. With this finding, they limited their analyses to the detection of conservation of adjacency of genes in the same DNA strand. Since the significance of conservation of adjacency increases with the phylogenetic distance of the genomes compared, Overbeek et al. *(3)* directly used phylogenetic distances as a score. However, selecting an appropriate threshold was a problem. Later, Ermolaeva et al. *(9)* proposed the excellent idea of using the conservation of adjacency of genes in opposite strands to calculate a confidence value. More recently, an approach using a simplified method similar to that presented by Ermolaeva et al. *(9)* was used to show that conservation of adjacency of paralogous genes is also useful for predicting operons and functional relationships of gene products *(10)*.

Another approach to conservation of gene order is to count the number of genomes in which genes are conserved next to each other *(11, 12)*. The main problem of such an approach is that conservation of adjacency in very closely related genomes counts as much as that among evolutionarily distant genomes. A more recent approach uses phylogenetic relationships to correct for this problem *(13)*. This chapter presents the simplified method mentioned in the preceding *(10)*, because the confidence values obtained in this method provide a direct measure of significance that is very easy to understand. Moreover, there is no proof yet that accounting for the number of genomes in which the genes are conserved produces any better results than conservation in evolutionarily distant genomes.

2. Systems, Software, and Databases

1. A UNIX-based operative system. This assumes that the user is working with a UNIX-based operating system. However, this method should be easily adapted to run under any other system.

2. The RefSeq bacterial genome database. The RefSeq database *(14, 15)* contains files with diverse information about each genome. The database can be downloaded using programs such as "wget" or "rsync." For instance, periodically running the command:

```
rsync -av –delete rsync://rsync.ncbi.nih.gov/
genomes/Bacteria/ \
LOCAL_GENOMES
```

would keep an updated directory "LOCAL_GENOMES" with all the information in the directory "genomes/Bacteria" of the NCBI rsync server. Here we use three files under each genome directory, those ending with ".gbk," with ".ptt" and ".rnt" (GBK, PTT, and RNT files). Although the GBK file generally contains all of the necessary information, the PTT and RNT files facilitate extracting gene neighbors.

3. BLASTPGP. To map the corresponding orthologous (homologous) genes it is necessary to compare proteins coded by the genes within all genomes. Identifying orthologs is important for any genome context method for inferring functional associations. Appropriate binaries of the BLASTPGP *(16)* program can be downloaded from the NCBI server at rsync://rsync.ncbi.nih.gov/blast/executables/LATEST/.

3. Methods

The method starts with the construction of files or databases of gene neighbors. For each gene in a given pair of neighbors in the genome of interest, the method verifies the existence of orthologs in an informative genome. If the orthologs exist, their adjacency is investigated. The following pseudocode summarizes this method:

```
GENE_NEIGHBOR_METHOD
1. for each informative_genome
2. count_conserved <- 0
3. conserved_list <- ""
4. for each NEIGHBORS(a, b) in genome_of_interest
```

```
5. if (ORTH(a) AND ORTH(b)) in informative_genome
6. if (NEIGHBORS(ORTH(a),ORTH(b))) in inform-
   ative_genome
7. ADD( a,b ) to conserved_list
8. count_conserved <- count_conserved + 1
9. return (informative_genome, count_conserved,
   conserved_list)
```

Notice that the results are returned for each informative genome. This is important in order to score the conservation. The detailed method that follows uses PERL code to exemplify each step. The programs and example files can also be downloaded from http://popolvuh.wlu.ca/gmh/Neighbor_Chapter/.

3.1. Learn Orthologs

Orthologs are defined as genes that diverge after a speciation event *(17)*. Such genes are sometimes colloquially referred to as the "same genes" in different species. Accordingly, orthologs are the appropriate genes to compare in the Neighbor method. The most common working definition of orthology is the bi-directional best hits, in which a given gene in the genome of interest finds a gene in an informative genome with the highest BLASTP scores, and, when used as a query, the gene in the informative genome also finds the gene in the genome of interest with the highest scores. However, the possibility of adjacently conserved paralogs (paralogs are genes that diverge after a duplication event) *(17)* was also discussed by Overbeek et al. *(3)*. Moreover, recent work has shown that operons do have a tendency toward producing paralog operons *(10, 18)*, and that strict detection of orthologs is not necessary for prediction of functional association *(10)*. Thus, this chapter uses conservation of uni-directional best hits; in other words, the best hits in the genome of interest as informative genes with no requirement for the informative genes to find the genes of interest as their best hits in the genome of interest. Detecting these top best hits is straightforward. A run of BLAST-PGP of the form:

```
blastpgp -i genome_of_interest -d informa-
tive_genome -e 1e-4 -F "m S" \
-I T -s T -z 5e8 -m 8 -o genome_of_interest.
informative_genome.blastpgp
```

produces a table of BLAST hits between the genome of interest and a given informative genome in the file "genome_of_interest.informative_genome.blastpgp." The -m 8 option instructs BLASTPGP to format the results into a table. The other options are: -e 1e-6, which sets the maximum E-value to 1e-6; -F "m S", which sets filtering of low information sequences during the blast search, but not during the alignment; -s T, which indicates a Smith-Waterman alignment to calculate the scores *(19)*; and -z

5e8, which fixes the search database to a size of 5e8 letters for the purpose of calculating E-values (**Note 1**).

BLAST presents results sorted from best to worst match. Thus, a subroutine in PERL that can get the best hits looks like this:

```
sub get_best_hits {
my ($genome_of_interest,$informative_
genome) = @_;
my %best_hits = ();
my %E_value = ();
my %bit_score = ();
my $blast_file
= "BLAST_RUNS/$genome_of_
interest.$informative_genome.blastpgp";
open( BLTBL,$blast_file );
while(<BLTBL>) {
  my ($query_id,$target_id,@stats) = split;

  # both the query and target have a com-
  plex name,
  # we only need the gi number to match
  the neighbor
  # table identifiers
  my ($query) = $query_id =~/gi\|(\d+)/;
  my ($target) = $target_id =~/gi\|(\d+)/;
  # the penultimate value is the E value
  my $E_value = $stats[$#stats - 1];

  # the last value is the bit score
  my $bit_score = $stats[$#stats];

  # now we actually learn the best hits
  if( $bit_score{$query} > 0 ) {
    if(
      ( $E_value{$query} == $E_value )
      && ( $bit_score{$query} == $bit_score )
    ) {
      $best_hits{$query} .= ",".$target;
    }
  }
  else {
    $E_value{$query} = $E_value;
    $bit_score{$query} = $bit_score;
    $best_hits{$query} = $target;
  }
}
close(BLTBL);
return(%best_hits);
}
```

3.2. Neighbors Database

The natural next step is to build a database of gene neighbors. The minimum information that this database should contain is a list of adjacent gene pairs and information on the strand on which each gene is found. To build this database, a convenient starting point is the RefSeq genomes database, available from the NCBI ftp server.

Several Refseq files could be used to obtain coordinates for each gene within the genome. Here we exemplify with the PTT (protein table) and RNT (robonucleotide table) files. The PTT file contains a table of protein-coding genes, whereas the RNT file contains a table of rRNA and tRNA genes (**Note 2**). The first column within these tables consists of the gene coordinates. As an example, a few lines of the PTT file are shown for the genome of *Escherichia coli* K12 *(20)*, accession "NC_000913", version "NC_000913.2 GI:49175990" (**Fig. 8.2**).

The first column in these tables corresponds to gene coordinates. Thus, the problem of forming pairs of adjacent genes becomes trivial. All that is needed is to sort the genes and associate each of them with the next gene in the list, formatting them into a table, or a database, of Gene Neighbors. The header of the resulting table might look like this:

Gene_a   Gene_b   Strands

Genes in the same strand have either "++" or "−−" in the "Strand" column, whereas genes in different strands have either "+−" (convergently transcribed), or "−+" (divergently transcribed) in this field (*see* **Note 3**).

If the genome is circular, a final pair should be formed with the last and the first genes in the table. The first line in the GBK file indicates whether the replicons are circular or linear (**Note 4**).

An example program in PERL that will output this table is:

```perl
#!/usr/bin/perl

$die_msg = "\tI need a genome to work
with\n\n";
$genome_of_interest = $ARGV[0] or die
$die_msg;

$die_msg = "\tNo $genome_of_interest
directory\n\n";
$genome_dir = "LOCAL_GENOMES/$genome_of_
interest";
opendir(GNMDIR, "$genome_dir") or die
$die_msg;

@ptt_files = grep { /\.ptt/ }
readdir(GNMDIR);
$results_dir = "NEIGHBORS";
mkdir($results_dir) unless ( -d $results_dir );

open( NGHTBL, ">$results_dir/$genome_of_
interest.nghtbl");
```

Escherichia coli K12, complete genome – 0..4639675
4237 proteins

Location	Strand	Length	PID	Gene	Synonym	Code	COG	Product
190..255	+	21	16127995	thrL	b0001	–	–	thr operon leader peptide
337..2799	+	820	16127996	thrA	b0002	E	C0G0460	bifunctional aspartokinase I/homoserine dehydrogenase I
2801..3733	+	310	16127997	thrB	b0003	E	C0G0083	homoserine kinase
3734..5020	+	428	16127998	thrC	b0004	E	C0G0498	threonine synthase
5234..5530	+	98	16127999	yaaX	b0005	–	–	hypothetical protein
5683..6459	–	258	16128000	yaaA	b0006	S	C0G3022	hypothetical protein
6529..7959	–	476	16128001	yaaJ	b0007	E	C0G1115	inner membrane transport protein
8238..9191	+	317	16128002	talB	b0008	G	C0G0176	transaldolase
9306..9893	+	195	16128003	mogA	b0009	H	C0G0521	molybdenum cofactor biosynthesis protein
9928..10494	–	188	16128004	yaaH	b0010	S	C0G1584	putative regulator, integral membrane protein
10643..11356	–	237	16128005	yaaW	b0011	S	C0G4735	hypothetical protein
10725..11315	+	196	16128006	htgA	b0012	–	–	positive regulator for sigma 32 heat shock promoters
11382..11786	–	134	16128007	yaaI	b0013	–	–	hypothetical protein
12163..14079	+	638	16128008	dnaK	b0014	0	C0G0443	molecular chaperone DnaK
14168..15298	+	376	16128009	dnaJ	b0015	0	C0G0484	chaperone with DnaK; heat shock protein
15445..16557	+	370	16128010	yi81_1	b0016	L	C0G3385	IS186 hypothetical protein
15869..16177	–	102	16128011	yi82_1	b0017	–	–	IS186 and IS421 hypothetical protein
16751..16960	–	69	16128012	mokC	b0018	–	–	regulatory peptide whose translation enables hokC (gef) expression
16751..16903	–	50	49175991	hokC	b4412	–	–	small toxic membrane polypeptide
17489..18655	+	388	16128013	nhaA	b0019	P	C0G3004	Na+/H antiporter, pH dependent
18715..19620	+	301	16128014	nhaR	b0020	K	C0G0583	transcriptional activator of cation transport (LysR family)
19811..20314	–	167	16128015	insB_1	b0021	L	C0G1662	IS1 protein InsB
20233..20508	–	91	16128016	insA_1	b0022	L	C0G3677	IS1 protein InsA
20815..21078	–	87	16128017	rpsT	b0023	J	C0G0268	30S ribosomal protein S20
21181..21399	+	72	16128018	yaaY	b0024	–	–	unknown CDS
21407..22348	+	313	16128019	ribF	b0025	H	C0G0196	hypothetical protein
22391..25207	+	938	16128020	ileS	b0026	J	C0G0060	isoleucyl-tRNA synthetase
25207..25701	+	164	16128021	lspA	b0027	M	C0G0597	signal peptidase II
25826..26275	+	149	16128022	fkpB	b0028	O	C0G1047	FKBP-type peptidyl-prolylcis-trans isomerase (rotamase)
26277..27227	+	316	16128023	ispH	b0029	I	C0G0761	4-hydroxy-3-methylbut-2-enyl diphosphate reductase
27293..28207	+	304	16128024	rihC	b0030	F	C0G1957	nucleoside hydrolase
28374..29195	+	273	16128025	dapB	b0031	E	C0G0289	dihydrodipicolinate reductase
29651..30799	+	382	16128026	carA	b0032	E	C0G0505	carbamoyl-phosphate synthase small subunit
30817..34038	+	1073	16128027	carB	b0033	E	C0G0458	carbamoyl-phosphate synthase large subunit
34195..34695	+	166	49175992	caiF	b0034	–	–	transcriptional regulator of cai operon
34781..35392	–	203	16128029	caiE	b0035	R	C0G0663	possible synthesis of cofactor for carnitine racemase and dehydratase
35377..36270	–	297	16128030	caiD	b0036	I	C0G1024	carnitinyl-CoA dehydratase
36271..37839	–	522	49175993	caiC	b0037	–	–	crotonobetaine/carnitine-CoA ligase

Fig. 8.2. A few lines of the PTT table of the genome of *Escherichia coli* K12. The first column of the PTT (protein-coding genes) and of the RNT (non-coding genes, those producing rRNAs and tRNAs) tables contains the gene coordinates. The second column contains the strand where the gene is found, which is useful for organizing the genes into stretches of adjacent genes in the same strand, called *directons*. The fourth column is the GI number (labeled here as a PID or protein identifier). This number is the best identifier for the protein-coding genes in a genome because it is unique. However, in the RNT tables this column is not the best identifier; the best identifiers seem to be the gene name (fifth column), and the synonym (sixth column). The table in the figure is formatted for display purposes, but the original PTT and RNT tables contain tab-separated plain text.

```perl
for my $ptt_file ( @ptt_files ) {

  # get proper name of the RNT and GBK files
  my $rnt_file = $ptt_file;
  my $gbk_file = $ptt_file;
  $rnt_file =~s/\.ptt/\.rnt/;
  $gbk_file =~s/\.ptt/\.gbk/;

  # Is the genome circular?
  # The information is in the first line
  of the GBK
  # file, which starts with the word "LOCUS"
  my $circular = "yes"; # make circular
  the default
  open (GBK, "$genome_dir/$gbk_file");
  while(<GBK>) {
    if(/^LOCUS/) {
      $circular = "no" if(/linear/i);
      last; # we do not need to read any
      further
    }
  }
  # now we read the table of protein coding
  genes
  # and their "leftmost" coordinate so we can
  # order them and find the neighbors
  my %strand = ();
  my %coords = ();
  my @ids = ();
  open(PTT, "$genome_dir/$ptt_file");
  while(<PTT>) {

    my @data = split;
    next unless($data[1] =~/^\+|\-$/);
    $gi = $data[3];
    $strand{$gi} = $data[1];
    my ($coord) = $data[0] =~/^(\d+)/;
    $coord{$gi} = $coord;
  }
close(PTT);

  # we verify that there is a table of rRNA
  and tRNA genes
  # if so, we get the genes
  if(-f "$genome_dir/$rnt_file") {
    open(RNT,"$genome_dir/$rnt_file");
    while (<RNT>) {
      my @data = split;
      next unless($data[1] =~/^\+|\-$/);
      # The identifier is not a GI
```

```
            # but I rather keep the variable
            names consistent
            # the best identifier for an 'RNA' gene is
            # the gene name (5th column)
            my $gi = $data[4];
            $strand{$gi} = $data[1];
            my ($coord) = $data[0] =~/^(\d+)/;
            $coord{$gi} = $coord;
        }
    }

    # now we build the table of direct neighbors
    my @ids = sort { $coord{$a} <=> $coord{$b}
    } keys %coord;
    for my $i( 0 .. $#ids ) {
        if (exists($strand{$ids[$i+1]})) {
            my $str = $strand{$ids[$i]}.$strand{$
            ids[$i+1]};
            print NGHTBL $ids[$i], "\
            t",$ids[$i+1], "\t",$str, "\n";
        }
    else {
        if ($circular eq "yes") {
            my $str = $strand{$ids[$i]}.$strand{$
            ids[0]};
    print NGHTBL $ids[$i],"\t",$ids[0],"\
    t",$str,"\n";
        }
      }
    }
}
close(NGHTBL);
```

and a subroutine that will read this table, learn the neighbors, and classify them as same-strand and opposite-strand neighbors is:

```
sub get_strands_of_neighbors {
    my $genome = $_[0];
    # we will learn the neighbors as hashes
    where the keys
    # are the neighbor pairs of genes and
    the values are
    # the strand situations (same strand or
    opposite strand
    my %strands_of = ();
    open( NGH,"NEIGHBORS/$genome.nghtbl");
    while(<NGH>) {
      my($gi,$gj,$strand) = split;
      my $neighbors =
      join(",",sort($gi,$gj));
```

```
              if (($strand eq "--") || ($strand eq
              "++")) {
                $strands_of{"$neighbors"} = "same";
              }
              elsif (($strand eq "-+") || ($strand
              eq "+-")) {
                $strands_of{"$neighbors"} = "opp";
              }
          }
          return(%strands_of);
      }
```

3.3. Putting Everything Together

As originally defined, the Gene Neighbor Method aims to find genes with conserved adjacency in evolutionarily distant genomes. However, Ermolaeva et al. *(9)* have obviated the need for a phylogenetic distance by using the genomes themselves to determine the significance of the conservation in the form of a confidence value. The idea behind the confidence value is that the proportion of conserved adjacencies in opposite strands represents conservation due to chance alone, or more properly, conservation due to short evolutionary distance and chance re-arrangement (**Note 5**). A simplified version of the confidence value calculated under the same assumption is:

$$C = 1 - 0.5 * \frac{P_{Opp}}{P_{Same}}$$

The confidence value (C) can be thought of as a positive predictive value (true-positives divided by the total number of predictions) for two genes to be conserved due to a functional interaction (they would be in the same operon) (**Note 6**). The value 0.5 in this expression is a prior probability for the genes to be in different transcription units. P_{Opp} is the count of pairs of orthologs conserved next to each other in opposite strands ("+−" and "−+" pairs of neighbor genes) divided by the total number of neighbors in opposite strands in the informative genome. P_{Same} is the count of orthologs conserved next to each other in the same strand ("++" and "−−") divided by the total number of neighbors in the same strand in the informative genome. Now, with all the necessary data, neighbors and best hits, and with a way of calculating a confidence value, the pseudocode above is modified as:

```
GENE_NEIGHBOR_METHOD
1. for each informative_genome
2. count_conserved <- 0
3. conserved_list <- ""
4. for each NEIGHBORS(a,b) in genome_of_
   interest
```

5. if (ORTH(*a*) AND ORTH(*b*)) in *informative_genome*
6. if (same-strand(ORTH(*a*),ORTH(*b*))) in *informative_genome*
7. ADD(*a,b*) to *conserved_same-strand*
8. count_same <- count_same + 1
9. else if (opposite-strand(ORTH(a), ORTH(b))
10. ADD(*a,b*) to *conserved_opposite-strand*
11. count_opposite <- count_opposite + 1
12. confidence <- 1-0.5*proportion(same)/proportion(opposite)
13. return (informative_genome, confidence, conserved_same-strand)

And a particular example program in PERL would be:

```perl
#!/usr/bin/perl

$genome_of_interest = "Escherichia_coli_K12";
@genomes = qw(
      Salmonella_typhi_Ty2
      Yersinia_pestis_KIM
      Rhizobium_etli_CFN_42
      Bacillus_subtilis
   );
$results_dir = "Confidence";
mkdir($results_dir) unless(-d $results_dir);

my %strands_of = get_strands_of_neighbors
($genome_of_interest);

open(CONF,">$results_dir/$genome_of_inter-
est.confidence");

for my $informative_genome (@genomes) {

  print $informative_genome,"\n";
  my %best_hits
      = get_best_hits( $genome_of_
      interest,$informative_genome);
  my %inf_strands_of
      = get_strands_of_neighbors ($informa-
      tive_genome);

  my $count_same = 0;
  my $count_opp = 0;
  my @predictions;

  for my $neighbors (keys %strands_of) {
      my ($gi,$gj) = split(/,/
      ,$neighbors);
```

```perl
# first see if there are any
orthologs
if (exists($best_hits{$gi})
  && exists($best_hits{$gj})) {

  # since there might be more than
  one ortholog, and
  # there might be more than one con-
  served pair,
  # we use a "flag" (count_conserv =
  "none") to
  # avoid "overcounting"
  my $count_conserv = "none";

  # now the actual verification of
  conservation
  for my $orth_i (split(/,/,$best_
  hits{$gi})) {
    for my $orth_j (split(/,/,$best_
    hits{$gj}) ) {
      my $test_neigh = join ",",
      sort($orth_i,$orth_j);
      if ($inf_strands_of{$test_
      neigh}
        eq $strands_of{$neighbors}) {
        $count_conserv = $strands_
        of{$neighbors};
      }
    }
  }
  # now we verify the flag and count
  any conservation
  if ($count_conserv eq "same") {
    $count_same++;
      push(@predictions,$neighbors);
    }
    elsif ($count_conserv eq "opp")
    {
    $count_opp++;
    }
  }
}
# now we also need to count the
number of genes in the same
# strand and those in opposite
strands in the informative genome
my $total_same = 0;
my $total_opp = 0;
```

```perl
for my $inf_ngh (keys %inf_strands_of) {
  if ($inf_strands_of{$inf_ngh} eq
  "same") {
     $total_same++;
  }
  elsif($inf_strands_of{$inf_ngh} eq
  "opp") {
     $total_opp++;
  }
}
# now we can calculate the confi-
dence value
my $P_same = $count_same/$total_same;
my $P_opp = $count_opp/$total_opp;
my $conf = 1 - 0.5 * ($P_opp/$P_same);
$conf = sprintf("%.2f",$conf);
print "CONFIDENCE = ",$conf,"\n";

# now print predictions with their
confidence values
for my $prediction (@predictions) {
  $prediction =~s/,/\t/;
  print CONF $prediction,"\
  t",$conf,"\t",$informative_
  genome,"\n";
}
}
```

When run, this program creates a single file with conserved neighbors, their confidence values, and the genome from which the value was obtained. At the same time, the program prints the following output to the display:

```
% ./neighbor-method.pl
Salmonella_typhi_Ty2
CONFIDENCE = 0.55
Yersinia_pestis_KIM
CONFIDENCE = 0.73
Rhizobium_etli_CFN_42
CONFIDENCE = 0.99
Bacillus_subtilis
CONFIDENCE = 1.00
```

The informative genomes are ordered evolutionarily from closest to farthest. As expected, the evolutionarily closest organism to *E. coli* K12 in this example, *Salmonella typhi* Ty2, gives the lowest confidence value, whereas the farthest gives the maximum confidence value. The threshold the author uses to accept predictions is a confidence value ≥ 0.95.

4. Notes

1. In order for BLASTPGP to run, the protein sequences found in the files ending with ".faa" (FAA file) have to be formatted into BLAST databases. Keep each genome separated so it is easy to update results when a new genome is published. The main caveat to this approach is that some prokaryotic genomes contain more than one replicon (*see* also **Note 4**). This means that there will be more than one FAA file for these genomes. It is better to have all the protein sequences in a single file. Thus, concatenate all the FAA files within the directory of each genome into a single file. A simple UNIX command that can do this job is:

   ```
   cat genome_of_interest/*.faa > FAADB/
   genome_of_interest.faa
   ```

   To build BLAST databases the command is:

   ```
   formatdb -i FAADB/genome_of_interest.faa -o T\
   -n BLASTDB/genome_of_interest
   ```

2. So far, no publications have taken into account the conservation of gene order of non-coding genes. Several tests should be run to determine whether the orthologs can be effectively mapped and whether the overall results would be improved.

3. It is also possible to allow gaps (i.e., intervening genes) between gene pairs. However, in the author's experience, allowing gaps neither improves nor worsens the results. This assessment is based on knowledge of the operons in *Escherichia coli* K12. However, allowing gaps might facilitate calculation of confidence values in very small genomes, in which the number of same- and opposite-strand genes might be too small. If gaps are used, it is important that the pairs of genes are part of the same stretch of genes in the same strand with no intervening genes in the opposite strand (such stretches are called *directons*). For opposite-strand genes it will be enough to confirm that they are in different strands. The extreme example is the same-directon vs. different-directon approach. The conservation to be evaluated would be that of two genes in the same directon, regardless of the number of genes in between. The control, or negative set, would consist of genes in different, yet adjacent, directions. This is very similar to a method that is now used at The Institute for Genomics Research (Maria Ermolaeva, personal communication), which is a simplified version of a method published by Ermolaeva et al. *(9)*. A program that will output a database of genes in the same directon, and genes in different directons, would be:

```perl
#!/usr/bin/perl
$genome_of_interest = $ARGV[0] or die "I
need a genome to work with\n\n";

$genome_dir = "LOCAL_GENOMES/$genome_of_
interest";
opendir(GNMDIR,"$genome_dir") or die $die_
msg;

@ptt_files = grep { /\.ptt/ }
readdir(GNMDIR);

$results_dir = "NEIGHBORS_DIRECTON";
mkdir($results_dir) unless ( -d $results_dir );

open(    NGHTBL,">$results_dir/$genome_of_
interest.nghtbl" );

PTT:
for my $ptt_file ( @ptt_files ) {

  # get proper name of the RNT and GBK files
  my $rnt_file = $ptt_file;
  my $gbk_file = $ptt_file;
  $rnt_file =~ s/\.ptt/\.rnt/;
  $gbk_file =~ s/\.ptt/\.gbk/;

  # Is the genome circular?
  # The information is in the first line
  of the "gbk"
  # file, which starts with the word "LOCUS"
  my $circular = "yes"; # make circular
  the default
  open ( GBK,"$genome_dir/$gbk_file" );
  while(<GBK>) {
    if( /^LOCUS/ ) {
      $circular = "no" if( /linear/i );
      last; # we do not need to read any
      further
    }
  }
# now we read the table os protein coding genes
# and their "leftmost" coordinate so we can
# order them and find the neighbors
my %strand = ();
my %coord = ();
open( PTT,"$genome_dir/$ptt_file" );
while(<PTT>) {
    my @data = split;
    next unless( $data[1] =~/^\+|\-$/ );
    my $gi = $data[3];
    $strand{$gi} = $data[1];
```

```perl
    my ($coord) = $data[0] =~/^(\d+)/;
    $coord{$gi} = $coord;
}
close(PTT);

if( -f "$genome_dir/$rnt_file" ) {
    open(RNT,"$genome_dir/$rnt_file");
    while (<RNT>) {
      my @data = split;
      next unless( $data[1] =~/^\+|\-$/ );

      # The identifier is not a GI
      # but I rather keep the variable
      names consistent
      # the best identifier for an 'RNA' gene is
      # the gene name (5th column)
      my $gi = $data[4];
      $strand{$gi} = $data[1];
      my ($coord) = $data[0] =~/^(\d+)/;
      $coord{$gi} = $coord;
    }
}
# we build directons: stretches of genes
in the same
# strand with no intervening gene in the
opposite
# strand
my @ids = sort { $coord{$a} <=> $coord{$b}
} keys %coord;
my @directon = ();
my $directon;
$prev_str = "none";
for my $gi( @ids ) {
    if ( $strand{$gi} eq $prev_str ) {
      $directon .= ",".$gi;
      $prev_str = $strand{$gi};
    }
    else {
      push( @directon,$directon ) if (
      defined $directon );
      $directon = $gi;
      $prev_str = $strand{$gi};
    }
}

# with circular genomes we make sure that
# we close the circle, meaning if first and last
# directon are in the same strand, they
form a single
# directon
```

```perl
if ( $strand{$ids[0]} eq $strand{$ids[$#ids]} ) {
   if ( $circular eq "yes" ) {
     $directon[0] = $directon.",".$directon[0];
   }
   else {
     push( @directon,$directon );
   }
}
else {
   push( @directon,$directon );
}
# now we do form pairs in same directon, and
# pairs in different directons
for my $i ( 0 .. $#directon ) {
   my @gi = split( /,/,$directon[$i] );
   # same directon
   my @expendable = @gi;
   while ( my $gi = shift @expendable ) {
     for my $gj ( @expendable ) {
       print NGHTBL $gi,"\t",$gj
         ,"\t",$strand{$gi}.$strand{$gj},"\n";
     }
   }
   ## different directon
   ## assuming circular replicons
   my $next_directon = "none";
   if ( $i < $#directon ) {
     $next_directon = $directon[$i+1];
   }
   else {
     if ( $circular eq "yes" ) {
$next_directon = $directon[0];
     }
     else {
       next PTT;
     }
   }
   my @gj = split( /,/,$next_directon );
   for my $gi ( @gi ) {
     for my $gj ( @gj ) {
       print NGHTBL $gi,"\t",$gj
         ,"\t",$strand{$gi}.$strand{$gj},"\n";
     }
   }
}
close(NGHTBL);
```

4. It is important to know that some of the prokaryotic genomes reported so far have more than one replicon, meaning more than one DNA molecule. Multi-replicon genomes can contain two or more chromosomes, mega-plasmids, and plasmids. All the published replicons are considered part of the genome, and thus the programs presented are designed to read all of the replicons under a given genome directory.

5. As stated, Overbeek et al. *(3)* noted that some divergently transcribed genes could be functionally related, but found that the proportion of conserved, divergently transcribed genes across evolutionarily distant species was very small. The main effect of this possibility is that the confidence value would be an *underestimate*. This is clear in the analyses presented by Ermolaeva et al. *(9)*, and in the particular examination of false positives presented by Janga et al. *(10)*, who found independent evidence that almost all of their false-positives had a functional relationship (*see* also **Note 6**). In these analyses, the confidence value of 0.95 seems to correspond to a positive predictive value (true-positives divided by the total number of predictions) of 0.98.

6. The relationship between the positive predictive value and the confidence value has been established *(9, 10)* using data on experimentally determined operons of *Escherichia coli* K12 from RegulonDB *(21)*. Another useful statistic is coverage (also called sensitivity: true-positives divided by the total number of truly related pairs). For protein-coding genes, the current estimate for most genomes is that 0.5 of all same-strand direct neighbors might be in the same operon. In *E. coli* K12, the total number of same-strand protein-coding genes is 2930. Thus, the total number of functionally related neighbors is approximately $2930/2 = 1465$. The maximum number of predictions for *E. coli* K12 compared against all the genomes in the current database is 640 at a confidence value ≥ 0.95. Thus, the estimated coverage is: $640 * 0.95 / 1465 = 0.41$. This coverage might be thought low, but the predictions are of excellent quality.

Acknowledgments

Research support from Wilfrid Laurier University, and as SHARCNET Chair in Biocomputing are acknowledged.

References

1. Dandekar, T., Snel, B., Huynen, M., et al. (1998) Conservation of gene order: a fingerprint of proteins that physically interact. *Trends Biochem Sci* 23, 324–328.

2. Overbeek, R., Fonstein, M., D'Souza, M., et al. (1999) Use of contiguity on the chromosome to predict functional coupling. *In Silico Biol* 1, 93–108.

3. Overbeek, R., Fonstein, M., D'Souza, M., et al. (1999) The use of gene clusters to infer functional coupling. *Proc Natl Acad Sci U S A* 96, 2896–2901.

4. Jacob, F., Perrin, D., Sanchez, C., et al. (1960) [Operon: a group of genes with the expression coordinated by an operator.]. *C R Hebd Seances Acad Sci* 250, 1727–1729.

5. Jacob, F., Perrin, D., Sanchez, C., et al. (2005) [The operon: a group of genes with expression coordinated by an operator.] *Comptes rendus biologies* 328, 514–520.

6. Mushegian, A. R., Koonin, E. V. (1996) Gene order is not conserved in bacterial evolution. *Trends Genet* 12, 289–290.

7. Bork, P., Dandekar, T., Diaz-Lazcoz, Y., et al. (1998) Predicting function: from genes to genomes and back. *J Mol Biol* 283, 707–725.

8. Korbel, J. O., Jensen, L. J., von Mering, C., et al. (2004) Analysis of genomic context: prediction of functional associations from conserved bidirectionally transcribed gene pairs. *Nat Biotechnol* 22, 911–917.

9. Ermolaeva, M. D., White, O., Salzberg, S. L. (2001) Prediction of operons in microbial genomes. *Nucleic Acids Res* 29, 1216–1221.

10. Janga, S. C., Moreno-Hagelsieb, G. (2004) Conservation of adjacency as evidence of paralogous operons. *Nucleic Acids Res* 32, 5392–5397.

11. Snel, B., Lehmann, G., Bork, P., et al. (2000) STRING: a web-server to retrieve and display the repeatedly occurring neighbourhood of a gene. *Nucleic Acids Res* 28, 3442–3444.

12. von Mering, C., Huynen, M., Jaeggi, D., et al. (2003) STRING: a database of predicted functional associations between proteins. *Nucleic Acids Res* 31, 258–261.

13. Zheng, Y., Anton, B. P., Roberts, R. J., et al. (2005) Phylogenetic detection of conserved gene clusters in microbial genomes. *BMC Bioinformatics* 6, 243.

14. Maglott, D. R., Katz, K. S., Sicotte, H., et al. (2000) NCBI's LocusLink and RefSeq. *Nucleic Acids Res* 28, 126–128.

15. Pruitt, K. D., Tatusova, T., Maglott, D. R. (2005) NCBI Reference Sequence (RefSeq): a curated non-redundant sequence database of genomes, transcripts and proteins. *Nucleic Acids Res* 33, D501–504.

16. Altschul, S. F., Madden, T. L., Schaffer, A. A., et al. (1997) Gapped BLAST and PSI-BLAST: a new generation of protein database search programs. *Nucleic Acids Res* 25, 3389–3402.

17. Fitch, W. M. (2000) Homology a personal view on some of the problems. *Trends Genet* 16, 227–231.

18. Gevers, D., Vandepoele, K., Simillion, C., et al. (2004) Gene duplication and biased functional retention of paralogs in bacterial genomes. *Trends Microbiol* 12, 148–154.

19. Schaffer, A. A., Aravind, L., Madden, T. L., et al. (2001) Improving the accuracy of PSI-BLAST protein database searches with composition-based statistics and other refinements. *Nucleic Acids Res* 29, 2994–3005.

20. Blattner, F. R., Plunkett, G., 3rd, Bloch, C. A., et al. (1997) The complete genome sequence of Escherichia coli K-12. *Science* 277, 1453–1474.

21. Salgado, H., Gama-Castro, S., Peralta-Gil, M., et al. (2006) RegulonDB (version 5.0): *Escherichia coli* K-12 transcriptional regulatory network, operon organization, and growth conditions. *Nucleic Acids Res* 34, D394–397.

Chapter 9

Phylogenetic Profiling

Shailesh V. Date and José M. Peregrín-Alvarez

Abstract

Phylogenetic profiles describe the presence or absence of a protein in a set of reference genomes. Similarity between profiles is an indicator of functional coupling between gene products: the greater the similarity, the greater the likelihood of proteins sharing membership in the same pathway or cellular system. By virtue of this property, uncharacterized proteins can be assigned putative functions, based on the similarity of their profiles with those of known proteins. Profile comparisons, when extended to the entire genome, have the power to reveal functional linkages on a genome-wide scale (the functional "interactome"), elucidating both known and novel pathways and cellular systems.

Key words: Phylogenetic profiles, phyletic patterns, protein–protein interactions, functional links, protein function assignments, BLAST.

1. Introduction

Proteins that are members of the same biochemical pathway or physical complex are likely to co-evolve and be co-inherited, if the function of the cellular system to which they belong is preserved during evolution. This basic evolutionary assumption suggests that protein components of a pathway or a system are likely to possess a pattern of conservation similar to the pattern displayed by other participating members, with regard to their presence or absence in different organisms. Conversely, matching conservation patterns between gene products, when found, are indicative of functional associations, and can be exploited to identify other members of the same system.

This logic is captured in the use of phylogenetic profiles, which are also sometimes referred to as phyletic patterns or

Jonathan M. Keith (ed.), *Bioinformatics, Volume II: Structure, Function and Applications, vol. 453*
© 2008 Humana Press, a part of Springer Science + Business Media, Totowa, NJ
Book doi: 10.1007/978-1-60327-429-6 Springerprotocols.com

phylogenomic profiles, for detecting functional associations between proteins *(1, 2)*. Phylogenetic profiles are representations of evolutionary signatures, which manifest themselves as a distribution of homologs of the query protein in any set of reference genomes (*see* **Table 9.1** for examples). Similar or matching phylogenetic profiles reveal functionally relevant linkages or interactions between components that make the system biologically viable.

Given their properties, phylogenetic profiles can be used to assign function to proteins that cannot be readily characterized by homology-based methods alone. The phylogenetic profile of an uncharacterized protein can be constructed and compared with profiles of proteins with known functional assignments. If a particular function, or more commonly, components of a particular pathway or system are over-represented in the set of matching profiles, the query protein can then be assigned a putative role as a member of the same system. The use of phylogenetic profiles can be further extended: profiles of uncharacterized proteins that match known components of a given pathway probably indicate the presence of new, previously unidentified components, whereas systems composed entirely of unknown proteins with matching profiles hint at novel cellular pathways *(3)*. Phylogenetic profile data have also been successfully used to explore patterns of evolution between pairs of genes *(4, 5)*, analyze sequence conservation *(5, 6)*, and improve genome annotation *(7)*.

This chapter describes the construction of protein phylogenetic profiles, and their use for investigating protein function and protein–protein interactions. The authors assume that readers are familiar with certain basic concepts of computer science and biology. Prospective users should be able to write simple computer programs for manipulating input and output data, create logical

Table 9.1
Constructing protein phylogenetic profiles

Genomes→ Proteins ↓	E. coli	S. cerevisiae	C. elegans	D. melanogaster	A. thaliana
P1	×	×	×	✓	✓
P2	✓	×	✓	✓	×
P3	✓	×	✓	✓	×

In the example, the presence of homologs of three proteins, *P1*, *P2*, and *P3*, is checked in five completely sequenced genomes. These presence/absence observations constitute the phylogenetic profile of each protein. Profiles for proteins *P2* and *P3* are identical, but different from that of *P1*, indicating the likelihood of proteins *P2* and *P3* being functionally linked. ✓, Reliable presence of a homolog; ×, absence of homolog. See **Section 3.1.2** for complete species names.

structures within their programs, and understand the biological context of the outlined procedure.

2. Materials

2.1. Hardware Requirements

A computer capable of running standalone Basic Local Alignment Search Tool (BLAST) software *(8)* is a primary requirement (*see* **Note 1**).

2.2. Software Requirements

1. Operating system: Given the high likelihood of users running their programs on a UNIX-based operating system such as Linux, the materials and methods described here implicitly assume that such a system is being employed. Some steps in the method absolutely depend on the feasibility of running custom-made computer programs (also referred to as computer scripts, or simply, scripts) within the confines of the operating system. The choice of the operating system may thus dictate which computer language is used when writing the programs. Commonly used languages such as PERL, C (or C++), Java, and Python are supported by a majority of the operating systems in use today, such as Linux, Windows, or the Macintosh operating system (MAC OS) (*see* **Note 2**).

2. BLAST software: BLAST software is available for download from the NCBI BLAST web site (http://www.ncbi.nlm.nih.gov/BLAST/download.shtml) or from the Washington University in St. Louis (WU-BLAST, http://blast.wustl.edu) (*see* **Note 3**).

3. BLAST results parser: A computer program is required for extracting necessary information from the BLAST search output. Importantly, regardless of the language or computer program used for this procedure, any such parser should be able to isolate details of a best matching sequence for every unique genome examined, given a query protein and some criteria (*see* **Note 4**). The parser program should be adapted or modified to use results from NCBI BLAST or WU-BLAST, based on the package selected.

4. Running serial BLAST jobs: If phylogenetic profiles are to be constructed for a significant number of proteins, it is advisable to use a computer program that is able to run BLAST searches serially, i.e., run BLAST searches sequentially for all proteins in the query set (*see* **Note 5**).

5. A script for comparing phylogenetic profiles: A computer program that is able to handle vectors as input, and measure

the similarity (or distance) between them, is also required. The program should be able to perform a mathematical/statistical operation on the profile set, and measure the distance between a pair of vectors.

3. Methods

3.1. Creating Phylogenetic Profiles

In the exercises described below, we will learn to construct phylogenetic profiles that are not merely binary representations of presence or absence (0/1 profiles) *(1)*, rather, we will use converted BLAST similarity scores over a fuzzy scale between 0 and 1 *(3)*. This procedure yields profiles of higher resolution, increasing the accuracy of similarity measurements during profile comparisons. By choosing a suitable BLAST expectation (E-) value cutoff, however, a user can easily convert high-resolution profiles built with BLAST scores to binary profiles, if need be.

3.1.1. Creating a Query Input File

Amino acid sequences of one or more query proteins are required to be present in a single file, in FASTA format (*see* **Note 6**). Here we will assume we want to construct and call this file *myQueryFile*. Profiles of the three example proteins are described in **Table 9.1**.

3.1.2. Assembling a Database of Reference Genomes

1. The first step in constructing and using protein phylogenetic profiles is to assemble a database of completely sequenced reference genomes, searchable by BLAST. Incompletely or partially sequenced genomes should not be included in this database, to avoid a false representation of absence of a match for the query protein.

2. For each included genome, a file containing amino acid sequences of all known or predicted open reading frames (ORFs) in FASTA format is required. These files can be obtained from NCBI, as well as repositories such as the Sanger Center (http://www.sanger.ac.uk) and the European Bioinformatics Institute (EBI; http://www.ebi.ac.uk/), or individual genome sequencing project web sites (*see* **Note 7**).

    For example, consider a database containing complete genome sequences of five organisms—one bacteria: *Escherichia coli (EC)*; and four eukarya: *Saccharomyces cerevisiae (SC)*, *Caenorhabditis elegans (CE)*, *Drosophila melanogaster (DM)*, and *Arabidopsis thaliana (AT)*. A FASTA file containing amino acid sequences of all ORFs for each of the genomes can be downloaded from NCBI (ftp://ftp.ncbi.nih.gov/genomes/; FASTA amino acid or *.faa* files; *see* **Note 8**).

3. FASTA amino acid files for all genomes are merged to produce a single file, which is then used for database construction. Let us call this file *myDatabase* in this example (*see* **Note 9**).

4. This database is then formatted with *formatdb*, a utility that is a part of the BLAST software bundle. In its simplest form, for the concatenated file of amino acid sequences, the command *formatdb –i myDatabase* can be used (*see* **Note 10**). The reference genome database is now ready for use.

3.1.3. Running BLAST

1. Based on the size of the query set, BLAST can be run individually, or in a serial manner (*see* **Note 5**). If the phylogenetic profile of only one protein is to be constructed, BLAST can be run individually as follows:

   /pathToBLASTExecutable/blastall –p blastp –i myQueryFile –d myDatabase -o myOutputFile –b 500 –e 1e-5

where

–p blastp	Query and subject are both amino acid sequences
–i myQueryFile	Name of the input file containing query protein sequences
–d myDatabase	Name of the reference genome database to use
–o myOutputFile	Name of the output file containing BLAST results
–b 500	Number of top scoring alignments to display
–e 1e-5	E-value cutoff to use

Two of the BLAST options listed in the preceding, *–b* and *–e*, should be modified based on the requirements of the user and the size of the reference database (*see* **Note 11**).

2. If phylogenetic profiles are to be constructed for a significant number of query proteins, say, for all proteins in the genome of *Escherichia coli*, BLAST has to be run serially for each protein and the output of individual queries stored separately (*see* **Notes 5** and **12**).

3.1.4. Parsing BLAST Output

1. BLAST results need to be parsed to extract the relevant information (*see* **Note 13**). Output of the parser program is required to capture two important aspects of the matching sequence: its identifier and the significance of the score (*see* **Note 14**).

Mock output for protein *P1* from a parser program is described in the following:

```
>query >subject raw_score: value | E-value:
value | query_start: value | query_end: value
| subject_start: value | subject_end: value
| match_length: value | identity_percent-
age: value | similarity_percentage: value
| query_length: value | subject_length:
value
```

```
>protein_P1    >protein_P52_athaliana   raw_
score: 300 | E-value: 1e-155 | query_start:
1 | query_end: 140 | subject_start: 1 | sub-
ject_end: 140 | match_length: 140 | identity_
percentage: 100 | similarity_percentage: 100
| query_length: 140 | subject_length: 140
```

```
>protein_P1    >protein_P8090_athaliana   raw_
score: 220 | E-value: 1e-138 | query_start:
1 | query_end: 105 | subject_start: 15 | sub-
ject_end: 155 | match_length: 105 | identity_
percentage: 78 | similarity_percentage: 91 |
query_length: 140 | subject_length: 244
```

```
>protein_P1     >protein_P679_dmelanogaster
raw_score: 132 | E-value: 1e-66 | query_
start: 22 | query_end: 80 | subject_start:
107 | subject_end: 165 | match_length: 58
| identity_percentage: 70 | similarity_
percentage: 88 | query_length: 140 | sub-
ject_length: 111
```

2. BLAST significance values can be transformed into more usable forms (scores). For instance, for each protein i, and its highest scoring match in a genome j, the value of each element p_{ij} in the profile vector can be denoted by:

$$p_{ij} = \frac{1}{\log E_{ij}},$$ where E_{ij} is the BLAST significance value.

In this transformation, values of $p_{ij} > 1$ are truncated to 1, to remove logarithm-induced artifacts. Users may wish to experiment with other transformations.

3.1.5. Generating Profiles from BLAST Data

This step entails identifying the best-matching sequences in each reference genome included in the database, using the parsed and transformed BLAST data. Using a computer program, significance values of the top hits for each genome are collected and retained (*see* **Note 15**). The output of the computer program is therefore a string or row of numbers representing the transformed significance values, attached to an identifier. For example:

Protein	EC	SC	CE	DM	AT
>P1	1.0	1.0	1.0	0.0	0.0
>P2	0.0	1.0	0.8	0.6	1.0
>P3	0.0	1.0	0.8	0.6	1.0

1.0 (absence) → 0.0 (presence)

For each protein *P1*, *P2*, and *P3*, the profile vector in the example is represented by six values, where the first value is the identifier and the remaining values indicate the strength of the sequence match in each reference genome. Profiles can thus be calculated for all proteins in the query set using the results of the BLAST search, and maintained in a file (*see* **Notes 16** and **17**).

3.2. Using Phylogenetic Profiles for Function Annotation

The degree of similarity between phylogenetic profiles is indicative of the degree of functional linkage between proteins. Using a computer program that can measure similarity between vectors in a pairwise manner, functional linkages can be determined between all profile vectors in an input file. The program should be able to compare all vectors against each other, resulting in $n(n - 1)/2$ total comparisons, excluding profiles of homologs (*see* **Note 18**).

3.2.1. Measuring Profile Similarity

The use of Euclidean distance or Pearson correlation as a measure of vector similarity is sufficient in many instances to identify matching profiles and draw meaningful conclusions. Other advanced measures for describing vector similarity, such as Mutual Information, Hamming distance, Jaccard coefficient, or the chance co-occurrence probability distribution have also been proposed *(9)*. The use of measures such as Mutual Information is recommended over simpler metrics for gauging profile vector similarity, in order to capture linear, as well as non-linear relationships present in the data set. Mutual information, an information theoretic measure, is greatest when there is perfect co-variation between the profiles of two genes and tends to zero as the profiles diverge. Mutual information (MI) for two proteins (A,B) is calculated as:

$$MI(A,B) = H(A) + H(B) - H(A,B)$$

where $H(A) = -\sum p(a)\ln p(a)$ represents the marginal entropy of the probability distribution $p(a)$ of gene A in each reference genome, summed over intervals in the probability distribution and $H(A,B) = -\sum\sum p(a,b)\ln p(a,b)$ represents the intrinsic entropy of the joint probability distribution of genes A and B.

Mutual information for each profile pair is calculated on histograms of the transformed p_{ij} values. For the proteins, *P1*, *P2*, and *P3* in our example (*see* **Section 3.1.5**), bins are constructed as follows:

Bins	P1	P2	P3
0	2	1	1
0.1	0	0	0
0.2	0	0	0
0.3	0	0	0
0.4	0	0	0
0.5	0	0	0
0.6	0	1	1
0.7	0	0	0
0.8	0	1	1
0.9	0	0	0
1	3	2	2

Here, $p(a_{Protein,\ Bin})$ for each bin is calculated as the number of elements in the bin, divided by the total number of elements in the profile. For instance,

$$p(P1_{Bin0.0}) = 2/5 = 0.4 \text{ and}$$
$$p(P1_{Bin0.0}) \ ln \ p(P1_{Bin0.0}) = -0.3665.$$

The intrinsic entropy $H(P1)$ is the sum of values from all bins for the protein *P1*. When calculating the joint entropy for two proteins (*P1*, *P2*), bins are constructed for each position in the profile vector.

$$\text{Bin } (1.0,0.0) = 1; \text{ Bin } (1.0,1.0) = 1; \ \ldots$$
$$\text{Bin } (0.0, \ 1.0) = 1$$
$$p(P1,P2_{Bin\ (1.0,0.0)}) = 1/5 = 0.2; \text{ and so on.}$$

The joint entropy $H(P1, P2)$ is calculated by summing values for each bin, and the intrinsic and joint entropy values are then used to calculate mutual information. It is important to note that in practice, log_2 values are used rather than natural log. The higher the mutual information score, the more likely it is that the proteins are functionally linked.

3.2.2. Characterizing Protein Function

Similarity measurements between profiles on a local or global scale reveal proteins that follow the same distribution in phylogenetic space as the query protein, indicating functional linkages that capture parts of known interaction networks and hint at unknown ones, and assist in assigning putative function to proteins based on their associations with characterized entities. (*See* **Fig. 9.1** for examples of protein annotation using links derived from phylogenetic profiles.)

When comparisons between profiles of all proteins are plotted as a graph, with proteins as nodes and functional links as edges, a network of interactions emerges, which describes the functional interactome of the species (*see* **Fig. 9.2** for an example).

3.3. Data Storage and Validation

The methods described in this chapter for constructing phylogenetic profiles involve a variety of input and output data, ranging from FASTA format sequence files and raw BLAST results to phylogenetic profiles themselves. It is advisable to maintain a copy of the raw data and data generated in intermediate steps, both for error checking and to regenerate results, if need be. These data can be maintained as simple text files (flat files), or stored in a database, depending on resource availability and user preference. Storing data in relational databases is ideal, as it allows coupling and association of additional biological knowledge, enhancing the value of the results.

3.3.1. Checking Input Data

Checking input and output data is one of the most important steps in the method described in the preceding sections. It is necessary to check and ensure that the data being used for profile construction are free of errors. Users should pay particular attention to their input data. Frequently, data obtained from sequencing centers may contain formatting as well as content errors. As the sequence database grows, a user should implement an error-checking method, perhaps encoded in a script, that scans the raw sequence files for characters and lines that do not relate to the protein sequence.

It is also important to ensure that raw data used for profile construction are exactly as provided by the sequencing centers, and no accidental changes have taken place. This can be achieved using the timestamp associated with the file, or tracking file status using the UNIX "stat" command, or better yet using a "checksum," generated using, say, the md5 method (the checksum utility is available for free as a part of GNU core utilities from http://www.gnu.org). Checksums are series of letters and numbers that are unique to the contents of a file. Changes in file content, even by a single byte, are reflected by a change in the checksum.

3.3.2. Sampling Intermediate Data for Errors

Intermediate data being generated during the process, such as raw BLAST data, or parsed BLAST results, should also be periodically sampled to ensure that the data are free of errors. Sampling raw BLAST results can reveal problems with the input data or the sequence database. Errors can also arise due to the use of incorrect regular expressions during parsing, use of delimiters that conflict with delimiters used by BLAST, or reliance on lines that are nonstandard output. The length of the profile vector should also be checked with relation to the sequence database, and should have a value for each sequence in the database.

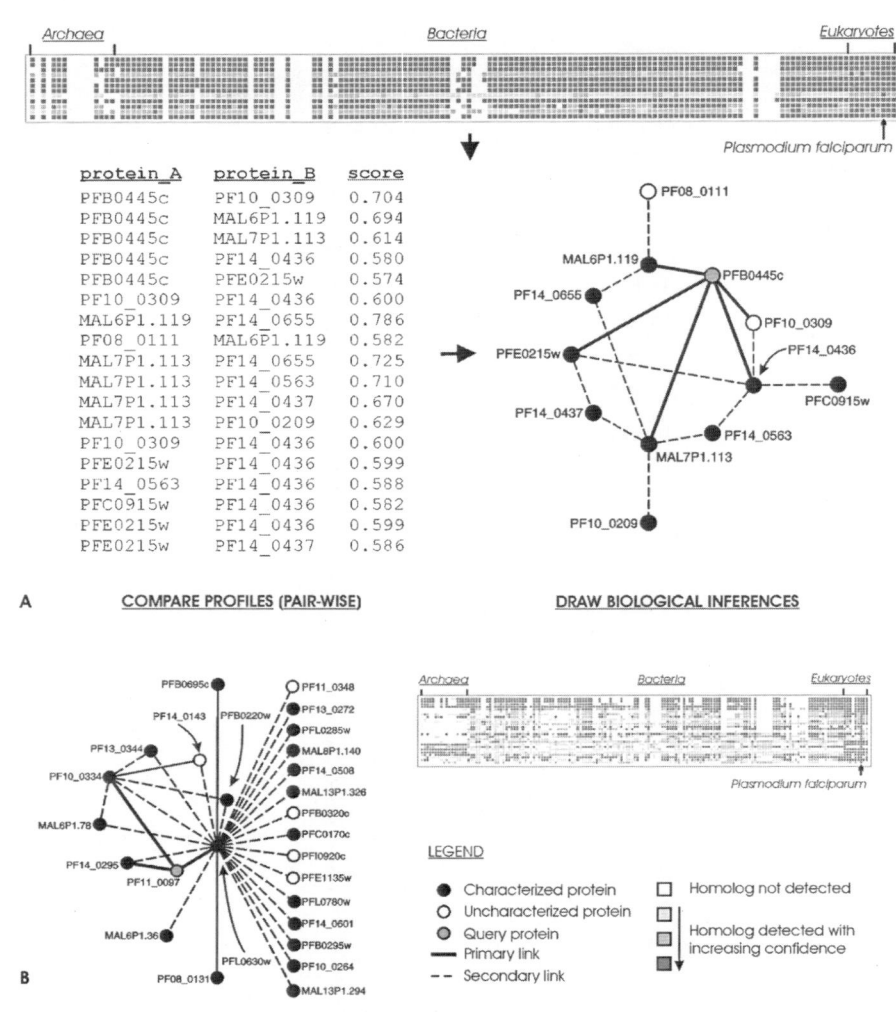

Fig. 9.1. Similarities between phylogenetic profiles reveal functional links that can be exploited to assign putative functions to proteins. Profiles are constructed by comparing a query amino acid sequence against a database of sequenced genomes, and pairwise similarity is measured between all profiles on a genome-wide scale, using mutual information. Primary links (proteins linked directly to the query protein forming an interaction core) and secondary links (proteins linked to the core) for proteins PFB0445 (**A**) and PF11_0097 (**B**) reveal sub-networks consisting of helicases and components of the oxidative phosphorylation machinery, respectively, and provide putative functional assignments for uncharacterized proteins in the network. PFB0445c, a helicase found in the human malarial parasite *Plasmodium falciparum*, is functionally linked with MAL6P1.119, MAL7P1.113, PF14_0436, and PFE0215w, (all of which are also annotated as helicases), when phylogenetic profiles for all proteins from this organism are compared on a genome-wide scale. These proteins are secondarily linked to even more helicases, providing a glimpse of a specific part of the functional interaction network of the parasite that groups together DNA-binding proteins. Based on these results, the two hypothetical proteins PF10_0309 and PF08_0111 included in the group can be confidently assigned DNA-binding roles. In another example, a search for profiles that match the profile of PF11_0097, the alpha subunit of succinyl-CoA synthetase, reveals connections to three other proteins: PF14_0295 (ATP-specific succinyl-CoA synthetase beta subunit, putative), PFL0630w (iron-sulfur subunit of succinate dehydrogenase), and PF10_0334 (flavoprotein subunit of succinate dehydrogenase), which describe two consecutive steps of the citric acid (TCA) cycle. Candidates secondarily linked to this core group include proteins such as MAL6P1.78 (para-hydroxybenzoate–polyprenyltransferase (4-hydroxybenzoate octaprenyltransferase), putative and PFL0780w (glycerol-3-phosphate dehydrogenase), which connect upstream

4. Notes

1. Running BLAST on computers with newer, faster processors is preferred over older machines. Although the computing power offered by older processors is adequate, a marked increase in speed is observed when computers with newer, more efficient processors and at least 256 MB of random access memory (RAM) are used. These observations also extend to computer clusters, in which the speed of a BLAST search is affected by the speed of each independent compute node running the process thread.

2. Users of the Windows operating system can take advantage of utilities such as Cygwin, which provide a UNIX-like environment for system and file manipulation. Current versions of the Macintosh operating system (version 10 or higher; OS X) provide users the ability to use a UNIX-style interface, simplifying the process of writing and executing programs.

3. The NCBI BLAST web site offers pre-compiled distributions (binaries) for all commonly used operating systems, including Linux, Windows, and the Macintosh operating system.

   For BLAST details *(8)*, see also: http://www.ncbi.nlm.nih.gov/Education/BLASTinfo/guide.html.

   A separate implementation of BLAST is also available from Washington University in St. Louis, referred to as WU-BLAST *(15)*. Users should note that results obtained from WU-BLAST are different in organization and content than NCBI BLAST. WU-BLAST developers claim more sensitivity, and significant differences can sometimes be observed between outputs of the two BLAST packages. We leave the choice of using WU-BLAST or NCBI BLAST up to the user. Depending on the distribution used, future steps in the protocol, such as selection of a BLAST results parser, should be modified accordingly.

Fig. 9.1. (continued) biosynthesis pathways to the TCA cycle. These functional linkages can be used to draw conclusions about uncharacterized proteins included within the group. Sequence analysis of the hypothetical protein PF14_0143, which is linked to the succinate dehydrogenase subunit PF10_0334, reveals the presence of the ABC1 and AarF domains within the protein. Yeast ABC1 is thought to be essential for ubiquinone biosynthesis *(10)*, and AarF is required for the production of ubiquinone in *Escherichia coli (11)*. These facts, together with links generated from phylogenetic profile analysis, suggest a putative role for PF10_0334 as a member of the oxidative phosphorylation machinery itself, or as a member of biosynthetic pathways that manufacture components involved in oxidative phosphorylation. Edges in the networks are not representative of link strength. Phylogenetic profile data for the malarial parasite were downloaded from the plasmoMAP web site (http://cbil.upenn.edu/plasmoMAP/ *(12)*, and the network layout and display done using Cytoscape *(13)* (http://www.cytoscape.org); images were optimized using CorelDRAW (http://www.corel.com).

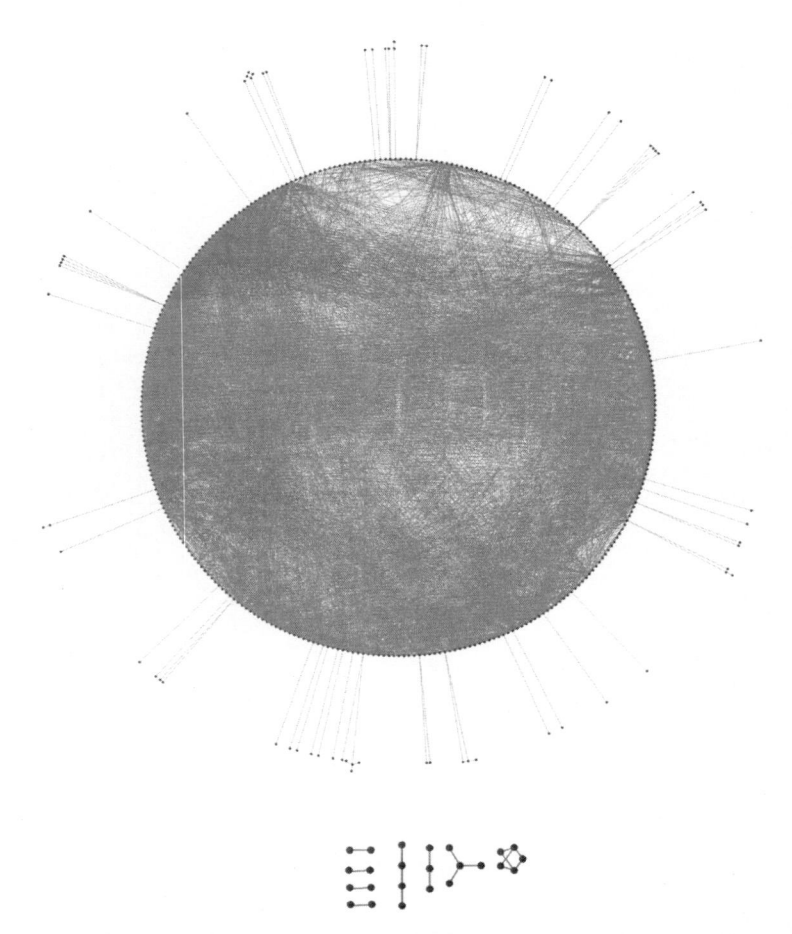

Fig. 9.2. A network of interacting proteins in P. falciparum. A genome-wide comparison of phylogenetic profile similarities for the malarial parasite can be used to reconstruct a network of functionally from all comparisons interacting proteins. Here, for purposes of illustration, the top 5,000 profile similarity scores from all comparisons were chosen, representing links between 408 unique proteins. Interaction networks are useful in understanding global relationships between gene and gene products. This network presents a global view of protein–protein interactions prevalent in the parasite, and can be further examined to understand properties such as connectivity and path length *(14)*. Some sub-networks of proteins that are independent of the main network are also seen.

4. Many BLAST result parsers are commonly available over the Internet, and indeed, users are free to write their own programs in any computer language. The open source effort BioPERL (http://www.bioperl.org) also provides result-parsing ability with its BLAST modules, if the PERL language is used for operation.

5. Running BLAST serially using a program is easy to achieve, the only requirement being that the program be able to store BLAST output (or parsed BLAST output) in such

a way that it can be identified as belonging to a particular query protein. This step becomes a necessity if a user plans to use a computer cluster for running BLAST searches. Serial BLAST searches can be run as a near parallel process, in which each search is designated to a specific node. In this scenario, a program for running BLAST serially can assume various degrees of complexity, the simplest being a script that breaks down the query set into equal parts and runs them on separate nodes, to a more complicated program that distributes jobs intelligently, based on node availability and finished searches.

6. For description of the FASTA and other file formats commonly used in sequence analysis, see http://www.ebi.ac.uk/help/formats_frame.html.

7. Genome sequencing projects often place restrictions on the use of pre-published sequence data for whole genome analysis. It is best to obtain permission from the principal investigators (PIs), before pre-published genomes are included in the user's reference database for constructing phylogenetic profiles.

8. If multiple files are present, such as individual chromosome files for eukaryotic organisms, they need to be merged to generate one FASTA amino acid file containing all ORFs for each organism.

9. Alternatively, FASTA amino acid files can be maintained separately for each genome, and used for comparison. However, this will affect BLAST E-values for sequence matches in the genome, especially near the threshold boundary (*see* **Note 11**).

10. Formatting the database file produces additional files with the same name, but with different extensions such as *.psq* and *.pin*, all of which are required by BLAST. *formatdb* has many other options that users are free to explore. Although the original file *myDatabase* is no longer required when BLAST is run, it is advisable to maintain it along with the database files, especially if there is minimal disk space cost. If multiple FASTA files are used for database construction (*see* **Note 9**), *formatdb* should be applied to all FASTA files.

11. By specifying the number of top alignments to display, an assumption is made as to the coverage of individual genomes in the database. If the database is very large, restricting the number of alignments will likely cause matches to proteins in some genomes to be excluded, and produce a truncated profile. This problem can be avoided by producing separate FASTA amino acid files for database construction (*see* **Note 12**), but this method should be used very carefully.

The *-e* option describes a significance value cutoff for pairwise alignments. In most sequence comparison studies, an E-value of 10^{-5} is considered the *de facto* threshold beyond which alignments cannot be trusted. This threshold is based on empirical observations of published literature by the authors. Users are free to choose a higher confidence threshold (with lower E-values) based on their experience or size of query sequence (see also BLAST description at http://www.ncbi.nlm.nih.gov/books/bv.fcgi?rid=handbook.chapter.610).

12. This can be accomplished using a computer program running on a single computer or a cluster of computers. BLAST results for all proteins can be collated or maintained separately for individual proteins.

13. For practical purposes, and to save disk space, it is advisable to combine running the BLAST search and parsing the output into a single step, using a computer program or a shell script. The original BLAST results can then be discarded (or saved, if needed, in a compressed form).

14. When parsing BLAST results, besides capturing the E-value, other fields such as the raw bit score, length of the match, sequence identity, and similarities, can also be extracted. Each of these parameters can be used in a filtering step to enhance accuracy during profile construction. The current protocol is restricted to using E-values as a parameter of presence/absence of a homolog in the subject genome.

15. This process is simplified by the fact that BLAST results are arranged in descending order of significance, and subsequent hits following the top hit can be safely ignored.

16. If the *-e* option for setting a significance cutoff threshold was not used during BLAST, a threshold describing presence or absence of the best matching sequence in each reference genome should be employed while constructing the profiles.

17. Pseudocode for constructing profiles (using PERL syntax):

```
read file containing genome names;

initialize genomeNameArray with genome names;

initialize $queryGeneId;
initialize $subjectGenomeId;
initialize %genomeNameHash;

open (myParsedOutputFile);
while(<myParsedOutputFile>){

  if (line is first line in the file){
    $queryGeneId = Id of query;
    $subjectGenomeId = name of subject genome;
```

```
$genomeNameHash{$subjectGenomeId} = "all
hit information";
next line;
}
if ((Id of query is not equal to previous
$queryGeneId) OR (endOfFile)){
print $queryGeneId;
for each $genome in $genomeNameArray {
  print information from $genomeNameHash
  {$genome};
}
reset $queryGeneId = Id of query;
reset %genomeNameHash to empty;
}
$subjectGenomeId = name of subject genome;
$genomeNameHash{$subjectGenomeId} = "all
hit information";
}
```

18. Similarity measurements between profiles of homologous proteins from the same organism are mostly inconsequential in nature, as homologs are expected to have the same phylogenetic profile. Therefore, such pairs should be excluded from analysis.

References

1. Pellegrini, M., Marcotte, E. M., Thompson, M. J., et al. (1999) Assigning protein functions by comparative genome analysis: protein phylogenetic profiles. *Proc Natl Acad Sci USA* 13, 4285–4288.

2. Gaasterland, T., Ragan, M. A. (1998) Microbial genescapes: phyletic and functional patterns of ORF distribution among prokaryotes. *Microb Comp Genom* 3, 199–217.

3. Date, S. V., Marcotte, E. M. (2003) Discovery of uncharacterized cellular systems by genome-wide analysis of functional linkages. *Nat Biotechnol* 21, 1055–1062.

4. Zheng, Y., Roberts, R. J., Kasif, S. (2002) Genomic functional annotation using co-evolution profiles of gene clusters. *Genome Biol* 10, RESEARCH0060.

5. Butland, G., Peregrin-Alvarez, J. M., Li, J., et al. (2005) Interaction network containing conserved and essential protein complexes in *Escherichia coli*. *Nature* 433, 531–537.

6. Peregrin-Alvarez, J. M., Tsoka, S., Ouzounis, C. A. (2003) The phylogenetic extent of metabolic enzymes and pathways. *Genome Res* 13, 422–427.

7. Mikkelsen, T. S., Galagan, J. E., Mesirov, J. P. (2005) Improving genome annotations using phylogenetic profile anomaly detection. *Bioinformatics* 21, 464–470.

8. Altschul, S. F., Gish, W., Miller, W., et al. (1990) Basic local alignment search tool. *J Mol Biol* 215, 403–410.

9. Wu, J., Kasif, S., DeLisi, C. (2003) Identification of functional links between genes using phylogenetic profiles. *Bioinformatics* 19, 1524–1530.

10. Do, T. Q., Hsu, A. Y., Jonassen, T., et al. (2001) A defect in coenzyme Q biosynthesis is responsible for the respiratory deficiency in *Saccharomyces cerevisiae* abc1 mutants. *J Biol Chem* 276, 18161–18168.

11. Macinga, D. R., Cook, G. M., Poole, R. K., et al. (1998) Identification and characterization of aarF, a locus required for production of ubiquinone in *Providencia stuartii* and *Escherichia coli* and for expression of 2'-N-acetyltransferase in *P. stuartii*. *J Bacteriol* 180, 128–135.

12. Date, S. V., Stoeckert, C. J. (2006) Computational modeling of the *Plasmodium*

falciparum interactome reveals protein function on a genome-wide scale. *Genome Res* 4, 542–549.

13. Shannon, P., Markiel, A., Ozier, O., et al. (2003) Cytoscape: a software environment for integrated models of biomolecular interaction networks. *Genome Res* 13, 2498–2504.

14. Barabasi, A. L., Oltvai, Z. N. (2004) Network biology: understanding the cell's functional organization. *Nat Rev Genet* 5, 101–113.

15. Lopez, R., Silventoinen, V., Robinson, S., et al. (2003) WU-Blast2 server at the European Bioinformatics Institute. *Nucleic Acids Res* 31, 3795–3798.

Chapter 10

Phylogenetic Shadowing
Sequence Comparisons of Multiple Primate Species

Dario Boffelli

Abstract

Comparisons between the sequence of the human genome and that of species at a variety of evolutionary distances from human have emerged as one of the most powerful strategies for identifying the functional coding and non-coding elements in the genome. Although the analysis of traits shared between human and distant relatives such as mouse or chicken can be effectively carried out through comparisons among the genomes of those organisms, analysis of traits specific to primates requires comparisons with our closest non-human primate relatives. In addition, comparisons of highly similar sequences simplify many computational problems in comparative genomics. This chapter describes a strategy for sequence comparisons between multiple primate species.

Key words: Genetics, comparative genomics, multiple primate comparisons, RankVISTA.

1. Introduction

With the goal of understanding the biology of *Homo sapiens*, non-human primates clearly constitute the most relevant model organisms. The recent divergence and consequent overall sequence conservation between individual members of this taxon make it difficult to distinguish functionally from neutrally conserved sequences on the basis of pairwise comparisons and have largely precluded the use of primates in comparative sequence studies. Accordingly, genomic sequence comparisons between distantly related species, such as human and mouse, have been used to identify genes and determine their intron-exon boundaries, as well as to identify regulatory elements present in the large non-coding

Jonathan M. Keith (ed.), *Bioinformatics, Volume II: Structure, Function and Applications, vol. 453*
© 2008 Humana Press, a part of Springer Science+Business Media, Totowa, NJ
Book doi: 10.1007/978-1-60327-429-6 Springerprotocols.com

fraction of the genome. However, sequences responsible for recently evolved functions among closely related species are likely to be missing in more distant species and unavailable for genome sequence comparisons. Of particular relevance is the fact that most primates differ in several ways from non-primates in their physiology, brain function, lifespan, diets, reproductive biology, and susceptibility to diseases. We developed phylogenetic shadowing to circumvent the problem of limited sequence divergence between primate species and enable primate sequence comparisons *(1)*. This approach is based on the simple premise that sufficient sequence variation is accumulated by extending the comparisons to several non-human primate species in addition to human. Although few differences can be detected between any pair of primate species, the regions of collective sequence variation detected among several primate species allow the identification of complementary regions of conservation (**Fig. 10.1**). These regions reveal putative functional regulatory elements, which can be prioritized for functional studies. To identify sequences conserved preferentially in primates, we search for sequences conserved in multiple primates but not between human and mouse. Two opposite evolutionary models can explain lack of detection of primate-conserved sequences in human-mouse comparisons: *(1)* the sequence appeared in the primate lineage, or *(2)* the sequence has been lost in the mouse

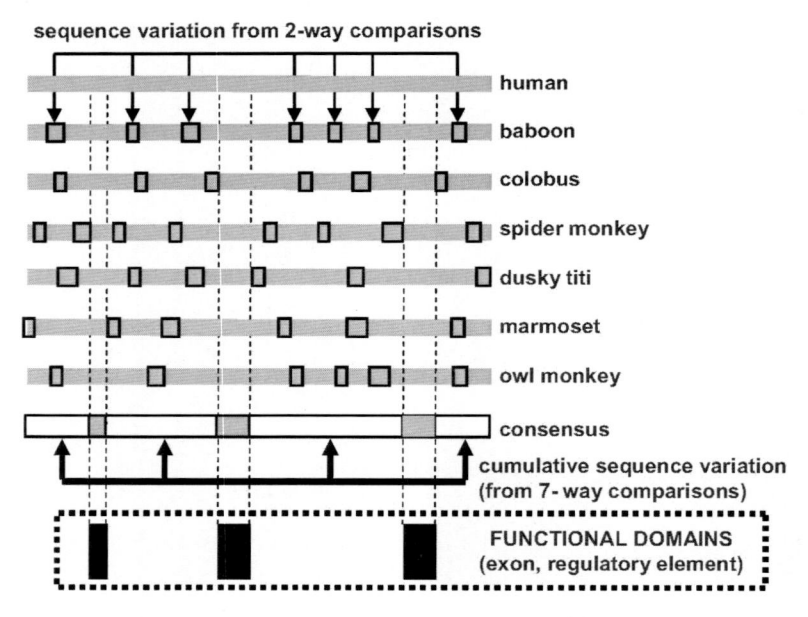

Fig. 10.1. Schematic representation of the phylogenetic shadowing approach. Few sequence differences *(cross-hatched boxes)* can be detected between any two pair of primate species: a consequence of primates' recent evolutionary divergence. To identify functional domains *(black boxes)*, the cumulative sequence variation obtained from the comparison of seven primate species is required.

lineage. To discriminate between these two possibilities, we use the dog sequence as outgroup (the dog lineage is ancestral to the human and mouse lineages): If a sequence missing in the mouse is also missing from the dog genome, we conclude that the sequence probably appeared in the primate lineage.

2. Materials

2.1. Sequence Data

1. Sequences of six anthropoid primate species are needed for phylogenetic shadowing (shown in **Fig. 10.2**). The human and Rhesus macaque genomes have been sequenced and are available from online genome browsers. (An example of how to retrieve these sequences is shown in **Section 3.1**) Sequenced genomes for the remaining species (colobus, marmoset, owl monkey, and dusky titi) are not available and the sequences of the relevant BAC clones need to be generated by the user at high-quality draft (6–7×). BAC libraries and clones for all the preceding species are available from BACPAC Resources (http://bacpac.chori.org/). Although the description of BAC clone sequencing is beyond the scope of this chapter, the generation and sequencing of 3 kb libraries (2, 3) and sequencing data assembly (4) are described in another volume in this series. Sequences submitted to

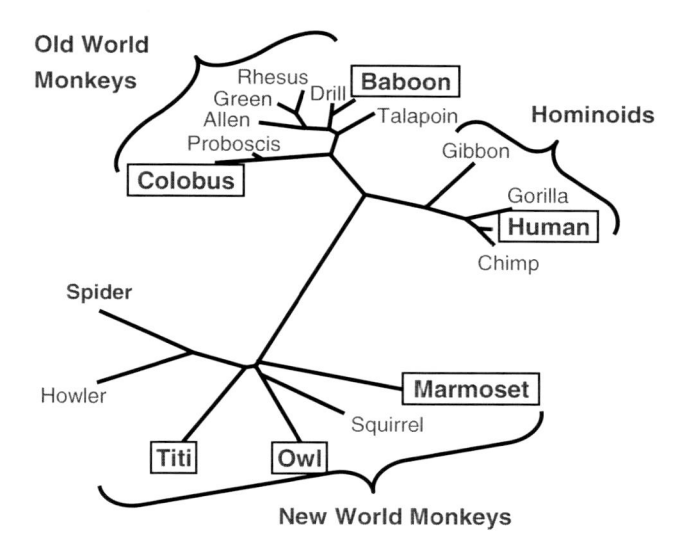

Fig. 10.2. Selection of primate species for phylogenetic shadowing. Sequencing of two old-world monkeys and three new-world monkeys in addition to the available human sequence (indicated by the boxes) captures at least 80% of the information that would be captured by sequencing all monkey species. This information is sufficient to identify non-coding sequences conserved among all primates analyzed.

the mVISTA server for the computation of the multiple sequence alignment need to be in FASTA format (*see* **Note 1**). User-generated sequences can be finished (i.e., 1 continuous sequence) or ordered/oriented draft (i.e., containing one or more contigs—sequence blocks—separated by a short string of "NNN," in the correct order and orientation in a single sequence). If using a word processor (e.g., Microsoft Word) to prepare your sequence files, make sure to save the files in plain text format (.txt). Other formats, such as .doc or .htm, will be rejected.

2. The annotation file for the human sequence helps visualizing known features such as exons and UTRs along the sequence (*see* **Section 3.2**). The format of the annotation file is as follows (**Fig. 10.3**): Each gene is defined by its start and end coordinates and the name, listed on one line. A greater than (>) or less than (<) sign should be placed before this line to indicate if the gene is transcribed on the plus or minus strand, although the numbering should be according to the plus strand. The exons are listed individually with the word "exon," after the start and end coordinates of each exon. UTRs are annotated in the same way as exons, with the word "utr" replacing "exon."

This chapter use sequences from the zeta/alpha-globin regulatory region to illustrate the steps involved in phylogenetic shadowing (human: chr16:43811-160260, May 2004 assembly; rhesus

```
> 43811  47443  C16orf33
  43811  43989  utr
  43990  44058  exon
  45459  45549  exon
  45777  45882  exon
  46539  46613  exon
  47086  47143  exon
  47144  47443  utr
< 48059  62591  RHBDF1
  48059  48338  utr
  48339  48758  exon
  48972  49125  exon
  49247  49347  exon
```

Fig. 10.3. Example annotation file. The entry for each gene is identified by a greater than (>) or less than (<) sign for the gene being transcribed on the plus or minus strand, respectively. The same line also contains the gene's start and end coordinates and the name listed on one line. All numbering is according to the plus strand. The exons are listed individually with the word "exon" following the start and end coordinates of each exon. UTRs are annotated in the same way as exons, with the word "utr" replacing "exon."

monkey: chr20:2423-137222, Jan 2006 assembly). The sequences for the zeta/alpha-globin regulatory region have been generated for the following non-human primates and can be retrieved from GenBank: colobus: accession #AC148220; marmoset: accession #AC146591; owl monkey: accession #AC146782; dusky titi: accession #AC145465).

2.2. Software

1. Several computational packages are now available for the quantitative comparative analysis of genome sequence alignments: RankVISTA, eShadow, GERP, and PhastCons. Here we use RankVISTA: http://genome.lbl.gov/vista/mvista/submit.shtml).

2. Sequences from available genomes can be retrieved from online genome browsers (UCSC Genome Browser: http://genome.ucsc.edu/cgi-bin/hgGateway and ENSEMBL: http://www.ensembl.org/index.html).

3. To generate the annotation file of the human sequence use GenomeVISTA: http://pipeline.lbl.gov/cgi-bin/GenomeVista.

4. To retrieve publicly available sequence data, use GenBank: http://www.ncbi.nlm.nih.gov/entrez/query.fcgi?db=Nucleotide.

3. Methods

The computation of the multiple primate comparison (phylogenetic shadowing) plot involves two steps: *(1)* generation of a multiple alignment of primate sequences for a genomic locus of interest using the mVISTA server, and *(2)* scanning of the multiple alignment to identify sequence blocks (usually a few hundreds base pairs in length) that are significantly more conserved than neutrally evolving sequences in that locus using the RankVISTA algorithm, available at the mVISTA server *(5)*. RankVISTA plots are based on the Gumby algorithm, which estimates neutral evolutionary rates from non-exonic regions in the multiple sequence alignment, and then identifies local segments of any length in the alignment that evolve more slowly than the background (**Note 2**).

3.1. Sequence Retrieval from Genome Browsers

Retrieve the sequences for species whose genomes are available (currently human and Rhesus macaque) from the UCSC Genome Browser as follows.

1. Select the genome for which you want to retrieve the sequence from the "Genome" drop-down menu and type the gene name or the genomic coordinates (if known) into the "position or search term" field.

2. Click the "Submit" button to open up the Genome Browser window to the requested location. Occasionally a list of several matches in response to a search will be displayed, rather than immediately displaying the Genome Browser window. When this occurs, click on the item in which you are interested and the Genome Browser will open to that location.

3. Click on the "DNA" button found in the blue menu bar at the top of the page to retrieve the DNA sequence for the region displayed in the Genome Browser.

4. Configure the retrieval and output format options in the "Get DNA in Window" page that appears.

5. Without changing the default settings, click on the "get DNA" button. Save the resulting DNA sequence using the "File → Save as" menu. Make sure to select "Text files" in the "Save as type" drop-down menu (**Note 3**).

3.2. Preparation of the Annotation File

The annotation file can be prepared manually or using the program GenomeVISTA. This program is used to align user-generated sequences to the human and other sequenced genomes. In addition to a pairwise conservation profile between the base (i.e., reference) genome and the user's sequence, GenomeVISTA returns the annotation of the base genome. Submission of the human sequence to be used for phylogenetic shadowing will enable you to retrieve the annotation for that sequence.

1. Upload from your computer to the GenomeVISTA server the human sequence for your region of interest as a plain text file in FASTA format using the "Browse" button.

2. Use "Human Genome" as the base genome. Enter your e-mail address, a name for your project and click on the "Submit query" button.

3. After a few minutes, you will receive an e-mail with a link to your results. On the results page, click on "Text Browser."

4. Write down the alignment coordinates (begin-end) of the submitted sequence relative to the human genome and then click on "Download RefSeq Genes" (**Fig. 10.4**).

5. Save the resulting annotation file on your computer using the "File → Save as" menu. Make sure to select "Text files" in the "Save as type" drop-down menu (**Note 3**). Note that the coordinates in this file are relative to the sequence of the human genome and not to the sequence in your human sequence file.

6. To adjust for this, use a spreadsheet program such as Microsoft Excel to subtract from all coordinates in the annotation file the beginning coordinate of the alignment to the human genome. All coordinates need to be >0. Inspect the

Human May 2004 chr16:43,811-178,231

<< >>

VISTA tracks on UCSC   VISTA Browser

Change Annotation: [Select Annotation ▾]

Download RefSeq genes   **Get CNS:** Human May 2004-HBA

Fig. 10.4. Retrieval of the human annotation file from the GenomeVISTA output. The alignment position of the human sequence submitted by the user is shown on the first line: the sequence aligns to chromosome 16, from 43,811 to 178,231, on the May 2004 Assembly of the Human Genome. The annotation file for this region (with coordinates relative to May 2004 assembly) can be retrieved by clicking on "Download RefSeq genes" *(arrow)*. To transform the coordinates from the May 2004 assembly to the user's human sequence, simply subtract the beginning coordinate on the May 2004 assembly (43,811, *underlined*) from all the coordinates in the annotation file.

resulting file and delete all exon annotation entries with zero or negative coordinates. Similarly, correct the start and end coordinates on each gene definition line if zero or negative numbers are present. Save the file again as "Text files."

3.3. Submission to mVISTA: Sequence Data Fields

To obtain a multiple sequence alignment for your sequences, submit your six sequences (the human and five non-human primate sequences) to the mVISTA server. Sequences can be uploaded in FASTA format from a local computer using the "Browse" button or, if available in GenBank, they can be retrieved by inputting the corresponding GenBank accession number in the "GEN-BANK identifier" field (*see* **Note 4**). Enter the human sequence as sequence#1.

3.4. Submission to mVISTA: Choice of Alignment Program

Three genomic alignments programs are available in mVISTA. "LAGAN" is the only program that produces multiple alignments of finished sequences, and is the most appropriate choice for phylogenetic shadowing *(6)*. Note that if some of the sequences are not ordered and oriented in a single sequence (*see* **Section 2.1.1**) your query will be redirected to AVID to obtain multiple pairwise alignment. "AVID" and "Shuffle-LAGAN" are not appropriate genomic aligners for phylogenetic shadowing as they produce only all-against-all pairwise alignments.

3.5. Submission to mVISTA: Additional Options

1. "Name": Select the names for your species that will be shown in the legend. It is advisable to use something meaningful, such as the name of an organism, the number of your experiment, or your database identifier. When using a GenBank identifier to input your sequence, it will be used by default as the name of the sequence.

2. "Annotation": If a gene annotation of the sequence is available, you can submit it in a simple plain text format to be displayed on the plot (*see* **Section 3.2**). Although to display the annotation on the RankVISTA plot you need to submit the annotation file for one species only, usually human, annotation files for all other species can also be submitted.

3. "RepeatMasker": Masking a base sequence will result in better alignment results. You can submit either masked or unmasked sequences. If you submit a masked sequence and the repetitive elements are replaced by letters "N," select the "one-celled/do not mask" option in the pull-down menu. mVISTA also accepts softmasked sequences, where repetitive elements are shown as lowercase letters while the rest of the sequence is shown in capital letters. In this case, you need to select "softmasked" option in the menu. If your sequences are unmasked, mVISTA will mask repeats with RepeatMasker. Select "human/primate" in the drop-down menu. If you do not want your sequence to be masked, select "one-celled/do not mask."

4. Leave the "Find potential transcription factor binding sites using rVISTA" and "Use translated anchoring in LAGAN/Shuffle-LAGAN" options unchecked.

3.6. Submission to mVISTA: Parameters for RankVISTA

1. The RankVISTA algorithm, used for the quantitative analysis of the multiple primate sequence comparisons, is run automatically on the alignment generated by mVISTA. The option "RankVISTA probability threshold $(0 < p < 1)$" tells the RankVISTA algorithm to ignore predictions with a p-value greater than that indicated in the box. The default setting of "0.5" means that all conserved sequences with a conservation p-value between 1 and 0.5 will not be reported.

2. If you know the phylogenetic tree relating the species you are submitting, enter it at "Pairwise phylogenetic tree for the sequences," otherwise LAGAN will calculate the tree automatically.

3. Click on "Submit" to send the data to the mVISTA server. If mVISTA finds problems with the submitted files, you will receive a message stating the type of problem; if not, you will receive a message saying that submission was successful. Several minutes after submitting your sequences, you will receive e-mail from vista@lbl.gov indicating your personal Web link to the location where you can access the results of your analysis.

3.7. Retrieval of the Results

Clicking on the link found in the body of the e-mail takes you to the results page. It lists every organism you submitted, and provides you with three viewing options using each organism as base. These three options are: *(1)* the "Text Browser," which provides all the detailed information—sequences, alignments, conserved sequence statistics, and RankVISTA results for multiple sequence comparisons. This is where you retrieve the coordinates of conserved regions predicted by phylogenetic shadowing. *(2)* The "Vista Browser," an interactive visualization tool that can be used to dynamically browse the resulting alignments and view a graphical presentation of RankVISTA results. *(3)* A PDF file, which is a static Vista plot of all pairwise alignments, and is not relevant to multiple primate comparisons. It is important to note that while mVISTA shows the results of all pairwise comparisons between one species chosen as the base (reference) sequence and all other submitted sequences, RankVISTA shows the result of the multiple (simultaneous) sequence comparisons of all submitted sequences and is independent of the choice of base sequence.

3.8. Text Browser

1. The Text Browser brings you to the results of your analysis in text format (**Fig. 10.5**). At the top of the page is a banner that displays the aligned organisms. The sequence listed in the darker header area is acting as base. (The choice of the base sequence is irrelevant for RankVISTA analysis.) This banner also lists the algorithm used to align your sequences. If you did not submit your own tree, you can click on "phylogenetic tree" to inspect the tree computed by MLAGAN and compare it with the one expected based on the known phylogeny of the species analyzed. Underneath is the navigation area, which shows the coordinates of the region currently displayed and offers a link to the Vista Browser (see the following) and a link to a list of all conserved regions found. Following that is the main table, which lists each pairwise alignment that was generated for the base organism. Each row is a separate alignment. Each column, except the last one, refers to the sequences that were submitted for analysis.

2. The last column (labeled "alignment") contains a link to the RankVISTA results and information pertaining to the whole alignment (**Fig. 10.5**, bottom panel). It also provides links to alignments in human readable and MFA (multi-fasta alignment) formats, a list of conserved regions from this alignment alone, and links to pdf plots of this alignment alone.

3.9. RankVISTA Text Browser

Clicking on the "RankVISTA" link in the alignment column takes you to a summary table of the RankVISTA analysis of multiple primate sequences. This is the primary result page for phylogenetic shadowing (**Fig. 10.6**). The table shows the start and end coordinates, relative to the sequence of the organism chosen as the base, for all regions predicted to be conserved across all primate species analyzed, and the length of the conserved regions. The *p*-value column shows the probability of seeing that level of conservation by chance in a neutrally evolving 10-kb segment of the base sequence, thus enabling the user to rank conserved sequences on the basis of

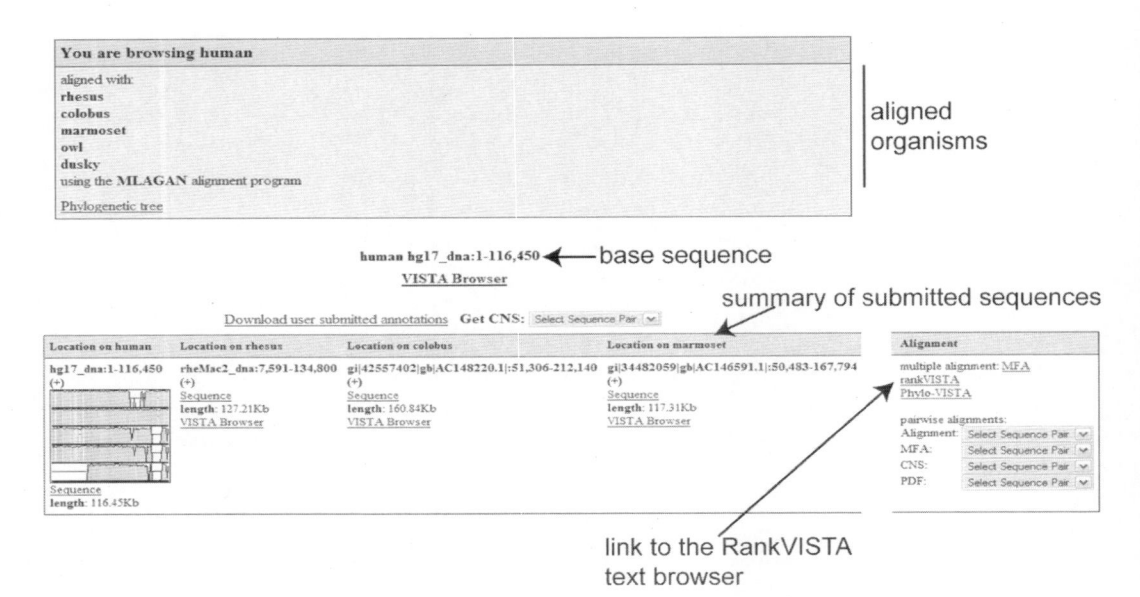

Fig. 10.5. mVISTA Text Browser. This is the entry table to the mVISTA results. The top panel presents a quick summary of your submission, including the list of submitted organisms and the program used to generate their multiple alignments. The middle section indicates the base sequence, used to visualize the alignments in the VISTA browser, accessible through the link right below. You can also access conserved sequences identified by pairwise comparisons through the "Get CNS" drop-down menu. The bottom panel shows a summary of all pairwise alignments between the base sequence and the other submitted sequences. At the left end of the panel, links to pairwise and multiple alignments are found. The link to the RankVISTA text browser is also found here.

```
****  RankVISTA conserved regions on hg17_dna:1-116,450  *****

start          end            length         p-value        type

4,527          4,939          413bp          0.18           exon

14,458         14,828         371bp          0.27           noncoding

31,592         31,980         389bp          0.3            exon

35,900         36,089         190bp          0.059          exon

36,163         36,954         792bp          0.46           noncoding

38,027         38,733         707bp          0.0014         noncoding

38,783         38,912         130bp          0.31           exon

39,049         39,405         357bp          0.074          noncoding

39,406         39,510         105bp          0.45           exon

40,067         40,475         409bp          0.25           noncoding

41,630         42,726         1097bp         0.00012        noncoding
```

Fig. 10.6. Example of RankVISTA results. This table, obtained from the Text Browser, contains the position (start and end, relative to the sequence shown on the first line of the table), the conservation probability (*p*-value) and the type (coding or non-coding) of conserved sequences identified through multiple primate comparisons.

their likelihood of conservation. The last column indicates whether the conserved sequence is coding or non-coding, and is based on the annotation file submitted to mVISTA.

3.10. VISTA Browser   Vista Browser is an interactive Java applet (*see* **Note 5**) designed to visualize pairwise and multiple alignments using the mVISTA (default) and RankVISTA scoring schemes, and to identify regions of high conservation across multiple species.

1. Clicking on the "Vista Browser" link will launch the applet with the corresponding organism selected as base (**Fig. 10.4**). The VISTA Browser main page displays all pairwise comparisons between the base sequence and all other submitted sequences using the mVISTA scoring scheme, which measures conservation based on the number of identical nucleotides (% conservation) in a 100 bp window. Multiple pairwise alignments sharing the same base sequence can be displayed simultaneously, one under another. The plots are numbered so that you can identify each plot in the list underneath the VISTA panel. The many additional features of the browser are described in detail in the online help pages, accessed by clicking on the "help" button in the top left corner of the browser.

2. To access the RankVISTA plot, click on the "1 more organism" drop-down menu found in the left panel of the browser ("Control Panel") and select "RankVISTA" (**Fig. 10.7**). The "Add curve" pop-up dialog prompts the user to set parameters for the RankVISTA plot. The "minimum Y" and "maximum Y" parameters set the range of displayed p-values, on a logarithmic scale. The default values of "0" and "5" instruct the server to display conserved regions with p-values ranging from 10^0 (=1) to 10^{-5} (note that the default "RankVISTA probability threshold" set at the mVISTA submission stage (*see* **Section 3.6**) instructs RankVISTA to cut off predictions at that given p-value). Both parameters can be adjusted after displaying the plot. The resulting RankVISTA plot is displayed in the Genome Browser below the pairwise comparisons.

3. The position of the plot can be reordered by selecting RankVISTA in the list underneath the VISTA panel and clicking on the "up" arrow. Conserved sequence regions predicted by RankVISTA are colored according to their annotation, with light-blue regions corresponding to annotated exons and pink regions to non-coding sequences (**Fig. 10.8**). Note that RankVISTA coloring is based on exon annotations of all aligned sequences (if annotation files were submitted for more than one species), not just the one currently used as the base. Consequently, an un-annotated region in the base sequence might still be colored as an exon because of annotations from other

navigation and utilities buttons

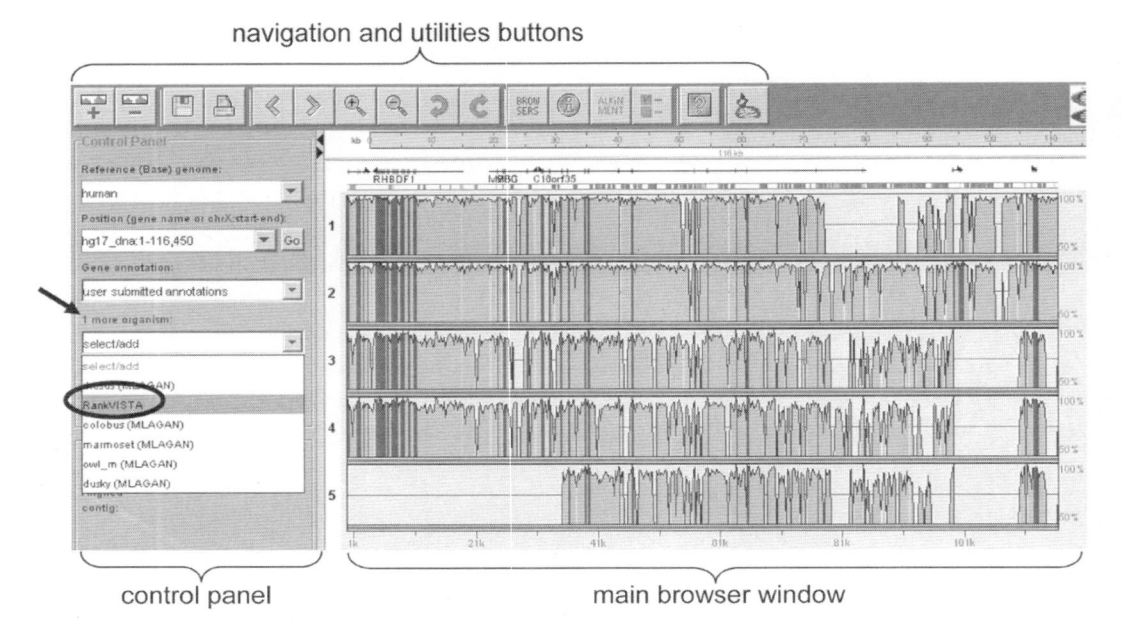

control panel   main browser window

Fig. 10.7. Adding the RankVISTA plot to the Genome Browser. This figure shows the main components of the VISTA Browser. To add the RankVISTA plot to the display, click on the "1 more organism" drop-down menu *(arrow)* in the "Control Panel" section and select "RankVISTA" *(circled)*.

Alignment details   Curve Parameters

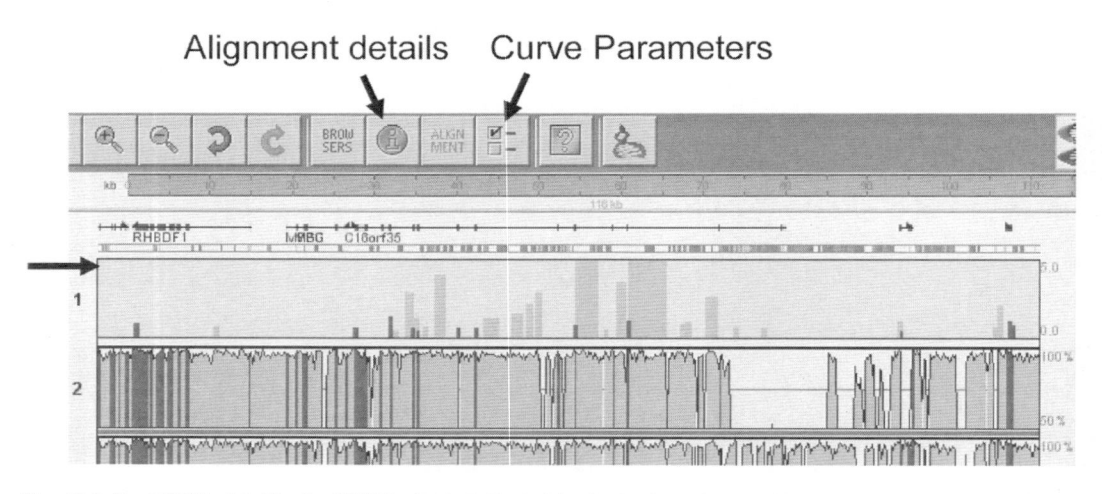

Fig. 10.8. RankVISTA plot. The RankVISTA plot is indicated by the horizontal arrow. The positions of conserved exons (coding and UTR) are shown by the blue *(dark gray here)* columns, while the pink *(light gray here)* columns identify conserved non-exonic regions. The height of the columns corresponds to the conservation p-value, on a logarithmic scale. The y-axis bounds can be modified by clicking on the "Curve parameters" button, and the coordinates of conserved exonic and non-exonic regions can be retrieved by clicking on "Alignment details," which takes you to the Text Browser (*see* **Section 3.9**).

sequences. In another deviation from the standard scheme, RankVISTA colors UTRs and coding exons the same, since they are treated identically by the underlying algorithm. The width of a predicted conserved region corresponds to its length, while the height corresponds to its conservation p-value.

4. To adjust the RankVISTA plot parameters, first select the plot by clicking on it. By clicking on the "Curve parameters" button (*see* **Fig. 10.8**), you can modify the y-axis bounds.

5. By clicking on the "Alignment details" button, you can quickly shift to the Text Browser (*see* **Section 3.8**) and retrieve the coordinates of conserved sequences.

6. To print the plot, click the "Print" button. The first time you do this, you will get a dialog box to confirm that you indeed requested that something be sent to the printer. This is a security measure in Java intended to handicap malicious code. Click "yes." A standard printing dialog box will appear. Proceed as you would with any other printing job.

7. To save the plot, click the "save as" button. In the menu that will appear, select the file type you want, adjust parameters such as image width if desired, and press "ok." If you have pop-up blocking software such as the Google toolbar or a later version of IE browser, you may need to hold down the CTRL key while clicking the OK button.

4. Conclusions

The RankVISTA analysis identifies coding and non-coding sequences conserved across multiple primate species. To determine whether these sequences are unique to primates or are shared with other mammalian species, a similar analysis should be repeated by comparing the human sequence with the orthologous sequence of non-mammals such as the mouse and dog, for which sequenced genomes are available, to identify which of the sequences conserved in primates are also conserved in these other mammals.

5. Notes

1. A sequence in FASTA format consists of a single-line description, followed by lines of sequence data. The first character of the description line is a greater than (>) symbol in the first column. All lines should be shorter than 80 characters. Make sure

to use meaningful names in the description line. Sequences have to be represented in the standard A/C/T/G nucleic acid code, with these exceptions: lowercase letters are accepted and are mapped into uppercase; and a single hyphen or dash can be used to represent a gap of indeterminate length. Any numerical digits in the query (=input) sequence should either be removed or replaced by N for unknown nucleic acid residue. Sequences retrieved from the UCSC Genome Browser are in FASTA format, and sequences available in GenBank can be retrieved in FASTA format.

2. RankVISTA uses the input alignment as its own training set to determine local background rates of sequence divergence. Consequently, small or grossly incomplete alignments should be avoided in accordance with these recommendations:

   a. The base sequence length should be at least 10 kb. Smaller alignments might be tolerated, but if RankVISTA detects an inadequate number of aligned positions, it will return no output.

   b. For the p-values to be meaningful, RankVISTA requires a reasonably complete alignment. As a rule of thumb, the number of "N" characters and spurious gap characters arising from missing sequence data should be <10% of the total number of characters in the alignment. For this reason, finished or high-quality draft sequences (6–7×) should be used with RankVISTA. The relative ranking of conserved regions by p-value would still be meaningful if this rule were violated. However, the p-value estimates would be systematically biased.

   c. RankVISTA's sensitivity in detecting non-exonic conservation can be increased by supplying exon annotations. The annotated regions are masked when estimating neutral evolutionary rates, resulting in a more accurate estimate of the background conservation level.

3. As always when saving documents, note in which folder you are saving your file (shown at the top of the "save as" window).

4. If you have problems submitting sequences using the GenBank accession number, download the sequence from GenBank in FASTA format to your computer and submit this to mVISTA.

5. VISTA Browser requires Java2. The browser checks automatically whether Java2 is installed and it returns an error message if Java2 is missing. For detailed instructions on downloading and installing Java2, visit http://pipeline.lbl.gov/vgb2help.shtml#install.

References

1. Boffelli, D., McAuliffe, J., Ovcharenko, D., et al. (2003) Phylogenetic shadowing of primate sequences to find functional regions of the human genome. *Science* 299, 1391–1394.

2. Roe, B. A. (2004) Shotgun library construction for DNA sequencing, in (Zao, S., Stodolsky, M., eds.), *Bacterial Artificial Chromosomes*, vol. 1. Humana Press, Totowa, NJ.

3. Predki, P. F., Elkin, C., Kapur, H., et al. (2004) Rolling circle amplification for sequencing templates, in (Zao, S., Stodolsky, M., eds.), *Bacterial Artificial Chromosomes*, vol 1. Humana Press, Totowa, NJ.

4. Schmutz, J., Grimwood, J., Myers, R. M. (2004) Assembly of DNA sequencing data, in (Zao, S., Stodolsky, M., eds.), *Bacterial Artificial Chromosomes*, vol 1. Humana Press, Totowa, NJ.

5. Martin, J., Han, C., Gordon, L. A., et al. (2004) The sequence and analysis of duplication-rich human chromosome 16. *Nature* 432, 988–994.

6. Brudno, M., Do, C. B., Cooper, G. M., et al. (2003) LAGAN and Multi-LAGAN: efficient tools for large-scale multiple alignment of genomic DNA. *Genome Res* 13, 721–731.

Chapter 11

Prediction of Regulatory Elements

Albin Sandelin

Abstract

Finding the regulatory mechanisms responsible for gene expression remains one of the most important challenges for biomedical research. A major focus in cellular biology is to find functional transcription factor binding sites (TFBS) responsible for the regulation of a downstream gene. As wet-lab methods are time consuming and expensive, it is not realistic to identify TFBS for all uncharacterized genes in the genome by purely experimental means. Computational methods aimed at predicting potential regulatory regions can increase the efficiency of wet-lab experiments significantly. Here, methods for building quantitative models describing the binding preferences of transcription factors based on literature-derived data are presented, as well as a general protocol for scanning promoters using cross-species comparison as a filter (phylogenetic footprinting).

Key words: Transcriptional regulation, transcription factor binding site, phylogenetic footprinting, PWM, PSSM.

1. Introduction

1.1. Transcriptional Regulation

Transcriptional regulation remains one of the most important processes to study for greater understanding of cellular biology. In experimental cell biology, the characterization of promoters and TFBS for specific genes remains the focus for many researchers. The finalizations of the genome projects have motivated further efforts in this area from both experimental and computational biologists.

The standard computational model describing the binding properties of transcription factors (TFs) is the position frequency matrix (PFM) *(1, 2)*. The PFM is constructed by counting nucleotide occurrences in the columns of an alignment of known TFBS

Jonathan M. Keith (ed.), *Bioinformatics, Volume II: Structure, Function and Applications, vol. 453*
© 2008 Humana Press, a part of Springer Science + Business Media, Totowa, NJ
Book doi: 10.1007/978-1-60327-429-6 Springerprotocols.com

collected from experiments. (*See* **Section 3.1.3** for details and examples.) Applied to a sequence, such a model can with high sensitivity predict sites that can be bound by the factor in question. In fact, the prediction is highly correlated with the *in vitro* binding properties of the TF *(3)*. However, the target of most gene regulation research is to identify binding sites that are functionally important *in vivo*, which is a greater challenge. Searching a sequence with a standard model as above is in most cases not selective enough to produce meaningful results: Functional sites can be found but will be indistinguishable from other predictions *(2)*. Although several approaches have been employed to increase selectivity (such as modeling the interaction between TFs) *(4)*, the application of cross-species comparison as a selection filter (phylogenetic footprinting) *(5, 6)* is arguably the most widely used method.

The goal of this chapter is to introduce the reader to the elementary methods for building models describing the DNA-binding properties of TFs, and to explain how these models can be used to scan genomic DNA for putative TFBS using cross-species comparison to increase predictive power. Readers should be aware that the methodology presented below is one of many possible ways of predicting regulatory elements. Due to limited space, discussions of alternative methods and the reasons for choosing certain analysis method are brief. Thus, for a deeper understanding, it is recommended to look further into other reviews of this area *(1, 2)*.

2. Assumed Knowledge

For understanding the concepts in this chapter fully, the reader is anticipated to have:

1. A basic understanding of fundamental molecular biology concepts and processes, in particular the commonly used definitions in transcriptomics and gene regulation science: the following textbooks and reviews provide a good starting point *(7–9)*.

2. A rudimentary understanding of probability theory. This is helpful to grasp the background of the models presented, in particular the construction of position weight matrices and the concept of information content. However, for casual users, most of the steps presented can be executed by a combination of Web tools such as MEME (http://meme.sdsc.edu) *(10)*, (http://jaspar.genereg.net) *(11)*, and (http://phylofoot.org/consite) *(12)*.

3. Knowledge of a programming language. This is essential for readers interested in applying these methods on larger scales. Perl is recommended since it is the primary programming language used in bioinformatics. In particular, many of the methods introduced in this chapter are included as methods in the open-source TFBS Perl programming module *(13)*.

3. Methods

In this section, a step-by-step guide is supplied for two important parts of the process of finding regulatory elements: *(1)* building a model for a transcription factor (TF) of interest, and *(2)* predicting regulatory targets for the TF. It is important to remember that many variants of these procedures are possible. Examples of the two processes are presented in **Sections 3.1.7** and **3.2.8**.

3.1. Construction of Models for TF-DNA Interactions

This section assumes that the goal is to construct a binding model for a specific TF using literature-derived data. For particularly well-studied factors, such models might already be available from open-access databases such as JASPAR *(11)* or commercial databases such as TRANSFAC *(14)*.

3.1.1. Selecting Sites for Building a Model for the Binding Properties of a Specific TF

In practice, sites used for building models from TFs are either based on individual functional sites extracted from several published articles or sets of sites identified by one or several rounds of *in vitro* site selection *(15)*. Extraction of sites from scientific literature is non-trivial since many different experimental techniques can be used for finding sites. Before the search, it is recommended to make a list of what type of evidence is required for accepting a site; for instance, is a simple gelshift enough or are mutations studies combined with reporter gene assays required? Sites should ideally be extracted with some flanking sequence (at least 5 bp on either side), as this will help in the subsequent alignment step. At a bare minimum, five sites or more are required to make a makeshift matrix model, whereas >20 sites in most cases are more than adequate for a representative model. If *in vitro* selection assays are used, the number of sites is usually not a problem. Instead, the danger lies in overselection; that is, the sites selected might not be representative of *in vivo* binding. It is hard to establish general rules applicable to every situation, but a rule of the thumb is to avoid studies with a large number of selection iterations in which all the sequences reported are close to identical.

3.1.2. Alignment of Sites

Once a set of sites that are judged to be reasonably representative are collected, it is necessary to align these for subsequent construction of models. To build a matrix model, an ungapped alignment is necessary. Although there are a multitude of alignment programs available, using a pattern finder such as MEME *(10)* is recommend as this tool automatically gives ungapped alignments and even calculates the count matrix (described in the next step). Pattern finders often have the option to ignore some of the sequences when finding patterns. In this specific context, this option is not recommended, since we already have defined all sites to be representative. As pattern finders are probabilistic, it is recommended to apply the same set of sequences several times to the algorithm to see that the solution is stable. Likewise, try different settings: in particular, vary the width of the pattern. Ideally, the resulting count matrix should have a clear sequence signature centered on the matrix and small or little signal in the edges. The best way to assess such properties by eye is by using graphical sequence logo representations, as described in **Section 3.1.5**.

3.1.3. Creating a Position Frequency Matrix

As described in the Introduction, the PFM is the model that will form the basis of all other analyses in this chapter. Each column in the alignment is assessed independently, counting the instances of each nucleotide A,C,G,T. The counts will form the corresponding column in the PFM $F = [f_{n,c}]$. (For clarity, F will have the same width as the alignment and one row corresponding to each nucleotide, indexed by c and n, respectively.) This can be done by hand or by using specific programs that often are coupled to specific alignment tools (e.g., the *prophecy* tool within the EMBOSS program suite; http://emboss.sourceforge.net/) *(16)*. As described, pattern finders will in general give such a PFM as the default output.

3.1.4. Conversion to PWM

The PFM can be considered as the primary database representation for TF binding models, but is not in itself optimized for scanning genome sequences (e.g., it is not normalized). Most researchers convert PFMs to position weight matrices, PWMs (also referred to as PSSMs). For the conversion, we must estimate the chance $p_{n,c}$ to draw nucleotide n in position c. To correct for small sample sizes (and to eliminate null values), a sampling collection known as a pseudocount is added in this process. There are several ways of defining the pseudocount function (*see* reference *(17)* for a review): Here it is simply the square root of the number of sites, N. Assuming that bases in the genome occur with approximately the same frequency ($p_A = p_C = p_G = p_T = 0.25$), we estimate $p_{n,c}$:

$$p_{n,c} = \frac{f_{n,c} + 0.25 * \sqrt{N}}{N + \sqrt{N}}$$

Given $p_{n,c}$, we can convert the cell values to the log-odds matrix which is the PWM. Conversion of the PFM $f_{n,c}$ to the PWM $W=[w_{n,c}]$ is made using the following equation:

$$w_{n,c} = \log_2 \left(\frac{p_{n,c}}{0.25} \right)$$

This operation can be made using the TFBS Perl module *(13)*. A quantitative score S for a potential site is calculated by summing up the relevant cell values in W, analogous to calculating the probability of observing the site using the relevant $p(n,c)$ values:

$$S = \sum_{c=1}^{d} w_{l_c,c}$$

where d is the number of columns in W and l_c the nucleotide found in position c in the potential site.

For sequences longer than the PWM, the PWM is slid over the sequence in one bp increments, producing one score per potential site. Given representative sites, this type of score is correlated to the binding strength of the factor *(1)*. As different matrices will have different score ranges, scores are normalized:

$$S_{norm} 100 * \frac{(S - S_{min})}{(S_{max} - S_{min})}$$

where S is a score obtained from some subsequence using the PWM W, S_{min} is the lowest possible score from W, and S_{max} is the highest possible score. A reasonable threshold for S_{norm} scores for identifying promising sites is >70–80%.

3.1.5. Visualization of the Model

Although the PWM can be considered the "workhorse" model representation that will be used in all sequence analysis, it is hard to assess PWMs by eye. The most popular way of representing profiles graphically is the sequence logo, as described in reference *(18)*, based on the concept of information content. Such representations are helpful throughout the modeling steps and not only as a "picture" of the final model. Sequence logos can be generated using web tools (such as WebLogo, http://weblogo. berkeley.edu), programming packages *(13)* or calculated "by hand" as in the following. Intuitively, the information content of a column in the PFM corresponds to how restricted the binding requirements in this position are. For instance, a column with only As will have the maximum information content (two bits), whereas a column with equal numbers of the four nucleotides will have an information content of 0 bits. The information content for a column c in F can be calculated as:

$$I_c = 2 + \sum_{n=A,C,G,T} p(f_{n,c}) \log_2 p(f_{n,c})$$

where F=[f_{n,c}]is the PFM constructed above from the alignment and:

$$p(f_{n,c}) = \frac{f_{n,c}}{N}$$

The total information content for F is the summed information content of all columns in F. Sequence logos are based on letter stacks corresponding to each column in F. The height of the stack is the information content of the column, whereas the individual heights of letters within the stack are proportional to their count distribution in the column.

3.1.6. Sharing of Models

A major problem in gene regulation bioinformatics is the lack of high-quality models for most TFs. Thus, sharing of constructed models is encouraged for the advancement of the whole field. The open-access JASPAR and PAZAR (http://www.cisreg.ca/) databases have interfaces for sharing both models and sites.

3.1.7. Example of Model Building

Sites for the FOXD1 factor were retrieved from an *in vitro* experiment *(19)*, and aligned using the AnnSpec program *(20)* (**Fig. 11.1A**), forming a PFM (*see* **Fig. 11.1B**) that was converted into a PWM (*see* **Fig. 11.1C**). The model was visualized as a sequence logo (*see* **Fig. 11.1D**). The model was deposited in the JASPAR database with the identifier MA0031 and will be part of the PAZAR database.

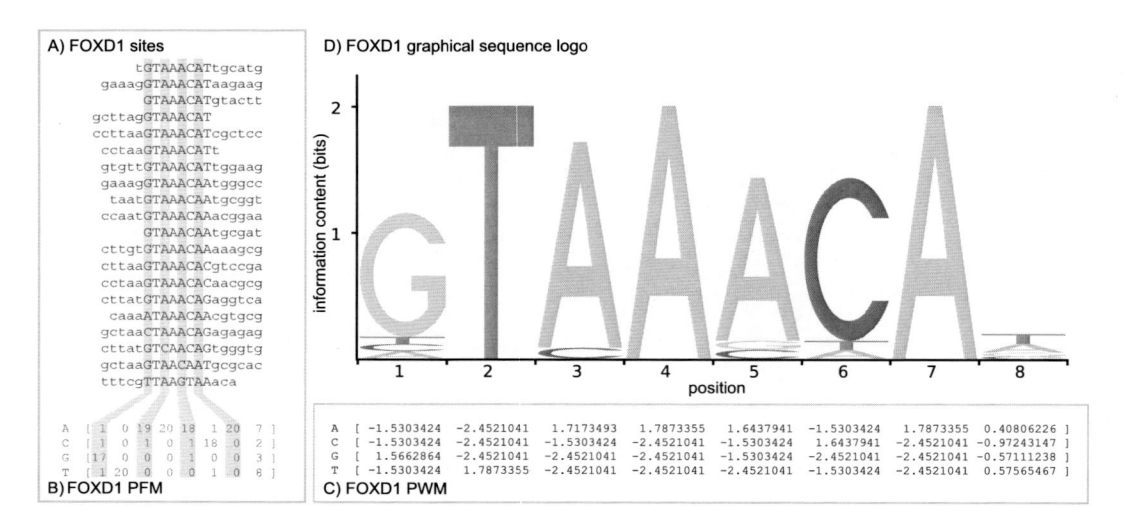

Fig. 11.1. Example of building and visualizing a model describing the binding preferences of the FOXD1 transcription factor. (**A**) FOXD1 sites as identified by Pierrou et al. *(19)*, aligned using AnnSpec. (**B**) The position frequency model (PFM) derived from the alignment. (**C**) The position weight matrix (PWM) derived from the PFM. (**D**) The PFM visualized as a sequence logo.

3.2. Predicting Regulatory Sequences with a Predefined Model

This section assumes that the goal is to find regulatory regions for factors whose binding specificity are modeled as PWMs (as in the preceding), responsible for regulating a specific gene. Furthermore, given the large amount of false predictions *(2)* using PWMs, phylogenetic footprinting between a pair of species is used to filter predictions.

3.2.1. Build Model

Construct or retrieve one or several PWM models of the TF(s) of interest. It is not advisable to have several models describing the binding specificity of the same factor in the analysis, as the results may be confusing.

3.2.2. Locate Genes and Promoters of Interest in Reference Species

Assuming that the reference species is human, use the UCSC browser (http://genome.ucsc.edu/) *(21)* to locate cDNAs corresponding to the gene of interest. Since phylogenetic footprinting will be employed in the following steps, locate promising candidate regions for analysis using both cDNA locations and conservation levels (at present, these are shown as Phastcons scores) *(22)*. Once a region is selected, it is possible to use the browser to extract the relevant genomic sequences for the next steps. This is a part of the analysis in which experience in looking at regulatory regions will help, as there are no simple rules to locate candidate regions. It is generally assumed that most regulatory regions lie upstream of the transcription start site. However, this is only a rough guide, as enhancers can be located several kbp away up or downstream of the gene, and regulatory regions can also be found in introns or even 3′ UTRs. At this time, the most accurate way of finding 5′ termini of genes is either by mapping full-length cDNA *(23)* to the genome (as in the UCSC *(21)* or Ensembl *(24)* browsers) or using sequenced 5′ ends of genes (CAGE) *(25)*.

3.2.3. Locating Orthologous Sequences

Once a candidate region in the reference species is selected, use the pre-generated whole-genome NET alignment track in the UCSC browser *(21)* to locate corresponding regions in the species of interest (*see* **Note 1** for additional comments). For analyzing the regulation of human genes, mouse is generally a good choice for phylogenetic footprinting *(5)*. (*See* **Note 2** for additional comments.) Once the region has been located in the other genome, the same procedure can be applied to extract the genomic sequence.

3.2.4. Aligning Promoters

Align the two sequences using an appropriate program. Choose an alignment program that is optimized for aligning non-coding genomic sequence and can handle stretches of low similarities. Examples include the global aligner LAGAN (http://lagan.stanford.edu/) *(26)* and the local aligners BlastZ *(27)* and TBA (http://www.bx.psu.edu/miller_lab) *(28)*. Aligners optimized for highly similar sequences such as BLAST *(29)* are not ideal for this type of analysis. Make sure that the alignment output format is compatible with the chosen tools in the next step before proceeding.

3.2.5. Identify Conserved Sub-sequences within the Alignment

The most common approach to selecting the conserved sub-sections of the genome is to use a sliding window approach. In short, a window with fixed width is slid over the alignments, counting the number of identical aligned nucleotides. The center nucleotide of the window is then labeled with the percent nucleotide identity calculated from the window. The window is then slid 1 bp further along the alignments. All nucleotides having a label higher than a certain threshold are considered conserved and used in the next step in the analysis. A typical choice of setting for human–mouse comparison is a 50-bp window and a conservation cutoff of 70% nucleotide identity. There are several applications that can perform this type of analysis, for instance ConSite *(12)* and rVista *(30)*, as well as programming modules *(13)*. An alternative approach is to segment the genome using multiple change-point analysis.

3.2.6. Scanning Conserved Sequences with PWM Models

Scan the conserved subsequences identified in the preceding using the scoring method presented in **Section 3.1.4** or another variant. This can either be done just using the reference sequence, but it is recommend to scan the sequences of both species at corresponding positions as implemented in the ConSite system *(12)*: in short, to accept a predicted site it must be predicted at overlapping positions in both genomes. The cutoff for the site scoring is dependent on the scoring system used and the model employed. If using the method presented here, typical site score cutoffs range from 70% to 80% (*see* **Note 3**). The TFBS Perl module has this scoring system implemented in a convenient system for power users *(13)*.

3.2.7. Interpretation of Results

Interpretation of the results of site searching can be daunting, since matrix-based models are prone to produce many spurious predictions, even though phylogenetic footprinting on average removes about 90% of the predictions compared with single sequence analysis *(5)*. It is generally beneficial to learn as much abut the gene and promoter as possible, or discuss the outcome with experimenters who are familiar with the gene in question and may be able to identify unlikely results. This type of information can often narrow the search by locating just a few factors of interest instead of searching with all available models. Access to expression data of both the factor(s) modeled and the target gene can help further, since if the factor and target are not co-expressed, the interaction is unlikely. In general, one should not expect to retrieve a single unambiguous answer by this type of analysis, especially if no prior data are available. However, in many cases the analysis can accelerate experimental elucidation of gene regulation by significantly reducing the search space to a few potential regulatory regions.

3.2.8. Example of Predicting a Functional Regulatory Region

The model describing the binding preferences of FOXD1 was applied to the proximal promoter of the human insulin receptor gene, located using the UCSC browser (**Fig. 11.2**). Corresponding mouse sequence was located using the NET alignment track within the browser. The sequences were aligned and analyzed as above using the ConSite webtool, using 80% as score threshold and 70% identity as conservation threshold within a 50-bp wide sliding window. The conservation profile showed a conserved region about 350 bp upstream of the transcription start site. With the given thresholds, only a single FOXD1 site was predicted within this region. Given that insulin insensitivity is coupled to the expression of forkhead proteins (the same type of transcription factors as FOXD1), the region and in particular the site is a promising candidate for targeted wet-lab experiments. The site was recently proved to be critical for the regulation of the insulin receptor (*31*).

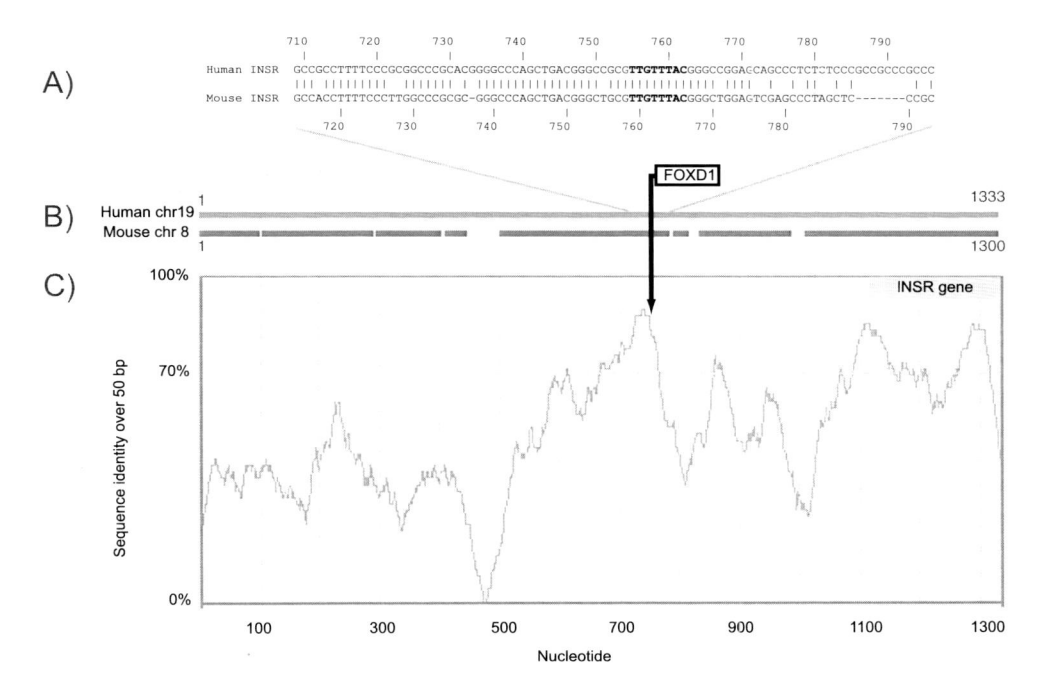

Fig. 11.2. Identification of a potential regulatory region and a putative binding site for FOXD1. Orthologous human and mouse promoter regions for the INSR gene were aligned and analyzed within the ConSite system, using the FOXD1 model defined in Fig. 11.1. (**A**) Detailed alignment of the region containing the predicted site. (**B**) Schematic plot of the whole alignment. (**C**) Conservation profile of the alignment, using a 50-bp sliding window. Only regions with >70% sequence identity were scanned with the FOXD1 model. The location of the sole predicted FOXD1 site is indicated by the arrow.

4. Notes

1. Finding orthologous genes: Finding orthologous genes and genome sequences is a major subfield in bioinformatics and has been reviewed extensively *(32)*. Technically, orthologous genes do not have to have the same function; however, the basis for phylogenetic footprinting is that genes and corresponding promoters have similar selective pressure. The concept of evolutionary turnover of both TFBS *(33)* and transcription start sites *(34)* between orthologous sequences adds further complications, although it is likely that such turnover events are rare. Therefore, it is necessary to be cautious in the analysis: Genes that have diverged in function are not ideal for phylogenetic footprinting analysis.

2. Ideal evolutionary distances for phylogenetic footprinting: Part of the selection of orthologous promoters is the choice of appropriate evolutionary distances for optimal filtering. It is obvious that the use of too closely related species does not give any significant effect: The results will be similar to scanning a single sequence. (Although *see* **Chapter 10** describing a different type of TFBS analysis, "phylogenetic shadowing," based on sets of similar sequences.) Conversely, too divergent species will not have retained enough promoter similarity for meaningful alignments, even if the TFBS in themselves are retained. However, species divergence is only a starting point, since different promoters evolve at different rates. For instance, for detection of enhancers of critical developmental genes, comparisons between species so diverged as human and fish are helpful *(35)*.

3. Thresholds settings for TF models: It is helpful to vary the thresholds for the models employed. Models with high information content might be overly selective, and it is often useful to lower the constraints in these cases. Some computational biologists prefer a *p*-value representation instead of a score-based cutoff. It is an open question in the field what system is the most effective. Other systems, such as the MATCH program, have preset thresholds for each matrix in the TRANSFAC database *(14)*.

Acknowledgments

Thanks to Ann Karlsson for comments on the text.

References

1. Stormo, G. D. (2000) DNA binding sites: representation and discovery. *Bioinformatics* 16, 16–23.

2. Wasserman, W. W., Sandelin, A. (2004) Applied bioinformatics for the identification of regulatory elements. *Nat Rev Genet* 5, 276–287.

3. Fickett, J. W. (1996) Quantitative discrimination of MEF2 sites. *Mol Cell Biol* 16, 437–441.

4. Wasserman, W. W., Fickett, J. W. (1998) Identification of regulatory regions which confer muscle-specific gene expression. *J Mol Biol* 278, 167–181.

5. Lenhard, B., Sandelin, A., Mendoza, L., et al. (2003) Identification of conserved regulatory elements by comparative genome analysis. *J Biol* 2, 13.

6. Wasserman, W. W., Palumbo, M., Thompson, W., et al. (2000) Human-mouse genome comparisons to locate regulatory sites. *Nat Genet* 26, 225–228.

7. Alberts, B., Johnson, A., Lewis, J., et al. (2002) *Molecular Biology of the Cell*. Garland Publishing, New York.

8. Kadonaga, J. T. (2004) Regulation of RNA polymerase II transcription by sequence-specific DNA binding factors. *Cell* 116, 247–257.

9. Lewin, B. (2004) *Genes VIII*. Pearsson Education, New York.

10. Bailey, T. L., Elkan, C. (1995) The value of prior knowledge in discovering motifs with MEME. *Proc Int Conf Intell Syst Mol Biol* 3, 21–29.

11. Vlieghe, D., Sandelin, A., De Bleser, P. J., et al. (2006) A new generation of JASPAR, the open-access repository for transcription factor binding site profiles. *Nucleic Acids Res* 34, D95–97.

12. Sandelin, A., Wasserman, W. W., Lenhard, B. (2004) ConSite: web-based prediction of regulatory elements using cross-species comparison. *Nucleic Acids Res* 32, W249–252.

13. Lenhard, B., Wasserman, W. W. (2002) TFBS: Computational framework for transcription factor binding site analysis. *Bioinformatics* 18, 1135–1136.

14. Matys, V., Kel-Margoulis, O. V., Fricke, E., et al. (2006) TRANSFAC and its module TRANSCompel: transcriptional gene regulation in eukaryotes. *Nucleic Acids Res* 34, D108–110.

15. Pollock, R., Treisman, R. (1990) A sensitive method for the determination of protein-DNA binding specificities. *Nucleic Acids Res* 18, 6197–6204.

16. Rice, P., Longden, I., Bleasby, A. (2000) EMBOSS: the European Molecular Biology Open Software Suite. *Trends Genet* 16, 276–277.

17. Durbin, R., Eddy, S. R., Krogh, A., et al. (2001) *Biological Sequence Analysis*. Cambridge Press, Cambridge, UK.

18. Schneider, T. D., Stephens, R. M. (1990) Sequence logos: a new way to display consensus sequences. *Nucleic Acids Res* 18, 6097–7100.

19. Pierrou, S., Hellqvist, M., Samuelsson, L., et al. (1994) Cloning and characterization of seven human forkhead proteins: binding site specificity and DNA bending. *Embo J* 13, 5002–5012.

20. Workman, C. T., Stormo, G. D. (2000) ANN-Spec: a method for discovering transcription factor binding sites with improved specificity. *Pac Symp Biocomput* 467–478.

21. Hinrichs, A. S., Karolchik, D., Baertsch, R., et al. (2006) The UCSC Genome Browser Database: update 2006. *Nucleic Acids Res* 34, D590–598.

22. King, D. C., Taylor, J., Elnitski, L., et al. (2005) Evaluation of regulatory potential and conservation scores for detecting cis-regulatory modules in aligned mammalian genome sequences. *Genome Res* 15, 1051–1060.

23. Carninci, P., Kasukawa, T., Katayama, S., et al. (2005) The transcriptional landscape of the mammalian genome. *Science* 309, 1559–1563.

24. Birney, E., Andrews, D., Caccamo, M., et al. (2006) Ensembl 2006. *Nucleic Acids Res* 34, D556–561.

25. Carninci, P., Sandelin, A., Lenhard, B., et al. (2006) Genome-wide analysis of mammalian promoter architecture and evolution, *Nat Genet* 38, 626–635.

26. Brudno, M., Do, C. B., Cooper, G. M., et al. (2003) LAGAN and Multi-LAGAN: efficient tools for large-scale multiple alignment of genomic DNA. *Genome Res* 13, 721–731.

27. Schwartz, S., Kent, W. J., Smit, A., et al. (2003) Human-mouse alignments with BLASTZ. *Genome Res* 13, 103–107.

28. Blanchette, M., Kent, W. J., Riemer, C., et al. (2004) Aligning multiple genomic sequences with the threaded blockset aligner. *Genome Res* 14, 708–715.

29. Altschul, S. F., Gish, W., Miller, W., et al. (1990) Basic local alignment search tool. *J Mol Biol* 215, 403–410.

30. Loots, G. G., Ovcharenko, I., Pachter, L., et al. (2002) rVista for comparative sequence-based discovery of functional transcription factor binding sites. *Genome Res* 12, 832–839.

31. Puig, O., Tjian, R. (2005) Transcriptional feedback control of insulin receptor by dFOXO/FOXO1. *Genes Dev* 19, 2435–2446.

32. Koonin, E. V. (2005) Orthologs, paralogs, and evolutionary genomics. *Annu Rev Genet* 39, 309–338.

33. Dermitzakis, E. T., Clark, A. G. (2002) Evolution of transcription factor binding sites in mammalian gene regulatory regions: conservation and turnover. *Mol Biol Evol* 19, 1114–1121.

34. Frith, M., Ponjavic, J., Fredman, D., et al. (2006) Evolutionary turnover of mammalian transcription start sites. *Genome Res* 16, 713–722.

35. Gomez-Skarmeta, J. L., Lenhard, B., Becker, T. S. (2006) New technologies, new findings, and new concepts in the study of vertebrate cis-regulatory sequences. *Dev Dyn* 235, 870–885.

Chapter 12

Expression and Microarrays

Joaquín Dopazo and Fátima Al-Shahrour

Abstract

High throughput methodologies have increased by several orders of magnitude the amount of experimental microarray data available. Nevertheless, translating these data into useful biological knowledge remains a challenge. There is a risk of perceiving these methodologies as mere factories that produce never-ending quantities of data if a proper biological interpretation is not provided.

Methods of interpreting these data are continuously evolving. Typically, a simple two-step approach has been used, in which genes of interest are first selected based on thresholds for the experimental values, and then enrichment in biologically relevant terms in the annotations of these genes is analyzed in a second step. For various reasons, such methods are quite poor in terms of performance and new procedures inspired by systems biology that directly address sets of functionally related genes are currently under development.

Key words: Functional interpretation, functional genomics, multiple testing, gene ontology.

1. Introduction

Genes operate within the cell in an intricate network of interactions that is only recently starting to be envisaged (1–3). It is a widely accepted fact that co-expressed genes tend to play common roles in the cell (4, 5). In fact, this causal relationship has been used to predict gene function from patterns of co-expression (6, 7).

In this scenario, a clear necessity exists for methods and tools that can help to understand large-scale experiments (microarrays, proteomics, etc.) and formulate genome-scale hypotheses from a systems biology perspective (8). Dealing with genome-scale data in this context requires the use of functional annotations of the genes, but this step must be approached from within a systems

Jonathan M. Keith (ed.), *Bioinformatics, Volume II: Structure, Function and Applications, vol. 453*
© 2008 Humana Press, a part of Springer Science + Business Media, Totowa, NJ
Book doi: 10.1007/978-1-60327-429-6 Springerprotocols.com

biology framework in which the collective properties of groups of genes are considered.

DNA microarray technology can be considered the dominant paradigm among genome-scale experimental methodologies. Although many different biological questions can be addressed through microarray experiments, three types of objectives are typically undertaken in this context: class comparison, class prediction, and class discovery *(9)*. The two first objectives fall in the category of supervised methods and usually involve the application of tests to define differentially expressed genes, or the application of different procedures to predict class membership on the basis of the values observed for a number of key genes. Clustering methods belong to the last category, also known as unsupervised analysis because no previous information about the class structure of the data set is used.

The extensive use of microarray technology has fueled the development of functional annotation tools that essentially study the enrichment of functional terms in groups of genes defined by the experimental values. Examples of such terms with functional meaning are gene ontology (GO) *(10)*, KEGG pathways *(11)*, CisRed motifs *(12)*, predictions of transcription factor binding sites *(13)*, Interpro motifs *(14)*, and others. Programs such as ontoexpress *(15)*, FatiGO *(16)*, GOMiner *(17)*, and others, can be considered representatives of a family of methods designed for this purpose *(18)*. These methods are used *a posteriori* over the genes of interest previously selected in a first step, in order to obtain some clues to the interpretation of the results of microarray experiments. Typical criteria for selection are differential expression (class comparison), co-expression (class discovery), or others. By means of this simple two-step approach, a reasonable biological interpretation of a microarray experiment can be reached. Nevertheless, this approach has a weak point: the list of genes of interest. This list is generally incomplete, because its definition is affected by many factors, including the method of analysis and the threshold imposed. In the case of class discovery analysis, the use of biological annotations has also been employed as a cluster validation criterion *(19)*.

Thus, the difficulties for defining repeatable lists of genes of interest across laboratories and platforms even using common experimental and statistical methods *(20)* has led several groups to propose different approaches that aim to select genes, taking into account their functional properties. The Gene Set Enrichment Analysis (GSEA) *(21, 22)*, although not free of criticisms *(23)*, pioneered a family of methods devised to search for groups of functionally related genes with a coordinate (although not necessarily high) over- or under-expression across a list of genes ranked by differential expression coming from microarray experiments. Different tests have recently been proposed with this aim for microarray data *(24–30)* and also for ESTs *(31)* and some of them are available in Web servers *(32, 33)*. In particular, the FatiScan procedure *(32, 33)*, which implements a segmentation test *(24)*, can deal with ordered lists of genes independently from the type of data that originated them. This interesting property allows its application to other types of data apart from microarrays. Also recently, biological information *(34, 35)* or phenotypic information *(36)* has been used as a constitutive part of clustering algorithms in the case of class discovery (clustering) analysis.

2. Methods

2.1. Threshold-Based Functional Analysis

The final aim of a typical genome-scale experiment is to find a molecular explanation for a given macroscopic observation (e.g., which pathways are affected by the deprivation of glucose in a cell, what biological processes differentiate a healthy control from a diseased case). The interpretation of genome-scale data is usually performed in two steps: In a first step genes of interest are selected (because they co-express in a cluster or they are significantly over- or under-expressed when two classes of experiments are compared), usually ignoring the fact that these genes are acting cooperatively in the cell and consequently their behaviors must be coupled to some extent (*see* **Note 1**). In this selection, stringent thresholds to reduce the false-positives ratio in the results are usually imposed. In a second step, the selected genes of interest are compared with the background (typically the rest of the genes) in order to find enrichment in any functional term. This comparison to the background is required because otherwise the significance of a proportion (even if high) cannot be determined. The procedure is illustrated in **Fig. 12.1** for the interpretation of either co-expressing genes found by clustering (*see* **Fig. 12.1A**) or genes selected by differential expressing among two pre-defined classes of experiments (*see* **Fig. 12.1B**).

This comparison is made by means of the application of tests such as the hypergeometric, χ^2, binomial, Fisher's exact test, and others. There are several available tools, reviewed in *(18)*. Among these methods, the most popular ones (more cited in the literature) are Onto-express *(15)* (http://vortex.cs.wayne.edu/ontoexpress/) and FatiGO *(16)* (http://www.fatigo.org). These tools use various biological terms with functional meaning such as GO *(10)*, KEGG pathways *(11)*, etc.

Although this procedure is the natural choice for analyzing clusters of genes, its application to the interpretation of differential gene expression experiments causes an enormous loss of information because a large number of false-negatives are obtained in

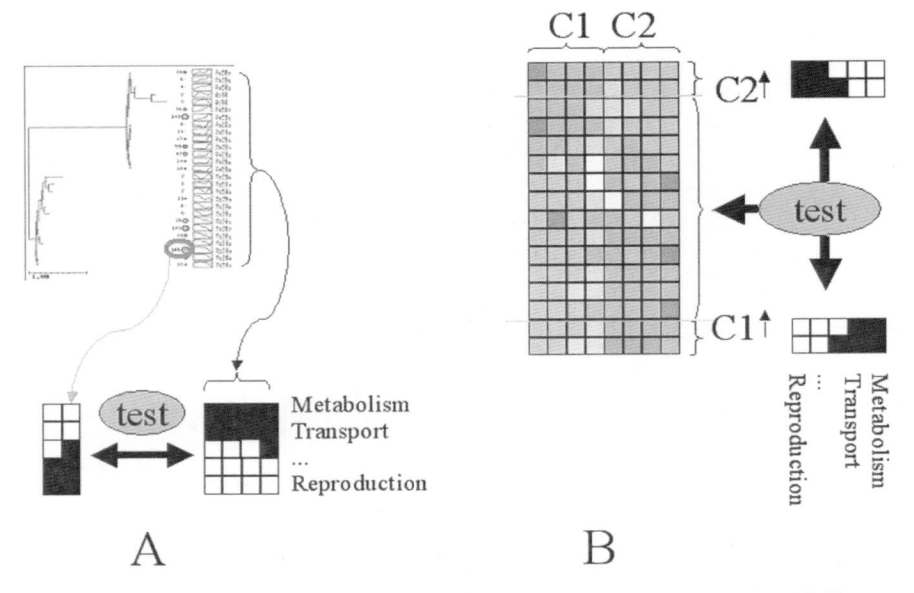

Fig. 12.1. Two-step procedure for the functional annotation of distinct microarray experiments. (**A**) The unsupervised approach: functional interpretation of clusters of co-expressing genes. A cluster of genes is selected for functional analysis and the genes inside it are checked for significant enrichment of functional terms with respect to the background (the rest of the genes). (**B**) The supervised approach: functional annotation of genes differentially expressed between two classes (C1 and C2). The differentially expressed genes are checked for enrichment of functional terms with respect to the background, the genes not showing differential expression between (**A**) and (**B**).

order to preserve a low ratio of false-positives (and the noisier the data the worse is this effect).

2.2. Threshold-Free Functional Analysis

From a systems biology perspective, this way of understanding the molecular basis of a genome-scale experiment is far from efficient. Methods inspired by systems biology focus on collective properties of genes. Functionally related genes need to carry out their roles simultaneously in the cell and, consequently, they are expected to display a coordinated expression. Actually, it is a long recognized fact that genes with similar overall expression often share similar functions *(4, 37, 38)*. This observation is consistent with the hypothesis of modularly behaving gene programs, where sets of genes are activated in a coordinated way to carry out functions. Under this scenario, a different class of hypotheses, not based on genes but on blocks of functionally related genes, can be tested. Thus, lists of genes ranked by any biological criteria (e.g., differential expression when comparing cases and healthy controls) can be used to directly search for the distribution of blocks of functionally related genes across the list without imposing any arbitrary threshold. Any macroscopic observation that causes this ranked list of genes will be the consequence of cooperative action of genes that are part of functional classes, pathways, etc. Consequently, each functional class "responsible" for the macroscopic observation will be found in the extremes of the ranking with highest probability. The previous imposition of a threshold based on the rank values that does not take into account the cooperation among genes is thus avoided under this perspective. **Fig. 12.2** illustrates this concept. Genes are arranged by differential expression between the classes C1 and C2. On the right part of the figure, labels for two different functional classes have been placed at the positions in the list where genes playing the corresponding roles are situated. Function A is completely unrelated to the experiment because it appears simultaneously over-expressed in class C1 and C2 and also in intermediate positions. Conversely, function B is predominantly performed by genes with high expression in class C2, but scarcely appears in C1. This observation clearly points to function B as one of the molecular bases of the macroscopic observation made in the experiment. Instead of trying to select genes with extreme values

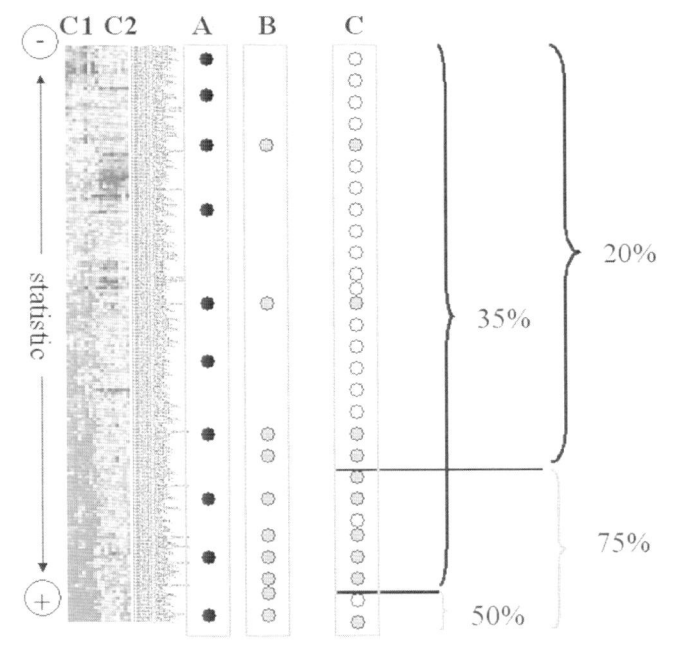

Fig. 12.2. Threshold-free procedure for the functional annotation of class comparison experiments. (**A**) Functional label unrelated to the experiment from which the rank of genes was obtained. (**B**) Functional label related to the experiment. (**C**) Schematic representation of two partitions of the segmentation test. In the first partition, 35% of the genes in the upper segment are annotated with the term C, whereas 50% of the genes in the lower segment are annotated with this term. This difference in percentages is not statistically significant. In the second partition, the 75% of the genes in the lower segment are annotated as C, whereas only 20% are annotated as C in the upper partition. These differences in the proportions are high enough to be considered significant after the application of a Fisher's exact test.

of differential expression, systems biology-inspired methods will directly search for blocks of functionally related genes significantly cumulated in the extremes of a ranked list of genes.

Other methods that have been proposed for this purpose, such as the GSEA *(21, 22)* or the SAFE *(39)* method, use a nonparametrical version of a Kolmogorov-Smirnov test. Other strategies are also possible, such as direct analysis of functional terms weighted with experimental data *(28)*, Bayesian methods *(29)*, or model-based methods *(26)*. Conceptually simpler and quicker methods with similar accuracy have also been proposed, such as the parametrical counterpart of the GSEA, the PAGE *(30)* or the segmentation test, Fatiscan *(24)*, which is discussed in the next section.

2.3. FatiScan: A Segmentation Test

A simple way of studying the asymmetrical distributions of blocks of genes across a list of them is to check if, in consecutive partitions, one of the parts is significantly enriched in any term with respect to the complementary part. **Figure 12.2C** illustrates this concept with the representation of ordered genes in which gray circles represent those genes annotated with a particular biological term and open circles represent genes with any other annotation. In the first partition, 35% of the genes in the upper segment are annotated with the term of interest, whereas 50% of the genes in the lower segment are annotated with this term. This difference in percentages is not statistically significant. However, in the second partition, the differences in the proportions are high enough to be considered significant (75% vs. 20%): The vast majority of the genes with the annotation are on the lower part of the partition.

The segmentation test used for threshold-free functional interpretation consists of the sequential application of the FatiGO *(16)* test to different partitions of an ordered list of genes. The FatiGO test uses a Fisher's exact test over a contingency table for finding significantly over- or under-represented biological terms when comparing the upper side with the lower side of the list, as defined by any partition. Previous results show that a number between 20 and 50 partitions often gives optimal results in terms of sensitivity and results recovered *(24)*. Given that multiple terms (T) are tested in a predefined number of partitions (P), the unadjusted p-values for a total of $T \times P$ tests must be corrected. The widely accepted FDR *(40)* can be used for this purpose. Nevertheless, carrying out a total of $T \times P$ tests would correspond to the most conservative scenario, in a situation in which no *a priori* functional knowledge of the system is available. Usually many terms can initially be discarded from the analysis due to prior information or just by common sense.

The FatiScan test has two fundamental advantages when compared to alternative methods based on Kolmogorov-Smirnov or related tests. On one hand, this method does not require an extreme non-uniform distribution of genes. It is able to find different types of asymmetries in the distribution of groups of genes across the list of data. On the other hand, and more importantly, this method does not depend on the original data from which the ranking of the list was derived. The significance of the test depends only on the ranking of the genes in the list and the strategy used for performing the partitions. This means that, in addition to DNA microarray data, this method can be applied to any type of genome-scale data in which a value can be obtained for each gene. FatiScan is available within the Babelomics package *(32, 33)* for functional interpretation of genome-scale experiments (http://www.babelomics.org).

2.4. Differential Gene Expression in Human Diabetes Samples

We have used data from a study of gene expression in human diabetes *(21)* in which a comparison between two classes (17 controls with normal tolerance to glucose versus 26 cases composed of 8 with impaired tolerance and 18 with type 2 diabetes mellitus, DM2) did not detect even a single gene differentially expressed. We ordered the genes according to their differential expression between cases and controls. A t-test, as implemented in the T-Rex tool from the GEPAS package *(41–43)* was used for this purpose (*see* **Note 2**). The value of the statistic was used as the ranking criteria for ordering the list. As in the original analysis *(21)*, we were unable to find individual genes with a significant differential expression (differentially expressed genes with an adjusted p-value < 0.05).

A total of 50 partitions of the ranked list were analyzed with the FatiScan algorithm for over- or under-expression of KEGG pathways and GO terms. The following KEGG pathways were found to be significantly over-expressed in healthy controls vs. cases: *oxidative phosphorylation, ATP synthesis,* and *Ribosome.* Contrarily, *Insulin signalling pathway* was up-regulated in diseased cases. When GO terms were analyzed, we found as significantly up-regulated in healthy controls: *oxidative phosphorylation* (GO:0006119), *nucleotide biosynthesis* (GO:0009165) (biological process ontology), *NADH dehydrogenase (ubiquinone) activity* (GO:0008137), *nuclease activity* (GO:0004518) (molecular function ontology), and *mitochondrion* (GO:0005739) (cellular component ontology). Some of the terms were redundant with the KEGG pathways, although here we have also the *ubiquinone* class, which does not appear in KEGG. Since FatiScan implements more functional terms, we also analyzed Swissprot keywords and found *Ubiquinone, Ribosomal protein, Ribonucleoprotein, Mitochondrion,* and *Transit peptide* as over-expressed in healthy controls vs. cases. Other alternative methods give similar results. *Oxidative phosphorylation* and *mitochondrion* are found by GSEA *(21)*, PAGE *(30)*, and other statistics *(27)*. *Nucleotide biosynthesis* can be assimilated to other datasets found by these three methods

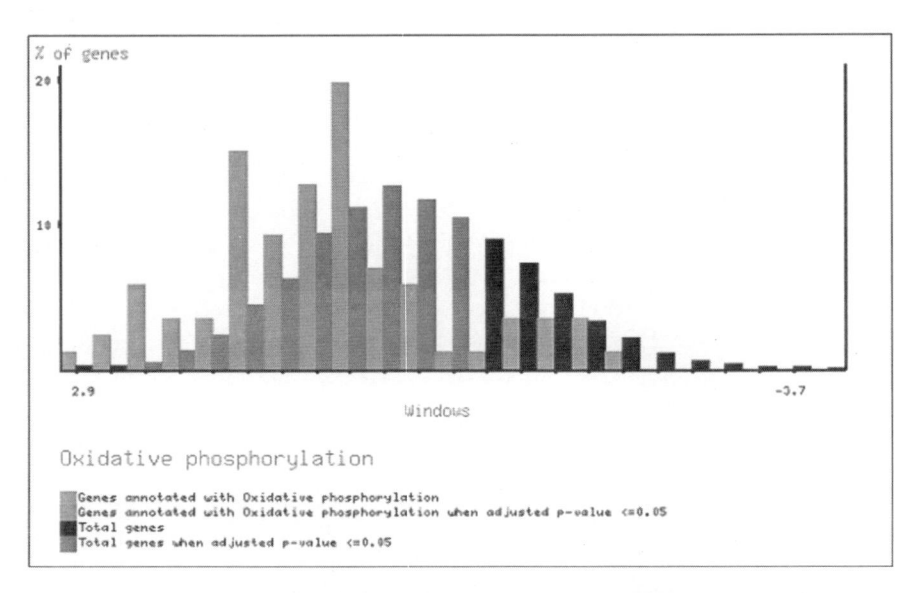

Fig. 12.3. Comparison between the background distribution of GO terms *(light gray bars)* and the distribution of *oxidative phosphorylation* GO term *(dark gray and black bars)*. The last distribution is clearly shifted towards highest values of the *t* statistic (horizontal axis), corresponding to high expression in healthy controls. The transition of colors black to gray makes reference to the values of the *t*-statistic for which the partitions were found to be significant.

(21, 27, 30) based on a set of functional categories developed by *(22)*. The rest of the terms were only found by FatiScan.

If, for example, the distribution of the GO term *oxidative phosphorylation* is compared to the background distribution of GO terms (**Fig. 12.3**) a clear trend to the over-expression of the complete pathway with high values of the *t*-statistic, corresponding to genes over-expressed in healthy controls, can be clearly observed.

3. Notes

1. Despite the fact that microarray technologies allow data on the behavior of all the genes of complete genomes to be obtained, hypotheses are still tested as if the problem consisted of thousands of independent observations. The tests applied involve only single genes and ignore all the available knowledge cumulated in recent years on their cooperative behavior, which is stored in several repositories (GO, KEGG, protein interaction databases, etc.). Since this process involves carrying out a large number of tests, severe corrections must be imposed to reduce the number of false-positives. Later, the genes selected using these procedures were functionally

analyzed in a second step. Much information is lost in both steps during this process. Recently, however, methods have been proposed that take into account the properties of the genes and address different biological questions not in a gene-centric manner, but in a function-centric manner for class comparison *(21, 22, 24–27, 29, 30)*, class assignation *(44)* and class discovery *(34–36)*.

A systems biology approach to the analysis of microarray data will in the future tend to use more information on the cooperative properties of the genes beyond their simple functional description. Thus it is expected that a deeper knowledge of the interactome or the transcriptional network of the genomes will contribute to more realistic biological questions being addressed by microarray experiments, resulting in more complete and accurate answers.

2. Given the number of steps necessary for the proper analysis of a microarray experiment (normalization, the analysis itself, and the functional interpretation), integrated packages are preferable in order to avoid problems derived from the change of formats. Among the most complete packages available on the web is GEPAS *(32, 41–43)* that offers various options for normalization, supervised and unsupervised analysis (http://www.gepas.org) and is coupled to the Babelomics suite *(32, 33)* for functional interpretation of genome-scale experiments (http://www.babelomics.org), in which various tests for two-steps of threshold-free functional interpretation are implemented.

Acknowledgments

This work is supported by grants from MEC BIO2005-01078, NRC Canada-SEPOCT Spain, and Fundación Genoma España.

References

1. Stelzl, U., Worm, U., Lalowski, M., Haenig, C., et al. (2005) A human protein-protein interaction network: a resource for annotating the proteome. *Cell* 122, 957–968.

2. Hallikas, O., Palin, K., Sinjushina, N., et al. (2006) Genome-wide prediction of mammalian enhancers based on analysis of transcription-factor binding affinity. *Cell* 124, 47–59.

3. Rual, J. F., Venkatesan, K., Hao, T., et al. (2005) Towards a proteome-scale map of the human protein-protein interaction network. *Nature* 437, 1173–1178.

4. Lee, H. K., Hsu, A. K., Sajdak, J., et al. (2004) Coexpression analysis of human genes across many microarray data sets. *Genome Res* 14, 1085–1094.

5. Stuart, J. M., Segal, E., Koller, D., et al. (2003) A gene-coexpression network for global discovery of conserved genetic modules. *Science* 302, 249–255.

6. van Noort, V., Snel, B., Huynen, M. A. (2003) Predicting gene function by conserved co-expression. *Trends Genet* 19, 238–242.

7. Mateos, A., Dopazo, J., Jansen, R., et al. (2002) Systematic learning of gene functional classes from DNA array expression data by using multilayer perceptrons. *Genome Res* 12, 1703–1715.

8. Westerhoff, H. V., Palsson, B. O. (2004) The evolution of molecular biology into systems biology. *Nat Biotechnol* 22, 1249–1252.

9. Golub, T. R., Slonim, D. K., Tamayo, P., et al. (1999) Molecular classification of cancer: class discovery and class prediction by gene expression monitoring. *Science* 286, 531–537.

10. Ashburner, M., Ball, C. A., Blake, J. A., et al. (2000) Gene ontology: tool for the unification of biology. The Gene Ontology Consortium. *Nat Genet* 25, 25–29.

11. Kanehisa, M., Goto, S., Kawashima, S., et al. (2004) The KEGG resource for deciphering the genome. *Nucleic Acids Res* 32, D277–280.

12. Robertson, G., Bilenky, M., Lin, K., et al. (2006) cisRED: a database system for genome-scale computational discovery of regulatory elements. *Nucleic Acids Res* 34, D68–73.

13. Wingender, E., Chen, X., Hehl, R., et al. (2000) TRANSFAC: an integrated system for gene expression regulation. *Nucleic Acids Res* 28, 316–319.

14. Mulder, N. J., Apweiler, R., Attwood, T. K., et al. (2005) InterPro, progress and status in 2005. *Nucleic Acids Res* 33, D201–205.

15. Draghici, S., Khatri, P., Martins, R. P., et al. (2003) Global functional profiling of gene expression. *Genomics* 81, 98–104.

16. Al-Shahrour, F., Diaz-Uriarte, R., Dopazo, J. (2004) FatiGO: a web tool for finding significant associations of Gene Ontology terms with groups of genes. *Bioinformatics* 20, 578–580.

17. Zeeberg, B. R., Feng, W., Wang, G., et al. (2003) GoMiner: a resource for biological interpretation of genomic and proteomic data. *Genome Biol* 4, R28.

18. Khatri, P., Draghici, S. (2005) Ontological analysis of gene expression data: current tools, limitations, and open problems. *Bioinformatics* 21, 3587–3595.

19. Bolshakova, N., Azuaje, F., Cunningham, P. (2005) A knowledge-driven approach to cluster validity assessment. *Bioinformatics* 21, 2546–2547.

20. Bammler, T., Beyer, R. P., Bhattacharya, S., et al. (2005) Standardizing global gene expression analysis between laboratories and across platforms. *Nat Methods* 2, 351–356.

21. Mootha, V. K., Lindgren, C. M., Eriksson, K. F., et al. (2003) PGC-1alpha-responsive genes involved in oxidative phosphorylation are coordinately downregulated in human diabetes. *Nat Genet* 34, 267–273.

22. Subramanian, A., Tamayo, P., Mootha, V. K., et al. (2005) Gene set enrichment analysis: a knowledge-based approach for interpreting genome-wide expression profiles. *Proc Natl Acad Sci U S A* 102, 15545–15550.

23. Damian, D., Gorfine, M. (2004) Statistical concerns about the GSEA procedure. *Nat Genet* 36, 663.

24. Al-Shahrour, F., Diaz-Uriarte, R., Dopazo, J. (2005) Discovering molecular functions significantly related to phenotypes by combining gene expression data and biological information. *Bioinformatics* 21, 2988–2993.

25. Goeman, J. J., Oosting, J., Cleton-Jansen, A. M., et al. (2005) Testing association of a pathway with survival using gene expression data. *Bioinformatics* 21, 1950–1957.

26. Goeman, J. J., van de Geer, S. A., de Kort, F., et al. (2004) A global test for groups of genes: testing association with a clinical outcome. *Bioinformatics*, 20, 93–99.

27. Tian, L., Greenberg, S. A., Kong, S. W., et al. (2005) Discovering statistically significant pathways in expression profiling studies. *Proc Natl Acad Sci U S A* 102, 13544–13549.

28. Smid, M., Dorssers, L. C. (2004) GO-Mapper: functional analysis of gene expression data using the expression level as a score to evaluate Gene Ontology terms. *Bioinformatics* 20, 2618–2625.

29. Vencio, R., Koide, T., Gomes, S., et al. (2006) BayGO: Bayesian analysis of ontology term enrichment in microarray data. *BMC Bioinformatics* 7, 86.

30. Kim, S. Y., Volsky, D. J. (2005) PAGE: parametric analysis of gene set enrichment. *BMC Bioinformatics* 6, 144.

31. Chen, Z., Wang, W., Ling, X. B., et al. (2006) GO-Diff: Mining functional differentiation between EST-based transcriptomes. *BMC Bioinformatics* 7, 72.

32. Al-Shahrour, F., Minguez, P., Tarraga, J., et al. (2006) BABELOMICS: a systems biology perspective in the functional annotation of genome-scale experiments. *Nucleic Acids Res* 34, W472–476.

33. Al-Shahrour, F., Minguez, P., Vaquerizas, J. M., et al. (2005) BABELOMICS: a suite of web tools for functional annotation and

analysis of groups of genes in high-throughput experiments. *Nucleic Acids Res*, 33, W460–464.

34. Huang, D., Pan, W. (2006) Incorporating biological knowledge into distance-based clustering analysis of microarray gene expression data. *Bioinformatics* 22, 1259–1268.

35. Pan, W. (2006) Incorporating gene functions as priors in model-based clustering of microarray gene expression data. *Bioinformatics* 22, 795–801.

36. Jia, Z., Xu, S. (2005) Clustering expressed genes on the basis of their association with a quantitative phenotype. *Genet Res* 86, 193–207.

37. Eisen, M. B., Spellman, P. T., Brown, P. O., et al. (1998) Cluster analysis and display of genome-wide expression patterns. *Proc Natl Acad Sci U S A* 95, 14863–14868.

38. Wolfe, C.J., Kohane, I. S., and Butte, A. J. (2005) Systematic survey reveals general applicability of "guilt-by-association" within gene coexpression networks. *BMC Bioinformatics* 6, 227.

39. Barry, W. T., Nobel, A. B., and Wright, F. A. (2005) Significance analysis of functional categories in gene expression studies: a structured permutation approach. *Bioinformatics* 21, 1943–1949.

40. Benjamini, Y., Yekutieli, D. (2001) The control of the false discovery rate in multiple testing under dependency. *Ann Stat* 29, 1165–1188.

41. Herrero, J., Al-Shahrour, F., Diaz-Uriarte, R., et al. (2003) GEPAS: A web-based resource for microarray gene expression data analysis. *Nucleic Acids Res* 31, 3461–3467.

42. Herrero, J., Vaquerizas, J. M., Al-Shahrour, F., et al. (2004) New challenges in gene expression data analysis and the extended GEPAS. *Nucleic Acids Res* 32, W485–491.

43. Vaquerizas, J. M., Conde, L., Yankilevich, P., et al. (2005) GEPAS, an experiment-oriented pipeline for the analysis of microarray gene expression data. *Nucleic Acids Res* 33, W616–620.

44. Lottaz, C., Spang, R. (2005) Molecular decomposition of complex clinical phenotypes using biologically structured analysis of microarray data. *Bioinformatics* 21, 1971–1978.

Chapter 13

Identifying Components of Complexes

Nicolas Goffard and Georg Weiller

Abstract

Identifying and analyzing components of complexes is essential to understand the activities and organization of the cell. Moreover, it provides additional information on the possible function of proteins involved in these complexes. Two bioinformatics approaches are usually used for this purpose. The first is based on the identification, by clustering algorithms, of full or densely connected sub-graphs in protein–protein interaction networks derived from experimental sources that might represent complexes. The second approach consists of the integration of genomic and proteomic data by using Bayesian networks or decision trees. This approach is based on the hypothesis that proteins involved in a complex usually share common properties.

Key words: Interacting network, clustering algorithm, integration, Bayesian network, decision tree.

1. Introduction

Protein complexes play a key role in many, if not all, cellular processes. They consist of proteins that interact with each other to form a physical structure and perform a specific function (e.g., the ribosome for gene translation, the proteasome for protein degradation, or ATP synthase for energy production). Identifying and analyzing components of complexes is essential to understand the activities and organization of the cell. It also may allow us to infer the possible function of an unknown protein belonging to a complex, as subunits of a multi-protein complex are generally involved in the same biological role.

For a few years now, huge experimental efforts have been carried out in order to characterize protein complexes for several model organisms (*Saccharomyces cerevisiae (1, 2)*, *Escherichia coli*

Jonathan M. Keith (ed.), *Bioinformatics, Volume II: Structure, Function and Applications, vol. 453*
© 2008 Humana Press, a part of Springer Science + Business Media, Totowa, NJ
Book doi: 10.1007/978-1-60327-429-6 Springerprotocols.com

(3), and *Drosophila melanogaster (4)*). These identifications were performed using techniques mainly based on purification by affinity coupled with mass spectrometry of entire multi-protein complexes *(5)*. However, the amount of data available is still limited. Therefore, several bioinformatics methods were developed in order to identify components of complexes, mostly in *Saccharomyces cerevisiae*, for which large genomic and proteomic datasets are available. Two of these approaches are presented here. The first one is based on the identification of molecular modules in protein–protein interaction networks derived from experimental sources that might represent complexes. Various clustering algorithms were developed to detect full (cliques) or densely connected (clusters) sub-graphs in those networks in order to predict protein complex membership *(6–10)*. However, these strategies rely on high throughput experimental data that are incomplete and frequently contain high rates of false-positives. To overcome this limitation, other genomic and proteomic data (e.g., essentiality, localization or expression) could be integrated in order to predict protein complexes *(11–13)*. This second approach is based on the hypothesis that proteins involved in the same complex share common properties. Indeed, they usually reside in the same cellular compartment and if a complex is essential, then its subunits should also be essential. (The *essentiality* of a protein is defined by the non-viability of the cell after transposon or gene deletion experiments.) Moreover, subunits of complexes have a significant tendency to be coordinated in terms of their mRNA expression levels *(14)*.

It is important to remember that the identification of complex components is a prediction of whether two proteins are in the same complex and not necessarily whether there is a physical contact between them.

2. Databases

1. Protein–protein interactions:

   a. The Biomolecular Interaction Network Database (BIND) *(15)*. Interaction data are available at http://bind.ca. Files containing BIND records are released daily (the file name defines its release date, i.e., yyyymmdd) and are available in plain-text, tab-delimited format. Files with names of the form *yyyymmdd.nrints.txt* contain a non-redundant list of interactions in BIND.

   b. The Database of Interacting Proteins (DIP) *(16)*. DIP datasets are available at http://dip.doe-mbi.ucla.edu/dip/, in XIN, tab-delimited and/or PSI-MI (MIF) formats.

Species-specific sets are obtainable. Registration is required to access these datasets; it is free for members of the academic community.

  c. IntAct *(17)*. IntAct interaction data files are available in HUPO PSI MI 1.0 XML format at ftp://ftp.ebi.ac.uk/pub/databases/intact/current/xml/, grouped by species.

2. Complex data:

  a. The MIPS (Munich Information Center for Protein Sequences) complexes catalog provides the currently known protein complexes in yeast *(18)*. These data are available at ftp://ftpmips.gsf.de/yeast/catalogs/complexcat/ and are described by two files. The file *complexcat.scheme* contains complex identifiers and definitions, whereas the file *complex_data_ddmmyyyy* describes components of complexes (ddmmyyyy represents the release date).

3. Essentiality data for *S. cerevisiae* are available from the MIPS database *(18)* at ftp://ftpmips.gsf.de/yeast/catalogs/gene_disruption/. It corresponds to the list of proteins having the same knockout phenotype. The file format is the same as for complexes (*gene_disruption.scheme* and *gene_disruption_data_ddmmyyyy*).

4. Subcellular localization information is available from the MIPS catalogs *(18)* for yeast at ftp://ftpmips.gsf.de/yeast/catalogs/subcell/. (*See* the preceding for the file description: *subcellcat.scheme* and *subcell_data_ddmmyyyy*.)

5. Expression data are available from the Stanford Micro-array Database *(19)*, which stores and provides results from publications at http://smd.stanford.edu.

3. Methods

3.1. Predicting Molecular Complexes in Protein Interaction Networks

With the development of proteomics techniques, it is now possible to create maps of protein–protein interaction networks. The identification of modules in these maps enables the prediction of components of complexes. Several algorithms have been developed for this purpose.

In order to identify all fully connected sub-graphs (cliques) (**Fig. 13.1A**), Spirin and Mirny *(7)* described a method enumerating all cliques. It is also possible to identify components of complexes from interaction networks by detecting clusters even if they are not fully connected (*see* **Fig. 13.1B**). In this case, it is necessary to identify highly connected sub-graphs that have

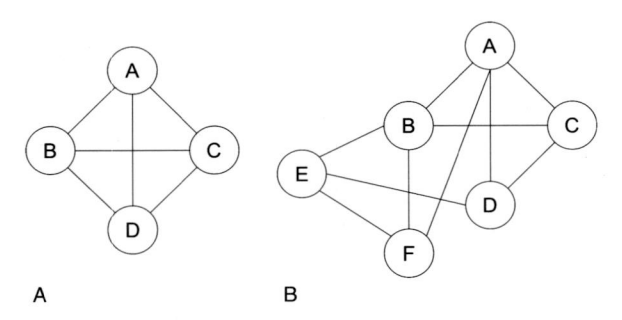

Fig. 13.1. Examples of modules in interaction networks. Protein–protein interactions can be represented as graphs in which nodes and links correspond to proteins and interactions. Modules in these networks might represent protein complexes. (**A**) Fully connected graph (clique) representing a protein complex with four subunits (*ABCD*). All proteins are interacting with all the proteins in the complex. (**B**) Densely connected graph (cluster) representing a complex with six subunits (*ABCDEF*).

more interactions within themselves and fewer with the rest of the network. Bader and Hogue *(8)* developed a clustering algorithm named Molecular Complex Detection (MCODE) to identify these modules and Spirin and Mirny proposed an approach based on an optimization Monte Carlo procedure to find a set of *n* vertices that maximizes the function of connection density in the cluster *(7)*.

3.1.1. Graph Processing

1. A non-redundant list of protein–protein interactions is generated with the databases available (*see* **Notes 1** and **2**).

2. A protein–protein network is represented as a graph, in which vertices (nodes) are proteins and edges (links) are interactions between these proteins (*see* **Note 3**).

3.1.2. Identify Fully Connected Sub-graphs (Cliques)

1. This method starts from the smallest size of clusters. To identify cliques with size 4, all pairs of edges are considered successively. For each pair of binary interactions (e.g., A–B and C–D), all possible pairs are tested to check if there are edges between them (*A* and *C*, *A* and *D*, *B* and *C*, and *B* and *D*). If all edges are present, it is a clique (*ABCD*).

2. For every clique of size 4 (*ABCD*), all known proteins are considered successively. For every protein *E*, if there is an edge between E and all proteins in the clique, then *ABCDE* is a clique with size 5.

3. This search is continued until the biggest clique is found.

4. The resulting cliques include many redundant cliques because cliques with size *n* contain *n* cliques with size *n-1*. Thus, to find all non-redundant sub-graphs, it is necessary to remove all cliques included in a larger one.

3.1.3. Detect Densely Connected Regions

The density of connections in the cluster is measured by the parameter $Q = 2m/(n(n-1))$, where n is the number of proteins in the cluster and m is the number of interactions between them.

3.1.3.1. Clustering Algorithm: Molecular Complex Detection

1. All proteins are weighted based on their local network density using the highest k-core of the protein neighborhood. A k-core is a graph in which every node has degree at least k and the highest k-core of a graph is the most densely connected sub-graph that is also a k-core. The core-clustering coefficient of a protein (p) is defined as the density of the highest k-core of the immediate neighborhood of p that includes p. Here the *immediate neighborhood* of p consists of p and the proteins that directly interact with p. The final weight given to a protein is the product of the protein core-clustering coefficient and the highest k-core level (k_{max}) in the immediate neighborhood of the protein.

2. This step starts from the complex with the highest weighted protein. Recursively, proteins whose weight is above a given threshold are included in the complex. This threshold is a given percentage less than the weight of the initial protein. If a protein is included, its neighbors are recursively checked in the same manner to determine whether they should be included in the complex. A protein is not checked more than once, so complexes cannot overlap.

3. This process stops once no more proteins can be added to the complex based on the given threshold and is repeated for the next highest weighted unchecked protein in the network. The protein weight threshold parameter defines the density of the resulting complex. A post-processing step could be applied to filter these complexes (*see* **Note 4**).

4. Resulting complexes are scored and ranked. The density of the complex score is defined as the product of the complex sub-graph, $C = (V,E)$ and the number of proteins in the complex sub-graph ($D_C \times |V|$). This ranks larger, denser complexes higher.

3.1.3.2. Monte Carlo Simulation

1. This method starts with a connected set of n proteins randomly selected on the graph and proceeds by "moving" these nodes along the edges of the graph to maximize Q.

2. For each protein pair A,B from this set, the shortest path L_{AB} between A and B in the sub-graph is calculated. The sum of all shortest paths L_{AB} for this set is denoted as L_0.

3. At every time step, one of n proteins is randomly selected, and one protein is picked at random out from among its neighbors. The new sum of all shortest paths, L_1, is calculated

for the sub-graph obtained by replacing the original node with this neighbor.

   a. If $L_1 < L_0$, the replacement takes place with probability 1.

   b. If $L_1 > L_0$, the replacement takes place with the probability $\exp{-(L_1 - L_0)/T}$, where T is the effective temperature.

4. This process is repeated until the original set converges to a complete sub-graph, or for a predetermined number of steps, after which the tightest sub-graph (the sub-graph corresponding to the smallest L_0) is recorded.

3.2. Integration of Genomic and Proteomic Data

Another approach to identifying components of complexes is based on the integration of genomic and proteomic data, using the fact that proteins involved in a complex share common properties *(11)*.

To predict membership of protein complexes for individual genes, two main methods have been developed, both of which combine different data sources corresponding to protein-pair features. They are based on learning methods and use already known complexes from the hand-curated MIPS complexes catalog *(18)* as a training set. Jansen et al. *(12)* developed a Bayesian approach to combine multiple sources of evidence for interactions (e.g., binary interactions, co-localization, co-essentiality, and co-expression) by comparing them with reference datasets (gold-standards). Zhang et al. *(13)* proposed a method based on a probabilistic decision tree, built from a reference set of complexes, for a list of features.

3.2.1. Bayesian Networks

1. Gold standard datasets are generated to evaluate each feature considered:

   a. Positive gold standard (pos): proteins that are in the same complex. This set is built from the MIPS complex catalog.

   b. Negative gold standard (neg): proteins that do not interact. This set is generated from proteins in separate subcellular compartments.

2. The features f considered have to be binned into discrete categories (*see* **Note 5**).

3. According to the gold standards, the conditional probabilities $P(f \mid pos)$ and $P(f \mid neg)$ are computed for each feature f.

4. Protein-pair features are compared to the gold standards using a likelihood ratio. It is defined as the fraction of gold standard positives having a feature f divided by the fraction of negatives having f. Thus, using Bayesian networks, for n features:

$$L(f_1...f_n) = \frac{P(f_1...f_n \mid pos)}{P(f_1...f_n \mid neg)}$$

If the features are conditionally independent, the Bayesian network is called the naïve network, and L can be simplified to:

$$L(f_1...f_n) = \prod_{i=1}^{n} L(f_i) = \prod_{i=1}^{n} \frac{P(f_i \mid pos)}{P(f_i \mid neg)}$$

5. The combined likelihood ratio is proportional to the estimated odds that two proteins are in the same complex, given multiple sources of information. Therefore, a protein pair is predicted as belonging to the same complex if its combined likelihood ratio exceeds a particular cutoff ($L > L_{cut}$).

3.2.2. Probabilistic Decision Tree

1. Features are represented as binary variables and are organized hierarchically (*see* **Note 6**).

2. A decision tree is generated by successively partitioning all protein pairs in the reference set according to the values (0 or 1) of their particular features. At the start, all protein pairs in the reference set are in a single root node. Recursively, each node N is partitioned into two sub-nodes based on the feature f that gives the greatest reduction of entropy (i.e., maximizes the probability that both proteins belong to the same complex).

3. This decision tree enables all protein pairs to be scored according to their probability of being involved in the same complex. Each protein pair tested is successively assigned to the nodes according to its features. At the end of this process, the protein pair is assigned to a leaf node N and its score corresponds to the probability that a protein pair assigned to this node is part of a known complex.

4. Notes

1. Some bioinformatics tools such as AliasServer *(20)* are available to handle multiple aliases used to refer to the same protein.

2. In order to improve the reliability of this dataset, only those pairs described by at least two distinct experiments should be retained. However, that can appreciably reduce the number of interactions.

3. Proteins with a large number of interactions can be removed in order to avoid spurious clusters with low statistical significance.

4. A post-processing stage can be applied. It consists of filtering complexes that do not contain at least a 2-core (graph of minimum degree 2) or in adding proteins to the resulting

complexes according to certain connectivity criteria. For every protein in the complex, its neighbors are added to the complex if they have not yet been checked and if the neighborhood density is higher than a given threshold.

5. Genomic features have to be binned into discrete categories.

   a. Proteins are divided according to whether they are essential or non-essential (based on their knock-out phenotype).

   b. The localization data can be separate into five compartments: nucleus, cytoplasm, mitochondria, extracellular, and transmembrane.

   c. For mRNA expression, a Pearson correlation is computed for each protein pair across all datasets. Protein pairs are then binned according to the interval in which their correlation coefficient lies.

   d. In principle, this approach can be extended to a number of other features related to interactions (e.g., phylogenetic occurrence, gene fusions, and gene neighborhood).

6. Each feature is mapped to one or more binary variables. For the expression correlation coefficient, intervals are used to convert it into binary groups. These features are hierarchically organized with an edge from attribute i to attribute j, indicating that any protein pair with attribute i also has attribute j.

References

1. Gavin, A. C., Bosche, M., Krause, R., et al. (2002) Functional organization of the yeast proteome by systematic analysis of protein complexes. *Nature* 415, 141–147.

2. Ho, Y., Gruhler, A., Heilbut, A., et al. (2002) Systematic identification of protein complexes in Saccharomyces cerevisiae by mass spectrometry. *Nature* 415, 180–183.

3. Butland, G., Peregrin-Alvarez, J. M., Li, J., et al. (2005) Interaction network containing conserved and essential protein complexes in Escherichia coli. *Nature* 433, 531–537.

4. Veraksa, A., Bauer, A., Artavanis-Tsakonas, S. (2005) Analyzing protein complexes in Drosophila with tandem affinity purification-mass spectrometry. *Dev Dyn* 232, 827–834.

5. Mann, M., Hendrickson, R. C., Pandey, A. (2001) Analysis of proteins and proteomes by mass spectrometry. *Annu Rev Biochem* 70, 437–473.

6. Rives, A. W., Galitski, T. (2003) Modular organization of cellular networks. *Proc Natl Acad Sci U S A* 100, 1128–1133.

7. Spirin, V., Mirny, L. A. (2003) Protein complexes and functional modules in molecular networks. *Proc Natl Acad Sci U S A* 100, 12123–1218.

8. Bader, G. D., Hogue, C. W. (2003) An automated method for finding molecular complexes in large protein interaction networks. *BMC Bioinformatics* 4, 2.

9. Bu, D., Zhao, Y., Cai, L., et al. (2003) Topological structure analysis of the protein-protein interaction network in budding yeast. *Nucleic Acids Res* 31, 2443–2450.

10. Asthana, S., King, O. D., Gibbons, F. D., et al. (2004) Predicting protein complex membership using probabilistic network reliability. *Genome Res* 14, 1170–1175.

11. Jansen, R., Lan, N., Qian, J., et al. (2002) Integration of genomic datasets to predict protein complexes in yeast. *J Struct Funct Genomics* 2, 71–81.

12. Jansen, R., Yu, H., Greenbaum, D., et al. (2003) A Bayesian networks approach for predicting protein-protein interactions from genomic data. *Science* 302, 449–453.

13. Zhang, L. V., Wong, S. L., King, O. D., et al. (2004) Predicting co-complexed protein pairs using genomic and proteomic data integration. *BMC Bioinformatics* 5, 38.

14. Jansen, R., Greenbaum, D., Gerstein, M. (2002) Relating whole-genome expression data with protein-protein interactions. *Genome Res* 12, 37–46.

15. Alfarano, C., Andrade, C. E., Anthony, K., et al. (2005) The Biomolecular Interaction Network Database and related tools 2005 update. *Nucleic Acids Res* 33, D418–424.

16. Salwinski, L., Miller, C. S., Smith, A. J., et al. (2004) The Database of Interacting Proteins: 2004 update. *Nucleic Acids Res* 32, D449–451.

17. Hermjakob, H., Montecchi-Palazzi, L., Lewington, C., et al. (2004) IntAct: an open source molecular interaction database. *Nucleic Acids Res* 32, D452–455.

18. Mewes, H. W., Frishman, D., Guldener, U., et al. (2002) MIPS: a database for genomes and protein sequences. *Nucleic Acids Res* 30, 31–34.

19. Ball, C. A., Awad, I. A., Demeter, J., et al. (2005) The Stanford Micro-array Database accommodates additional micro-array platforms and data formats. *Nucleic Acids Res* 33, D580–582.

20. Iragne, F., Barre, A., Goffard, N., et al. (2004) AliasServer: a web server to handle multiple aliases used to refer to proteins. *Bioinformatics* 20, 2331–2332.

Chapter 14

Integrating Functional Genomics Data

Insuk Lee and Edward M. Marcotte

Abstract

The revolution in high throughput biology experiments producing genome-scale data has heightened the challenge of integrating functional genomics data. Data integration is essential for making reliable inferences from functional genomics data, as the datasets are neither error-free nor comprehensive. However, there are two major hurdles in data integration: heterogeneity and correlation of the data to be integrated. These problems can be circumvented by quantitative testing of all data in the same unified scoring scheme, and by using integration methods appropriate for handling correlated data. This chapter describes such a functional genomics data integration method designed to estimate the "functional coupling" between genes, applied to the baker's yeast *Saccharomyces cerevisiae*. The integrated dataset outperforms individual functional genomics datasets in both accuracy and coverage, leading to more reliable and comprehensive predictions of gene function. The approach is easily applied to multicellular organisms, including human.

Key words: Data integration, function prediction, guilt-by-association, gene association, functional coupling, data correlation, data heterogeneity.

1. Introduction

The ultimate goal of functional genomics is to identify the relationships among all genes of an organism and assign their physiological functions. This goal is certainly ambitious, but seems somewhat more attainable when considering the remarkable advances in automation of molecular biology and innovations in high throughput analysis techniques, which are already producing enormous sets of functional data. Such data include micro-array analyses of gene expression *(1, 2)*, protein interaction maps using yeast two-hybrid *(3–8)*, affinity purification of protein complexes *(9–11)*, synthetic

Jonathan M. Keith (ed.), *Bioinformatics, Volume II: Structure, Function and Applications, vol. 453*
© 2008 Humana Press, a part of Springer Science+Business Media, Totowa, NJ
Book doi: 10.1007/978-1-60327-429-6 Springerprotocols.com

lethal screening *(12, 13)*, and many others, including computational methods that predict gene functions using comparative genomics approaches. These methods are introduced in earlier chapters of this book: prediction of gene function by protein sequence homology (*see* **Chapter 6**), Rosetta Stone proteins or gene fusions (*see* **Chapter 7**), gene neighbors (*see* **Chapter 8**), and phylogenetic profiling (*see* **Chapter 9**)—provide related information with relatively low cost. All of these data enable the prediction of gene function through guilt-by-association—the prediction of a gene's function from functions of associated genes. For example, if we observe a gene associated with other genes known to be involved in ribosomal biogenesis, we might infer the gene is involved in ribosomal biogenesis as well.

Although functional genomics data are accumulating rapidly, the assignment of functions to the complete genome or proteome is still far from complete. One complicating factor is the fact that all functional analyses (both experimental and computational) contain errors and systematic bias. For example, yeast two-hybrid methods can detect associations between physically interacting genes only, whereas genetic methods only occasionally do *(14, 15)*. Integration of diverse functional genomics data can potentially overcome both errors and systematic bias. In practice, data integration improves prediction of gene function by guilt-by-association, generally resulting in stronger inferences and larger coverage of the genome *(16–23)*.

Nevertheless, successfully integrating data is not trivial due to two major problems: heterogeneity and correlation among the data to be integrated. This chapter discusses an approach for integrating heterogeneous and correlated functional genomics data for more reliable and comprehensive predictions of gene function, applying the method to genes of a unicellular eukaryotic organism, the yeast *Saccharomyces cerevisiae*.

2. Methods

2.1. Standardization of Heterogeneous Functional Genomics Data

A major hurdle in data integration is the heterogeneity of the data to be integrated. All data to be integrated must be assessed for relevance and informativeness to the biological hypothesis that one wants to test. In practice, this means choosing a common quantitative test of relevance to apply to each dataset, allowing comparison of the datasets by a single unified scheme. After this data standardization, integration becomes a much easier task. Note that this process is not strictly necessary—many classifiers do not require it—but it greatly simplifies later interpretation of results.

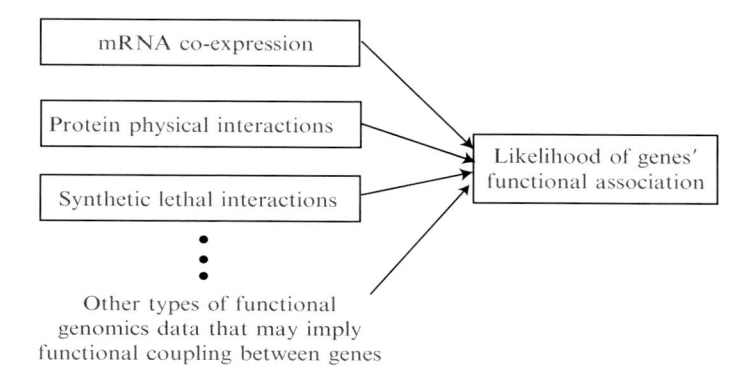

Fig. 14.1. Schematic for applying functional genomics data to estimate functional coupling between genes. Different functional genomics data imply different types of gene associations with varying confidence levels. To predict functional associations, these diverse data are re-interpreted as providing likelihoods of functional associations, and then combined into a single, integrated estimate of the observed coupling between each pair of genes.

The first step in a data integration scheme is choosing the biological hypothesis we want to test. In the prediction of gene functions by guilt-by-association, we are often interested in whether two given genes are functionally associated or not. Although we may be interested in a more specific type of relationship (e.g., protein physical interactions, genetic interactions, pathway associations), we illustrate here a more general notion of "functional" associations, which implicitly includes all of these more specific associations (**Fig. 14.1**). This can be defined more precisely as participation in the same cellular system or pathway.

2.2. Constructing a Reference Set for Data Evaluation

Standardization of heterogeneous data is carried out by evaluating them using a common benchmarking reference. Here, the reference set consists of gene pairs with verified functional associations under some annotation scheme (prior knowledge). The positive reference gene associations are generated by pairing genes that share at least one common annotation and the negatives by pairing annotated genes that do not share any common annotation. The quality of the reference set—both its accuracy and extensiveness—is critical to successful data evaluation. It is also important to keep in mind that the reference set must be consistent throughout the entire data evaluation and integration.

Several different annotations can be used to generate the reference set of functional associations. For example, reference sets might consist of gene pairs sharing functional annotation(s), sharing pathway annotation(s), or found in the same complex(es). The Kyoto Encyclopedia of Genes and Genomes (KEGG) pathway annotations, Gene Ontology (GO), and Munich Information Center for Protein Sequences (MIPS) complex annotation (which is available only for *Saccharomyces cerevisiae*) are useful annotation sets to generate reference sets (*see* **Note 1**). A reference set generated from genes sharing or not sharing KEGG pathway annotation is used for the discussion in this chapter. We define two genes to be functionally associated if we observe at least one KEGG pathway term (or pathway code) annotating both genes.

2.3. A Unified Scoring System for Data Integration

One way to compare heterogeneous functional genomics data is to measure the likelihood that the pairs of genes are functionally associated conditioned on the data, calculated as a log-likelihood score:

$$LLR = \ln\left(\frac{P\big(I|D\big)/P\big(\sim I|D\big)}{P(I)/P(\sim I)}\right),$$

where $P(I|D)$ and $P(\sim I|D)$ are the frequencies of functional associations observed in the given dataset *(D)* between the positive *(I)* and negative *(~I)* reference set gene pairs, respectively. $P(I)$ and $P(\sim I)$ represent the prior expectations (the total frequencies of all positive and negative reference set gene pairs, respectively). A score of zero indicates interaction partners in the data being tested are no more likely than random to be functionally associated; higher scores indicate a more informative dataset for identifying functional relationships.

2.4. Mapping Data-Intrinsic Scores into Log-Likelihood Scores

Many data come with intrinsic scoring schemes, which can easily be converted to log-likelihood scores. **Figure 14.2** describes such a mapping for using 87 DNA micro-array datasets measuring mRNA expression profiles for different cell cycle time points *(1)*. Genes co-expressed under similar temporal and spatial conditions are often functionally associated. The degree of co-expression can be measured as the Pearson correlation coefficient (*PCC*) of the two genes' expression profiles. Gene pairs are sorted by the calculated *PCC*, and then binned into equal-sized sets of gene pairs, starting first with the higher Pearson correlation coefficients (*see* **Note 2**). For each set of gene pairs in a given bin, $P(I|D)$ and $P(\sim I|D)$ are calculated. These probabilities correspond to a given degree of co-expression within this dataset. The log-likelihood score is calculated from these probabilities, along with $P(I)$ and $P(\sim I)$, the unconditional probabilities calculated from the reference gene pairs. If the PCC provides a significant correlation with the log-likelihood score, we then define a regression model with which we can map all data-intrinsic scores (PCC scores) into the standardized scores (log-likelihood scores) (*see* **Note 3**). In this way, the learned relationship between co-expression and functional coupling can be extended to all un-annotated pairs. The datasets re-scored using log-likelihood scores will be used for integrating the datasets.

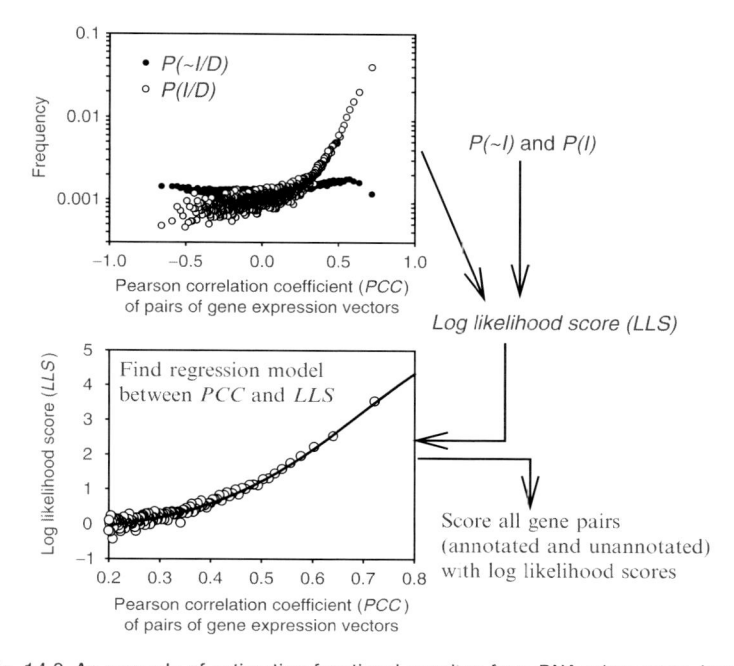

Fig. 14.2. An example of estimating functional coupling from DNA micro-array–based mRNA expression data. Many functionally associated genes tend to be co-expressed through the course of an experiment. Thus, the Pearson correlation coefficient of two genes' expression vectors shows a positive correlation with their tendency to share pathway annotation. Here, yeast mRNA expressions patterns across the cell cycle (1) are compared to their participation in the same KEGG (24) pathways, plotted for all annotated gene pairs as a function of each pair's Pearson correlation coefficient. The frequencies of gene pairs sharing pathway annotation *(P(I|D))* are calculated for bins of 20,000 gene pairs. In contrast, the frequencies of gene pairs not sharing pathway annotation *(P(~I|D))* show no significant correlation with the correlation in expression. The ratio of these two frequencies, corrected by the unconditional frequencies (*P(I)* and *P(~I)*), provides the likelihood score of belonging to the same pathway for the given condition. In practice, we calculate the natural logarithm for the likelihood score to create an additive score (log-likelihood score). Using a regression model for the relationship, we can score all gene pairs (not just the annotated pairs) with log-likelihood scores, indicating the normalized likelihood of functional coupling between genes. (Adapted from Lee, I., Date, S. V., et al. (2004) A probabilistic functional network of yeast genes. *Science* 306, 1555–1558.)

2.5. Integration of Correlated Data Using a Simple Weighted Sum Approach

If datasets to be integrated are completely independent, integration is simple: We can use a naïve Bayes approach, simply adding all available log-likelihood scores for a given gene pair to give the pair's integrated score. However, this assumption of independence among datasets is often unrealistic. We often observe strong correlations among functional genomics data. Integrating correlated datasets using formal models, such as Bayesian networks *(25)*, requires defining the degree of correlation among the datasets. The complexity of this correlation model increases exponentially with the number of datasets. Therefore, defining a correlation model among datasets is often computationally challenging or impractical. If the dataset correlation is relatively weak, the naïve Bayes approach provides reasonable performance in integration. As we accumulate more and more functional genomics data to be integrated, however, this convenient and efficient assumption becomes more troublesome.

An alternative approach that is simple but still accounts for data correlation is a variant of naïve Bayes with one additional parameter accounting for the relative degree of correlation among datasets. In this weighted sum method, we first collect all available log-likelihood scores derived from the various datasets and lines of evidence, then add the scores with a rank-order determined weighting scheme. The weighted sum (WS) score for the functional linkage between a pair of genes is calculated as:

$$WS = \sum_{i=1}^{n} \frac{L_i}{D^{(i-1)}},$$

where L represents the log-likelihood score for the gene association from a single dataset, D is a free parameter roughly representing the relative degree of correlation between the various datasets, and i is the rank index in order of descending magnitude of the n log-likelihood scores for the given gene pair. The free parameter D ranges from 1 to $+\infty$, and is chosen to optimize overall performance (accuracy and coverage, *see* **Note 4**) on the benchmark. When $D = 1$, WS represents the simple sum of all log-likelihood scores and is equivalent to a *naïve* Bayesian integration. We might expect D to exhibit an optimal value of 1 in the case that all datasets are completely independent. As the optimal value of D increases, WS approaches the single maximum value of the set of log-likelihood scores, indicating that the various datasets are entirely redundant (i.e., no new evidence is offered by additional datasets over what is provided by the first set). **Figure 14.3** illustrates the performance using different values of D in integrating datasets with different degrees of correlation. Datasets from similar types of functional genomics studies are often highly correlated. The integration of highly correlated DNA micro-array datasets from different studies *(1)* is illustrated in **Fig. 14.3A**. Here, the assumption of data independence ($D = 1$) provides high scores for a limited portion of the proteome. However, accounting for partial data correlation (e.g., $D = 2$) provides significantly increased coverage of the proteome in identifying functionally associated gene pairs for a reasonable cost of likelihood. In fact, the assumption of complete correlation ($D = +\infty$) among the different gene expression datasets provides a very reasonable trade-off between accuracy and coverage in identifying functionally associated gene pairs. In contrast, **Fig. 14.3B** shows the integration of 11 diverse functional genomics datasets, described in full in Lee et al. *(21)*. This integration is optimal with

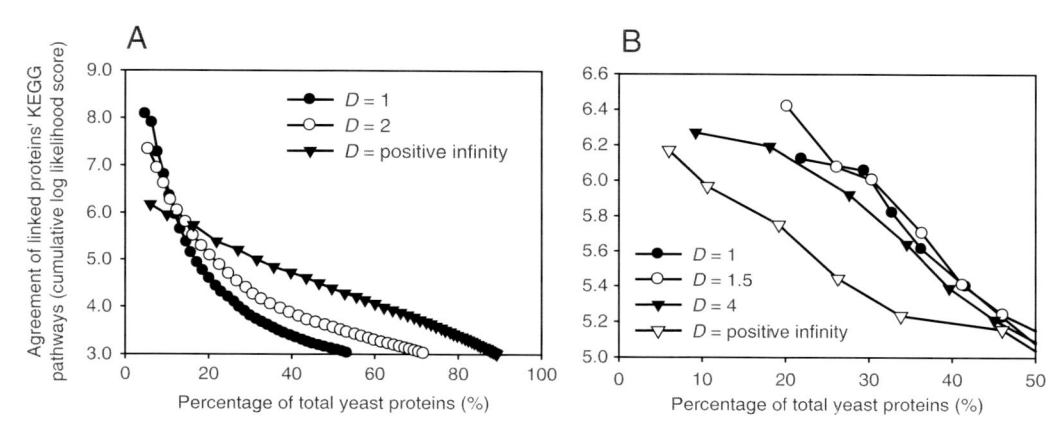

Fig. 14.3. Effects of data correlation on data integration. Here (**A**) 12 different DNA micro-array datasets or (**B**) 11 diverse functional genomics datasets from Lee et al. (21) are integrated by the weighted sum method, applying different values of D. The quality of the integration is assessed by measuring coverage (percentage of total protein-coding genes covered by gene pairs in the data) and accuracy (cumulative log-likelihood scores measured using a reference set of KEGG pathway annotations), with each point indicating 2,000 gene pairs. Integration of DNA micro-array datasets with naïve Bayes approach ($D = 1$) shows high likelihood scores for the top-scoring gene functional pairs, but a rapid decrease in score with increasing proteome coverage. In contrast, integration assuming higher ($D = 2$) or complete correlation ($D =$ positive infinity) provides a dramatic improvement in coverage for a reasonable cost of likelihood. For the integration of 11 diverse functional genomics datasets, the naïve Bayes approach ($D = 1$) shows reasonable performance. However, the best performance is observed for $D = 1.5$. These examples illustrate that better performance in data integration is achieved by accounting for the appropriate degree of correlation among the datasets—similar types of datasets are often highly correlated, requiring high values of D, whereas more diverse types of data can be relatively independent, requiring low values of D or naïve Bayes for the optimal integration.

neither complete independence ($D = 1$) nor complete dependence ($D = + \infty$), integration with $D = 1.5$, accounting for intermediate dependence, achieves optimal performance.

Integrated data generally outperforms the individual datasets. A precision-recall curve (*see* **Note 5**) for the 11 individual datasets and the integrated set demonstrates that data integration improves performance of identifying functional associations in terms of both recall and precision (**Fig. 14.4**).

2.6. Inference of New Gene Function by Guilt-by-Association: A Case Study of PRP43

The gene associations arising from integrating different functional genomics datasets often generate new biological hypotheses. For example, PRP43 was initially implicated only in pre-mRNA splicing *(26)*. Interestingly, many genes involved in ribosomal biogenesis are strongly associated with PRP43 in an integrated gene network *(21)*. Among the 5 genes most strongly associated with PRP43, as ranked by the log-likelihood scores of their associations, three are known to be involved in ribosomal biogenesis (**Table 14.1**). Three recent experimental studies have validated this association, demonstrating that PRP43 is a regulator of both pre-mRNA splicing and ribosomal biogenesis *(27–29)*.

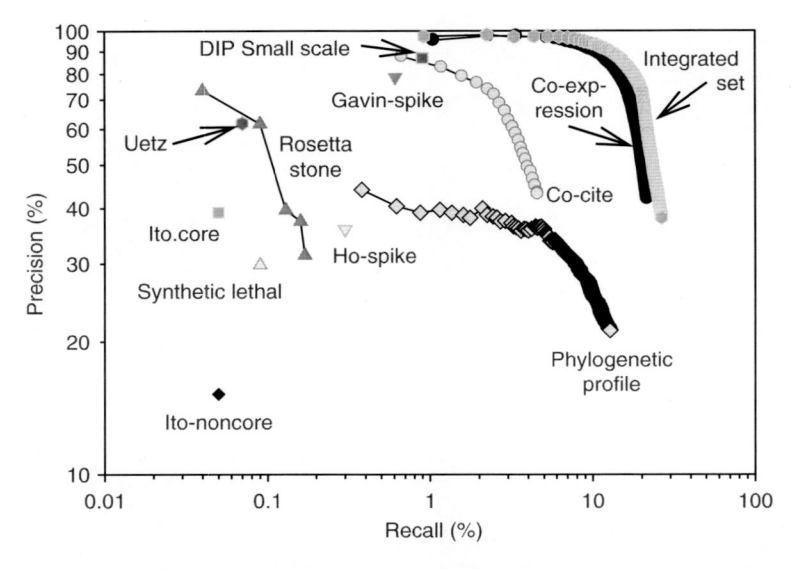

Fig. 14.4. A comparison of the quality of gene functional associations found using 11 diverse functional genomics datasets and an integrated dataset. The predictive power of 11 diverse functional genomics datasets and the integrated dataset *(21)* are assessed by a recall-precision curve (*see* **Note 5**). Measurements are carried out for bins of 2,000 gene pairs. The integrated dataset outperforms all individual datasets, with data integration improving the prediction of functional associations in both accuracy and coverage. Assessment curves are plotted with logarithm of both recall and precision for visualization purpose; thus, the predictive powers are significantly different between co-expression and integrated dataset with linear scale.

Table 14.1
The five genes most strongly associated with Prp43

Rank	Name	Cellular location[a]	Cellular function[b]
1	ERB1	nucleolus	rRNA processing
2	RRB1	nucleolus	ribosome biogenesis
3	LHP1	nucleus	tRNA processing
4	URA7	cytosol	CTP biosynthesis, phospholipid biosynthesis, pyrimidine base biosynthesis
5	SIK1	small nucleolar ribonucleo protein complex	rRNA modification, 35S primary transcript processing, processing of 20S pre-rRNA

[a]Annotated by Gene Ontology cellular component.
[b]Annotated by Gene Ontology biological process.

3. Notes

1. These annotations are hierarchically organized, and choosing different levels of the annotation hierarchy may generate quite different evaluations for the same dataset. Generally speaking, top-level annotations provide extensive coverage but low information specificity (resolution), whereas low-level annotations decrease coverage but increase information specificity. Therefore, the choice of appropriate levels of hierarchical annotation must be considered carefully in order to achieve the optimal trade-off between coverage and specificity. KEGG pathway and GO biological process annotations are available for yeast from ftp://ftp.genome.jp/pub/kegg/pathways/sce/sce_gene_map.tab and http://www.geneontology.org/ontology/process.ontology, respectively. CYGD (the comprehensive yeast genome database) functional categories from MIPS are available at ftp://ftpmips.gsf.de/yeast/catalogues/complexcat/). CYGD lists yeast cellular complexes and their member proteins. Similar data are available for many other organisms. It is striking that the current yeast annotation and reference sets are quite nonoverlapping *(30)*. For example, fewer than half of the KEGG pathway database associations are also contained in the Gene Ontology (GO) annotation set. The low overlap is primarily due to different data mining methods and to inclusion bias among the annotation sets. However, this provides the opportunity to generate more comprehensive reference sets by combination of different annotation sets.

2. Here, an appropriate choice of bin size is important. Generally, a minimum of 100 annotated gene pairs per bin is recommended to obtain statistically reliable frequencies. Overly large bin sizes decrease the resolution of evaluation. Binning must start with the more significant values first. For Pearson correlation coefficient scores, positive values tend to be more meaningful than negative values. (We observe significant signals with Pearson correlation coefficient > 0.3 in **Fig. 14.2.**) Thus, we rank gene pairs with decreasing Pearson correlation coefficients (starting from $+1$) and bin in increasing increments of x pairs ($x = 20,000$ in **Fig. 14.2**).

3. Regression models may suffer from noisy data. For microarray data, gene pairs with negative Pearson correlation coefficient scores are often un-correlated with log-likelihood scores of gene functional associations. As most of the $\sim$18 million of yeast gene pairs belong to this group of noisy data, taking only gene pairs with positive Pearson correlation coefficients generally gives an improved regression model.

4. The ability to identify functional associations is assessed by measuring accuracy for a given cost of coverage. To control for systematic bias (the dataset may predict well for only certain gene groups, e.g., associations between ribosomal proteins), we measure the coverage of total genes as the percentage of all protein-coding genes represented in the dataset. Accuracy is defined as the cumulative log-likelihood score of the dataset. The area under this coverage-accuracy curve line indicates relative performance. We select the value of D that maximizes the area under the coverage-accuracy curve.

5. One formal way to evaluate data coverage and accuracy is by plotting a recall-precision curve. *Recall* (defined as the percentage of positive gene associations in the reference set correctly predicted as positive gene associations in the dataset) provides a measure of coverage and *precision* (defined as the percentage of predicted positive gene associations in the dataset confirmed as true positive gene associations by the reference set) provides a measure of accuracy. The evaluation method should be able to identify any possible over-fitting during data integration, which occurs when the training process simply learns the training set, rather than a more generalized pattern. Over-fitting is tested using a dataset that is completely independent from the training set. The simple way of making an independent test set is to leave out some fraction (e.g., ~30%) of the original training set as a separate test set. However, this reduces the size of the training set, thus decreasing training efficiency. An alternative method is randomly splitting the original training set into k subsets, then iterating training k times, using one subset for the testing and all others for the training. The average measurements from this iterative process, called k-fold cross-validation, provide a fairly unbiased evaluation with minimal loss of training power (*see* **Chapter 15**, **Section 3.4.3** and **Note 3** for more details).

Acknowledgments

This work was supported by grants from the N.S.F. (IIS-0325116, EIA-0219061, 0241180), N.I.H. (GM06779-01), Welch (F1515), and a Packard Fellowship (E.M.M.).

References

1. Gollub, J., Ball, C. A., Binkley, G., et al. (2003) The Stanford Micro-array Database: data access and quality assessment tools. *Nucleic Acids Res* 31, 94–96.

2. Barrett, T., Suzek, T. O., Troup, D. B., et al. (2005) NCBI GEO: mining millions of expression profiles-database and tools. *Nucleic Acids Res* 33, D562–566.

3. Uetz, P., Giot, L., Cagney, G., et al. (2000) A comprehensive analysis of protein-protein interactions in *Saccharomyces cerevisiae*. *Nature* 403, 623–627.

4. Ito, T., Chiba, T., Ozawa, R., et al. (2001) A comprehensive two-hybrid analysis to explore the yeast protein interactome. *Proc Natl Acad Sci USA* 98, 4569–4574.

5. Giot, L., Bader, J. S., Brouwer, C., et al. (2003) A protein interaction map of *Drosophila melanogaster*. *Science* 302, 1727–1736.

6. Li, S., Armstrong, C. M., Bertin, N., et al. (2004) A map of the interactome network of the metazoan *C. elegans*. *Science* 303, 540–543.

7. Rual, J. F., Venkatesan, K., Hao, T., et al. (2005) Towards a proteome-scale map of the human protein-protein interaction network. *Nature* 437, 1173–1178.

8. Stelzl, U., Worm, U., Lalowski, M., et al. (2005) A human protein-protein interaction network: a resource for annotating the proteome. *Cell* 122, 957–968.

9. Gavin, A. C., Bosche, M., Krause, R., et al. (2002) Functional organization of the yeast proteome by systematic analysis of protein complexes. *Nature* 415, 141–147.

10. Ho, Y., Gruhler, A., Heilbut, A., et al. (2002) Systematic identification of protein complexes in Saccharomyces cerevisiae by mass spectrometry. *Nature* 415, 180–183.

11. Bouwmeester, T., Bauch, A., Ruffner, H., et al. (2004) A physical and functional map of the human TNF-alpha/NF-kappa B signal transduction pathway. *Nat Cell Biol* 6, 97–105.

12. Tong, A. H., Evangelista, M., Parsons, A. B., et al. (2001) Systematic genetic analysis with ordered arrays of yeast deletion mutants. *Science* 294, 2364–2368.

13. Tong, A. H., Lesage, G., Bader, G. D., et al. (2004) Global mapping of the yeast genetic interaction network. *Science* 303, 808–813.

14. Wong, S. L., Zhang, L. V., Tong, A. H., et al. (2004) Combining biological networks to predict genetic interactions. *Proc Natl Acad Sci USA* 101, 15682–15687.

15. Kelley, R., Ideker, T. (2005) Systematic interpretation of genetic interactions using protein networks. *Nat Biotechnol* 23, 561–566.

16. Mellor, J. C., Yanai, I., Clodfelter, K. H., et al. (2002) Predictome: a database of putative functional links between proteins. *Nucleic Acids Res* 30, 306–309.

17. Troyanskaya, O. G., Dolinski, K., Owen, A. B., et al. (2003) A Bayesian framework for combining heterogeneous data sources for gene function prediction (in Saccharomyces cerevisiae). *Proc Natl Acad Sci USA* 100, 8348–8353.

18. Jansen, R., Yu, H., Greenbaum, D., et al. (2003) A Bayesian networks approach for predicting protein-protein interactions from genomic data. *Science* 302, 449–453.

19. von Mering, C., Huynen, M., Jaeggi, D., et al. (2003) STRING: a database of predicted functional associations between proteins. *Nucleic Acids Res* 31, 258–261.

20. Bowers, P. M., Pellegrini, M., Thompson, M. J., et al. (2004) Prolinks: a database of protein functional linkages derived from coevolution. *Genome Biol* 5, R35.

21. Lee, I., Date, S. V., Adai, A. T., et al. (2004) A probabilistic functional network of yeast genes. *Science* 306, 1555–1558.

22. Gunsalus, K. C., Ge, H., Schetter, A. J., et al. (2005) Predictive models of molecular machines involved in Caenorhabditis elegans early embryogenesis. *Nature* 436, 861–865.

23. Myers, C. L., Robson, D., Wible, A., et al. (2005) Discovery of biological networks from diverse functional genomic data. *Genome Biol* 6, R114.

24. Kanehisa, M., Goto, S., Kawashima, S., et al. (2002) The KEGG databases at GenomeNet. *Nucleic Acids Res* 30, 42–46.

25. Jensen, F. V. (2001) *Bayesian Networks and Decision Graphs*. Springer, New York.

26. Martin, A., Schneider, S., Schwer, B. (2002) Prp43 is an essential RNA-dependent ATPase required for release of lariat-intron from the spliceosome. *J Biol Chem* 277, 17743–17750.

27. Lebaron, S., Froment, C., Fromont-Racine, M., et al. (2005) The splicing ATPase prp43p is a component of multiple preribosomal particles. *Mol Cell Biol* 25, 9269–9282.

28. Leeds, N. B., Small, E. C., Hiley, S. L., et al. (2006) The splicing factor Prp43p, a DEAH box ATPase, functions in ribosome biogenesis. *Mol Cell Biol* 26, 513–522.

29. Combs, D. J., Nagel, R. J., Ares, M., Jr., et al. (2006) Prp43p is a DEAH-box splicco-some disassembly factor essential for ribosome biogenesis. *Mol Cell Biol* 26, 523–534.

30. Bork, P., Jensen, L. J., von Mering, C., et al. (2004) Protein interaction networks from yeast to human. *Curr Opin Struct Biol* 14, 292–299.

Section III

Applications and Disease

Chapter 15

Computational Diagnostics with Gene Expression Profiles

Claudio Lottaz, Dennis Kostka, Florian Markowetz, and Rainer Spang

Abstract

Gene expression profiling using micro-arrays is a modern approach for molecular diagnostics. In clinical micro-array studies, researchers aim to predict disease type, survival, or treatment response using gene expression profiles. In this process, they encounter a series of obstacles and pitfalls. This chapter reviews fundamental issues from machine learning and recommends a procedure for the computational aspects of a clinical micro-array study.

Key words: Micro-arrays, gene expression profiles, statistical classification, supervised machine learning, gene selection, model assessment.

1. Introduction

In clinical micro-array studies, tissue samples from patients are examined using micro-arrays measuring gene expression levels of as many as 50,000 transcripts. Such high-dimensional data, possibly complemented by additional information about patients, provide novel opportunities for molecular diagnostics through automatic classification.

For instance, Roepman et al. *(1)* describe a study on head and neck squamous cell carcinomas. In this disease, treatment strongly depends on the presence of metastases in lymph nodes near the neck. However, diagnosis of metastases is difficult. More than half of the patients undergo surgery unnecessarily, whereas 23% remain under-treated. Roepman et al. show that treatment based on micro-array prediction can be significantly more accurate: In a validation cohort, under-treatment would have been

Jonathan M. Keith (ed.), *Bioinformatics, Volume II: Structure, Function and Applications, vcl. 453*
© 2008 Humana Press, a part of Springer Science + Business Media, Totowa, NJ
Book doi: 10.1007/978-1-60327-429-6 Springerprotocols.com

completely avoided while the rate of unnecessary surgery would have dropped from 50% to 14%.

From a statistical point of view, the major characteristic of micro-array studies is that the number of genes is orders of magnitude larger than the number of patients. For classification this leads to problems involving overfitting and saturated models. When blindly applying classification algorithms, a model rather adapts to noise in the data than to the molecular characteristics of a disease. Thus, the challenge is to find molecular classification signatures that are valid for entire disease populations.

The following briefly describes the machine learning theory, as it is needed for computational diagnostics using high-dimensional data. Software solutions are suggested in **Section 2** and a procedure for a clinical micro-array study in **Section 3**. The last section describes pitfalls and points out alternative strategies.

1.1. The Classification Problem

Classification is a well-investigated problem in machine learning. This chapter gives a brief overview of the most fundamental issues. Refer to *(2–6)* for further reading on the more theoretical concepts of machine learning, and to *(7, 8)* for an in-depth description of their application to micro-array data.

The task in classification is to determine classification rules that enable discrimination between two or more classes of objects based on a set of features. Supervised learning methods construct classification rules based on training data with known classes. They deduce rules by optimizing a classification quality criterion, for example, by minimizing the number of misclassifications. In computational diagnostics, features are gene expression levels, objects are tissue samples from patients, and object classes are phenotypic characteristics of patients. Phenotypes can include previously defined disease entities, as in the leukemia study of Haferlach et al. *(9)*; risk groups, as in the breast cancer studies of van't Veer et al. *(10)*; treatment response, as in the leukemia study by Cheok et al. *(11)*; or disease outcome, as in the breast cancer study of West et al. *(12)*. In this context, classification rules are called *diagnostic signatures*.

Study cases are always samples from a larger disease population. Such a population comprises all patients who had a certain disease, have it now, or will have it in the future. We aim for a diagnostic signature with good performance not only on the patients in the study, but also in future clinical practice. That is, we aim for a classification rule that generalizes beyond the training set. Different learning algorithms determine signatures of different complexity. We illustrate signature complexity using a toy example in which diagnosis of treatment response is based on the expression levels of only two genes (**Fig. 15.1**). A linear signature corresponds to a straight line, which separates the space defined by the two genes into two parts, holding good and bad

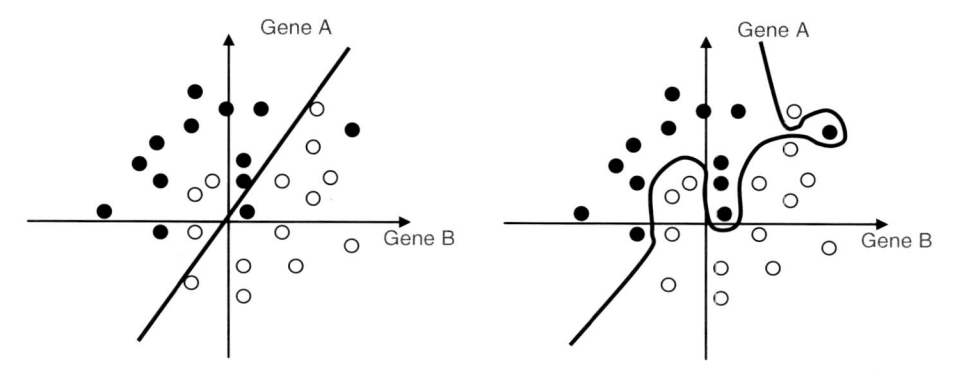

Fig. 15.1. Overfitting. The linear boundary *(left-hand side)* reflects characteristics of the disease population, including an overlap of the two classes. The complex boundary *(right-hand side)* does not show any misclassifications but adapts to noise. It is not expected to perform well on future patient data.

responders, respectively. When expression levels of many genes are measured, linear signatures correspond to hyperplanes separating a high-dimensional space into two parts. Other learning algorithms determine more complex boundaries. In micro-array classification, however, the improvement achieved by sophisticated learning algorithms is controversial *(13)*. Complex models are more flexible to adapt to the data. However, they also adapt to noise more easily and may thus miss the characteristic features of the disease population. Consider **Fig. 15.1**, in which black data points represent bad responders and white data points represent good responders. A linear signature is shown in the left panel of **Fig. 15.1**, whereas a complex boundary is drawn in the right panel. The linear signature reflects the general tendency of the data but is not able to classify perfectly. On the other hand, the complex boundary never misclassifies a sample, but it does not appear to be well supported by the data. When applying both signatures to new patients, it is not clear whether the complex boundary will outperform the linear signature. In fact, experience shows that complex signatures often do not generalize well to new data, and hence break down in clinical practice. This phenomenon is called *overfitting*.

1.2. The Curse of Dimensionality

In micro-array studies, the number of genes is always orders of magnitude larger than the number of patients. In this situation, overfitting also occurs with linear signatures. To illustrate why this is a problem, we use another simplistic toy example: Two genes are measured in only two patients. This is the simplest scenario in which the number of patients does not exceed the number of genes. We want to construct a linear signature that can discriminate between the two classes, say good and bad responders, each represented by one patient. This is the same problem as

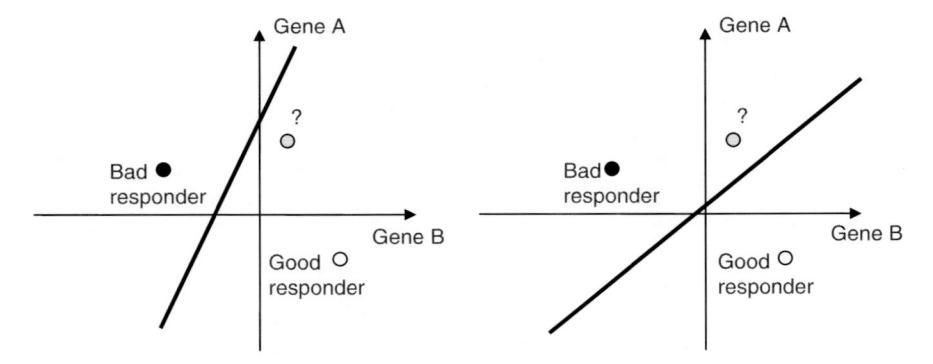

Fig. 15.2. Ill-posed classification problem. Even linear signatures are underdetermined when the number of patients does not exceed the number of genes. The black and white datapoints represent the training data. The linear signatures in both panels are equally valid but classify the novel data point *(gray)* differently.

finding a straight line separating two points in a plane. Clearly, there is no unique solution (**Fig. 15.2**). Next, think about a third point, which does not lie on the line going through the first two points. Imagine it represents a new patient with unknown diagnosis. The dilemma is that it is always possible to linearly separate the first two points such that the new one lies on the same side as either one of them. No matter where the third point lies, there is always one signature with zero training error, which classifies the new patient as a good responder, and a second signature, equally well supported by the training data, which classifies him as a bad responder. The two training patients do not contain sufficient information to diagnose the third patient. We are in this situation whenever there are at least as many genes as patients. Due to the large number of genes on micro-arrays this problem is inherent in gene expression studies.

The way out of the dilemma is *regularization*. Generally speaking, regularization means imposing additional criteria in the signature building process. A prominent method of regularization is *gene selection*, which restricts the number of genes contributing to the signature. This can be biologically motivated, since not all genes carry information on disease states. In many cases, gene selection improves the predictive performance of a signature *(14)*. Furthermore, adequate selection of genes opens the opportunity to design smaller and hence cheaper diagnostic micro-arrays or marker panels *(15)*.

Often genes are selected independently of the classification algorithm according to univariate selection criteria. This is called *gene filtering (16)*. In the context of classification, selection criteria reflecting the correlation with the class labels are generally used. Hence one favors genes with low expression levels in one class and high expression levels in the other. Popular choices are variants of the *t*-statistic or the non-parametric Wilcoxon rank sum statistic.

1.3. Calibrating Model Complexity

As illustrated in the previous sections, regularization is essential in the process of determining a good diagnostic signature. Aggressive regularization, however, can be as harmful as too little of it; hence, regularization needs to be calibrated. One way to do so is to vary the number of genes included in the signature: Weak regularization means that most genes are kept, whereas strong regularization removes most genes.

With little regularization, classification algorithms fit very flexible decision boundaries to the data. This results in few misclassifications on the training data. Nevertheless, due to overfitting, this approach can have poor predictive performance in clinical practice. With too much regularization, the resulting signatures are too restricted. They have poor performance on both the study patients and in future clinical practice. We refer to this situation as *underfitting*. The problem is schematically illustrated in **Fig. 15.3**, in which two error rates are compared. First there is the training error, which is the number of misclassifications observed on data from which the signature was learned. In addition, there is a test error, which is observed on an independent test set of additional patients randomly drawn from the disease population. Learning algorithms minimize the training error, but the test error measures whether a signature generalizes well.

In order to learn signatures that generalize well, we adapt model complexity so that test errors are minimized. To this end, we have to estimate test errors on an independent set of patients, the calibration set, which is not used in signature learning. To calibrate regularization, we learn signatures of varying complexity, evaluate them on the calibration set, and pick the signature that performs best.

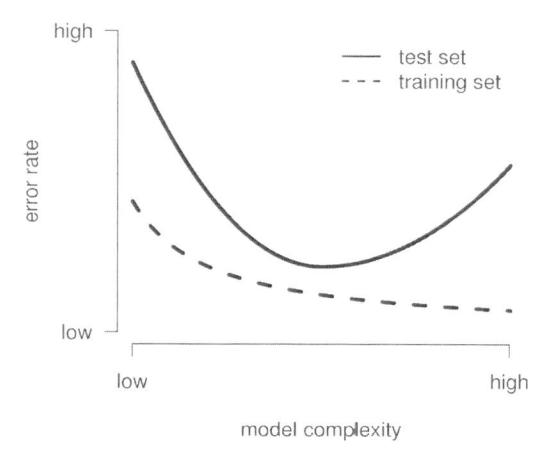

Fig. 15.3. Overfitting-underfitting trade off. The x-axis codes for model complexity and the y-axis for error rates. The dashed line displays the training error, the solid line the test error. Low complexity models (strong regularization) produce high test errors (underfitting) and so do highly complex models (weak regularization, overfitting).

1.4. Evaluation of Diagnostic Signatures

Validation of a diagnostic signature is important, because the errors on a training set do not reflect the expected error in clinical practice. In fact, the validation step is most critical in computational diagnosis and several pitfalls are involved. Estimators can be overly optimistic (biased), or they can have high sample variances. It also makes a difference whether we are interested in the performance of a fixed signature (which is usually the case in clinical studies), or in the power of the learning algorithm that builds the signatures (which is usually the case in methodological projects). The performance of a fixed signature varies due to the random sampling of the test set, whereas the performance of a learning algorithm varies due to sampling of both training and test set.

In computational diagnostics, we are usually more interested in the evaluation of a fixed diagnostic signature. The corresponding theoretical concept is the *conditional error rate*, also called *true error*. It is defined as the probability of misclassifying new patients given the training data used in the study. The true error is not obtainable in practice, since its computation involves the unknown population distribution. Estimates of the true error rate, however, can be obtained by evaluating the diagnostic signature on independent test samples.

2. Software

Most of the techniques we describe can be found in packages contributed to the statistical computing environment R *(17, 18)* and are available from http://cran.R-project.org. LDA, QDA, and many other classification methods are implemented in package *MASS* from the *VR* bundle, DLDA is contained in package *supclust*. Support vector machines are part of package *e1071*; the entire regularization path for the SVM can be computed by *svmpath*. The nearest centroid method is implemented in package *pamr*. The package *MCRestimate* from the Bioconductor project *(19)* implements many helpful functions for nested cross-validation.

3. Methods

This section describes our choice of tools for the development of a diagnostic signature. In addition to expression profiles, patients have an attributed class label reflecting a clinical phenotype. The challenge is to learn diagnostic signatures on gene expression data that enable prediction of the correct clinical phenotype for new patients.

3.1. Notation

We measure p genes on n patients. The data from the micro-array corresponding to patient i is represented by the *expression profile* $x^{(i)} \in \Re^p$. We denote the *label* indicating the phenotype of patient i by $y_i \in \{-1, +1\}$. In this setting, the class labels are binary and we restrict our discussion to this case. However, the discussed methods can be extended to multi-class problems dealing with more than two clinical phenotypes. The profiles are arranged as rows in a *gene expression matrix* $\underline{X} \in \Re^{n \times p}$. All labels together form the vector $y = (y_i)_{i=1,\ldots,n}$. The pair $(\underline{X}, y)$ is called a data set D. It holds all data of a study in pairs of observations $\{(x^{(i)}, y_i)\}$, $i = 1,\ldots,n$. The computational task is to generate a mathematical model f relating x to y.

Although we never have access to the complete disease population, it is convenient to make it part of the mathematical formalism. We assume that there is a data-generating distribution $P(X, Y)$ on $\Re^p \times \{-1, +1\}$. $P(X, Y)$ is the joint distribution of expression profiles and associated clinical phenotypes. The patients who enrolled for the study, as well as new patients who need to be diagnosed in clinical practice, are modeled as independent samples $\{(x^{(i)}, y_i)\}$ drawn from P. We denote a diagnostic signature, or *classification rule*, by $f: \Re^p \rightarrow \{-1, +1\}$.

3.2. Diagonal Linear Discriminant Analysis

Various learning algorithms have been suggested for micro-array analysis. Many of them implement rather sophisticated approaches to model the training data. However, Wessels et al. *(13)* report that in a comparison of the most popular algorithms and gene selection methods, simple algorithms perform particularly well. They have assessed the performance of learning algorithms using six datasets from clinical micro-array studies. Diagonal linear discriminant analysis (DLDA) combined with univariate gene selection achieved very good results. This finding is in accordance with other authors *(14)*. Thus, we recommend and describe this combination of methods.

Diagonal linear discriminant analysis (DLDA) is based on a comparison of multivariate Gaussian likelihoods for two classes. The conditional density $P(x \mid y = c)$ of the data given membership to class $c \in \{-1, +1\}$ is modeled as a multivariate normal distribution $N(\mu^c, \Sigma_c)$ with class mean μ^c and covariance matrix Σ_c. The two means and covariance matrices are estimated from the training data. A new point is classified to the class with higher likelihood. Restrictions on the form of the covariance matrix control model complexity: *Quadratic discriminant analysis* (QDA) allows different covariance matrices in both classes; *linear discriminant analysis* (LDA) assumes that they are the same, and diagonal linear discriminant analysis (DLDA) additionally restricts the covariance matrix to diagonal form. The parameters needed by DLDA are therefore class-wise mean expression values for each gene and one pooled variance per gene.

To derive the decision rule applied in DLDA, consider the multivariate Gaussian log-likelihood $N(\mu^c, \Sigma)$ for class with diagonal covariance matrix Σ. The vector μ^c contains the mean gene expression values μ^c_i for class c. It is also called the *class centroid*. The covariance matrix Σ contains pooled gene-wise variances σ_i. The log-likelihood $L_c(x)$ can then be written in the form:

$$L_c(x) = -\frac{1}{2} \cdot \sum_{i=1}^{p} \log\left(2\pi\sigma_i^2\right) - \frac{1}{2} \cdot \sum_{i=1}^{p} \frac{\left[x_i - \mu_i^c\right]^2}{\sigma_i^2}$$

The first term of $L_c(x)$ does not depend on the class and is therefore neglected in DLDA. DLDA places patients into the class for which the absolute value of the second term is minimized.

3.3. Univariate Gene Selection

Diagonal linear discriminant analysis can be directly applied to micro-array data. Nevertheless, *gene selection* considerably improves its performance. In gene selection, we impose additional regularization by limiting the number of genes in the signature. A simple way to select informative genes is to rank them according to a univariate criterion measuring the difference in mean expression values between the two classes. We suggest a regularized version of the *t*-statistic used for detecting differential gene expression. For gene i it is defined as:

$$d_i = \frac{\mu_i^{-1} - \mu_i^{+1}}{\sigma_i + \sigma_o},$$

where σ_o denotes the statistic's regularization parameter, the so called *fudge factor*. It is typically set to the median of all σ_i, and ensures that the statistic does not grow exceedingly when low variances are underestimated. Only top-ranking genes in the resulting list are chosen for classification.

3.4. Generation of Diagnostic Signatures

This section suggests a simple procedure to generate and validate a diagnostic signature within a clinical micro-array study, which is illustrated in **Fig. 15.4**.

3.4.1. Preprocess Your Data

First, normalize your micro-array data in order to make the expression values comparable. A detailed discussion of micro-array data preprocessing is given in **Chapter 4, Volume 1**. All methods presented there are equally valid in computational diagnostics. From now on, we assume that gene expression profiles $x^{(i)}$ are normalized and on an additive scale (log transformed).

3.4.2. Divide Your Data into a Training and a Test Set

In order to allow an unbiased evaluation of the generated diagnostic signature, we suggest separating the study data right from the start into two parts: the training set L for learning and the test set E for evaluation. The entire learning process must be executed on the training data only. The evaluation of the resulting diagnostic signature must be performed afterward using only the test set.

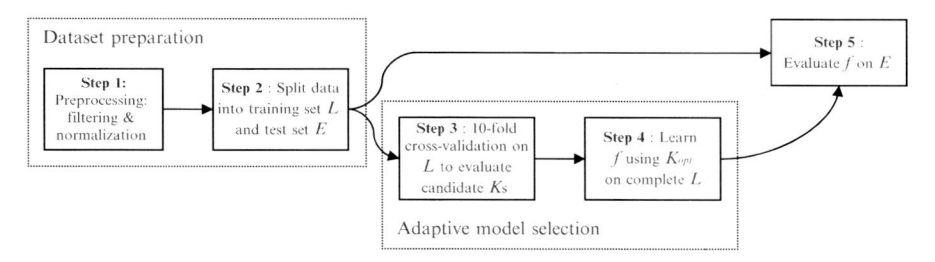

Fig. 15.4. Overview of the suggested procedure to generate a diagnostic signature. K represents the regularization level (e.g., number of genes selected), K_{opt} denotes the best regularization level found in cross validation. L represents the training set and E the test set.

The training set's exclusive purpose is to learn a diagnostic signature. Unclear or controversial cases can and should be excluded from learning. You may consider focusing on extreme cases. For instance Liu et al. improve their outcome prediction by focusing the learning phase on very short- and long-term survivors *(20)*.

The test set's exclusive purpose is to evaluate the diagnostic signature. Since you want to estimate the performance of a signature in clinical practice, the test set should reflect expected population properties of the investigated disease. For example, if the disease is twice as common in women as in men, the gender ratio should be close to 2:1 in the test set.

The size of the training and test sets has an impact on both the performance of the signature and the accuracy of its evaluation. Large training sets lead to a better performance of the signature, whereas large test sets lead to more accurate estimates of the performance. Actually, small test sets result in unreliable error estimation due to sampling variance (*see* **Note 1**). We recommend splitting the data in a ratio of 2:1.

3.4.3. Find the Best Regularization Level

Apply univariate gene filtering and DLDA to learn signatures with varying complexity and estimate their performance on independent data. Your best choice for the regularization level is the one leading to the best performance. *See* **Note 2** for alternative learning algorithms and feature selection schemes.

To estimate performance on independent data we recommend 10-fold cross validation. Partition the training set into 10 bins of equal size. Take care to generate bins that are balanced with respect to the classes to be discriminated. That is, classes should have the same frequency in the bins as in the complete training data. Use each bin in turn as the calibration set, and pool the other bins to generate the learning set. In each iteration, select genes according to the regularized t-statistics d_i and the regularization level K to be evaluated, learn a signature by

1. Choose regularization level K.
2. Separate the training set L into 10 bins $L_1, ..., L_{10}$
3. For i in 1 to 10 do
    a. Select features S according to K from $D[L{-}L_i]$
    b. Learn $f_i(K)$ on $D[L{-}L_i, S]$
    c. Evaluate $f_i(K)$ on $D[L_i, S]$
4. Average error rates determined in step 3.c

Fig. 15.5. Algorithmic representation of the 10-fold cross-validation procedure. K is the regularization level to be evaluated. $D[L]$ denotes the study data restricted to patients in the set L. $D[L, S]$ represents the study data restricted to the patient set L and the gene set S. The operator "-" is used for set difference.

applying DLDA on the restricted learning set, and compute the number of misclassifications in the calibration bin. The cross-validation error is the sum of these errors. Use this estimate for performance on independent data to determine the optimal amount of regularization *(21, 22)*. See **Fig. 15.5** for an algorithmic representation of the cross-validation procedure.

A straightforward method for finding a good set of candidate values for K is *forward filtering*. Starting with the most discriminating genes, include genes one by one until the cross-validation error reaches an optimum, stop iterating and set the optimal regularization level K_{opt} to the value of K that produced the smallest cross-validation error.

3.4.4. Learn the Diagnostic Signature

For the optimal level of regularization K_{opt}, compute the diagnostic signature f on the complete training set L. This is the final signature.

3.4.5. Evaluate Your Signature

Apply your final signature to the test set to estimate the misclassification rate. Note that the result is subject to sampling variance (*see* **Note 3** for more information on sampling variance).

3.4.6. Document Your Signature

Diagnostic signatures eventually need to be communicated to other health care centers. It should be possible for your signature to be used to diagnose patients worldwide, at least if the same array platform is used to profile them. Therefore, you should provide a detailed description of the signature that you propose. The list of genes contributing to the signature is not enough. The mathematical form of both the classification and the preprocessing models needs to be specified together with the values of all parameters. For DLDA, communication of the centroid for each group and the variance for each gene is necessary to specify the signature completely. Although exact documentation of the classification signature is crucial, involvement of genes in a signature should not be interpreted biologically (*see* **Note 4**).

4. Notes

1. Pitfalls in signature evaluation: There are several pitfalls leading to overly optimistic estimations, for instance, using too small or unbalanced validation sets *(36)*. When using a single independent test set for evaluation of diagnostic signatures, only training data is used for gene selection, classifier learning, and adaptive model selection. The final signature is then evaluated on the independent test set. Unfortunately, this estimator can have a substantial sample variance, due to the random selection of patients in the test set. This is especially the case if the test set is small. Thus, good performance in small studies can be a chance artifact. For instance, Ntzani et al. *(37)* have reviewed 84 micro-array studies carried out before 2003 and observed that positive results were reported strikingly often on very small datasets.

   Another prominent problem is the selection bias caused by improperly combining cross-validation with gene selection *(12, 35, 36)*. Gene selection has a strong impact on the predictive performance of a signature. It is an essential part of the signature-building algorithm. There are two possible ways to combine gene selection with cross-validation: Either apply gene selection to the complete dataset and then perform cross-validation on the reduced data, or perform gene selection in every single step of cross-validation anew. We call the first alternative *out-of-loop gene selection* and the second *in-loop gene selection*. In-loop gene selection gives the better estimate of generalization performance, whereas the out-of-loop procedure is overoptimistic and biased toward low error rates. In out-of-loop gene selection, the genes selected for discriminative power on the whole dataset bear information on the samples used for testing. In-loop gene selection avoids this problem. Here, genes are only selected on the training data of each cross-validation iteration and the corresponding test sets are independent.

   The impact of overoptimistic evaluation through out-of-loop gene selection can be very prominent. For example, Reid et al. *(38)* report the re-evaluation of two public studies on treatment response in breast cancer. They observe classification errors of 39% and 46%, respectively, when applying in-loop gene selection. Using the overoptimistic out-of-loop gene selection, error rates are underestimated at 25% and 24%, respectively. Similarly, Simon et al. *(36)* describe a case in breast cancer outcome prediction in which the out-of-loop cross-validation method estimates the error rate to 27%, whereas the in-loop cross-validation method estimates an error rate of 41%. Nevertheless, Ntzani et al. *(37)* report that 26% of 84 reviewed micro-array studies published before 2003 provide overoptimistic error rates due to out-of-loop gene selection. Seven of these studies have been re-analyzed by Michiels et al. *(39)*. They determine classification rates using nested cross-validation loops and average over many random cross-validation partitionings. In five of the investigated datasets, classification rates no better than random guessing are observed.

2. Alternative classification algorithms: Several authors have observed that complex classification algorithms quite often do not outperform simple ones like diagonal linear discrimination on clinical micro-array data *(13, 14, 23, 24)*. We report two more algorithms, which have shown good performance on micro-array data.

   The first one is a variant of DLDA called Prediction Analysis of Micro-arrays (PAM) *(25, 26)*. An outstanding feature of PAM is *gene shrinkage*. Filtering uses a hard threshold: when selecting k genes, gene $k + 1$ is thrown away even if it bears as much information as gene k. Gene shrinkage is a smoother, continuous, soft-thresholding method. PAM is an application of nearest centroid classification, in which the class centroids are shrunken in the direction of the overall centroid. For each gene i the value δ_{ic} measures the distance of the centroid for class c to the overall centroid in units of its standard deviation. Each δ_{ic} is then reduced by an amount Δ in absolute value and is set to zero if its value becomes negative. With increasing Δ, all genes lose discriminative power and more and more of them will fade away. Genes with high variance vanish faster than genes with low variance. A link between PAM and classical linear models is discussed in Huang et al. *(27)*.

   Support vector machines (SVMs) *(2, 28, 29)* avoid the ill-posed problem shown in **Fig. 15.2** in **Section 1.2** by fitting a maximal (soft) margin hyperplane between the two classes. In high-dimensional problems there are always several perfectly separating hyperplanes, but there is only one separating hyperplane with maximal distance to the nearest training points of either class. *Soft margin SVMs* trade off the number of misclassifications with the distance between hyperplane and nearest datapoints in the training set. This trade-off is controlled by a tuning parameter C. The maximal margin hyperplane can be constructed by means of inner products between training examples. This observation is the key to the second building block of SVMs: The inner product $x_i^T x_j$ between two training examples x_i and x_j is replaced by a non-linear *kernel function*. The use of kernel functions implicitly maps the data into a high-dimensional space, where the maximal margin hyperplane is constructed. Thus, in the original input space, boundaries between the classes can be complex when choosing non-linear kernel functions. In micro-array data, however, choosing a linear kernel and thus deducing linear decision boundaries usually performs well.

A gene selection algorithm tailored to SVMs is *recursive feature elimination (30)*. This procedure eliminates the feature (gene) contributing least to the normal vector of the hyperplane before retraining the support vector machine on the data, excluding the gene. Elimination and retraining are iterated until maximal training performance is reached. In high-dimensional micro-array data, eliminating features one by one is computationally expensive. Therefore, several features are eliminated in each iteration.

3. Alternative evaluation schemes: The signature evaluation suggested in **Section 3.4** yields an estimate of the signature's misclassification rate, but does not provide any information on its variability. In order to investigate this aspect, you can apply methods designed to evaluate signature generating algorithms. In order to evaluate the performance of a learning algorithm, the sampling variability of the training set has to be taken into account. One approach is to perform the partitioning into training and test set (L and E) many times randomly and execute the complete procedure illustrated in **Fig. 15.4** for each partitioning. This yields a distribution of misclassification rates reflecting sample variability of training and test sets.

More effective use of the data can be made via cross-validation. Together with the procedure described in Section 3.4., two nested cross-validation loops are needed *(22, 23)*. Braga-Neto and Dougherty *(31)* advise averaging cross-validation errors over many different partitionings. Ruschhaupt et al. *(23)* and Wessels et al. *(13)* implement such complete validation procedures and compare various machine learning methods.

In the leave-one-out version of cross-validation, each sample is used in turn for evaluation, whereas all other samples are attributed to the learning set. This evaluation method estimates the expected error rate with almost no bias. For big sample sizes, it is computationally more expensive than 10-fold cross-validation and suffers from high variance *(7, 32, 33)*. Efron et al. *(34)* apply *bootstrap smoothing* to the leave-one-out cross-validation estimate. The basic idea is to generate different *bootstrap replicates*, apply leave-one-out cross-validation to each and then average results. Each bootstrap replicate contains n random draws from the original dataset (with replacement so that samples may occur several times). A result of this approach is the so-called "0.632+ estimator." It takes the possibility of overfitting into account and reduces variance compared to the regular cross-validation estimates. Ambroise et al. *(35)* have found it to work well with gene expression data.

Cross-validation and bootstrap smoothing error rates, as well as error rates determined by repeated separation of data into training and test set, refer to the classification algorithm, not to a single signature. In each iteration, a different classifier is learned based on different training data. The cross-validation performance is the average of the performance of different signatures. Nevertheless, cross-validation performance can be used as an estimate of a signature's expected error rate.

4. Biological interpretation of diagnostic signatures: The methodology in the preceding was presented in the classification context only. In addition, you might be tempted to interpret the genes driving the models biologically, but this is dangerous. First, it is unclear how exactly regularization biases the selection of signature genes. Although this bias is a blessing from the diagnostic perspective, this is not the case from a biological point of view. Second, signatures are generally not unique: Although outcome prediction for breast cancer patients has been successful in various studies (*40–42*), the respective signatures do not overlap at all. Moreover, Ein-Dor, L. et al. *(43)* derived a large number of almost equally performing non-overlapping signatures from a single dataset. Also, Michiels et al. *(39)* report highly unstable signatures when resampling the training set.

This is not too surprising considering the following: The molecular cause of a clinical phenotype might involve only a small set of genes. This primary event, however, has secondary influences on other genes, which in turn deregulate more genes, and so on. In clinical micro-array analysis we typically observe an avalanche of secondary or later effects, often involving thousands of differentially expressed genes. Although complicating biological interpretation of signatures, such an effect does not compromise the clinical usefulness of predictors. On the contrary, it is conceivable that only signals enhanced through propagation lead to a well generalizing signature.

References

1. Roepman, P., Wessels, L. F., Kettelarij, N., et al. (2005) An expression profile for diagnosis of lymph node metastases from primary head and neck squamous cell carcinomas. *Nat Genet* 37, 182–186.

2. Schölkopf, B., Smola, A. J. (2001) *Learning with Kernels*. MIT Press, Cambridge, MA.

3. Ripley, B. D. (1996) *Pattern Recognition and Neural Networks*. Cambridge University Press, Cambridge, UK.

4. Devroye, L., Györfi, L., Lugosi, L. (1996) *A Probabilistic Theory of Pattern Recognition*. Springer, New York.

5. Hastie, T., Tibshirani, R., Friedman, J. (2001) *The Elements of Statistical Learning: Data Mining, Inference, and Prediction*. Springer-Verlag, New York.

6. Duda, R. O., Hart, P. E., Stork, D. G. (2001) *Pattern Classification*. Wiley, New York.

7. McLachlan, G. J., Do, K. A., Ambroise, C. (2004) *Analyzing Micro-array Gene Expression Data*. Wiley, New York.

8. Terry Speed (ed.) (2003) *Statistical Analysis of Gene Expression Micro-array Data*. Chapman & Hall/CRC, Boca Raton, FL.

9. Haferlach, T., Kohlmann, A., Schnittger, S., et al. (2005) A global approach to the diagnosis of leukemia using gene expression profiling. *Blood* 106, 1189–1198.

10. van't Veer, L. J., Dai, H., van de Vijver, M. J., et al. (2002) Gene expression profiling predicts clinical outcome of breast cancer. *Nature* 415, 530–536.

11. Cheok, M. H., Yang, W., Pui, C. H., et al. (2003) Treatment-specific changes in gene expression discriminate in vivo drug response in human leukemia cells. *Nat Genet* 34, 85–90.

12. West, M., Blanchette, C., Dressman, H., et al. (2001) Predicting the clinical status of human breast cancer by using gene expression profiles. *Proc Natl Acad Sci U S A* 98, 11462–11467.

13. Wessels, L. F., Reinders, M. J., Hart, A. A., et al. (2005) A protocol for building and evaluating predictors of disease state based on micro-array data. *Bioinformatics* 21, 3755–3762.

14. Dudoit, S., Fridlyand, J., Speed, T. (2002) Comparison of discrimination methods for the classification of tumors using gene expression data. *J Amer Stat Assoc* 97, 77–87.

15. Jäger, J., Weichenhan, D., Ivandic, B., et al. (2005) Early diagnostic marker panel determination for micro-array based clinical studies. *SAGMB* 4, Art 9.

16. John, G. H., Kohavi, R., Pfleger, K. (1994) *Irrelevant Features and the Subset Selection Problem Morgan Kaufmann Publishers* International Conference on Machine Learning, San Francisco CA, USA pp. 121–129.

17. Ihaka, R., Gentleman, R. (1996) R: a language for data analysis and graphics. *J Comput Graphical Stat* 5, 299–314.

18. Team, R. D. C. (2005), *R Foundation for Statistical Computing*. Vienna. A Language and Environment for Statistical Computing, R Foundation for Statistical Computing, Vienna, Austria

19. Gentleman, R. C., Carey, V. J., Bates, D. M., et al. (2004) Bioconductor: Open software development for computational biology and bioinformatics. *Gen Biol* 5, R80.

20. Liu, Li, Wong (2005) Use of extreme patient samples for outcome prediction from gene expression data. *Bioinformatics*.

21. Stone, M. (1974) Cross-validatory choice and assessment of statistical predictions. *J Roy Stat Soc Series B (Method)* 36, 111–147.

22. Geisser, S. (1975) The predictive sample reuse method with applications. *J Amer Stat Assoc* 70, 320–328.

23. Ruschhaupt, M., Huber, W., Poustka, A., et al. (2004) A compendium to ensure computational reproducibility in high-dimensional classification tasks. *Stat Appl Gen Mol Biol* 3, 37.

24. Dudoit, S. (2003) *Introduction to Multiple Hypothesis Testing*. Biostatistics Division, California University Berkeley CA, USA.

25. Tibshirani, R., Hastie, T., Narasimhan, B., et al. (2003) Class prediction by nearest shrunken centroids, with applications to DNA micro-arrays. *Statist Sci* 18, 104–117.

26. Tibshirani, R., Hastie, T., Narasimhan, B., et al. (2002) Diagnosis of multiple cancer types by shrunken centroids of gene expression. *Proc Natl Acad Sci U S A* 99, 6567–6572.

27. Huang, X., Pan, W. (2003) Linear regression and two-class classification with gene expression data. *Bioinformatics* 19, 2072–2078.

28. Vapnik, V. (1998) *Statistical Learning Theory*. Wiley, New York.

29. Vapnik, V. (1995) *The Nature of Statistical Learning Theory*. Springer, New York.

30. Guyon, I., Weston, J., Barnhill, S., et al. (2002) Gene selection for cancer classification using

support vector machines. *Machine Learning* 46, 389–422.

31. Braga-Neto, U. M., Dougherty, E. R. (2004) Is cross-validation valid for small-sample micro-array classification? *Bioinformatics* 20, 374–380.

32. Kohavi, R. (1995) *IJCAI* 1137–1145.

33. Hastie, T., Tibshirani, R., Friedman, J. (2001) *The Elements of Statistical Learning.* Springer, New York.

34. Efron, B., Tibshirani, R. (1997) Improvements on cross-validation: the 632+ bootstrap method. *J Amer Stat Assoc* 92, 548–560.

35. Ambroise, C., McLachlan, G. J. (2002) Selection bias in gene extraction on the basis of micro-array gene-expression data. *Proc Natl Acad Sci U S A* 99, 6562–6566.

36. Simon, R., Radmacher, M. D., Dobbin, K., et al. (2003) Pitfalls in the use of DNA micro-array data for diagnostic and prognostic classification. *J Natl Cancer Inst* 95, 14–18.

37. Ntzani, E. E., Ioannidis, J. P. A. (2003) Predictive ability of DNA micro-arrays for cancer outcomes and correlates: an empirical assessment. *Lancet* 362, 1439–1444.

38. Reid, J. F., Lusa, L., De Cecco, L., et al. (2005) Limits of predictive models using micro-array data for breast cancer clinical treatment outcome. *J Natl Cancer Inst* 97, 927–930.

39. Michiels, S., Koscielny, S., Hill, C. (2005) Prediction of cancer outcome with micro-arrays: a multiple random validation strategy. *Lancet* 365, 488–492.

40. van de Vijver, M. J., He, Y. D., van't Veer, L. J., et al. (2002) A gene-expression signature as a predictor of survival in breast cancer. *N Engl J Med* 347, 1999–2009.

41. Sorlie, T., Tibshirani, R., Parker, J., et al. (2003) Repeated observation of breast tumor subtypes in independent gene expression data sets. *Proc Natl Acad Sci U S A* 100, 8418–8423.

42. Ramaswamy, S., Ross, K. N., Lander, E. S., et al. (2003) A molecular signature of metastasis in primary solid tumors. *Nat Genet* 33, 49–54.

43. Dor, L. E., Kela, I., Getz, G., et al. (2005) Outcome signature genes in breast cancer: is there a unique set? *Bioinformatics* 21, 171–178.

Chapter 16

Analysis of Quantitative Trait Loci

Mario Falchi

Abstract

Diseases with complex inheritance are characterized by multiple genetic and environmental factors that often interact to produce clinical symptoms. In addition, etiological heterogeneity (different risk factors causing similar phenotypes) obscure the inheritance pattern among affected relatives and hamper the feasibility of gene-mapping studies. For such diseases, the careful selection of quantitative phenotypes that may represent intermediary risk factors for disease development (intermediate phenotypes) is etiologically more homogeneous than the disease per se. Over the last 15 years quantitative trait locus mapping has become a popular method for understanding the genetic basis for intermediate phenotypes. This chapter provides an introduction to classical and recent strategies for mapping quantitative trait loci in humans.

Key words: Quantitative traits, linkage, association, review.

1. Introduction

The last century was characterized by a shift of biological science from an observational to a predictive point of view. Advances in molecular biology techniques together with theoretical work in population genetics are increasingly enabling the elucidation of the connection between genotypes and phenotypes. The main remarkable progress has been achieved by human genetic research, motivated by the impact that this scientific field could have on public health and disease prevention. Quantitative genetics, historically led by plant and animal geneticists, is now receiving popular growing support in human genetics as the appropriate tool with which to tackle the challenge of unravelling the genetic determinants for common disorders.

Jonathan M. Keith (ed.), *Bioinformatics, Volume II: Structure, Function and Applications, vol. 453*
© 2008 Humana Press, a part of Springer Science + Business Media, Totowa, NJ
Book doi: 10.1007/978-1-60327-429-6 Springerprotocols.com

The last 20 years provided great successes in dissecting the molecular bases of different diseases in humans through the powerful strategy of positional cloning, in which genes are identified and cloned on the basis of their chromosomal location only. Positional cloning generally begins with linkage analysis, a statistical method that, by studying the co-segregation in families of a disease (or a trait) with genetic markers of known position in the genome, enables identification of the chromosomal location of specific genes involved in the studied disease (or trait). Most of the statistical theory behind linkage analysis has been described in the first 50 years of the twentieth century. In 1910, Thomas Hunt Morgan *(1)* observed that loci on the same chromosome have a non-zero probability of being co-inherited (linkage). Some years later, Haldane *(2)* proposed the application of linkage methods to map genes in human. Haldane and Smith *(3)* and Barnard *(4)*, using the statistical technique of maximum-likelihood analysis developed by Fisher *(5, 6)*, proposed the logarithm of Odds (LOD) score analysis of human pedigrees. In 1955, Morton *(7)* established the guidelines for the LOD score analysis interpretation, building the basis of positional cloning studies. The ensuing computational enhancement begun by Simpson *(8)*, Elston and Stewart *(9)*, and Ott *(10)*, and the development of the first set of molecular markers for whole-genome spanning by Botstein *(11)*, determined the beginning of a new era for gene mapping in human.

The positional cloning approach has been extremely successful in identifying more than 1,500 genes causing human diseases or traits, as reported in the OMIM catalog (online mendelian inheritance in man: http://www.ncbi.nlm.nih.gov/entrez/query.fcgi?db=OMIM). However, these successes have been principally limited to simple mendelian genes, characterized by a close "one-to-one" correspondence between genotypes and observed phenotypes. The success in monogenic disorder studies and the identification of a subset of genes underlying more common disorders, such as BRCA-1 for breast cancer or APP for Alzheimer's disease, generated a strong optimism for an equivalent success in identifying genes for more common disorders, such as obesity, schizophrenia, or hypertension. However, the positional cloning strategy is failing in identifying the numerous genes of smaller effect that underlie the most common, familial diseases. For these traits, different contributions of genes, environmental factors, and their interactions may often lead to similar phenotypes (complex traits). This etiological heterogeneity obscures the inheritance pattern of common diseases, thus hampering gene-mapping studies.

A powerful approach to reduce the etiological heterogeneity in a genetic mapping study is to focus on the quantitative phenotypes (intermediate phenotypes) that are risk factors for a specific disease. For example, body mass index (BMI), serum lipid levels, blood pressure, and other intermediate phenotypes are likely to be involved in the pathogenesis of ischemic heart disease. Therefore, the resulting outcome of disease onset can be an insensitive marker of different underlying processes, each of them under a different degree of genetic and environmental control and etiologically more homogeneous than the disease itself.

In other—and more obvious—circumstances, the disease status is defined by an arbitrary threshold of a quantitative measurement, such as BMI values for obesity, or bone mineral density measurements for osteoporosis. Instead of using such arbitrary thresholds, the direct analysis of quantitative measurements may more objectively explain the variation observed among individuals.

The importance of genetic mapping for quantitative traits is not only related to the discovery of genetic factors for susceptibility to common diseases. Quantitative phenotypes include individual differences in morphology, physiology, and behavior and represent the main source of variations both between and within populations. The identification of these genetic determinants will also further a major goal in biological research: to clarify the mechanisms of life preservation, evolution, and the dynamics of ontogenesis.

There is a growing availability of user-friendly software for the statistical analysis of genetic data. The most comprehensive catalogue is maintained at the Rockefeller University: http://linkage.rockefeller.edu/soft/list.html

Most of them are easy to use and extensively documented, and are often accompanied by step-by-step guides.

The purpose of this chapter is to provide an introduction to the principles underlying quantitative trait genetic analysis, in order to help the reader better understand limitations and advantages of classic and new approaches, and the theory behind the software developed to support their application.

2. Underlying Principles

In 1865, the monk Gregor Mendel *(12)* first reconsidered the validity of the blending theory of inheritance, or pangenesis, which had been described by Hippocrates (ca 400 BC) and was generally accepted by Mendel's contemporary scientists. However, the particulate inheritance of discontinuous characters demonstrated by Mendel through his garden pea *(Pisum Sativum)* studies was popularized only at the beginning of the twentieth century, as a result of which the concept of gene became accepted as the atomic unit of genetic inheritance.

Nevertheless, the mendelian segregation laws appeared inaccurate in explaining the familiar resemblance of some quantitative measures, as first observed by Galton's studies on the heritability of anthropometric traits. Since the offspring phenotype was in the average midway between the parents' phenotypes, the pangenesis theory looked more suitable in explaining the inheritance of such quantitative traits. It was the theoretical work of Fisher *(13)*, published in 1918, that unified the two schools of thought by introducing the concept of polygenic inheritance. Therefore, although the mendelian discrete phenotypes were determined by variations at a single genetic locus (monogenic), the observed variability of the phenotypes studied by the biometrical school was attributable to variations at multiple genetic loci (polygenic). For those traits, the individual phenotypic value is determined by the sum of each contributing loci, each of them inherited according to the mendelian laws of segregation.

A polymorphic locus whose alleles affect the variation of a quantitative phenotype is named a quantitative trait locus (QTL). Generally, quantitative traits are influenced by the segregation of multiple QTLs, which effects are sensible to the genetic background and to the external and sexual environments. By definition, a single QTL can either overlap a single gene or encompass a cluster of genes involved in the trait, if they cannot be separated by recombinations or through functional studies and considered as separated QTLs.

2.1. Polygenic Inheritance

Although many different alleles may exist for a gene, the effect of a single gene determining a quantitative trait can be simply described by a diallelic gene through the classical biometrical model *(14)*.

Consider a single gene with two possible alleles, A and B, having population frequencies p and q = (1 − p), respectively. Assuming random mating in the population, the genotypic frequencies of the three possible genotypes (AA, AB, and BB) are p^2, 2pq, and q^2, respectively. Define *genotypic value* as the average phenotype associated with a given genotype. We can arbitrarily fix, without loss of generality, the genotypic value of the homozygotes AA to 0. If we assume the allele B increases the phenotypic value, we can define $2a$ as the genotypic value of the homozygote BB. The heterozygote AB has genotypic value $a + d$, where d measures the degree of dominance between the two alleles and represents the departure from complete additivity. Indeed, when d is equal to 0 there is no dominance and each copy of the B allele provides an amount a to the phenotypic value (**Fig. 16.1**). If A is dominant over B, $d < 0$, whereas if B is dominant over A, $d > 0$.

Assuming $d = 0$ (additive model) and p = q = 0.5 (equal allele frequencies), the distribution of the quantitative phenotype due to this single gene would have three values in the population, whose frequencies are determined by the three genotypes' frequencies (**Fig. 16.2A**).

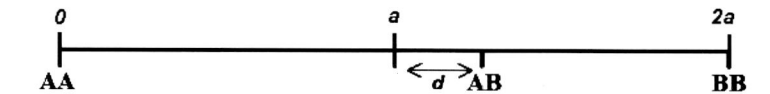

Fig. 16.1. Biometrical genetic model for a single diallelic gene.

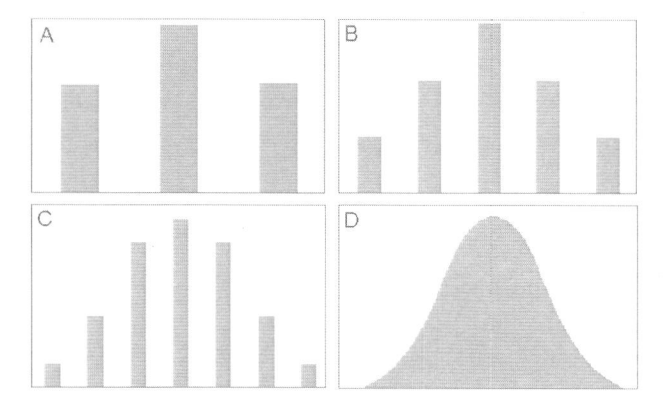

Fig. 16.2. Quantitative trait distributions in a population determined by an increasing number of diallelic genes, assuming only additive effects, no interactions and equal allele frequencies. (**A**) 1; (**B**) 2; (**C**) 3; (**D**) *n*.

If a second independent gene having equal characteristics influences the trait, the population will show nine genotypic combinations at the two genes, whose frequencies are determined by the product of the genotype frequency at each gene. Consider, for instance, a second gene with equally frequent alleles C and D. Assume that both the alleles B and D have $a = 1$ and $d = 0$, and the genotypic value for both AA and CC is 0. Then each copy of B and D alleles cause an increment of 1 of the trait over the AACC genotype (which has value 0). Under this context, the expected multi-locus frequencies and trait values in the population are given in **Table 16.1**. Since some of the two-gene genotypes determine the same genotypic value (e.g., ABCC and AACD), the two-locus model determines only five distinct phenotypic values in the population (*see* **Fig. 16.2B**). The number of genotypic combinations and genotypic values at *n* loci increases as additional genes are included in this simplified model (*see* **Fig. 16.2C**). Therefore, according to the polygenic model, under the influence of a large number of genes, each of them segregating according to the mendelian laws of inheritance, the distribution of the genotypic values in the population will tend to a normal distribution (*see* **Fig. 16.2D**).

In the presence of environmental influences the phenotypic value of an individual carrying a particular genotypic configuration can show a deviation from the genotypic value.

**Table 16.1
Genotypes, genotypic frequencies, and corresponding phenotypic values for a quantitative trait determined by two genes**

Genotype	Frequency	Trait value
AACC	0.0625	0
ABCC	0.125	1
BBCC	0.0625	2
AACD	0.125	1
ABCD	0.25	2
BBCD	0.125	3
AADD	0.0625	2
ABDD	0.125	3
BBDD	0.0625	4

Assuming no genotype-environment interaction, the environmental deviations are assumed to have the same effect on each genotype, following in the population a normal distribution with expected average of zero. Therefore, a normally distributed phenotype can result even if the trait is oligogenic; that is, affected by only a few genes.

The theoretical viewpoint has been that hundreds of genes determine a quantitative trait. However, if this were true, efforts to map a QTL would be in vain. Early QTL mapping revealed a much simpler genetic architecture for many quantitative traits *(15–17)*. Thus, even if many genes determine the value of a quantitative trait, some of them are likely to have a major effect (major QTL), and hence their chromosomal locations could be potentially determined.

2.2. Resemblance Between Relatives

In classical quantitative genetics, assuming independence between genetic and environmental effects, the observed phenotypic value (p) of an individual is determined as the sum of unobserved genotypic (g) and environmental (e) values.

$$p = g + e \qquad [1]$$

The genotypic value can be further decomposed into the additive (a), dominance (d), and epistatic values (i).

$$g = a + d + i \qquad [2]$$

The additive value, as shown in the previous paragraph, depends on the average effect of the alleles at each locus, the dominant value on the interaction between alleles within each locus, whereas the epistatic effect depends on the interaction between genotypes at different loci. Given a large number of genetic clones, the genotypic value can be calculated as the average phenotypic value and the environmental value as the difference between the phenotype and the estimated genotypic value. These values cannot be calculated using a human population sample—which is typically composed of different types of relatives—because the genotypic values between individuals are differentiated in sizes. However, through these differences it is possible to infer the amount of phenotypic variation determined by each component.

Consider the simple case of a single diallelic locus influencing the trait (having allele frequencies p and q = [1 − p]). The mean effect of the gene to the population mean of the trait is the sum of the product of the genotypic values $(2a, a+d, 0)$ and the genotypic frequencies:

$$(p^2, 2pq, q^2): 2ap^2 + 2pq(a+d) + 0q^2 = a + a(p-q) + 2pqd.$$

The total genetic variance can be obtained by applying the formula for variances:

$$\sigma_X^2 = \sum_{i=1}^{n} f_i(x_i - \mu)^2$$

where f_i denotes the i-genotype frequency, x_i the i-genotypic value, and μ the population mean. The total genetic variance is:

$$\sigma_g^2 = p^2[2a - a - a(p-q) - pqd]^2 + 2pq[d + a - a - a(p-q) - pqd]^2$$
$$+ q^2[0 - a(p-q) - pqd]^2$$

This can be simplified to:

$$\sigma_g^2 = p^2[2q(a - dp)]^2 + 2pq[a(q-p) + d(1-2pq)]^2$$
$$+ q^2[-2p(a + dq)]^2$$

And further simplified to:

$$\sigma_g^2 = 2pq[a - (p-q)d]^2 + (2pqd)^2$$

These two terms correspond to the additive and dominance variances (*see* also *(14)* for further details) in the population:

$$\sigma_a^2 = 2pq[a - (p-q)d]^2$$
$$\sigma_d^2 = (2pqd)^2 \tag{3}$$

A property of variances is that if two variables are independent, the variance of their sum equals the sum of their single variances:

$$\sigma_{(X+Y)}^2 = \sigma_X^2 + \sigma_Y^2$$

Therefore, since we assume independence of the genotypic and environmental factors, according to Equation 1 the phenotypic variance σ_p^2 can be decomposed into the sum of the genetic and environmental variances:

$$\sigma_p^2 = \sigma_g^2 + \sigma_e^2$$

and from Equation 2 the genotypic variance can be decomposed into:

$$\sigma_g^2 = \sigma_a^2 + \sigma_d^2 + \sigma_i^2$$

For simplicity we will hereafter ignore the epistatic effects. The covariance between the phenotypic values of a large number of relative pairs can be used to estimate the contribution of the genetic and environmental components of variance to the phenotypic variance. The covariance of the phenotypic values for any pairs of relatives (i,j) is:

$$\mathrm{Cov}\,(p_i, p_j) = \mathrm{Cov}\,(g_i + e_i, g_j + e_j)\,.$$

Assuming g and e are independent (no genotype-environment interactions), we have that:

$$\mathrm{Cov}\,(p_i, p_j) = \mathrm{Cov}\,(g_i, g_j) + \mathrm{Cov}\,(e_i, e_j)$$

The sharing of alleles of the same genes generates the covariance of the genotypic values between individuals. For example, if two relatives share two alleles at the gene influencing the trait, they would share both the additive and the dominance variance $\sigma_a^2 + \sigma_d^2$.

$$\mathrm{Cov}\,(g_i, g_j) = \sigma_a^2 + \sigma_d^2$$

If they share only one allele, the dominance variance cannot be observed and they share half of the additive variance $\sigma_a^2/2$ as only the additive effect of one allele can be observed.

$$\mathrm{Cov}\,(g_i, g_j) = \frac{1}{2}\sigma_a^2$$

When we do not have any information on the genes involved in the trait and/or their genotypic configuration, the expectation of gene sharing over the genome is used to model the covariance due to the expected shared genetic effects between relatives, and estimate the overall additive and dominance variance components (due to all loci influencing the trait):

$$\mathrm{Cov}\,(g_i, g_j) = r\sigma_a^2 + \delta\sigma_d^2$$

Here r is the coefficient of relatedness and δ is the fraternity coefficient of two related individuals. The coefficient of relatedness r is the expected fraction of alleles shared by two individuals, and is equal to two times the kinship coefficient (2ϕ), when neither individual is inbred *(18)*. The kinship coefficient is defined as the probability that, picking from the same location in the genome an allele at random for each relative, the two chosen alleles are identical-by-descent (IBD). Two alleles are IBD when they descend from a single ancestral copy, and are therefore the same exact copy of a gene.

The fraternity coefficient δ is defined as the probability that two relatives can share two alleles IBD for a locus.

In a non-inbred population, a parent and an offspring share one allele IBD at each locus in the genome. Picking an allele at random from a locus of a parent and an offspring, the coefficient of kinship is $0.5^2 = 0.25$. Their fraternity coefficient is 0.

Two full siblings have probability 0.5 to share 1 allele IBD, 0.25 probabilities to share 0 alleles IBD, and 0.25 probabilities to share 2 alleles IBD (the latter is their fraternity coefficient). If they share 1 allele IBD, as just observed for a parent and an offspring the probability to pick the same allele at random is 0.25. If they share 2 alleles IBD, this probability is 0.5. The expected proportion of alleles shared IBD is therefore:

$$0.25 \times 0.5 + 0.5 \times 0.25 = 0.25$$

Different methods can be used to estimate the components of variance for quantitative traits from the phenotypic covariance of relative pairs *(19)*.

The simplest methods are parent–offspring regression and the analysis of variance (ANOVA). To evaluate a parent–offspring regression, the trait of interest is measured in both parents and all their offspring. The mean scores of the offspring are plotted against the mean score of the two parents (mid-parent value) as shown in the scatter plot of **Fig. 16.3**.

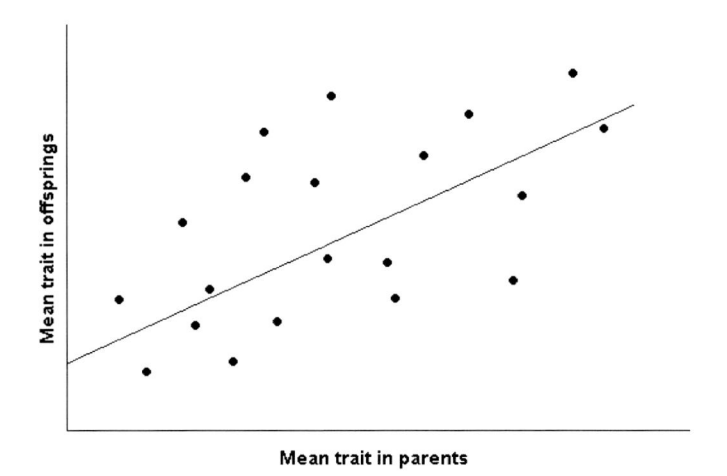

Mean trait in parents

Fig. 16.3. Parent–offspring regression. Scatter plot and regression line showing the relationship between the mid-parental values (x-axis) and the mean trait value in the offspring (y-axis).

Linear regression is a statistical technique of drawing the best fitting straight line through these points, which produces an equation for the line in the form:

$$Y = a + bX$$

The slope b of the regression line is Cov (X, Y)/Var (X)—in this case the covariance between the mid-parent trait value and the offspring divided by the variance of the trait in the parental generation. The slope of the regression line on the mid-parent value is:

$$\text{Cov } (X, Y)/\text{Var } (X) = \frac{\sigma_a^2}{\sigma_p^2}$$

This value, representing the proportion of the phenotypic variance explained by the additive genetic variance, is described in the following as being the "narrow heritability" of a trait. The slope of the regression line reflects how much of the phenotypic differences that we find in parents are retrieved in their offspring. The steeper the slope, the higher is the proportion of phenotypic variance passed on from parents to offspring that is genetically additive.

Offspring–parent regression is not often used in practice. It requires data on two generations, and uses only this kind of data.

ANOVA can be applied to sib and/or half-sib families, monozygotic and dizygotic twins. This technique partitions the total variance observed in a dataset into within- and between-group variances to estimate the relative contributions of the genetic and environmental components. It is more flexible than parent–offspring regression and provides unbiased estimators, even if the data are not normally distributed. However, ANOVA is not ideal for populations for which one has information on many relationship types and unbalanced family sizes.

A more flexible alternative for estimation of variance components is maximum likelihood (ML), or more often restricted maximum likelihood (REML) variance-component estimation (20, 21). Assuming a particular distribution, generally multivariate normal, for the phenotype, the methods allow partitioning the variance into its basic genetic and non-genetic components, using a sample of individuals of known relationship. Because the analysis can use the information from all types of relative pairs in the data, without concern for balanced numbers of families of restricted relationship types, the information available in the data are used more efficiently than in ANOVA methods.

In ML and REML the aim is to find the set of model parameters that maximizes the likelihood of the data. To this purpose, different algorithms have been developed. An overview is given by Meyer (22).

2.3. Heritability

Once the variance components have been estimated, the global influence of the genetic effects on the phenotype variability can be summarized through a statistical descriptor called heritability.

The more common summary statistic is the narrow sense heritability of a phenotype:

$$h^2 = \frac{\sigma_a^2}{\sigma_p^2}$$

which is the fraction of the total variance due to additive genetics effects and is specific to the population and environment for which it has been evaluated. Another measure of the genetic influence on the phenotype is the broad sense heritability, which represents the fraction of the phenotypic variability due to the total genetic effects:

$$H^2 = \frac{\sigma_G^2}{\sigma_p^2}$$

It is typically assumed that the additive effects are the primary contributors to the trait and most studies in humans ignore dominance effects. Moreover, from Equation 3, the dominance effect contributes to the additive variance. Even when there is considerable dominance, σ_a^2 is usually larger than σ_d^2. It is, in fact, possible to have a high additive variance even when all loci follow a dominant model. Therefore, a high additive variance does not necessarily indicate that there are any loci that follow a strictly additive model. However, in human studies, dominance variance components should be considered in the inbred case, as a pair of relatives are more likely to share two copies of the same allele at the same locus *(23–25)*.

A small narrow heritability (or a heritability = 0) only indicates that the influence of additive variance on the phenotype is small (or = 0), but does not imply the trait is not under genetic influence. Furthermore, a trait can be genetically determined even if all the phenotypic variance within a population is environmental. For example, the number of fingers on a human hand is genetically determined, but the heritability of number of fingers in humans is almost certainly very low (if the amount of variation in the genotype for a trait is negligible, the heritability is close to zero even if genetic factors determine a trait).

The heritability value also depends on the distribution of allele frequencies at the genes determining the trait variability—as suggested by Equation 3—therefore, it is strictly specific to the population for which it has been evaluated, and comparing trait heritability in different populations is meaningless. Nevertheless, observing a high heritability for a trait is sufficient justification to set up a mapping study. Indeed, even though a high heritability does not guarantee the success of a mapping study, there are more chances to detect a major locus for the trait if a strong genetic component is observed in the sample.

A popular way to estimate the heritability of a trait in a population compares the resemblance between twins. Since identical monozygotic twins (MZ) share all genes while dizygotic twins (DZ) share half of the genome, the narrow heritability is approximately twice the difference in the trait correlation r between MZ and between DZ twins:

$$h^2 = 2\left(r\left(MZ\right) - r\left(DZ\right)\right)$$

2.4. Major Quantitative Trait Locus Effect

The effect of a major QTL, a locus with major effect on a quantitative trait can be described through the biometrical model described in **Section 2.1.** for the single gene model. The genotypic values a and d are now characterized in the context of the normal distribution of a trait in a population, as shown in **Fig. 16.4** for an additive QTL with alleles A and B.

In a simple model, the phenotypic values are assumed to be normally distributed around each genotypic value because of the independent effects of the polygenic and environmental components. By definition, the average genotypic effects of a major QTL stand out against the remaining polygenic and environmental background effects, although the phenotypic distribution around each genotypic value can partially overlap with the others.

The population variance due to the segregation of the QTL is:

$$\sigma_g^2 QTL = \sigma_a^2 QTL + \sigma_d^2 QTL$$

In practice, many mapping studies in outbred populations consider only the QTL additive variance. Indeed, moderate dominance effects, as those observed in outbred populations, are likely to be absorbed by the additive variance *(26)*.

The additive heritability of the QTL, or QTL effect, is defined as the proportion of the total phenotype variability due to the QTL.

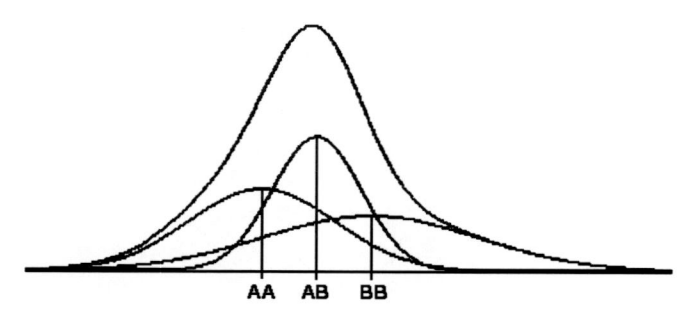

Fig. 16.4. Illustration of the genotypic effects for a major diallelic quantitative trait locus (QTL). Distribution of the quantitative trait in the population (upper curve) as sum of the distributions of the phenotypic values around each QTL genotype.

$$h^2_{QTL} = \frac{\sigma^2_a QTL}{\sigma^2_p} \qquad [4]$$

Here, the total phenotypic variance σ^2_p is the sum of the QTL effect, the environmental effects, and the polygenic effects (due to all the other genes influencing the trait).

The QTL additive component, as previously shown for the single locus effect, is:

$$\sigma^2_a QTL = 2pq\,[a - (p - q)d\,]^2 \qquad [5]$$

Therefore the QTL effect is mainly a function of both the displacement between the genotypic means and the allele frequencies.

When a novel QTL is identified through a mapping study, confirmation in another independent sample is usually requested in order to corroborate the results. Equations 4 and 5 highlight some straightforward implications for the success of a replication study. From Equation 4 it is clear that, even neglecting all kinds of interactions, the same QTL can show different effects among populations under different environmental and genetic background influences, as they contribute in determining the total phenotypic variance. On the other hand, Equation 5 suggests that different QTL effects might be observed between two populations sharing the same environment and genetic background, if they have different allele frequencies at the QTL.

In general, no two human samples can be genetically identical, as a result of evolutionary history, and environmental and culture variations cannot be avoided in natural populations. Consequently, the prospects for replication are often very poor and the fact that a QTL effect is observed in a population sample only implies that it is easier to observe its effect in that specific sample.

These problems are exacerbated for QTL with small displacement. According to Equation 5, a rare QTL with a large displacement can show the same effect as a more common QTL with a smaller displacement. However, the genotypic effects of a QTL with large displacement are distinguishable, as each genotype corresponds with slight ambiguity to a different mean of the quantitative trait. This category of QTLs is also referred to as mendelian, because the segregation of the trait level within the families closely follows the segregation of the QTL. An example is the apolipoprotein (a)—a risk factor for cardiovascular diseases—whose quantitative variation results from the effects of alleles at a single locus *(27)*. A mendelian QTL can be easily identified by a mapping study, and its effects are more likely to be consistent among independent samples.

When the displacement is small the QTL genotypic effects overlap to a large extent. Therefore, different genotypes at the

QTL correspond to the same phenotypic values, hampering their identification and replication through mapping studies.

3. QTL Mapping

The main approaches to QTL mapping are the *candidate gene* and the *marker locus* approaches. In the candidate gene approach, a trait is tested against the functional polymorphism within a gene selected on the basis of a biological hypothesis. Selecting a candidate gene *a priori* is sometimes difficult and does not allow the identification of new QTLs. However, this approach is also used to screen the candidate genes in a chromosomal interval previously identified by a marker locus approach. This is a hypothesis-driven approach and conforms to rigorous epidemiological principles.

The marker locus approach tests a trait against a number of polymorphic genetic markers at known position in the genome to identify those chromosomal regions that are likely to contain a gene for the trait. These approaches are hypothesis-free, and less well suited to pure epidemiological science. However, they are more popular, as they allow the identification of several genes for mendelian traits, and are beginning to show some successes also in the identification of QTLs for complex traits *(28, 29)*.

Only a small proportion of the human genome codes for proteins. About 1% of the genome presents sequence variants. The most important structural variants used in genetic mapping are short tandem repeats (or microsatellites) and single nucleotide polymorphisms (SNPs). Microsatellites consist of multiple repeats of short sequences (generally 2 to 8 base pairs, bp) and alleles are differentiated by the number of repeats. Microsatellites have a very high mutation rate; therefore, there is a high probability of observing a large number of heterozygous loci in an individual genome. Despite the efforts expended to automate the genotyping process, allele scoring is prone to errors, as it requires an accurate assessment of the size of the PCR product through gel electrophoresis.

SNPs are variations at a single nucleotide and are very frequent in the genome; currently more than 10 millions SNPs in the human genome are known. They do not require gel electrophoresis sizing and are suitable for high throughput genotyping at low cost. SNPs are less informative than microsatellites; however, the large number and the ease of typing allows generating very dense maps suitable for mapping studies.

3.1. Linkage Studies

Methods for linkage analysis are based on the analysis of the meiotic recombination events in families. Meiosis is the process by which diploid germ cells divide to produce haploid gametes for sexual reproduction. In meiotic cell division the germ cells undergo two rounds of chromosomal division following a single round of chromosome replication. In prophase of meiosis I, chromosomes first replicate into sister chromatids, and then search for and align with their homologous partner. Recombination between homologous chromosomes is initiated during this stage of meiosis, and is mediated by a cohort of enzymes that accumulate at the site of recombination giving rise to the mechanism of crossing over. Two alleles at different loci on the same chromosome can be separated by an odd number of crossings over (**Fig. 16.5**).

The probability of crossing over between two loci is proportional to the distance between them. Therefore, close loci on the same chromosome are more likely to segregate together than two distant loci. The recombination fraction θ between two loci is defined as the probability that the two alleles at these loci are recombinant, and hence derives from different parental chromosome. The values of θ between two loci can assume every value between 0, when no recombinations can be observed between two loci, to $\theta = 0.5$ when they segregate independently.

The principal method for the study of mendelian traits is the so-called LOD-score method and relies on the analysis

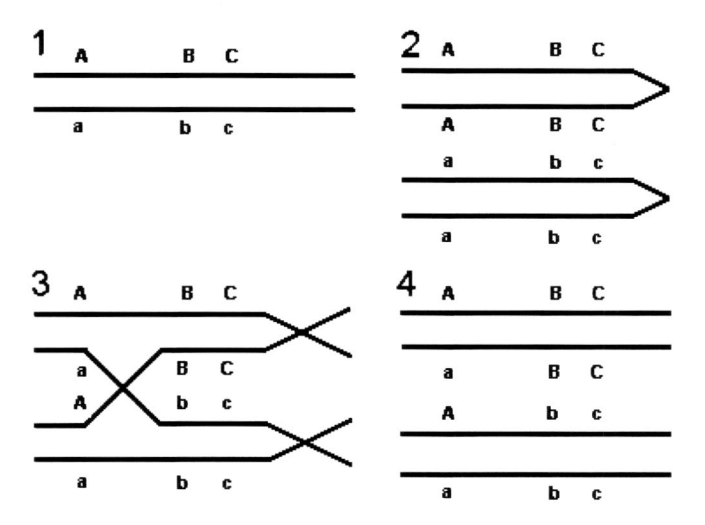

Fig. 16.5. Recombination is generated by an odd-number of crossings over. After chromosomes duplicate (1), the sister chromatids pair up (2) and a crossing–over takes place between loci A and B (3). This results in recombination between the locus pairs A–B and A–C, whereas no recombination is observed between the relatively closer loci B and C (4).

of recombination between a hypothetical QTL and a map of genetic markers with known location in the human genome. This method, also called parametric or model-based, requires the specification of a genetic model for the QTL, including allele frequencies, mean effect of the genotypes, and genetic variance in the population. The relative location of a QTL within a map of markers is estimated by maximizing the likelihood of the genetic model parameters at different values of recombination fraction between 0 and 0.5 *(9, 30, 31)*. Significance of linkage is evaluated by comparing the likelihood of the model assuming a value of $\theta < 0.5$ with a model that assumes the QTL and the map interval segregate independently ($\theta = 0.5$). Conventionally, significance of linkage for a given θ is declared when the $\log_{10}$ of the ratio of the two models' likelihoods is >3 *(7)*. This value takes into account testing with multiple markers and corresponds to an asymptotic *p*-value of 0.0001. The 1-LOD-support interval, identified by moving along the chromosome from the highest test statistic until the LOD score decreases by 1 unit, approximately corresponds to a 95% confidence interval of the QTL localization.

In more recent studies, higher LOD-score thresholds (in the range 3.3–4.0, depending on the study design) have been shown to provide a genome-wide significance level of 0.05, when accounting for the multiple tests that are performed in a genome-wide search *(32)*. However, the specification of a trait model for non-mendelian loci is particularly complex, and a wrong specification of the model parameters might lead to the overestimation of the recombination fraction. Moreover, these methods are computationally challenging, although Markov chain Monte Carlo (MCMC) methods *(33)* have provided new tools able to fit such complex models even in extended families.

The high parameterization of LOD-score methods and the difficulty of obtaining correct parameter estimates has driven the development of various nonparametric, or model-free, methods. These models do not make any assumption on the mode of inheritance for the trait and are based on the estimated proportion of alleles shared IBD among individuals in the tested chromosomal interval.

In model-free methods, information derived from observation of the meiotic segregation process is used to estimate, for each pair of individuals, the proportion of alleles shared IBD. Using a map of polymorphic markers, the same marker alleles can be observed for different individuals since each marker has a limited number of alleles; for example, SNPs typically have only two alleles. The probability for two individuals to share the same SNP alleles identical from a molecular point of view—or identical-by-state (IBS)—is very high. However, no assumptions can be made on the genetic content of the chromosomal regions flanking a marker (or comprised in a markers map interval).

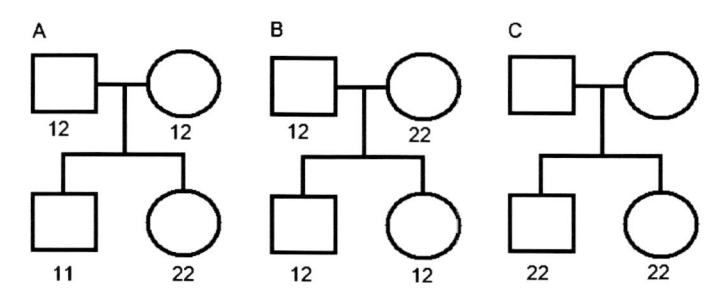

Fig. 16.6. Difference between identity-by-state (IBS) and identity-by-descent (IBD). The two sibs in pedigree A share 0 alleles IBS and 0 alleles IBD. In pedigree B the two sibs share 2 alleles IBS. Allele 1 is IBD since it has been transmitted at both of them from the father. However, it is impossible to tell if allele 2 is IBS or IBD, because the mother is homozygous and could have transmitted the same allele to both sibs or a different copy to each of them, and these events can occur with the same probability. In pedigree C the two sibs share all the alleles IBS, but it is impossible to tell how many alleles they share IBD. This probability can be calculated based on its population frequency, and thus the probability of each parent carrying one or two copies of the same allele.

If two individuals share the same allele IBS, whose segregation can be traced through a common ancestor, they are also likely to share the same genetic content around the marker (**Fig. 16.6**). In order to obtain better estimates of the IBD status, the recombination events are observed taking into account multiple markers at the same time. Non-informative marker segregation can be determined probabilistically from the segregation pattern of the adjacent markers given their recombination fractions. Several algorithms have been developed (9, 30, 31) for estimating IBD sharing among each pair of individuals using the available family members and population genotype frequencies.

The non-parametric method tests whether individuals who show similar trait values also share genomic regions at IBD levels higher than those expected by chance. Individuals with dissimilar trait values should show a decreased level of IBD close to the trait locus. Given π_0, π_1, π_2—the probabilities of two non-inbred individuals sharing, respectively, 0, 1, and 2 alleles IBD at a chromosomal region—the expected proportion of alleles IBD is:

$$\hat{\pi} = \pi_1 / 2 + \pi_2.$$

The two widely used approaches to QTL mapping are regression based (RB) and variance component (VC) linkage analysis.

3.1.1. Regression Based Linkage Analysis

The most popular non-parametric model for linkage analysis of quantitative traits is the Haseman and Elston (HE) regression method for sib-pair data *(34)*. The HE method tests the linear regression relationship between the squared difference in trait values $Y = (X_1 - X_2)^2$ of a sibling pair (the dependent variable) and the proportion $\hat{\pi}$ of alleles shared IBD between them at a position in the genome (the independent variable). The idea is that if the genomic location is linked with a QTL, small differences in the trait should be associated with high IBD sharing, and vice versa.

Assuming only additive effects, for the pairs (i, j):

$$Y_{ij} = \left(X_i - X_j\right)^2 = b_0 + b_{ij}\,\hat{\pi}_{ij} + e_{ij}$$

where b_0 is the intercept, b_{ij} the slope and e_{ij} an error term.

The slope of the regression line is $-2\sigma_a^2 QTL$ if the recombination fraction between the marker and the QTL is equal to zero. If the recombination fraction is different from zero, the regression coefficient is $-2(2\Psi - 1)\sigma_a^2 QTL$, where $\Psi = \theta^2 (1 - \theta)^2$ and $(2\Psi - 1)$ is the correlation between the proportion of alleles IBD at the chromosomal location and the proportion of alleles IBD at the QTL.

Under the hypothesis of no linkage, the regression coefficient is 0, while under linkage it is less than 0. The null hypothesis is easily tested by a standard one-sided Student's t-test.

Using the square of trait differences ignores information inherent in the complete bivariate distribution of the individuals' trait values and results in a lost of power. Several authors have subsequently proposed modification of the original test to improve its power, using the squared sum of the trait values together with the squared difference in two separated regressions (*see (35)* for a survey on revised HE methods). Further extensions have been proposed to analyze extended families with multiple sib-pairs *(36, 37)*. Indeed, the squared sum and squared differences of the trait values of different pairs belonging to the same family are correlated. Recently Sham and colleagues *(38)* proposed to reverse the dependent and independent variables, modeling the proportions of alleles shared IBD as a function of the trait values. This method allows the analysis of general pedigrees.

Regression methods are computationally fast, allowing the application of resampling techniques that make them robust to violation of normality of the trait distribution and permit estimation of confidence intervals for the parameters. These approaches are consequently suited for the analysis of selected samples, such as extreme sib-pairs, which are expected to increase the power to detect linkage for trait loci with low heritability *(39)*.

However, for unselected samples the RB methods do not necessarily have more statistical power than the VC *(40)*.

3.1.2. Variance Component Linkage Analysis

The ML variance-component framework has been introduced in **Section 2.2.** as a method for partitioning the phenotypic variance into its basic genetic and non-genetic components, using a sample of individuals of known relationship.

In order to extend the framework to allow for the effect of a QTL in a location of the genome, the covariance structure of the data is modeled by adding a variance component $\sigma_g^2 QTL$ conditional on the IBD sharing of all relative pairs. The genetic covariance for the pair of individuals (i,j) can be rewritten as:

$$Cov\left(g_i, g_j\right) = r\sigma_a^2 + \delta\sigma_d^2 + \pi\sigma_a^2 QTL + \pi_2\sigma_d^2 QTL$$

where $\hat{\pi}$ (the estimated proportion of alleles shared IBD in a location of the genome) is used to model the additive variance due to the QTL, and $\hat{\pi}_2$ (the estimated probability of sharing two alleles IBD) models the QTL dominance variance. Since it is usual to assume an additive model:

$$Cov\left(g_i, g_j\right) = r\sigma_a^2 + \delta\sigma_d^2$$

The VC approach for human QTL mapping was first proposed by Goldgar *(41)*, and further developed by Amos *(42)* in order to analyze several kinds of common relative pairs. The VC was successively extended to accommodate pedigrees of any size and structure *(43)*. VC can jointly model covariate effects along with variance components and provides flexible modeling capabilities. Additional variance components can be incorporated in the model to allow shared environmental effects, dominance effects, gene–gene and gene–environment interactions. The linkage is evaluated by a likelihood ratio test, presented as a LOD score, between a model estimating all components of variance and a reduced model in which one or more components of variance are required to equal zero. For instance, the presence of a QTL in a given chromosomal region is evaluated by comparing the likelihood of a model estimating $\sigma_a^2 QTL$ and the likelihood of a reduced model in which $\sigma_a^2 QTL = 0$. When the test considers a single component of variance, the likelihood ratio test is asymptotically distributed as a $1/2:1/2$ mixture of a χ^2 variable and a point mass at zero *(44)*.

For complex traits, the simultaneous modeling of QTLs in an oligogenic model has been suggested to improve the power of the analysis, at the same time reducing the number of false-positives. When n QTLs are modeled together, the genetic covariance for the pair of individuals (i,j) can be rewritten as:

$$Cov\left(g_i, g_j\right) = \Sigma_{i=1}^n \hat{\pi}\sigma_a^2 QTL + r\sigma_a^2$$

The null model can then be evaluated by constraining one or more QTLs to have zero effect. When multiple components of variance are jointly tested, the resulting likelihood ratio test statistic has a more complex asymptotic distribution that continues to be a mixture of χ^2 distributions. Given the relative low power for the sample sizes commonly used in human genetics linkage studies, it has been shown that QTL effect estimates can be upwardly biased for several reasons *(45)*. This problem can reduce the power of oligogenic models if the total variance attributable to the QTLs jointly considered in the model would exceed the trait genetic variance *(46)*.

The trait values of related individuals are assumed to follow a multivariate normal distribution *(47, 48)* and when this assumption holds the VC method is very powerful. When this assumption is violated the probability of false-positives (or type I error) can be inflated to a great degree *(49)*. Some solutions have been suggested, for example in *(50, 51)*. Methods based on permutation tests, as proposed for instance in *(52)*, allow one to obtain robust empirical *p*-values. For instance, permutation tests in sib-pairs can be evaluated by shuffling the trait values between pairs and repeating the VC analysis a large number of times (e.g., 10,000 times). The estimate of the empirical significance is determined by the proportion of the permuted samples in which the test statistic exceeds the one observed in the actual data set. However, these methods entail an enormous increase in the computational load.

3.1.3. Interactions

Complex models in linkage analysis might encompass gene–gene and gene–environment interactions.

In classical mendelian genetics, interaction between loci (or epistasis) refers to the masking of genotypic effects at one locus by genotypes of another. The usage of the term in quantitative genetics is much broader, and refers to any statistical interaction between genotypes at two or more loci *(53, 54)*.

Some studies of epistasis among major or candidate genes have found epistasis in the expression of complex traits of medical importance in humans, including type II diabetes *(55)* and inflammatory bowel disease *(56)*.

However, the sample sizes of human studies will probably need to be considerably larger to achieve enough power to detect epistatic effects when the trait gene is unmeasured. Epistatic effects are generally smaller than the additive and dominance effects and difficult to distinguish *(57, 58)*; hence, their effect can be absorbed by the other variance components *(59)*. Consequently, a test for interaction between two unmeasured covariates is less powerful than those for the main effect of each component *(60)*. Evaluations of epistasis are not included as a standard procedure in the genetic analysis of complex traits in human populations.

For analogous reasons, estimation of the gene–environment components of variance of a trait in human is difficult, even though some studies have found gene–environment interactions for complex traits, for example, *see (61)*. Genetic studies in animal breeding experiments have shown that gene–environment interactions are extremely common *(19)*. In human quantitative genetic analyses, it is often not possible to distinguish environmental and genetic influences unless the specific genotype and environmental factors are explicitly measured. Monozygous twins are a potential resource for detecting genotype–environment interactions (Jinks & Fulker, 1970). From a public health perspective, identifying gene–environment interactions involved in the development of a disease would in principle allow targeted prevention measures for those people having particularly high genetic risk.

3.1.4. Multivariate Models

Linkage of two correlated traits to a single chromosomal region is often taken as further support that a major gene controlling both traits exists in that region. Several authors have suggested that multi-trait analysis can increase the power of QTL mapping, providing finer precision in the localization, and allowing one to test for linkage versus pleiotropy when a QTL appears to affect two distinct traits *(62)*.

Amos and co-workers *(63)* compared HE and VC methods in a multivariate context. They demonstrated that the power of multivariate models is high when polygenic and major gene correlations in traits were opposite in sign.

3.2. Association Studies

3.2.1. Linkage Disequilibrium

Linkage disequilibrium (LD), the tendency for alleles of linked loci to co-occur non-randomly on chromosomal haplotypes, is a useful phenomenon that can be exploited to localize and identify a variant influencing a quantitative trait.

Consider loci A and B with alleles A_1, A_2, and B_1, B_2, with frequencies p_A, q_A, p_B, and q_B, respectively. Under linkage equilibrium, the expected frequency of the haplotype A_1B_1 is $p_A p_B$, the product of the allelic frequencies in the population. If the observed haplotype frequency is significantly different from the expectation the two loci are in linkage disequilibrium.

The amount of linkage disequilibrium D *(64)* is measured as:

$$D = P_{AB} - P_A P_B$$

Since the parameter D depends also on the allele frequencies, its value is of little use for measuring the strength of and comparing levels of LD. The two most common measures of LD based on D are the absolute value of D′ and r^2. The absolute value of D′ is determined by dividing D by its maximum possible value, given the allele frequencies at the two loci, $D' = |D/D_{max}|$, r^2 is equal to D^2 divided by the product of the allele frequencies at the two loci, $r^2 = D^2/(p_A q_A p_B q_B)$.

Both these measures equal 0 when there is independence between the two loci, and 1 in the case of complete disequilibrium. The correlation coefficient between the two loci, r^2, is the measure of choice for quantifying and comparing LD in a gene-mapping context and there is a simple relationship between this measure and the sample size required to detect association between the marker locus and the variant. Moreover, r^2 is less inflated in small samples than D', and intermediate values of r^2 are more interpretable than intermediate values of D'.

An overview of the different measures of LD and their utility in LD mapping can be found in Devlin and Risch *(65)*.

When a new mutation occurs on a chromosome, the alleles at any locus of the chromosome will be in complete LD with the mutation. When the chromosome is transmitted through the generations, the recombinations between the mutant allele and the other loci will restore linkage equilibrium between the mutant allele at all but closely linked loci. This phenomenon is exploited in LD mapping, as individuals sharing the mutant allele will also have a genomic fragment in LD with the original mutation. The length of the genomic fragment depends on the average amount of recombination per generation in that particular region of the genome, the number of generations that have passed since the original mutation, and the population size *(14, 66, 67)*.

3.2.2. Linkage Disequilibrium Mapping

The relatively small number of meiotic events observable in families limits the resolution of QTL linkage studies *(68)*. LD mapping provides one possible strategy for narrowing the chromosomal regions previously identified by a linkage study. In addition, LD studies are expected to have more power than linkage studies, particularly for traits having low heritability *(69, 70)*. The limited successes of linkage analysis for complex traits, together with advances in genotyping technology, are shifting the attention of mapping studies in humans toward LD studies. These studies focus their attention on the trait–allele relationship instead of on IBD-sharing in families. Therefore, the success of an LD study is influenced by the chance of selecting those markers whose allele frequencies match the trait allele frequencies *(71, 72)*.

QTL association designs can be broadly categorized into population- and family-based designs. Population-based association studies use a sample of unrelated individuals from a natural population. In the simplest design the population sample is stratified by marker genotype, and association is inferred if there is a significant difference in trait mean between the different genotypic classes. For instance, the measured-genotype test *(73, 74)* compares by ANOVA the mean quantitative-trait values of individuals who have either specific alleles or specific genotypes. A comparison among regression, ANOVA, and maximum-likelihood analyses to detect LD between a marker locus and QTL in samples from a random mating population can be found in Luo et al. *(75)*. Novel developments of population-based LD-mapping methods are in Fan et al. *(76)*.

The measured genotype approach can be also implemented in a variance component framework and represents a powerful test to detect association in families *(77)* by testing for genotype-specific differences in the means of traits while allowing for non-independence among family members. However, spurious marker-trait associations can be detected as a consequence of population admixture *(78)*. For instance, spurious association can arise in a sample recruited for a population derived from the union of two or more genetically distinct populations, if the populations have different allele frequencies at the marker locus and also different allele frequencies at the trait locus, or different means for the trait. All the population-based tests are also affected by the presence of population substructures in the sample and should be used with ethnically homogeneous samples. A suitable way to detect population stratification is testing for Hardy-Weinberg equilibrium at the genotyped markers *(79)*.

Several solutions have been proposed to allow association studies in the presence of population stratification. For instance, since stratification acts on the whole genome and not only locally, a genomic control *(80)* that uses unlinked markers from the genome has been suggested to create appropriate corrections for population-based association tests.

Most of the so-called family-based designs derive from the transmission disequilibrium test (TDT) method, originally developed for binary traits *(81)*. The TDT has been extended to quantitative trait analysis *(82–86)* and can be used in the presence of population stratification. These quantitative TDT-like tests examine whether subjects with different genotypes within the same family have different trait means. Because all family members belong to the same subpopulation, significant within-family differences cannot be the result of admixture and are thus indicative of association between the trait and the tested polymorphism(s). Family-based methods are not without their own drawbacks. Recruiting family members can be difficult and more genotyping resources are required for family-based studies. Moreover, in some settings, family-based designs are less powerful than population-based designs *(87)*, even for the same number of individuals sampled. A review of family-based designs is provided by Laird and Lange *(88)*.

The HapMap project, the SNP Consortium *(89)*, and Perlegen Sciences *(90)* have already identified over 5 million SNPs using different human populations. Since it is not practical to identify a different panel of SNPs for each study, these data allow researchers to select a set of validated markers likely to be informative for their own sample. The emerging literature on strategies for LD mapping is partly driven by the current economical and experimental conditions. Although it is possible to identify and genotype all the candidate variants for a small number of candidate genes, for marker-based approaches in large chromosomal regions (or even genome-wide) it is often necessary to reduce the potentially available set of genotypic variants to a smaller but still informative subset. Therefore, a number of potential approaches have been proposed to maximize the amount of information that can be extracted from a chromosomal region using a reasonably small number of genotyped markers. This aim can be achieved by a careful marker selection and/or retrieving the hidden information between the markers by means of multi-locus haplotype data. The informativeness of markers might be increased by using highly polymorphic markers such as microsatellites (91), or by using a set of tag-SNPs likely to be a good proxy for the common variants in the genome. Microsatellites have a higher degree of heterozygosity than SNPs, on average ~70% vs. ~20%. Even if microsatellites are less frequent than SNPs in the genome, they have the advantage of showing a longer LD range because of their high mutation rate (and therefore younger age). However, when using multi-allelic markers, the degrees of freedom of LD tests increases considerably with the number of alleles, although methods have been proposed in order to deal with this problem (92). This problem can be exacerbated for multi-locus models using haplotype data. Indeed, some authors suggest that using multiple markers simultaneously could increase the power of detection and improve the accuracy in localization of QTLs (93, 94). On the other hand, other authors have observed that uncertainty in haplotype determination may introduce bias into the subsequent analysis, making haplotypes less powerful than the corresponding unphased genotypes (95, 96).

Some multi-locus models do not assume that haplotype phases are known, for instance, Haplotype Trend Regression developed by Zaykin et al. (97) (which is very close to the method of Schaid et al.) (98). Other models assume that haplotype phases are known, for example, the approach proposed by Meuwissen and Goddard (93), which uses mixed models to model the haplotype effect.

Phase could be derived using either molecular techniques, or by typing additional relatives, but both methods are costly and often impractical, and in some cases, phase ambiguity may remain. Alternatively, the phase could be estimated from a population sample using a reconstruction algorithm such as the EM algorithm (described for example in Excoffier and Slatkin (99)), or PHASE (100). For a detailed coverage of recent computational methods on this subject, see Niu (101).

Another strategy to select a subset of maximally informative markers has built upon the early observation that a particular extent of correlation and structure characterizes the haplotype pattern of the human genome (102–104). Large projects, such as the International HapMap Consortium, have been founded with the purpose of detecting the common patterns of DNA sequence variation by comparing genetic sequences among people within and among populations. There is an emerging literature on methods for choosing an optimal subset of SNPs (tag-SNPs) that could capture most of the signals from the untyped markers in an association study *(105–107)*. The proposed strategy is to genotype the subset of SNPs that are in strong LD with the largest number of common SNPs. This subset of tag-SNPs would then serve as a proxy for the untyped SNPs. The utility of tag-SNPs in association studies derives from the hypothesis that common variants play an important role in the etiology of common diseases *(108)*. This hypothesis has been supported by some examples of genes in which a single common allele increases the risk for a disease, for example, APOE locus with Alzheimer's disease and heart disease *(109)*, PPARγ and type-2 diabetes *(110)*, and a recent finding for the complement factor H and age-related macular degeneration *(111)*. However, the validity of this strategy has been questioned *(112, 113)*.

Improvements in genotyping technology have enabled typing large numbers of SNPs, which opens the possibility of genome-wide LD mapping for complex traits *(114)*, thus dispensing with traditional linkage mapping. LD studies have increasingly reported positive results *(29, 111, 115)*, but the number of replications is very low *(116)*. The main reasons are the variability in study design, population stratification, and the stringent threshold for individual tests as a consequence of multiple testing when using a large number of markers. Assuming a transmission disequilibrium test with one million SNPs, Risch and Merikangas *(114)* suggested a pointwise significance of 5×10^{-8} after Bonferroni correction for a genome-wide Type I error rate of 0.05. Extremely large sample sizes are needed to achieve a similar level of significance, especially for the small QTL effects expected for complex traits. However, the suggested threshold is over-conservative, as the allelic configuration at the various loci is not independent because of the background LD among the markers. The problem can be circumvented by assessing an empirical significance, for instance, through the application of permutation tests. This procedure can be extremely computationally intensive, since the empirical significance is assessed by repeating the initial analysis in a large number of permuted datasets. A promising approach to correct for multiple comparisons is the false discovery rate, proposed by Benjamini and Hochberg *(117)*. This approach controls, when correcting for multiple comparisons, the proportion of wrongly rejected null hypotheses over all the rejections. Weller et al. *(118)* proposed the first application of this procedure for QTL mapping. A review of false discovery rate for QTL mapping is provided by Benjamini and Yekutieli *(119)*.

References

1. Morgan, T. H. (1910) Sex limited inheritance in *Drososphila*. *Science* 32, 120–122.

2. Haldane, J. B. S. (1936) A provisional map of a human chromosome. *Nature* 137, 398–400.

3. Haldane, J. B. S., Smith, C. A. B. (1947) A new estimate of the linkage between the genes for color-blindness and hemophilia in man. *Ann Eugen* 14, 10–31.

4. Barnard, G. A. (1949) Statistical Inference. *J Roy Stat Soc Ser B* 11, 115–139.

5. Fisher, R. A. (1922) On the mathematical foundations of theoretical statistics. *Phil Trans Roy Soc Lond* Ser A, 222, 309–368.

6. Fisher, R. A. (1922) The systematic location of genes by means of crossover observations. *Am Nat* 56, 406–411.

7. Morton, N. E. (1955) Sequential tests for the detection of linkage. *Am J Hum Genet* 7, 277–318.

8. Simpson, H. R. (1958) The estimation of linkage on an electronic computer. *Ann Hum Genet* 22, 356–361.

9. Elston, R. C., Stewart, J. (1971) A general model for the genetic analysis of pedigree data. *Hum Hered* 21, 523–542.

10. Ott, J. (1974) Estimation of the recombination fraction in human pedigrees: efficient computation of the likelihood for human linkage studies. *Am J Hum Genet* 26, 588–597.

11. Botstein, D., White, R. L., Skolnick, M., et al. (1980) Construction of a genetic linkage map in man using restriction fragment length polymorphisms. *Am J Hum Genet* 32, 314–331.

12. Mendel, G. (1866) Versuche uber Pflanzenhybriden. *Verh Naturforsh Ver Brunn* 4, 3–44.

13. Fisher, R. A. (1918) The correlation between relatives on the supposition of Mendelian inheritance. *Trans Roy Soc Edinburgh* 52, 399–433.

14. Falconer, D. S., Mackay, T. F. C. (1996) *Introduction to Quantitative Genetics*. Longman, Essex, UK.

15. Hilbert, P., Lindpaintner, K., Beckmann, J. S., et al. (1991) Chromosomal mapping of two genetic loci associated with blood-pressure regulation in hereditary hypertensive rats. *Nature* 353, 521–529.

16. Jacob, H. J., Lindpaintner, K., Lincoln, S. E., et al. (1991) Genetic mapping of a gene causing hypertension in the stroke-prone spontaneously hypertensive rat. *Cell* 67, 213–224.

17. Paterson, A. H., Lander, E. S., Hewitt, J. D., et al. (1988) Resolution of quantitative traits into Mendelian factors by using a complete linkage map of restriction fragment length polymorphisms. *Nature* 335, 721–726.

18. Pamilo, P. (1989) Comparison of relatedness estimators. *Evolution* 44, 1378–1382.

19. Lynch, M., Walsh, B. (1998) *Genetics and Analysis of Quantitative Traits*. Sinauer Associates, Sunderland, MA.

20. Hartley, H. O., Rao, J. N. (1967) Maximum-likelihood estimation for the mixed analysis of variance model. *Biometrika* 54, 93–108.

21. Patterson, H. D., Thompson, R. (1971) Recovery of interblock information when block sizes are unequal. *Biometrika* 58, 545–554.

22. Meyer, K. (1990) 4th World Congress on Genet. Applied to Livest. pp. 403, Prod. Edinburgh.

23. Cockerham, C. C., Weir, B. S. (1984) Covariances of relatives stemming from a population undergoing mixed self and random mating. *Biometrics* 40, 157–164.

24. Harris, D. L. (1964) Genotypic covariances between inbred relatives. *Genetics* 50, 1319–1348.

25. Jacquard, A. (1974) *The Genetic Structure of Populations*. Springer-Verlag, New York.

26. Pratt, S. C., Daly, M. J., Kruglyak, L. (2000) Exact multipoint quantitative-trait linkage analysis in pedigrees by variance components. *Am J Hum Genet* 66, 1153–1157.

27. Boerwinkle, E., Leffert, C. C., Lin, J., et al. (1992) Apolipoprotein(a) gene accounts for greater than 90% of the variation in plasma lipoprotein(a) concentrations. *J Clin Invest* 90, 52–60.

28. Falchi, M., Forabosco, P., Mocci, E., et al. (2004) A genomewide search using an original pairwise sampling approach for large genealogies identifies a new locus for total and low-density lipoprotein cholesterol in two genetically differentiated isolates of Sardinia. *Am J Hum Genet* 75, 1015–1031.

29. Herbert, A., Gerry, N. P., McQueen, M. B., et al. (2006) A common genetic variant is associated with adult and childhood obesity. *Science* 312, 279–283.

30. Lander, E. S., Green, P. (1987) Construction of multi-locus genetic linkage maps in humans. *Proc Natl Acad Sci U S A* 84, 2363–2367.

31. Sobel, E., Lange, K. (1996) Descent graphs in pedigree analysis: applications to haplotyping, location scores, and marker-sharing statistics. *Am J Hum Genet* 58, 1323–1337.

32. Lander, E., Kruglyak, L. (1995) Genetic dissection of complex traits: guidelines for interpreting and reporting linkage results. *Nat Genet* 11, 241–247.

33. Heath, S. C. (1997) Markov chain Monte Carlo segregation and linkage analysis for oligogenic models. *Am J Hum Genet* 61, 748–760.

34. Haseman, J. K., Elston, R. C. (1972) The investigation of linkage between a quantitative trait and a marker locus. *Behav Genet* 2, 3–19.

35. Feingold, E. (2002) Regression-based quantitative-trait-locus mapping in the 21st century. *Am J Hum Genet* 71, 217–222.

36. Elston, R. C., Buxbaum, S., Jacobs, K. B., et al. (2000) Haseman and Elston revisited. *Genet Epidemiol* 19, 1–17.

37. Ghosh, S., Reich, T. (2002) Integrating sibship data for mapping quantitative trait loci. *Ann Hum Genet* 66, 169–182.

38. Sham, P. C., Purcell, S., Cherny, S. S., et al. (2002) Powerful regression-based quantitative-trait linkage analysis of general pedigrees. *Am J Hum Genet* 71, 238–253.

39. Risch, N., Zhang, H. (1995) Extreme discordant sib pairs for mapping quantitative trait loci in humans. *Science* 268, 1584–1589.

40. Yu, X., Knott, S. A., Visscher, P. M. (2004) Theoretical and empirical power of regression and maximum-likelihood methods to map quantitative trait loci in general pedigrees. *Am J Hum Genet* 75, 17–26.

41. Goldgar, D. E. (1990) Multipoint analysis of human quantitative genetic variation. *Am J Hum Genet* 47, 957–967.

42. Amos, C. I. (1994) Robust variance-components approach for assessing genetic linkage in pedigrees. *Am J Hum Genet* 54, 535–543.

43. Almasy, L., Blangero, J. (1998) Multipoint quantitative-trait linkage analysis in general pedigrees. *Am J Hum Genet* 62, 1198–1211.

44. Self, S. G., Liang, K.-Y. (1987) Asymptotic properties of maximum likelihood estimators and likelihood ratio tests under nonstandard conditions. *J Am Stat Assoc* 82, 605–610.

45. Goring, H. H., Terwilliger, J. D., Blangero, J. (2001) Large upward bias in estimation of locus-specific effects from genomewide scans. *Am J Hum Genet* 69, 1357–1369.

46. Falchi, M., Andrew, T., Snieder, H., et al. (2005) Identification of QTLs for serum lipid levels in a female sib-pair cohort: a novel application to improve the power of two-locus linkage analysis. *Hum Mol Genet* 14, 2971–2979.

47. Lange, K. (1978) Central limit theorems for pedigrees. *J Math Biol* 6, 59–66.

48. Lange, K., Boehnke, M. (1983) Extensions to pedigree analysis. IV. Covariance components models for multivariate traits. *Am J Med Genet* 14, 513–524.

49. Allison, D. B., Neale, M. C., Zannolli, R., et al. (1999) Testing the robustness of the likelihood-ratio test in a variance-component quantitative-trait loci-mapping procedure. *Am J Hum Genet* 65, 531–544.

50. Blangero, J., Williams, J. T., Almasy, L. (2000) Robust LOD scores for variance component-based linkage analysis. *Genet Epidemiol* 19, S8–14.

51. Sham, P. C., Zhao, J. H., Cherny, S. S., et al. (2000) Variance-Components QTL linkage analysis of selected and non-normal samples: conditioning on trait values. *Genet Epidemiol* 19, S22–28.

52. Iturria, S. J., Williams, J. T., Almasy, L., et al. (1999) An empirical test of the significance of an observed quantitative trait locus effect that preserves additive genetic variation. *Genet Epidemiol* 17, S169–173.

53. Cheverud, J. M., Routman, E. J. (1995) Epistasis and its contribution to genetic variance components. *Genetics* 139, 1455–1461.

54. Cockerham, C. (1954) An extension of the concept of partitioning hereditary variance for analysis of covariance among relatives when epistasis is present. *Genetics* 39, 859–882.

55. Cox, N. J., Frigge, M., Nicolae, D. L., et al. (1999) Loci on chromosomes 2 (NIDDM1) and 15 interact to increase susceptibility to diabetes in Mexican Americans. *Nat Genet* 21, 213–215.

56. Cho, J. H., Nicolae, D. L., Gold, L. H., et al. (1998) Identification of novel susceptibility loci for inflammatory bowel disease on chromosomes 1p, 3q, and 4q: evidence for epistasis between 1p and IBD1. *Proc Natl Acad Sci U S A* 95, 7502–7507.

57. Eaves, L. J. (1994) Effect of genetic architecture on the power of human linkage studies to resolve the contribution of quantitative trait loci. *Heredity* 72, 175–192.

58. Mitchell, B. D., Ghosh, S., Schneider, J. L., et al. (1997) Power of variance component linkage analysis to detect epistasis. *Genet Epidemiol* 14, 1017–1022.

59. Purcell, S., Sham, P. C. (2004) Epistasis in quantitative trait locus linkage analysis: interaction or main effect? *Behav Genet* 34, 143–152.

60. Breslow, N., Day, N. (1987) *The Design and Analysis of Cohort Studies*. IARC Scientific Publications, Lyon, France.

61. Yaffe, K., Haan, M., Byers, A., et al. (2000) Estrogen use, APOE, and cognitive decline: evidence of gene-environment interaction. *Neurology* 54, 1949–1954.

62. Lebreton, C. M., Visscher, P. M., Haley, C. S., et al. (1998) A nonparametric bootstrap method for testing close linkage vs. pleiotropy of coincident quantitative trait loci. *Genetics* 150, 931–943.

63. Amos, C., de Andrade, M., Zhu, D. (2001) Comparison of multivariate tests for genetic linkage. *Hum Hered* 51, 133–144.

64. Lewontin, R. C., Kojima, K. (1960) The evolutionary dynamics of complex polymorphisms. *Evolution* 14, 450–472.

65. Devlin, B., Risch, N. (1995) A comparison of linkage disequilibrium measures for fine-scale mapping. *Genomics* 29, 311–322.

66. Hartl, D. L., Clark, A. G. (1997) *Principles of Population Genetics*. Sinauer Associates, Sunderland, MA.

67. Hill, W. G., Robertson, A. (1968) Linkage disequilibrium in finite populations. *Theor Appl Genet* 38, 231.

68. Boehnke, M. (1994) Limits of resolution of genetic linkage studies: implications for the positional cloning of human disease genes. *Am J Hum Genet* 55, 379–390.

69. Mackay, T. F. (2001) The genetic architecture of quantitative traits. *Annu Rev Genet* 35, 303–339.

70. Sham, P. C., Cherny, S. S., Purcell, S., et al. (2000) Power of linkage versus association analysis of quantitative traits, by use of variance-components models, for sibship data. *Am J Hum Genet* 66, 1616–1630.

71. Abecasis, G. R., Cookson, W. O., Cardon, L. R. (2001) The power to detect linkage disequilibrium with quantitative traits in selected samples. *Am J Hum Genet* 68, 1463–1474.

72. Schork, N. J., Nath, S. K., Fallin, D., et al. (2000) Linkage disequilibrium analysis of biallelic DNA markers, human quantitative trait loci, and threshold-defined case

and control subjects. *Am J Hum Genet* 67, 1208–1218.

73. Boerwinkle, E., Chakraborty, R., Sing, C. F. (1986) The use of measured genotype information in the analysis of quantitative phenotypes in man. I. Models and analytical methods. *Ann Hum Genet* 50, 181–194.

74. Boerwinkle, E., Visvikis, S., Welsh, D., et al. (1987) The use of measured genotype information in the analysis of quantitative phenotypes in man. II. The role of the apolipoprotein E polymorphism in determining levels, variability, and covariability of cholesterol, betalipoprotein, and triglycerides in a sample of unrelated individuals. *Am J Med Genet* 27, 567–582.

75. Luo, Z. W., Tao, S. H., Zeng, Z. B. (2000) Inferring linkage disequilibrium between a polymorphic marker locus and a trait locus in natural populations. *Genetics* 156, 457–467.

76. Fan, R., Jung, J., Jin, L. (2006) High-resolution association mapping of quantitative trait loci: a population-based approach. *Genetics* 172, 663–686.

77. Hopper, J. L., Mathews, J. D. (1982) Extensions to multivariate normal models for pedigree analysis. *Ann Hum Genet* 46, 373–383.

78. Spielman, R. S., Ewens, W. J. (1996) The TDT and other family-based tests for linkage disequilibrium and association. *Am J Hum Genet* 59, 983–989.

79. Tiret, L., Cambien, F. (1995) Departure from Hardy-Weinberg equilibrium should be systematically tested in studies of association between genetic markers and disease. *Circulation* 92, 3364–3365.

80. Devlin, B., Roeder, K. (1999) Genomic control for association studies. *Biometrics* 55, 997–1004.

81. Spielman, R. S., McGinnis, R. E., Ewens, W. J. (1993) Transmission test for linkage disequilibrium: the insulin gene region and insulin-dependent diabetes mellitus (IDDM). *Am J Hum Genet* 52, 506–516.

82. Abecasis, G. R., Cardon, L. R., Cookson, W. O. (2000) A general test of association for quantitative traits in nuclear families. *Am J Hum Genet* 66, 279–292.

83. Allison, D. B. (1997) Transmission-disequilibrium tests for quantitative traits. *Am J Hum Genet* 60, 676–690.

84. Fulker, D. W., Cherny, S. S., Sham, P. C., et al. (1999) Combined linkage and association sib-pair analysis for quantitative traits. *Am J Hum Genet* 64, 259–267.

85. Laird, N. M., Horvath, S., Xu, X. (2000) Implementing a unified approach to family-based tests of association. *Genet Epidemiol* 19, S36–42.

86. Rabinowitz, D. (1997) A transmission disequilibrium test for quantitative trait loci. *Hum Hered* 47, 342–350.

87. Risch, N., Teng, J. (1998) The relative power of family-based and case-control designs for linkage disequilibrium studies of complex human diseases I. DNA pooling. *Genome Res* 8, 1273–1288.

88. Laird, N. M., Lange, C. (2006) Family-based designs in the age of large-scale gene-association studies. *Nat Rev Genet* 7, 385–394.

89. Sachidanandam, R., Weissman, D., Schmidt, S. C., et al. (2001) A map of human genome sequence variation containing 1.42 million single nucleotide polymorphisms. *Nature* 409, 928–933.

90. Hinds, D. A., Stuve, L. L., Nilsen, G. B., et al. (2005) Whole-genome patterns of common DNA variation in three human populations. *Science* 307, 1072–1079.

91. Tamiya, G., Shinya, M., Imanishi, T., et al. (2005) Whole genome association study of rheumatoid arthritis using 27 039 microsatellites. *Hum Mol Genet* 14, 2305–2321.

92. Terwilliger, J. D. (1995) A powerful likelihood method for the analysis of linkage disequilibrium between trait loci and one or more polymorphic marker loci. *Am J Hum Genet* 56, 777–787.

93. Meuwissen, T. H., Goddard, M. E. (2000) Fine mapping of quantitative trait loci using linkage disequilibria with closely linked marker loci. *Genetics* 155, 421–430.

94. Zeng, Z. B. (1994) Precision mapping of quantitative trait loci. *Genetics* 136, 1457–1468.

95. Clayton, D., Chapman, J., Cooper, J. (2004) Use of unphased multi–locus genotype data in indirect association studies. *Genet Epidemiol* 27, 415–428.

96. Morris, A. P., Whittaker, J. C., Balding, D. J. (2004) Little loss of information due to unknown phase for fine-scale linkage-disequilibrium mapping with single-nucleotide-polymorphism genotype data. *Am J Hum Genet* 74, 945–953.

97. Zaykin, D. V., Westfall, P. H., Young, S. S., et al. (2002) Testing association of statistically inferred haplotypes with discrete and continuous traits in samples of unrelated individuals. *Hum Hered* 53, 79–91.

98. Schaid, D. J., Rowland, C. M., Tines, D. E., et al. (2002) Score tests for association between traits and haplotypes when linkage phase is ambiguous. *Am J Hum Genet* 70, 425–434.

99. Excoffier, L., Slatkin, M. (1995) Maximum-likelihood estimation of molecular haplotype frequencies in a diploid population. *Mol Biol Evol* 12, 921–927.

100. Stephens, M., Smith, N. J., Donnelly, P. (2001) A new statistical method for haplotype reconstruction from population data. *Am J Hum Genet* 68, 978–989.

101. Niu, T. (2004) Algorithms for inferring haplotypes. *Genet Epidemiol* 27, 334–347.

102. Daly, M. J., Rioux, J. D., Schaffner, S. F., et al. (2001) High-resolution haplotype structure in the human genome. *Nat Genet* 29, 229–232.

103. Gabriel, S. B., Schaffner, S. F., Nguyen, H., et al. (2002) The structure of haplotype blocks in the human genome. *Science* 296, 2225–2229.

104. Reich, D. E., Cargill, M., Bolk, S., et al. (2001) Linkage disequilibrium in the human genome. *Nature* 411, 199–204.

105. Cardon, L. R., Abecasis, G. R. (2003) Using haplotype blocks to map human complex trait loci. *Trends Genet* 19, 135–140.

106. Johnson, G. C., Esposito, L., Barratt, B. J., et al. (2001) Haplotype tagging for the identification of common disease genes. *Nat Genet* 29, 233–237.

107. Stram, D. O., Haiman, C. A., Hirschhorn, J. N., et al. (2003) Choosing haplotype-tagging SNPS based on unphased genotype data using a preliminary sample of unrelated subjects with an example from the Multiethnic Cohort Study. *Hum Hered* 55, 27–36.

108. Reich, D. E., Lander, E. S. (2001) On the allelic spectrum of human disease. *Trends Genet* 17, 502–510.

109. Corder, E. H., Saunders, A. M., Strittmatter, W. J., et al. (1993) Gene dose of apolipoprotein E type 4 allele and the risk of Alzheimer's disease in late onset families. *Science* 261, 921–923.

110. Altshuler, D., Hirschhorn, J. N., Klannemark, M., et al. (2000) The common PPARgamma Pro12Ala polymorphism is associated with decreased risk of type 2 diabetes. *Nat Genet* 26, 76–80.

111. Klein, R. J., Zeiss, C., Chew, E. Y., et al. (2005) Complement factor H polymorphism in age-related macular degeneration. *Science* 308, 385–389.

112. Terwilliger, J. D., Hiekkalinna, T. (2006) An utter refutation of the "Fundamental Theorem of the HapMap". *Eur J Hum Genet* 14, 426–437.

113. Terwilliger, J. D., Weiss, K. M. (2003) Confounding, ascertainment bias, and the blind quest for a genetic 'fountain of youth'. *Ann Med* 35, 532–544.

114. Risch, N., Merikangas, K. (1996) The future of genetic studies of complex human diseases. *Science* 273, 1516–1517.

115. Maraganore, D. M., de Andrade, M., Lesnick, T. G., et al. (2005) High-resolution whole-genome association study of Parkinson disease. *Am J Hum Genet* 77, 685–693.

116. Clayton, D. G., Walker, N. M., Smyth, D. J., et al. (2005) Population structure, differential bias and genomic control in a large-scale, case-control association study. *Nat Genet* 37, 1243–1246.

117. Benjamini, Y., Hochberg, Y. (1995) Controlling the false discovery rate: a practical and powerful approach to multiple testing. *J Roy Stat Soc* Ser B, 57, 289–300.

118. Weller, J. I., Song, J. Z., Heyen, D. W., et al. (1998) A new approach to the problem of multiple comparisons in the genetic dissection of complex traits. *Genetics* 150, 1699–1706.

119. Benjamini, Y., Yekutieli, D. (2005) Quantitative trait Loci analysis using the false discovery rate. *Genetics* 171, 783–790.

Chapter 17

Molecular Similarity Concepts and Search Calculations

Jens Auer and Jürgen Bajorath

Abstract

The introduction of molecular similarity analysis in the early 1990s has catalyzed the development of many small-molecule-based similarity methods to mine large compound databases for novel active molecules. These efforts have profoundly influenced the field of computer-aided drug discovery and substantially widened the spectrum of available ligand-based virtual screening approaches. However, the principles underlying the computational assessment of molecular similarity are much more multifaceted and complex than it might appear at first glance. Accordingly, intrinsic features of molecular similarity analysis and its relationship to other methods are often not well understood. This chapter discusses critical aspects of molecular similarity, an understanding of which is essential for the evaluation of method development in this field. Then it describes studies designed to enhance the performance of molecular fingerprint searching, which is one of the most intuitive and widely used similarity-based methods.

Key words: Computational drug discovery, database mining, fingerprints, fingerprint profiling and scaling, molecular descriptors, molecular diversity and dissimilarity, molecular similarity, pharmacophores, similarity metrics, similarity searching, structure–activity relationships, virtual compound screening.

1. Introduction

The introductory sections provide a brief perspective on the development of approaches to search databases for active compounds and the position of molecular similarity analysis in this context. They also introduce similarity searching using molecular fingerprints, which is the subject of the methodological part of this chapter.

Jonathan M. Keith (ed.), *Bioinformatics, Volume II: Structure, Function and Applications, vol. 453*
© 2008 Humana Press, a part of Springer Science+Business Media, Totowa, NJ
Book doi: 10.1007/978-1-60327-429-6 Springerprotocols.com

1.1. Database Searching

Since the late 1970s and the 1980s, database searching for pharmaceutically relevant molecules was dominated by three-dimensional (3D) pharmacophore and docking methods (1, 2). A pharmacophore is the spatial arrangement of atoms or groups in a molecule that are responsible for its biological activity. As such, pharmacophore searching is a ligand-based method and, for the most part, a hypothesis-driven approach, since it usually requires a model as input. By contrast, docking methods attempt to place small molecules into the active site of a target protein, predict their binding conformations, and rank them according to their probability of being "true" ligands. Thus, docking depends on detailed knowledge of 3D target protein structures and their binding sites. Large-scale docking calculations are often referred to as structure-based virtual screening, as opposed to ligand-based virtual screening (3).

During the 1970s, two-dimensional (2D) substructure searching was also developed as an approach to detect molecules that share structural fragments (4). In these calculations, a pre-defined 2D structure is used as a query and database molecules containing this exact substructure are identified, irrespective of other groups they might contain. Searching of databases with conventional 2D molecular graph representations as input was also attempted (5), but these searches became computationally infeasible for larger compound sets due to the complexity of sub-graph matching procedures. Only recently have such problems been circumvented by the introduction of simplified graph representations for database searching (6).

1.2. Molecular Similarity

Molecular similarity analysis was introduced during the early 1990s and triggered by the formulation of the "similarity property principle," which simply states that "similar molecules should have similar biological properties" (7). Computationally, "similarity" is often quantified by considering both structural and physicochemical features of compounds through the application of various molecular descriptors. Such descriptors are mathematical models of molecular structure and properties (8). The similarity property principle is intuitive and perhaps sounds almost trivial, but it spawned the development of new types of database mining methods that are today as widely used as the more "traditional" ligand- and structure-based approaches mentioned above (9). As discussed in the following, however, consequences of the similarity property principle and implications for method development are far from being simplistic.

1.3. Fingerprint Searching

Fingerprints are bit string representations of molecular structure and properties and are among the most popular similarity search tools (8, 9). A fingerprint encodes various molecular descriptors as pre-defined bit settings. These descriptors can be recorded as binary features (e.g., does the molecule contain features a, b, c...; yes/no) *(9)*. Alternatively, descriptor values can be numerically encoded. Depending on the dimensionality of the descriptors used, one distinguishes between 2D or 3D fingerprints (although their bit strings are one-dimensional). Furthermore, fingerprints can be divided into keyed or hashed designs. In keyed fingerprints, each bit position is associated with a specific descriptor or binary feature, whereas in hashed designs descriptor values or feature patterns are mapped to overlapping bit segments. Fingerprints that record substructure patterns in molecules are prototypic 2D keyed designs *(10)*, whereas Daylight fingerprints that capture connectivity pathways through molecules represent exemplary 2D hashed designs *(11)*. In addition, pharmacophore fingerprints are 3D fingerprints and monitor possible pharmacophore patterns in molecules explored by systematic conformational search *(12)*. Pharmacophore fingerprints are typically keyed. Besides these prototypes, other types of fingerprints also exist. The length of fingerprint bit strings can vary dramatically from hundreds of bit positions for structural fragment fingerprints to thousands of bits for connectivity fingerprints and literally millions of bits for pharmacophore fingerprints. However, the length and complexity of fingerprints do not necessarily correlate with their search performance *(8)*.

In similarity searching, fingerprints are calculated for query and database compounds and then compared. This means the search proceeds in fingerprint vector space. Quantitative pairwise comparison of bit string overlap serves as a criterion for molecular similarity and is based on the calculation of similarity coefficients *(13)*. Molecules are considered similar if calculated similarity coefficients exceed a pre-defined threshold value.

Different types of fingerprint search calculations are described and analyzed in this chapter in order to illustrate one of the most widely used methodologies for global molecular similarity analysis. In the accompanying chapter by Eckert and Bajorath, a mapping algorithm is described in detail that presents a conceptually different approach for the evaluation of global molecular similarity. Similarity searching and compound mapping are methodologically distinct but complement each other in similarity-oriented compound database analysis.

2. Molecular Similarity Concepts

On the basis of the similarity property principle, similar structures and properties of molecules are considered indicators of similar biological activities. As discussed in the following, several variations of this theme must be considered to fully appreciate the opportunities and limitations of molecular similarity analysis.

2.1. A Global Molecular Approach

Essentially all small-molecule–based computational screening methods rely on the evaluation of molecular similarity in some ways. However, a fundamental difference between *molecular similarity analysis* and other structure–activity correlation methods is the way molecules are considered in computational terms. Different from pharmacophore *(1)* or quantitative structure–activity relationship (QSAR) *(14)* analyses, methods for molecular similarity analysis, as we understand it today, do not attempt to focus on parts of a molecule that are thought or known to be crucial for activity. Rather, similarity analysis considers the whole molecule when assessing potential relationships to other molecules. This has both advantages and drawbacks. In contrast to pharmacophores and QSAR, molecular similarity analysis is not hypothesis-driven and is therefore in principle less prone to errors. On the other hand, molecular similarity calculations can be much influenced by structural features that do not contribute to binding interactions and are not critical for biological activity. In this context, it is worth noting that graph-matching algorithms, as mentioned, also attempted to assess similarity on a whole-molecule basis and thus represented an early form of similarity analysis, years before the similarity property principle was formulated. **Figure 17.1** summarizes major types of methods that explore (and exploit) global or local molecular similarity.

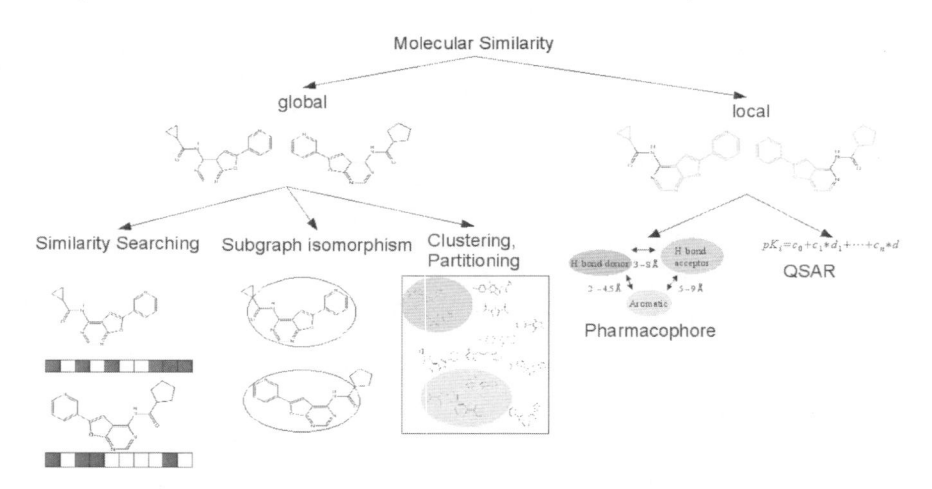

Fig. 17.1. Exploration of molecular similarity. The figure shows different types of similarity-based methods that are used to mine databases for active compounds and evaluate either global or local similarity between template and test compounds.

2.2. Molecular Similarity, Dissimilarity, and Diversity

What is the relationship between molecular similarity and diversity? Molecular diversity analysis aims at the design of large structure- and property-diverse compound sets but is not a preferred technique to study pairwise molecular relationships. Diversity analysis is most often applied as a design approach in an "inductive" manner. Typically, one applies diversity techniques to evenly populate a pre-defined chemical reference space with computationally generated compounds (or molecules selected from large pools) *(15)*. Diversity-oriented compound design is often based on the specification of chemical reactions for given sets of possible reagents in order to guide combinatorial chemistry efforts. Thus, achieving comprehensive coverage of chemical space with a limited number of molecules is a major goal of diversity design. By contrast, in similarity analysis, one tries to find compounds that are structurally and functionally related to templates and focuses on a rather limited region of chemical space. The inverse of molecular similarity analysis is not diversity but dissimilarity analysis, which aims to find molecules that are most different from a given compound or set of compounds *(16)*. Thus, like similarity analysis, dissimilarity calculations typically explore pairwise compound relationships. Major goals of dissimilarity analysis include the selection of a maximally diverse subset of compounds from a much larger set or the identification of compounds that are most dissimilar to an existing collection. Such tasks differ from diversity design and require the application of different algorithms *(16)*. In summary, it might be best to rationalize diversity in the context of library design and dissimilarity in the context of compound selection.

Similarity-based computational methods do not necessarily depend on exploring pairwise molecular relationships. For example, partitioning algorithms project compound sets into chemical reference spaces based on their descriptor coordinates *(17)*. Closeness in chemical space then implies similarity. However, unlike clustering methods, partitioning algorithms do not compare distances between molecules in a pairwise manner and are therefore applicable to very large compound databases. In addition, partitioning algorithms are also suitable for diversity selection or design.

2.3. Similarity beyond the Computational Point of View

Given its intuitive appeal, the similarity property principle is best viewed from a medicinal chemistry perspective *(18)*. It is well known that series of similar molecules, often called analogs, typically share similar activity. On the other hand, it is also known that analogs can exhibit dramatic differences in potency *(18)*, a situation that one explores and exploits in QSAR analysis *(14)*. This situation also has implications for approaches that are based on a whole-molecule view. With rare exceptions *(19)*, molecular similarity methods do not take compound potency into account and are thus more qualitative in nature. However, any similarity method must recognize analogs as (very) similar structures, irrespective of their relative potency. In some instances, the global molecular view applied in similarity analysis can cause significant problems and lead to failures.

2.4. Caveats

The fact that minor modifications can render active compounds nearly completely inactive, which is well known to medicinal chemists, presents a major caveat for molecular similarity analysis. **Figure 17.2** illustrates this point.

The two analogs shown in **Fig. 17.2** are very similar, but only one of them is active. Any meaningful similarity-based method must detect both compounds. However, had only the inactive analog been selected, the similarity search would have been considered a failure, although the approach succeeded in identifying a novel active structural motif. This situation illustrates an intrinsic limitation of the similarity property principle and its applications, which has been termed the "similarity paradox" *(20)*. Simply put, very similar structures can have dramatically different activity. How can one best tackle such difficulties in molecular similarity analysis? First, one can select (and test) as many analogs as possible, thereby reducing the chance of selecting the "wrong" ones. Another possibility is to isolate the core structure shared by a series of analogs. If a new active scaffold is indeed identified, it should display at least some weak intrinsic activity. Of course, the latter approach would very likely produce only novel hits with very low potency (which might not be sufficiently attractive for medicinal chemists to initiate a hit-to-lead optimization effort). Thus, one way or another, the similarity paradox presents in some cases a substantial challenge for molecular similarity analysis, which can only be addressed by application of well-designed compound selection strategies.

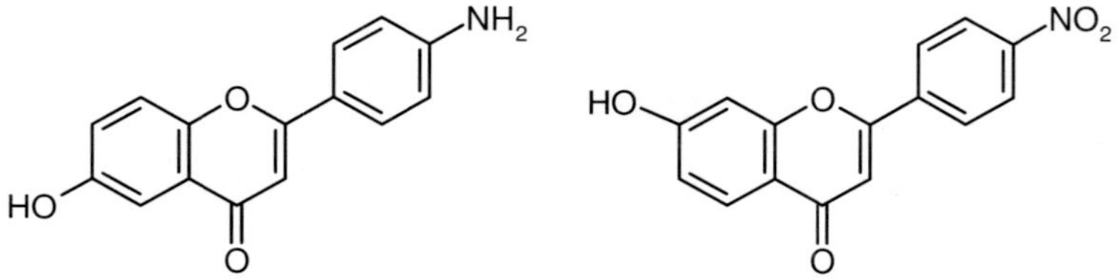

$$IC_{50} = 1.2\ \mu M \qquad\qquad IC_{50} > 2000\ \mu M$$

Fig. 17.2. Two compounds identified by similarity searching using a known protein kinase inhibitor as a template. These compounds share the same core structure (scaffold) and are very similar. However, the compound on the left has kinase inhibitory activity, whereas the compound on the right is inactive. IC_{50} gives the compound concentration for half-maximal enzyme inhibition.

Another critical aspect of similarity analysis is the potency and/or specificity relationship between known template compounds and newly identified hits. Since similarity analysis usually starts from molecules that have gone through lead optimization efforts (e.g., patented or published compounds), there is usually a price to pay for finding novel structures with similar activity: as one deliberately moves away from an optimized structural motif, novel hits are generally likely to be much less potent (or selective) than the templates. Optimized leads typically display potencies in the nanomolar range, whereas hits identified by similarity analysis are mostly active in the micromolar range. Thus, they often present another starting point for chemical optimization efforts.

2.5. Opportunities

A major attraction of molecular similarity analysis is its ability to identify different types of structures (also called chemotypes) having similar activity, a process often referred to as lead hopping *(21)*. As illustrated in **Fig. 17.3**, the ability of similarity-based methods to recognize remote similarity relationships should go beyond mere structural resemblance. This sets molecular similarity analysis apart from other ligand-based hit identification methods.

The examples shown in **Fig. 17.3** illustrate that structurally diverse compounds indeed share similar activity. Molecular similarity analysis identifies a spectrum of molecules that gradually diverge from the structures of templates. This spectrum includes closely related analogs and increasingly diverse structures. Following the similarity property principle, a key question is how far molecules can diverge from templates but still retain similar activity? In other words, we are challenged to outline an "activity radius" in chemical space.

Importantly, similarity analysis does not assume that similar compounds must have a similar binding mode or mechanism of

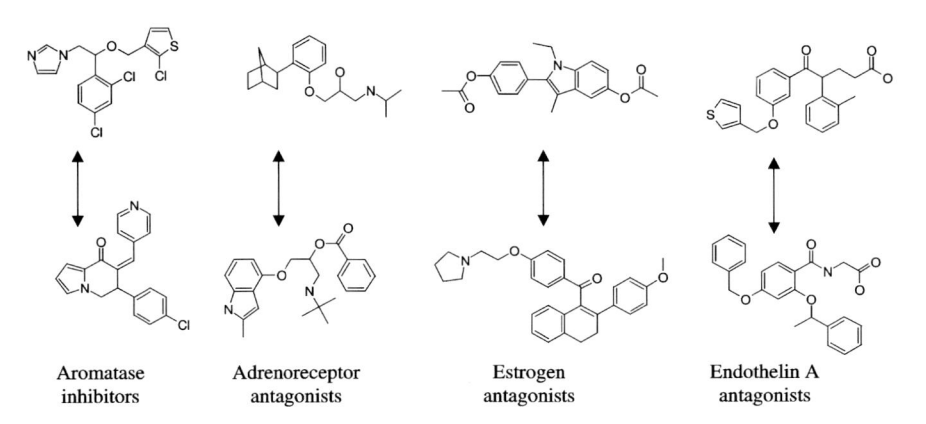

| Aromatase inhibitors | Adrenoreceptor antagonists | Estrogen antagonists | Endothelin A antagonists |

Fig. 17.3. Examples of remote similarity relationships (diverse structures – similar activity) identified using similarity-based methods.

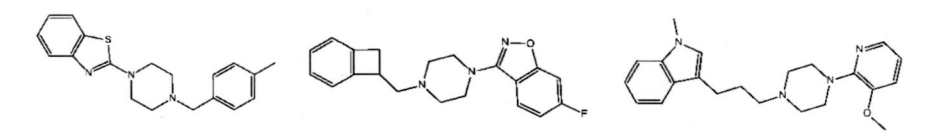

Fig. 17.4. Compounds with related structures and overlapping specificity found using known serotonin receptor ligands as templates. **Left to right:** dopamine receptor antagonist, serotonin and dopamine antagonist, serotonin receptor antagonist.

action. Thus, in some instances, molecules with overlapping specificities can be identified, as illustrated in **Fig. 17.4**.

The example shows that structural variations among similar compounds might produce a spectrum of activities, the detection of which is another attractive aspect of the approach. As mentioned, for practical purposes it is important to select a sufficient number of compounds from similarity calculations (e.g., 100 or so) to cover at least part of the molecular spectrum within or beyond an investigated activity radius.

3. Methods

Similarity searching with fingerprints relies on the correlation of bit-string representations of molecular structure and properties with biological activity. Fingerprints are designed to encode specific features of active molecules. Compared with compound classification methods, similarity searching has an intrinsic advantage: It can be applied when only a single template compound is available. However, fingerprints (and other similarity search tools) also benefit from the availability of multiple reference compounds. Accordingly, fingerprint methods utilizing multiple templates often increase search performance. Exemplary multiple-template-dependent search techniques are discussed in the following.

3.1. Exemplary Fingerprint Design

For our calculations, we use a 2D fingerprint termed MPMFP that was previously developed in our laboratory *(22)*. It is a relatively simple fingerprint consisting of only 175 bits: 114 bits encode structural fragments *(23)* and 61 binary chemical descriptors. The motivation and rationale of MPMFP design was based on the "mini-fingerprint" idea *(24)*. Previously, the generation of prototypic mini-fingerprints that were considerably smaller than many other 2D or 3D fingerprints had revealed that combining structural fragment descriptors and selected chemical property descriptors provided a solid design framework for achieving good performance in similarity searching *(24)*. Structural keys provide a well-established basis for the selection of structurally similar compounds that can then be further discriminated by use of more (and more diverse) information-carrying chemical descriptors. **Table 17.1** reports a possible classification scheme for descriptors *(25)*.

A simple way to classify descriptors is according to the dimensionality of the molecular representation from which they are calculated. The simplest representation is a chemical formula (1D) and molecular weight is an example of a 1D descriptor. Properties derived from molecular connectivity, for example connectivity indices, require a connection table or graph representation, which is a 2D drawing of a molecule. In addition, conformation-dependent descriptors include, for example, 3D shape descriptors or the van der Waals surface area. Brown and Martin *(25)* distinguish between two more types of descriptors. Type IV descriptors in **Table 17.1** account for biological properties and Type V descriptors are complex descriptors built from sets of others.

Since biological activity is strongly influenced by 3D properties such as molecular shape, it would be expected that 3D descriptors are generally more sensitive to biological activity than simpler representations. However, this is not necessarily the case. Numerous studies indicate that 2D information is often sufficient to produce accurate results *(25, 26)*. In addition, 2D representations are typically much less prone to errors than descriptors calculated from hypothetical 3D conformations (that have little probability to be bioactive conformations).

In order to be encoded in fingerprints, chemical descriptors must be transformed into a binary format, a process called discretization. For this purpose, a conventional approach is binning,

Table 17.1
Classification scheme for chemical descriptors

Type	Derived from	Example
I	Global (bulk) molecular properties	logP(o/w)
II	2D structure	Structural keys
III	3D structure	MolSurf
IV	Biological properties	AFP
V	Combination of different descriptors	BCUT

"logP(o/w)" is the logarithm of the octanol/water partition coefficient (a measure of hydrophobicity). "MolSurf" means molecular surface area and "AFP" affinity fingerprints (descriptors capturing experimental binding profiles of test compounds). "BCUT" stands for Burden-CAS-University of Texas descriptors. The classification scheme was adapted from Brown, R. D., Martin, Y. C. (1996) Use of structure-activity data to compare structure-based clustering methods and descriptors for use in compound selection. J *Chem Inf Comput Sci* 36, 572–584.

that is, the division of a value range into a number of intervals. A preferred binning scheme is the division of an original value range into intervals such that the frequency of database compounds found within each interval is the same. This process is referred to as equal frequency binning. MPMFP is based on a different discretization strategy. In this case, continuous chemical descriptors are transformed into a binary format through statistical analysis of their medians. The median of each descriptor is calculated from a screening database of compounds and used as a threshold in order to set the binary value to 0 (<median) or 1 (≥median). Perhaps surprisingly, this straightforward coding scheme has been shown to yield better performance than value range encoding of descriptors *(22)*. These findings could be rationalized. As single entities, binary-transformed descriptors have little predictive value, but multiple binary descriptors in combination were shown to become highly discriminatory *(22)*.

One of the most important aspects of fingerprint design is descriptor selection. From the thousands of chemical descriptors that are available, one ultimately has to choose a subset that is particularly suitable for discriminating similar from non-similar compounds. For MPMFP, a set of 150 descriptors was analyzed for their chemical information content in compound databases using our adaptation of the Shannon entropy approach for database profiling *(27, 28)*. This pre-selection step identified a set of 129 descriptors having at least detectable but often significant information content. However, chemical descriptors are often correlated, so including all descriptors without further pre-processing would count some features multiple times, thus favoring certain chemical properties over others, only because of statistical effects and not for chemical reasons. In order to minimize such correlation effects, pairwise correlation coefficients were computed and from each pair with a correlation coefficient >0.8, the descriptor with lower information content was removed (**Note 1**). Finally, a set of 61 descriptors remained. These descriptors were then combined with 114 structural keys, consistent with the principles of mini-fingerprint design, producing MPMFP containing a total of 175 bits. In summary, our design efforts were guided by four principles: *(1)* exploit the previously observed predictive power of combined structural and chemical descriptors, *(2)* select descriptors with high chemical information content and limited correlation, *(3)* encode descriptors in a binary-transformed format, and *(4)* generate a fingerprint of relatively small size.

3.2. Similarity Metrics

In order to calculate the similarity between two compounds, fingerprint overlap must be quantified. A chosen metric should assign a value of 0 for completely distinct fingerprints and a value of 1 for identical ones. The most prominent metric for fingerprint comparison is the Tanimoto coefficient (Tc; also known as the Jaccard coefficient) *(13)* that is defined for binary vectors A and B as:

$$Tc(A,B) = \frac{N_{AB}}{N_A + N_B - N_{AB}}$$

where N_A is the number of bits set to 1 in fingerprint A, N_B the number of bits set to 1 in fingerprint B, and N_{AB} the number of common bits. This formulation is known as the "binary" Tc.

For application of MPMFP, the binary Tc had to be modified to also take bits set to 0 into account because for median-based binary descriptors, a bit set to 0 carries as much information as a bit set to 1 (*see* **Note 2**).

3.3. Fingerprint Profiling

Similarity searching using fingerprints is based on the premise that compounds with similar activity should also produce similar bit strings. When multiple template compounds are available, their fingerprints can reveal additional information, as we have been able to show *(29)*. Comparing bit strings of several compounds with similar activity identifies bit settings that are shared by all compounds. These bit patterns are often characteristic for a compound class *(29)*. In order to compare fingerprint settings for different activity classes, the concept of "fingerprint profiles" was introduced *(29, 30)*. A profile records the relative frequency for bit settings in fingerprints of compounds belonging to an activity class. Fingerprint profiles are calculated by counting the number of bits set to 1 at each bit position and dividing the sum by the number of compounds per class. **Figure 17.5** compares profiles for three activity classes. As can be seen, bit frequencies differ significantly, especially in the first half of the profile.

Considering the special binary transformation of property descriptors encoded in MPMFP, two profiles of equal relevance can be generated, one counting bits set to 1 and the other

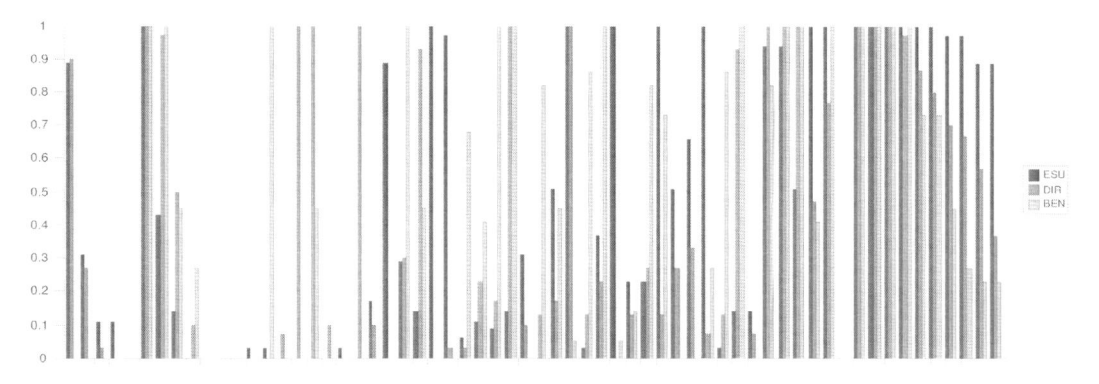

Fig. 17.5. Exemplary fingerprint profiles. Reported are relative bit frequencies for three activity classes (black, dark gray, light gray) obtained with a mini-fingerprint.

recording unset bits. From these complementary profiles, the most characteristic bit settings for an activity class can be derived. These so-called consensus bits are shared by the majority of compounds in an activity class and are thus likely to indicate descriptors producing signature values.

3.4. Profile Scaling

A straightforward way to make use of the activity-class-specific information of fingerprint profiles is to emphasize consensus bit positions in similarity search calculations. This process involves applying scaling factors to these bits. The comparison of scaled fingerprints sharing these bit settings produces higher Tc or avTc values than calculations in the absence of scaling *(30)*. A profile cut-off value (PCV) determines the minimum frequency for a bit setting to be scaled. Every bit setting with a frequency higher than PCV is scaled by a profile scaling factor (PSF) >1 (i.e., the bit is counted PSF times). Frequencies below the cut-off value are not scaled and are submitted as regular bit values to Tc calculations. In order to further discriminate between lower and higher bit frequencies, PSF is linearly weighted within the interval [PCV, 1.0] (*see* **Note 3**). Scaling can be applied to any keyed fingerprint, but the MPMFP fingerprint has been shown to benefit most in a study comparing different keyed designs, also including a fingerprint consisting of 166 structural keys as a standard *(30)*. In summary, fingerprint profiling and scaling increases the information content of similarity search calculations by determining and emphasizing activity class–dependent bit signatures (**Note 4**).

3.5. Centroid and Nearest Neighbor Techniques

The scaling procedure described in the preceding utilizes fingerprint profiles as a basis for emphasizing selected bit settings at the level of similarity calculations. Another approach is to use a profile or an ensemble of fingerprints directly as a search query, which is conceptually similar to the generation and application of protein sequence profiles for detecting distant relationships *(31)*. This idea was followed by Schuffenhauer et al. *(32)*, who introduced a centroid method for fingerprint searching with multiple reference structures. In the centroid approach, an average fingerprint vector is calculated over all template molecules and then compared to database compounds using an appropriate similarity coefficient (*see* **Note 5**).

Schuffenhauer et al. also investigated a nearest neighbor method *(32)*. In this case, the Tanimoto similarity of a database compound is separately calculated against all reference compounds and values of nearest neighbors are selected for averaging (**Note 6**). For practical purposes, it should be noted that computational requirements of fingerprint profile scaling and the centroid method differ from those of the nearest neighbor approach (*see* **Note 7**). **Figure 17.6** presents a summary of the different types of fingerprint search calculations presented herein.

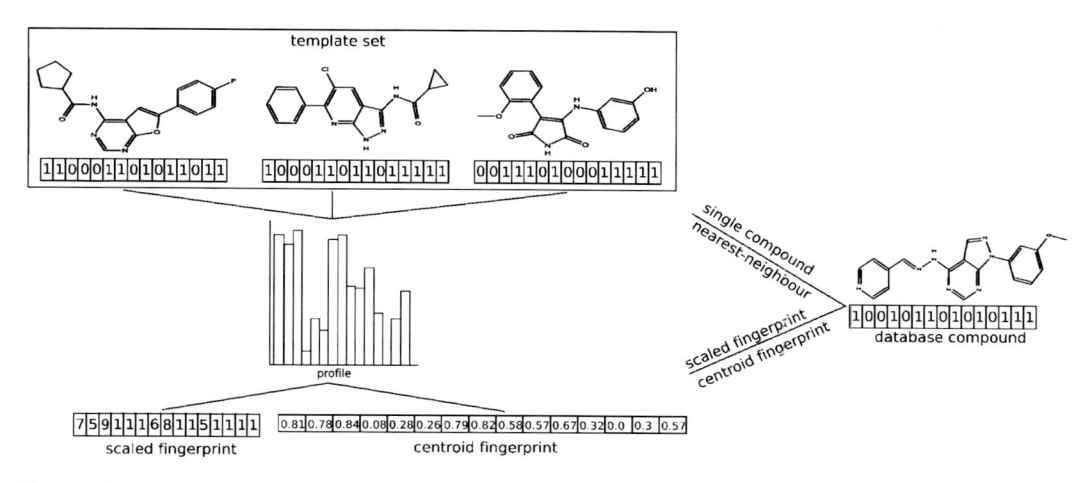

Fig. 17.6. Summary of different fingerprint search calculations. The fingerprint of a database compound can be compared with the fingerprint of one or more template molecules, and nearest neighbor analysis can be carried out on the basis of such pairwise comparisons. Furthermore, a profile generated from the fingerprints of all template molecules can be used for fingerprint scaling or the generation of a centroid fingerprint. The fingerprint of a database compound is then compared with the scaled fingerprint of a template molecule or the centroid fingerprint.

3.6. Case Study

Previously, we showed that profile scaling significantly improved MPMFP search performance relative to calculations using single template molecules. Here we present a comparison of the different techniques for fingerprint searching based on multiple templates, as discussed in the preceding.

3.6.1. Dataset

Calculations were carried out on a dataset consisting of 5,000 randomly selected synthetic compounds and 12 different activity classes that are listed in **Table 17.2**. This background database was used previously when studying three of our activity classes (H3H, BEN, and CAE) *(30)*. These classes are also investigated here. The remaining nine classes were assembled from the MDL Drug Data Report (MDDR), a database collecting biologically relevant compounds from the patent literature or chemistry publications *(33)*.

3.6.2. Calculation Protocols

For each "experiment," the activity classes were randomly divided into two halves. One half was used to compute the fingerprint profiles and the other added to the background database as potential hits. All 5,000 background compounds were considered inactive and potential false-positives. All calculations were carried out with 50 randomly selected compound sets, each of which was used once as the template (profile) set and once as the hit set. In addition, for profile scaling, in which a query compound had to be selected, every compound from the profile set was taken once as the query (which resulted in approximately 18,000 search calculations for profile scaling). In these calculations, the profile cut-off value (PCV) was set to 0.95 and the scaling factor (PSF)

Table 17.2
Compound activity classes

Class	Activity	N	min	max	avg	dev
ESU	Inhibitor of estrone sulfatase activity	35	0.29	1.00	0.61	0.11
DIR	Dihydrofolate reductase inhibitor	30	0.34	0.95	0.57	0.12
MEL	Melatonin agonist	25	0.31	0.95	0.55	0.11
XAN	Xanthine oxidase inhibitor	35	0.12	1.00	0.56	0.18
CHO	Cholesterol esterase inhibitor	30	0.13	0.98	0.48	0.18
INO	Inosine monophosphate dehydrogenase inhibitor	35	0.34	1.00	0.65	0.11
KRA	Kainic acid receptor antagonist	22	0.30	1.00	0.55	0.17
LAC	Lactamase (beta) inhibitor	29	0.15	1.00	0.44	0.18
DD1	Dopamine (D1) agonist	30	0.23	1.00	0.57	0.14
BEN	Benzodiazepine receptor ligands	22	0.51	1.00	0.67	0.13
CAE	Carbonic anhydrase II inhibitors	22	0.45	1.00	0.64	0.13
H3E	H3 antagonists	21	0.41	1.00	0.66	0.15

For each class, similarity statistics for pairwise compound comparisons are reported. The minimum *(min)*, maximum *(max)*, and average *(avg)* Tanimoto coefficient were computed with MACCS structural keys and the Tc standard deviation *(dev)* was calculated. *N* is the number of active compounds per class *(see* also **Note 8**).

to 10. For nearest neighbor searching, only one template compound was selected (1-NN; *see* **Note 6**).

3.6.3. Results

For all calculations, we determined the numbers of correctly identified hits and false positives above the avTc value that maximizes a scoring function relating the numbers of correctly and incorrectly identified compounds to each other (**Note 9**). Another way to assess the performance of similarity searches is to compute cumulative recall rates that report the number of hits contained in a certain percentage of the ranked list of database compounds. We report here the recall rate for 1% of the database (i.e., the top-ranked 50 compounds), which is often used as a measure of search performance *(34, 35)*. The search results for each class are given in **Table 17.3**. As one would expect, search performance

Table 17.3
Results obtained for each of the three similarity search methods

Class Centroid	avTc	Actives	TP	TP%	FP	Score	Recall
BEN	0.76	6.34	6.34	100	0.00	0.58	10.82
CAE	0.76	9.66	7.74	80	1.92	0.52	10.38
H3E	0.76	8.38	8.08	96	0.30	0.73	10.50
CHO	0.76	2.48	2.18	88	0.30	0.12	9.74
DD1	0.76	9.46	9.46	100	0.00	0.63	13.80
DIR	0.76	8.28	7.32	88	0.96	0.42	11.60
ESU	0.76	6.28	6.28	100	0.00	0.36	15.56
INO	0.76	14.56	10.58	73	3.98	0.38	13.18
KRA	0.76	1.48	1.36	92	0.12	0.11	6.04
LAC	0.76	1.10	1.10	100	0.00	0.08	8.62
MEL	0.76	10.12	6.18	61	3.94	0.19	11.60
XAN	0.76	12.80	9.62	75	3.18	0.37	13.08
1-NN							
BEN	0.81	10.08	9.08	90	1.00	0.74	10.91
CAE	0.81	7.13	7.13	100	0.00	0.67	10.29
H3E	0.81	7.67	7.67	100	0.00	0.74	10.49
CHO	0.81	9.17	8.17	89	1.00	0.48	12.36
DD1	0.81	8.14	8.14	100	0.00	0.55	14.82
DIR	0.81	5.67	5.17	91	0.50	0.32	11.93
ESU	0.81	5.75	5.75	100	0.00	0.34	16.54
INO	0.81	10.49	9.49	90	1.00	0.49	14.63
KRA	0.81	6.15	6.15	100	0.00	0.58	10.13
LAC	0.81	5.68	5.68	100	0.00	0.40	11.55
MEL	0.81	5.94	4.48	75	1.46	0.24	10.07
XAN	0.81	12.27	11.0	90	1.27	0.56	14.39

(continued)

Table 17.3 (continued)
Results obtained for each of the three similarity search methods

Profile scaling

BEN	0.88	7.41	7.05	95	0.37	0.61	10.88
CAE	0.88	11.53	8.18	71	3.35	0.44	10.59
H3E	0.88	9.19	8.60	94	0.59	0.77	10.49
CHO	0.88	2.38	2.17	91	0.21	0.13	6.24
DD1	0.88	3.93	3.93	100	0.00	0.26	13.76
DIR	0.88	6.73	5.86	87	0.87	0.34	13.05
ESU	0.88	3.85	3.65	95	0.20	0.20	15.17
INO	0.88	11.25	7.03	63	4.22	0.19	11.67
KRA	0.88	5.40	2.64	49	2.76	0.12	4.58
LAC	0.88	2.80	2.67	95	0.14	0.18	6.46
MEL	0.88	9.16	4.17	45	4.99	0.06	8.77
XAN	0.88	8.36	5.42	65	2.94	0.18	9.89

"Actives" reports the number of molecules above the (MPMFP-specific) *"avTc"* threshold value and *"TP"* and *"FP"* the number of correctly identified true- and false-positives, respectively. *"Recall"* gives the number of true positives among the top ranked 50 database compounds.

varies over activity classes and also among the compared methods. Furthermore, optimum avTc threshold values differ and range from 0.76 (centroid) to 0.88 (profile scaling). Clearly, some of the activity classes are in general more difficult to treat than others (e.g., MEL, XAN), but overall all three methods produce encouraging results. This becomes evident when the numbers of correctly identified hits (true-positives) at optimum similarity threshold values are compared with the false-positive rates. With a few exceptions, the percentage of detected true-positives is consistently high, whereas false-positive rates are moderate to low. In the majority of cases, the recall for the three methods was comparable in magnitude. These observations indicate that recall rates are not a "high-resolution" measure for the comparison of search performance, as also suggested by Bender and Glen *(36)*. Therefore, a scoring function that takes hits and false-positives into account (**Note 9**) should provide a more appropriate performance criterion. Applying this scoring scheme, 1-NN performs overall best, followed by centroid searching and profile scaling. This is mainly because 1-NN produces the lowest false-positive rates (e.g., no false-positives for six of 12 classes), which is a pre-requisite for achieving high scores. The overall best performance of profile

Table 17.4
Averages over all activity classes

Method	avTc	Actives	TP	TP%	FP	Score	Recall
Centroid	0.76	7.58	6.35	88	1.23	0.38	11.24
1-NN	0.81	7.85	7.33	94	0.52	0.51	12.34
Scaling	0.88	6.83	5.12	79	1.72	0.29	10.13

"*Actives*" reports the number of molecules above the (MPMFP-specific) "*avTc*" threshold value and "*TP*" and "*FP*" the number of correctly identified (true- and false-positives, respectively). "*Recall*" gives the number of true positives among the top ranked 50 database compounds.

searching using one nearest neighbor is also interesting from the point of view that it is conceptually the simplest of the three methods compared here. **Table 17.4** reports the results of the different methods averaged over all classes. This summary shows that differences in search performance are overall subtle at the level of compound selection. For example, on average, 1-NN detects about one more hit than profile scaling and one fewer false-positive per search calculation. Thus, for similarity search applications, all three approaches have considerable potential. It should also be considered that the relative performance of these techniques is much influenced by the intrinsic diversity of the template compounds. For increasingly diverse compound sets, 1-NN is not expected to generally produce superior results.

3.7. Conclusions

In this chapter, we have discussed aspects of molecular similarity that are important for the design of methods to analyze and compare structure–activity relationships of small molecules. The development of such approaches is currently a major topic in chemoinformatics research. As defined herein, molecular similarity analysis considers similarity as a global molecular property, in contrast with pharmacophore or QSAR methods, which focus on the analysis of local similarities between molecules such as specific arrangements of atoms or functional groups. This implies that much less prior knowledge about activity determinants is required for the application of molecular similarity analysis than for pharmacophore or QSAR modeling. This is a major attraction of molecular similarity analysis and also one of its major limitations because the whole-molecule view renders it vulnerable to structure–activity relationships in which small chemical changes cause large effects on potency or other biologically relevant molecular properties. Nevertheless, for compound database mining and hit identification in drug discovery, methodologies that evaluate molecular similarity from a global point of view have become very popular. Among these, similarity searching using molecular fingerprints is one of the most widely used and intuitive approaches.

Fingerprints are bit string representations of molecular structure and properties. They are applied to facilitate pairwise similarity-based molecular comparisons. Quantitative assessment of fingerprint overlap is used as a measure of molecular similarity. The basic ideas underlying fingerprint searching are well in accord with the similarity property principle. The fingerprint methods presented herein are advanced similarity search techniques designed to make use of the information provided by multiple template structures. Similarity search benchmark calculations typically assess search performance based on compound retrieval statistics over a number of different activity classes. Such benchmark statistics provide a meaningful initial comparison of different search tools and techniques but do not provide information about search performance on specific sets of compounds, for example, molecules having similar activity but increasingly diverse structures. Regardless, the results of our case study show that conceptually different fingerprint search techniques using multiple template compounds produce promising hit rates and moderate to low false-positive rates. In addition to the methods discussed herein, there are a number of other possibilities to tune fingerprint search techniques for application on single or multiple template structures.

4. Notes

1. Descriptor selection can also be supported by approaches other than information content analysis or calculation of correlation coefficients. For example, statistical techniques such as principal component analysis or partial least squares regression can be applied in order to identify descriptors that significantly account for property variability within a molecular data set.

2. In order to adjust the Tanimoto coefficient to equally account for bits set to 1 or 0, we have introduced a modified Tc version (avTc) accounting for the average of Tc calculations focusing on 1 or 0 bit settings:

$$avTc(A,B) = \frac{Tc(A,B) + Tc(A^0, B^0)}{2}$$

Here $Tc(A,B)$ is the conventional binary Tanimoto coefficient calculated for bits set to 1 and $Tc(A^0, B^0)$ the corresponding coefficient calculated for bits set to 0.

3. For a bit frequency f > PCV, linear weighting produces an effective scaling factor of f*PSF. More complex scaling functions, e.g., exponential functions, have also been evaluated but have not generated notable improvements in search performance.

4. Outline of a profile-scaled similarity search with a set of known active compounds as input and an arbitrarily chosen query compound from this set:

   a. Calculate the MPMFP fingerprint for each active compound.

   b. Count the frequencies of set and unset bits at each position and generate two profiles (one counting bits set to 1 and the other counting bits to 0).

   c. For similarity searching, bits set with a frequency above a threshold PCV form the set of scaled consensus bit positions.

   d. Randomly select a query compound from the profile set.

   e. When computing the similarity of a database compound to the query compound, count each non-scaled bit as either 0 or 1 and scaled bit positions as f*PSF.

   f. Calculate the scaled avTc value.

5. The centroid approach corresponds, in a pictorial view, to calculating the center in chemical space for the cluster formed by all template compounds. Thus, the centroid vector contains real numbers (rather than 0 or 1 bit settings). Because the binary Tanimoto coefficient is not applicable for real-valued vectors, a suitable similarity measure for comparing such vectors must be defined. However, since the binary Tc is only a special case of the general Tanimoto coefficient (Tc_G), this metric can be used. For two vectors of real values, Tc_G is defined as:

$$Tc_G(A, B) = \frac{\sum_{i=1}^{n} A_i B_i}{\sum_{i=1}^{n} A_i^2 + \sum_{i=1}^{n} B_i^2 - \sum_{i=1}^{n} A_i B_i}$$

For binary vectors, this formulation is reduced to the binary Tc. In a centroid-based search, a maximal similarity value of 1.0 could, in principle, only be obtained if the ensemble consists of identical fingerprints. Otherwise, the centroid vector contains real values within the range [0.0, 1.0], whereas the fingerprint of a database compound that is compared to the centroid is a binary vector of elements {0,1}.

6. The nearest neighbor approach compares each database compound to all reference structures, calculates binary Tc values, and selects k-nearest-neighbors (k-NN) for averaging. In

their calculations, Schuffenhauer et al. obtained overall best result when using only one nearest neighbor (1-NN); that is, the Tc value of the most similar template compound *(32)*. Thus, in this case, the final similarity value is the maximum Tc (and no average is calculated).

7. Nearest neighbor searching is computationally more expensive than the centroid approach, as it requires N computations (with N being the number of reference structures) for every database compound. By contrast, for centroid searching, and also profile scaling, only one calculation is required.

8. A Tc value of 1.00 means that the fingerprints of two compared molecules are identical and that the two molecules are very similar but not necessarily identical. For example, depending on the descriptors encoded in a fingerprint, closely related analogs might produce identical fingerprint settings.

9. The scoring function used here was $S = (TP - FP)/N$, where TP is the number of true-positives, FP the number of false-positives, and N the total number of compounds per activity class. This scoring scheme was previously applied for the analysis of fingerprint scaling calculations *(30)*.

References

1. Gund, P. (1977) Three-dimensional pharmacophore pattern searching. in (Hahn, F. E., ed.), *Progress in Molecular and Subcellular Biology*, vol 5. Springer-Verlag, Berlin.

2. Kuntz, I. D., Blaney, J. M., Oatley, S. J., et al. (1982) A geometric approach to macromolecule-ligand interactions. *J Mol Biol* 161, 269–288.

3. Kitchen, D. B., Decornez, H., Furr, J. R., et al. (2004) Structure-based virtual screening and lead optimization: methods and applications. *Nat Rev Drug Discov* 3, 935–949.

4. Cramer, R. D. III, Redl, G., Berkoff, C. E. (1974) Substructural analysis: a novel approach to the problem of drug design. *J Med Chem* 17, 533–535.

5. Barnard, J. M. (1993) Substructure searching methods. Old and new. *J Chem Inf Comput Sci* 33, 532–538.

6. Gillet, V. J., Willett, P., Bradshaw, J. (2003). Similarity searching using reduced graphs. *J Chem Inf Comput Sci* 43, 338–345.

7. Johnson, M. A., Maggiora, G. M. (Eds.) (1990) *Concepts and Applications of Molecular Similarity*. Wiley, New York.

8. Bajorath, J. (2001) Selected concepts and investigations in compound classification, molecular descriptor analysis, and virtual screening. *J Chem Inf Comput Sci* 41, 233–245.

9. Bajorath, J. Integration of virtual and high-throughput screening (2002) *Nat Rev Drug Discov* 1, 337–346.

10. Barnard, J. M., Downs, G. M. (1997) Chemical fragment generation and clustering software. *J Chem Inf Comput Sci* 37, 141–142.

11. James, C. A., Weininger, D. (2006) *Daylight Theory Manual.* Daylight Chemical Information Systems, Inc., Irvine, CA.

12. Bradley, E. K., Beroza, P., Penzotti, J. E., et al. (2000) A rapid computational method for lead evolution: description and application to α_1-adrenergic antagonists. *J Med Chem* 43, 2770–2774.

13. Willett, P., Barnard, J. M., Downs, G. M. (1998) Chemical similarity searching. *J Chem Inf Comput Sci* 38, 983–996.

14. Esposito, E. X., Hopfinger, A. J., Madura, J. D. (2004) Methods for applying the quantitative structure-activity relationship paradigm. *Methods Mol Biol* 275, 131–214.

15. Martin, Y. C. (2001) Diverse viewpoints on computational aspects of molecular diversity. *J Comb Chem* 3, 231–250.

16. Willett, P. (1999) Dissimilarity-based algorithms for selecting structurally diverse sets of compounds. *J Comput Biol* 6, 447–457.

17. Stahura, F. L., Bajorath, J. (2003) Partitioning methods for the identification of active molecules. *Curr Med Chem* 10, 707–715.

18. Kubinyi, H. (1998) Similarity and dissimilarity: a medicinal chemist's view. *Perspect Drug Discov Design* 9/10/11, 225–252.

19. Godden, J. W., Stahura, F. L., Bajorath, J. (2004) POT-DMC: a virtual screening method for the identification of potent hits. *J Med Chem* 47, 4286–4290.

20. Bajorath, J. (2002) Virtual screening: methods, expectations, and reality. *Curr Drug Discov* 2, 24–28.

21. Cramer, R. D., Jilek, R. J., Guessregen, S., et al. (2004) "Lead hopping." Validation of topomer similarity as a superior predictor of biological activities. *J Med Chem* 47, 6777–6791.

22. Xue, L., Godden, J. W., Stahura, F. L., et al. (2003) Design and evaluation of a molecular fingerprint involving the transformation of property descriptor values into a binary classification scheme. *J Chem Inf Comput Sci* 43, 1151–1157.

23. *MACCS Structural Keys*. MDL Information Systems, Inc., San Leandro, CA.

24. Xue, L., Godden, J. W., Bajorath, J. (2003) Mini-fingerprints for virtual screening: design principles and generation of novel prototypes based on information theory. *SAR QSAR Env Res* 14, 27–40.

25. Brown, R. D., Martin, Y. C. (1996) Use of structure-activity data to compare structure-based clustering methods and descriptors for use in compound selection. J *Chem Inf Comput Sci* 36, 572–584.

26. Xue, L., Bajorath, J. (2002) Accurate partitioning of compounds belonging to diverse activity classes. *J Chem Inf Comput Sci* 42, 757–764.

27. Shannon, C. E. Mathematical theory of communication (1948). *Bell Syst Tech J* 27, 379–423.

28. Godden, J. W., Bajorath, J. (2002) Chemical descriptors with distinct levels of information content and varying sensitivity to differences between selected compound databases identified by SE-DSE analysis. *J Chem Inf Comput Sci* 42, 87–93.

29. Godden, J. W., Xue, L., Stahura, F. L., et al. (2000) Searching for molecules with similar biological activity: analysis by fingerprint profiling. *Pac Symp Biocomput* 8, 566–575.

30. Xue, L., Godden, J. W., Stahura, F. L., et al. (2003) Profile scaling increases the similarity search performance of molecular fingerprints containing numerical descriptors and structural keys. *J Chem Inf Comput Sci* 43, 1218–1225.

31. Gribskov, M., McLachlan, A. D., Eisenberg, D. (1987) Profile analysis: detection of distantly related proteins. *Proc Natl Acad Sci U S A* 84, 4355–4358.

32. Schuffenhauer, A., Floersheim, P., Acklin, P., et al. (2003) Similarity metrics for ligands reflecting the similarity of the target proteins. *J Chem Inf Comput Sci* 43, 391–405.

33. *MDL Drug Data Report*. MDL Information Systems Inc., San Leandro, CA.

34. Bender, A., Mussa, H. Y., Glen, R. C., et al. (2004) Similarity searching of chemical databases using atom environment descriptors (MOLPRINT 2D): evaluation of performance. *J Chem Inf Comput Sci* 44, 1708–1718.

35. Hert, J., Willett, P., Wilton, D. J., et al. (2005) Enhancing the effectiveness of similarity-based virtual screening using nearest neighbour information. *J Med Chem* 8, 7049–7054.

36. Bender, A., Glen, R. C. (2005) A discussion of measures of enrichment in virtual screening: comparing the information content of descriptors with increasing levels of sophistication. *J Chem Inf Model* 45, 1369–1375.

Chapter 18

Optimization of the MAD Algorithm for Virtual Screening

Hanna Eckert and Jürgen Bajorath

Abstract

The approach termed *Determination and Mapping of Activity-Specific Descriptor Value Ranges (MAD)* is a conceptually novel molecular similarity method for the identification of active compounds. MAD is based on mapping of compounds to different (multiple) activity class-selective descriptor value ranges. It was recently developed in our laboratory and successfully applied in initial virtual screening trials. We have been able to show that selected molecular property descriptors can display a strong tendency to respond to unique features of compounds having similar biological activity, thus providing a basis for the approach. Accordingly, a crucial step of the MAD algorithm is the identification of activity-sensitive descriptor settings. The second critical step of MAD is the subsequent mapping of database compounds to activity-sensitive descriptor value ranges in order to identify novel active molecules. This chapter describes the optimization of the MAD approach and evaluates the second-generation algorithm on a number of different compound activity classes with a particular focus on the recognition of remote molecular similarity relationships.

Key words: Compound activity classes, compound class-specific features, descriptors, descriptor distribution analysis, descriptor signature value ranges, mapping algorithms, molecular similarity, scoring functions, search performance, virtual screening.

1. Introduction

Virtual screening methods are generally divided into structure-dependent *(1–3)* and ligand-based *(4–6)* approaches. Ligand-based virtual screening uses small molecules with known activity as templates for database analysis and the identification of novel hits by application of various computational approaches that assess molecular similarity *(6, 7)* or evaluate structure–activity relationships *(8, 9)*. Similarity-based virtual screening methods

Jonathan M. Keith (ed.), *Bioinformatics, Volume II: Structure, Function and Applications, vol. 453*
© 2008 Humana Press, a part of Springer Science+Business Media, Totowa, NJ
Book doi: 10.1007/978-1-60327-429-6 Springerprotocols.com

can roughly be divided into similarity search tools such as molecular fingerprints *(10, 11)* or graph queries *(12)* and compound classification methods such as clustering or partitioning algorithms *(13, 14)* and support vector machines *(15)*. Classification methods generally depend on the use of molecular structure and property descriptors *(11, 16)* for the generation of chemical reference spaces *(4, 11)*. Among classification methods, mapping algorithms introduced by our group present a unique class, as they use sets of known active compounds to generate activity-dependent consensus positions in chemical reference space *(17)* or combine activity-selective descriptor value ranges *(18)* in order to identify database compounds most likely to be active.

Descriptor selection for ligand-based virtual screening applications presents a significant problem *(4)*, since literally thousands of different chemical descriptors are available *(16)*. Mapping methods attempt to facilitate descriptor selection by identifying descriptor subsets from large pools that selectively respond to specific compound activities. In MAD, this is achieved by applying a scoring function that compares descriptor value distributions in screening databases with those in classes of active compounds. On this basis, descriptors displaying significant deviations are selected. In our original implementation of MAD *(18)*, these deviations were detected by comparing sizes of value ranges occurring in active and database compounds.

This chapter presents the optimization of MAD for virtual screening applications. Optimization efforts aim at further increasing hit and compound recovery rates and the potential of identifying remote similarity relationships (i.e., compounds having diverse structures yet similar activity), a process often referred to as "lead hopping" *(19)*. We introduce an advanced technique to identify activity-sensitive descriptors by comparing relative frequencies of descriptor values instead of value range sizes. Activity-specific value ranges for compound mapping are then obtained from class value ranges of the selected descriptors. However, dependent on individual compound class compositions, a decision is made whether statistical outlier values are removed and/or value ranges are slightly extended (in order to further increase the potential for lead hopping). This procedure can be rationalized as a class-oriented optimization step. Finally, database compounds are ranked either using a simple similarity metric (relying only on the number of matched value ranges) or by a more complex scoring scheme introducing additional descriptor score-dependent weighting factors.

The optimized version of MAD was compared to our original implementation in systematic virtual screening trials on six different compound activity classes that were previously not tested with MAD. The results of these calculations confirmed that our optimization strategy led to a significant further increase in MAD performance in virtual screens.

2. Methods

The following describes and compares methodological details of descriptor scoring, determination of activity-specific value ranges, and compound ranking as implemented in original and optimized MAD, and virtual screening results.

2.1. Original MAD Implementation

In a preliminary step, descriptor values are calculated for an activity class and the compound database that is used as the source for screening. Then descriptors are scored by comparing the sizes of their value ranges in the database and activity class. This is done in order to reveal significant deviations, if they exist. Such deviations are indicative of class-specific descriptor value ranges. When comparing the sizes of descriptor value ranges one has to take into account that descriptor values of large compound sets (e.g., a screening database) are usually not evenly distributed but follow a bell-shaped distribution with "frayed" borders. Therefore, it is advisable to use a database sub-range and not the entire database range for size comparison in order to avoid a biasing influence of extreme values. Hence, for each descriptor, the minimum ($dbMin$), 25%-quantile ($q^{0.25}$), 75%-quantile ($q^{0.75}$), and maximum ($dbMax$) of its database value distribution are determined to distinguish the range of the central 50% database values (termed $centralRange$) from the range of the highest 25% and lowest 25% of its values. For compounds of an activity class, only the minimum ($classMin$) and maximum ($classMax$) value of each descriptor are calculated. On this basis, one can distinguish between the following three scoring categories (*see* also **Note 1**):

1. A class value range falls within the $centralRange$ or overlaps with it ($classMax \geq q^{0.25}$ and $classMin \leq q^{0.75}$). Then the descriptor score is calculated as:

$$score = \frac{q^{0.75} - q^{0.25}}{classMax - classMin} \qquad [1]$$

2–3. A class value range completely falls below ($classMax < q^{0.25}$) or above ($classMin > q^{0.75}$) the $centralRange$. Then we calculate the descriptor score as follows:

$$score = \frac{q^{0.25} - dbMin}{classMax - classMin} * 2 \qquad [2]$$

$$score = \frac{dbMax - q^{0.75}}{classMax - classMin} * 2 \qquad [3]$$

The factor 2 for categories 2 and 3 is included to account for the fact that the value range of only 25% of the database molecules is used here, whereas in 1 the value range of 50% of the molecules is used. Applying the above scoring scheme, scores of 0 or greater are produced. Descriptors obtain scores smaller than 1 if the class value range is greater than the range of the central 50% database compounds. This usually corresponds to a situation where more than 50% of the database molecules match the class value range of the descriptor, and this descriptor would not be regarded as an activity-sensitive one. Descriptor scores >1 are generally obtained when fewer than half of the database molecules match the class value range. For compound mapping, the top-scoring descriptors are selected in descending order down to a score of 1, but no more than 70 descriptors are selected. This cut-off value is empirically chosen based on the results of various benchmark calculations.

In order to identify new active compounds, database molecules are mapped to slightly extended class value ranges of the selected descriptors and ranked by using a simple similarity function that divides the number of matched class value ranges by their total number. The range extension is applied with the purpose of increasing the ability to recognize compounds with similar activity but diverse structures (i.e., lead hopping potential), which represents one of the premier goals of ligand-based virtual screening analysis. The extent of value range expansion is determined by a simple function depending on the number of active compounds (termed *numBaits*) and the size of the class value range:

$$\delta = \frac{classMax - classMin}{numBaits - 1}$$

The obtained range fraction is added at both ends of the original class range leading to the expanded range [*classMin* − δ, *classMax* + δ] for compound mapping. **Figure 18.1** illustrates the different stages of the MAD algorithm, as described in the preceding.

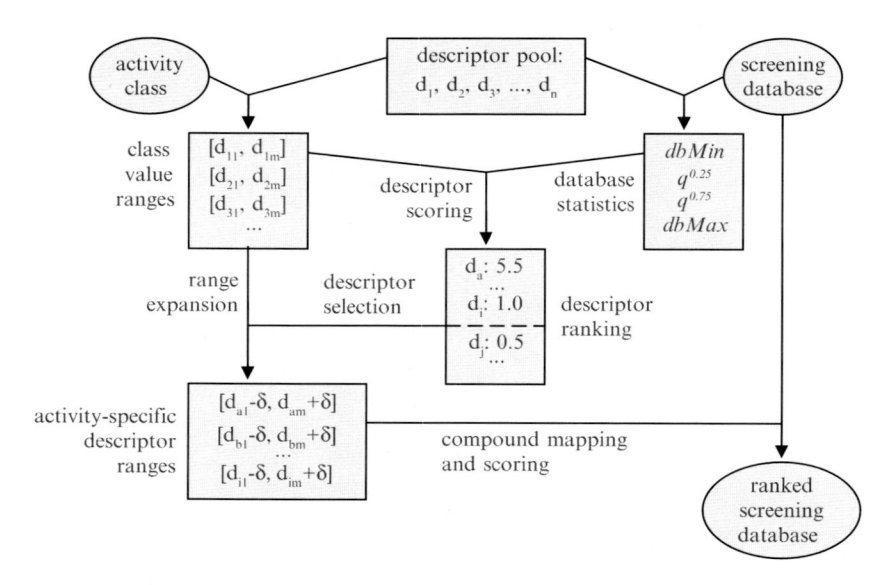

Fig. 18.1. Flowchart of MAD analysis. A descriptor statistic of a screening database and class value ranges for an activity class provide the basis for scoring and ranking a pool of descriptors. The top-scoring descriptors are selected and a controlled range expansion is applied to their class value ranges. Finally, database compounds are mapped to these expanded class ranges and ranked using a similarity function.

In benchmark calculations, MAD produced promising results. When utilized in simulated virtual screening situations with six different compound activity classes, hit and recovery rates between 10–25% and 25–75%, respectively, were achieved *(18)*. These rates were significant and compared very well to those obtained with different reference methods (*see* **Note 2**).

2.2. Refinement of the Descriptor Scoring Scheme

An initial focal point for improvement of the original MAD method was the descriptor scoring function. The scoring function is used to identify activity-sensitive descriptors for compound mapping and it is thus critically important for the success of the approach. The idea behind MAD's descriptor scoring scheme is the assessment of the specificity of descriptor value ranges adopted by a compound activity class: A class value range is considered to be increasingly specific the fewer database compounds match it. The preceding descriptor scoring function only approximately accounts for this situation because comparing value range sizes of database and active compounds only reflects selectivity tendencies of descriptors, not the number of database compounds that actually match the value range of an activity class. Consequently, we designed a new descriptor scoring function that directly relates descriptor scores to mapping probabilities of class ranges:

$$score = 100 - P(X \in [classMin, \ classMax]) * 100$$

The scoring function determines the fraction of the database that matches the class range and calculates scores on this basis: the fewer database compounds, the higher the score. When applying this function, a maximum descriptor score of 100 can be achieved (i.e., no database compound matches the class range) and a minimum score of 0 (i.e., all compounds match the class range). In order to be selected, a descriptor must have a minimum score of 50. The mapping probabilities required for score calculations are obtained by applying an approximation (*see* **Note 3**). **Figure 18.2** illustrates principal differences between original and revised score calculations. In the example given in this figure, a descriptor displaying a clear class-specific tendency would only be selected if the new scoring function was applied.

2.3. Refinement of Value Range Expansion

Next we optimized the value range expansion function for compound mapping. Class value ranges are extended to favor the recognition of molecules with similar activity but increasingly diverse structures. For this purpose, MAD in its original implementation made use of a naïve expansion function linearly depending on a class value range size. Since most descriptors have uneven DB value distributions, which often display significant differences, the original expansion scheme might sometimes lead to unpredictable effects in compound mapping. Consequently,

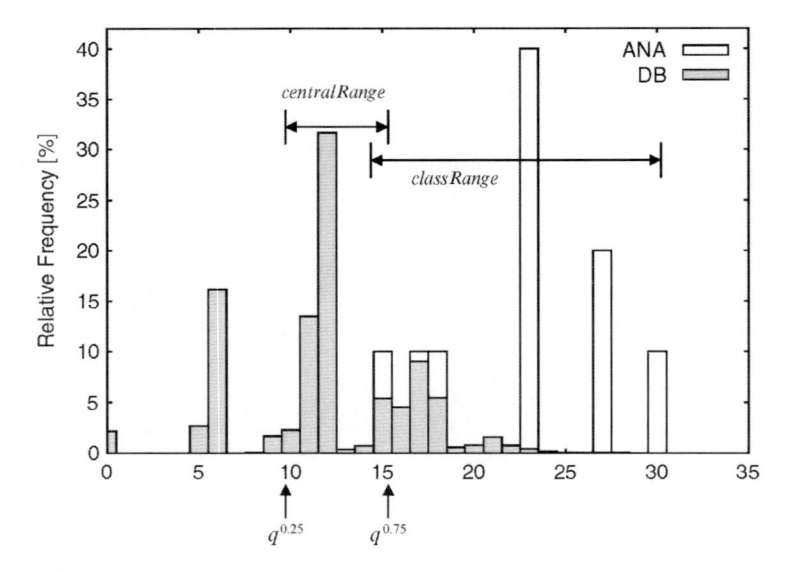

Fig. 18.2. Original versus revised descriptor scoring. The value distributions of a screening database (termed DB) and an activity class, angiotensin II antagonists (ANA), for descriptor a_aro (number of aromatic atoms in a molecule) are shown. On the horizontal axis, $q^{0.25}$ and $q^{0.75}$ mark the positions of the 25% and 75% quantiles of DB. Double-headed arrows indicate which value ranges are used for calculation of the original scoring function. Descriptor a_aro belongs to category "a" of the original scoring function (as defined); that is, the class value range overlaps with the central range of DB. The potential drawback of the original scoring function is clearly evident in this example. The size of the class value range (i.e., 15) is compared to the size of the central range (i.e., 5) because both ranges overlap, albeit only marginally. This leads to the score of 0.33 and, accordingly, the descriptor would not be selected for compound mapping. At the same time, it is obvious that descriptor a_aro has class-specific descriptor value settings for ANA because value distribution characteristics of ANA and DB are strikingly different. By contrast, the relative frequency of database descriptor values falling within the class range [15, 30] is 29%, leading to a revised descriptor score of 71 and the selection of a_aro for compound mapping.

we designed two alternative expansion functions that make the extent of value range extension (Δp) dependent on the mapping probability p of an activity class:

$$Function\ 1: \Delta p = \frac{1-p}{100 * p + 10}$$

$$Function\ 2: \Delta p = \frac{1-p}{8}$$

In both cases, for increasing mapping probabilities p, the additional mapping probability Δp decreases. These two alternative functions differ in the extent of expansion. Function 1 only leads to a significant expansion in the presence of low mapping probabilities (0–30%), whereas Function 2 constantly reduces the magnitude of expansion with increasing mapping probabilities. Thus, Function 1 is similar to a correction function that expands too small class value ranges to an adequate degree. The constant 10 in Function 1 ensures that the minimum mapping probability following expansion is 10%, whereas in Function 2 the constant 8 is used corresponding to a minimal mapping probability of 12.5%. Given Δp, class value ranges are extended both on the lower and upper end (each by $\Delta p/2$) by including the next database descriptor values until the total additional mapping probability is reached (*see* also **Note 4**).

2.4. Handling Structurally Diverse Compound Activity Classes

For virtual screening applications aiming at recognition of remote similarity relationships, very similar structures (e.g., an analog series) are usually not suitable as templates. Rather, template sets should include diverse structures having different scaffolds. Such compound activity classes with distinct diversity often display more widespread descriptor class value ranges and thus broader ranges for compound mapping. In many instances, this increases the probability of false-positive recognitions (inactive compounds) during virtual screening trials. However, a larger class range is often simply the result of single compounds producing "outlier" values. Therefore, omitting the minimum and maximum descriptor values obtained for a compound set during the calculation of class value ranges is an effective option to prevent compound mapping to artificially large ranges. Thus, although value range expansion can favor correct compound recognition, it is also necessary to examine calculated class value ranges for the effects of outliers and control their sizes.

2.5. Parameter Optimization

For individual activity classes, it is often advisable to balance parameter settings for MAD calculations with respect to value range expansion and statistical outliers of bait-set descriptor values prior to virtual screening trials. For this purpose, a set of known active compounds is divided into a *training set* used for calculation of class value ranges and a *test set* that is added to the source database. Therefore, test calculations can be carried out either without value range expansion or using expansion Functions 1 or 2 and, in addition, either with unmodified bait descriptor value sets or after removing the minimum and maximum descriptor values from each set. For these benchmark calculations, compounds in training and test sets should also be exchanged. Best parameters are determined by comparing the different distributions of descriptor scores obtained in these calculations and the resulting hit and recovery rates. Such individually optimized parameter settings are then used for a compound activity class in virtual screening applications.

2.6. Introduction of Descriptor-Dependent Scaling Factors for Compound Ranking

The compound ranking function of MAD has also been further refined. As described, database compounds are mapped to combinations of selected activity-sensitive descriptor value ranges. In the first implementation of MAD, mapped database compounds are ranked by dividing the number of matched value ranges by the total number of selected descriptors. This represents a simple similarity function in which every selected descriptor is considered equivalent. We now introduce descriptor score-dependent weighting factors that put more emphasis on descriptors with clearly defined class-specific value settings. A continuous weighting function was found to further increase search performance in test calculations:

$$w = \left(\frac{score}{50} \right)^5$$

For instance, applying this function, a scaling factor of 1 is obtained for descriptors with score of 50 (minimum score for descriptors to be selected), a factor of ~2.5 for a score of 60, a factor of ~5.4 for a score of 70, and finally a factor of 32 for the maximum (hypothetical) score of 100.

2.7. Application Example

In order to test the optimized version of MAD in simulated virtual screening trials, we carried out a comparative study using the original MAD implementation as our reference. We assembled six different activity classes containing between 30 and 45 compounds from a database of bioactive compounds *(20)*. These classes are reported in **Table 18.1**. Virtual screening trials were carried out in a publicly available compound database containing ~1.44 million molecules (*see* **Note 5**). In our calculations, 155 1D and 2D descriptors implemented in the Molecular Operating Environment (MOE) *(21)* were used as a basis set and their values were calculated for all active and database compounds. For each of the six activity classes, 100 sets of 10 compounds each were randomly selected. Each set was taken once as the "bait" set (i.e., template compounds for virtual screening) for the calculation of descriptor scores, selection of descriptors, and determination of specific value ranges for compound mapping. The remaining active compounds (between 20 and 35) were added to the database as potential hits, whereas all database compounds were considered inactive (and thus potential false-positives). As performance measures for comparing the original and refined MAD approach, hit rates (HR; number of selected active compounds relative to selected database molecules) and recovery rates (RR; number of selected active compounds relative to the total number of potential database hits) were calculated for selection sets consisting of either 10 and 50 compounds and averaged over the 100 independent trials. **Table 18.1** summarizes the results of this study.

Table 18.1
Comparison of original and optimized MAD in virtual screening trials

Activity class	ADC	Method	RR_10	HR_10	RR_50	HR_50	Optimized parameters
ANA	35	Original	17.3	60.4	31.8	22.2	—
		Optimized	20.1	70.3	39.4	27.6	+, 2
		Scaled	21.5	75.3	45.9	32.1	
DD1	20	Original	23.1	46.1	36.5	14.6	—
		Optimized	29.0	58.0	38.9	15.6	+, 2
		Scaled	31.2	62.3	43.0	17.2	
GLY	24	Original	25.7	61.7	40.1	19.2	—
		Optimized	26.1	62.5	40.4	19.4	−, 1
		Scaled	30.7	73.6	51.6	24.8	
INO	25	Original	5.5	13.8	11.8	5.9	—
		Optimized	7.4	19.1	12.4	6.2	+, 0
		Scaled	14.6	36.4	22.7	11.3	
SQS	32	Original	8.4	26.7	17.2	11.0	—
		Optimized	15.1	48.2	23.3	14.9	+, 0
		Scaled	15.7	50.3	24.1	15.4	
XAN	25	Original	6.6	16.5	19.2	9.6	—
		Optimized	17.9	44.9	25.5	12.8	+, 1
		Scaled	20.0	50.0	28.2	14.1	

"Activity Class" reports the names of the six compound activity classes studied in our calculations: angiotensin II antagonists (ANA), dopamine D1 agonists (DD1), glycoprotein IIb/IIIa receptor antagonists (GLY), inosine monophosphate dehydrogenase inhibitors (INO), squalene synthetase inhibitors (SQS), xanthine oxidase inhibitors (XAN). "ADC" reports the number of active database compounds (potential hits) and "RR_10," "HR_10," "RR_50," and "HR_50" the recovery and hit rates (in %) for selection sets of 10 or 50 compounds. Under "Method," "original" stands for the original MAD implementation and "optimized" for MAD as described herein with optimized parameter settings but no descriptor scaling; in "scaled" calculations, descriptor scaling for compound mapping was also applied. "Optimized Parameters" indicates which parameter settings produced the best results for the refined MAD approach: "+" or "−" indicate whether or not minimum and maximum descriptor values were removed and "0," "1," and "2" indicate the chosen expansion function; "0" means that class value ranges were not expanded.

As can be seen in **Table 18.1**, improvements in both hit and recovery rates are consistently achieved for the six activity classes studied here using optimized MAD without descriptor scaling and further improvements are obtained when the descriptor score-dependent scaling function is used. In some cases, improvements in averaged hit and recovery rates are fairly small, but in others (e.g., classes SQS or XAN) they are substantial. A characteristic feature of MAD calculations is the very good search performance for small compound selection sets and optimized MAD further improves these results. For HR_10, absolute improvements range from ~12% (GLY) to ~34% (XAN) and for RR_10, improvements range from ~5% (GLY) to ~13% (XAN). As one would expect, best parameter settings vary dependent on the activity class. For six activity classes, four different preferred parameter settings are observed. Thus, parameter optimization is an important step. Furthermore, scaling of descriptor contributions can also cause significant effects. For HR_10, absolute improvements as a consequence of scaling range from 2% (SQS) to 17% (INO).

Our findings show that introduction of the new descriptor scoring function leads to consistently better search performance of MAD for the six activity classes studied here. The most plausible explanation for these observations is that the new scoring function selects descriptors that are not chosen by the original function. The results shown in **Fig. 18.3** confirm this view. For activity class SQS, only 27 descriptors are selected by both functions, whereas 35 other descriptors are selected exclusively by the new function. These descriptors were not recognized as activity-sensitive by the original function.

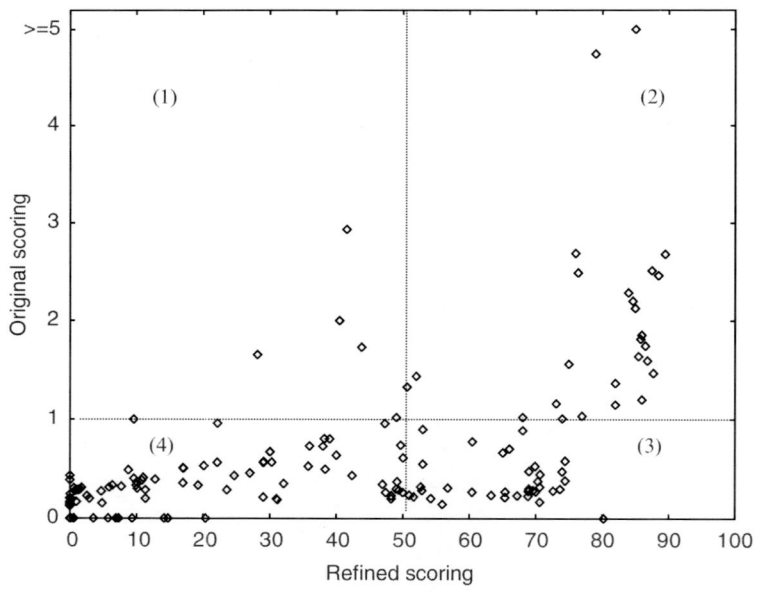

Fig. 18.3. Comparison of different descriptor scoring functions. As an example, descriptor scores obtained with the original scoring function and the one reported herein are compared for activity class SQS. Scores of the refined scoring function range from 0 to 100, whereas scores of the original function have no fixed upper limit. Applying the original scoring function, 35 descriptors with scores ≥1 are selected. These descriptors are located above the horizontal line in areas (1) and (2). When using the refined scoring function almost twice as many (i.e., 62) descriptors with scores ≥50 are selected. These descriptors are seen on the right side of the vertical line in (2) and (3). Only 27 descriptors are selected by both functions and are in area (2). Descriptors located in (4) are not selected by either function.

In the case of SQS, the original scoring function selects only 35 descriptors, whereas the new function selects a total of 62. Thus, compound mapping using optimized MAD should have greater discriminatory power, which is consistent with the observed improvements in search performance.

2.8. Perspective

With MAD, we have introduced an approach belonging to a novel class of methods, termed mapping algorithms, which add to the spectrum of currently available molecular similarity-based methods. Among similarity-based methods, mapping algorithms such as MAD or DMC (17) are unique. MAD could be easily adopted for large-scale virtual screening applications and its computational efficiency is comparable to partitioning methods. A characteristic feature of both MAD and DMC is their high level of search performance for compound selection sets that are much smaller in size than those produced with many other similarity-based methods, which is consistent with the high stringency of mapping calculations on the basis of multiple descriptor contributions.

The optimization of MAD for virtual screening reported in this chapter provides an instructive example of the way predictive performance can be improved by refinement of algorithmic details and systematic parameter adjustments. In particular, the results reported herein demonstrate the crucial role of the descriptor scoring function for the success of the MAD approach. This is expected because MAD relies on the detection of activity class-sensitive descriptors, which is a key feature of this novel methodology. It is well known that the performance of similarity-based methods, irrespective of their details, generally shows a strong compound activity class dependence. Thus, we cannot expect that current or future refinements of MAD will improve search performance in each and every case. In fact, for some activity classes, MAD's original descriptor scoring function already detects the most relevant activity-sensitive descriptors and thus further improvements cannot be expected. In other cases, such as the compound classes tested in this study, optimized MAD recognizes more activity-selective descriptors, which together with its more refined compound mapping function leads to further improved hit and recovery rates.

3. Notes

1. Prior to score calculations, a slight correction of the class value range [classMin, classMax] has to be applied when its size is zero (i.e., when all active compounds adopt the same descriptor value), simply because division by zero is not possible. In this case, we extend the class range by use of ε,

defined as a fraction of the standard deviation of the descriptor database distribution:

$$\varepsilon = \frac{stdDev}{200}$$

Thus, the magnitude of the correction is made dependent on the individual standard deviation of each affected descriptor. We add ε to both sides of the original class range, thereby making its size greater than zero. The constant 200 for calculation of ε was empirically chosen and ensures that affected descriptors achieve top scores within the typically observed scoring range.

2. Hit rates measure the content of novel hits within a compound selection of a given size (e.g., 10 active compounds within 100 selected databases compounds correspond to a hit rate of 10%). Recovery rates give the number of identified hits relative to the total number of active compounds in the database (e.g., if 10 of 100 potential hits are selected, the recovery rate is 10%). Although hit rates serve as a measure of virtual screening performance in benchmark calculations and "real-life" applications, recovery rates can usually only be reported for simulated virtual screening trials.

3. Descriptor value distributions in the source database are determined and stored in a preliminary step: For descriptors having a discrete value distribution with up to about 100 different values, the relative frequency of each descriptor value is calculated. For descriptors with a continuous value range or a discrete value distribution containing significantly more than 100 or so values, we divide the value distribution into sub-intervals, each having an overall relative frequency for database compounds of 1% at most. Mapping probabilities required for descriptor scoring are then obtained by addition of relative frequencies of discrete values x_i for descriptors with discrete distributions:

$$P\left(X \in [classMin, classMax]\right) = \sum_{classMin \le x_i \le classMax} P\left(X = x_i\right)$$

or otherwise by addition of matching probabilities of sub-intervals $[x_{i1}, x_{i2}]$:

$$p\left(X \in [classMin, classMax]\right) = \sum_{\substack{classMin \le x_{i2} \\ and \\ classMax \ge x_{i1}}} P\left(X \in [x_{i1}, x_{i2}]\right)$$

that fall within or overlap with the class value range. The maximum possible absolute error due to the introduction of this approximation is 0.02, but the error is usually negligible or zero for discrete value distributions.

4. Given Δp and the simplified value distribution described in **Note 3**, value range expansion is actually done by including the next discrete value or sub-interval alternately on the lower and upper end until the total additional mapping probability Δp is achieved.

5. The precursor of our source database was the publicly available compound collection ZINC *(22)*, containing ~2.01 million molecules in predicted 3D conformational states, from various vendor sources. For each compound, we calculated the values of 155 1D and 2D molecular descriptors, and used them to create a "2D-unique" version of ZINC by identifying compounds with duplicate descriptor settings. This led to the removal of ~0.57 million ZINC compounds. Thus, the 2D-unique version representing the source database for our calculations contained ~1.44 million compounds.

References

1. Brooijmans, N., Kuntz, I. D. (2003) Molecular recognition and docking algorithms. *Annu Rev Biophys Biolmol Struct* 32, 335–373.

2. Shoichet, B. K. (2004) Virtual screening of chemical libraries. *Nature* 432, 862–865.

3. Kitchen, D. B., Decornez, H., Furr, J. R., et al. (2004) Structure-based virtual screening and lead optimization: methods and applications. *Nature Rev Drug Discov* 3, 935–949.

4. Bajorath, J. (2002) Integration of virtual and high-throughput screening. *Nature Rev Drug Discov* 1, 882–894.

5. Green, D. V. (2003) Virtual screening of virtual libraries. *Prog Med Chem* 41, 61–97.

6. Stahura, F. L., Bajorath, J. (2005) New methodologies for ligand-based virtual screening. *Curr Pharm Des* 11, 1189–1202.

7. Johnson, M., Maggiora, G. M., eds. (1990) *Concepts and Applications of Molecular Similarity.* John Wiley & Sons, New York.

8. Martin, Y. C. (1992) 3D database searching in drug design. *J Med Chem* 35, 2145–2154.

9. Esposito, E. X., Hopfinger, A. J., Madura, J. D. (2004) Methods for applying the quantitative structure-activity relationship paradigm. *Methods Mol Biol* 275, 131–214.

10. Xue, L., Godden, J. W., Bajorath, J. (2003) Mini-fingerprints for virtual screening: design principles and generation of novel prototypes based on information theory. *SAR QSAR Env. Res.* 14, 27–40.

11. Bajorath, J. (2001) Selected concepts and investigations in compound classification, molecular descriptor analysis, and virtual screening. *J Chem Inf Comput Sci* 41, 233–245.

12. Gillet, V. J., Willett, P., Bradshaw, J. (2003) Similarity searching using reduced graphs. *J Chem Inf Comput Sci* 43, 338–345.

13. Stahura, F. L. and Bajorath, J. (2003) Partitioning methods for the identification of active molecules. *Curr Med Chem* 10, 707–715.

14. Pearlman, R. S., Smith, K. M. (1998) Novel software tools for chemical diversity. *Perspect Drug Discov Design* 9, 339–353.

15. Jorissen, R. N., Gilson, M. K. (2005) Virtual screening of molecular databases using a support vector machine. *J Chem Inf Model* 44, 549–561.

16. Todeschini, R., Consonni, V. (2000) Handbook of molecular descriptors, in (Mannhold, R., Kubinyi, H., Timmerman, H., eds.), *Methods and Principles in Medicinal Chemistry, vol.* 11. Wiley, New York.

17. Godden, J. W., Furr, J. R., Xue, L., et al. (2004) Molecular similarity analysis and virtual screening by mapping of consensus positions in binary-transformed chemical descriptor spaces with variable dimensionality. *J Chem Inf Comput Sci* 44, 21–29.

18. Eckert, H., Bajorath, J. (2006) Determination and mapping of activity-specific descriptor value ranges (MAD) for the identification of active compounds. *J Med Chem* 49, 2284–2293.

19. Cramer, R. D., Jilek, R. J., Guessregen, S., et al. (2004) "Lead hopping." Validation of topomer similarity as a superior predictor of biological activities. *J Med Chem* 47, 6777–6791.

20. Molecular Drug Data Report (MDDR). MDL Information Systems, Inc., San Leandro, CA.

21. MOE (Molecular Operating Environment), Chemical Computing Group, Inc., Montreal, Quebec, Canada.

22. Irwin, J. J., Shoichet, B. K. (2005) ZINC: a free database of commercially available compounds for virtual screening. *J Chem Inf Model* 45, 177–182.

Chapter 19

Combinatorial Optimization Models for Finding Genetic Signatures from Gene Expression Datasets

Regina Berretta, Wagner Costa, and Pablo Moscato

Abstract

The aim of this chapter is to present combinatorial optimization models and techniques for the analysis of microarray datasets. The chapter illustrates the application of a novel objective function that guides the search for high-quality solutions for sequential ordering of expression profiles. The approach is unsupervised and a metaheuristic method (a memetic algorithm) is used to provide high-quality solutions. For the problem of selecting discriminative groups of genes, we used a supervised method that has provided good results in a variety of datasets. This chapter illustrates the application of these models in an Alzheimer's disease microarray dataset.

Key words: Combinatorial optimization, integer programming, gene selection, feature selection, gene ordering, microarray data analysis, Alzheimer's disease.

1. Introduction

Microarray technologies are revolutionizing life sciences with their ability to simultaneously measure thousands of gene expression values. However, it is well known that microarray technologies, although extremely powerful, provide measurements that are prone to errors. Consequently, techniques that help to analyze microarray information should be relatively robust to measurement noise and provide good solutions in reasonable time. The analysis of microarray datasets requires both unsupervised and supervised methods *(1, 2)* (*see* also **Chapter 12**). If we have class labels associated with each experiment, we can apply supervised techniques when the final objective is, for example, building a classifier *(3)*. In the case of unsupervised methods, the goal is

Jonathan M. Keith (ed.), *Bioinformatics, Volume II: Structure, Function and Applications, vol. 453*
© 2008 Humana Press, a part of Springer Science + Business Media, Totowa, NJ
Book doi: 10.1007/978-1-60327-429-6 Springerprotocols.com

generally to find groups of co-expressed genes using clustering/ordering algorithms *(4, 5)*.

This chapter presents combinatorial optimization models and algorithms for problems related to unsupervised and supervised methodologies applied to microarray data analysis. In the case of supervised methods, we present mathematical models for *feature set selection* problems. As an example of an unsupervised method, we describe a gene ordering approach, which will be addressed using a metaheuristic approach (a memetic algorithm). We apply both methodologies to an Alzheimer's disease microarray dataset contributed by Brown et al. *(6)*. We have previously applied the techniques described here to the classification of different types of cancer *(7)* and, using a different discretization method, we have applied it to the same dataset used here *(8)* (*see* **Note 1**). In this book, **Chapter 20** illustrates the application of these techniques to a rodent model of Parkinson's disease, another microarray dataset which is available in the public domain.

In terms of computational complexity, our mathematical formulations of the problems of interest to life scientists are based on NP-complete problems. Such problems are deemed "mathematically intractable" and the search for exact polynomial-time solutions may be ultimately an impossible quest. However, recent advances in exact techniques (Branch-and-Cut and other variants) and several metaheuristic methods have provided insights into how to address optimization problems which were considered practically intractable a few years ago *(9)*.

In the bioinformatics literature, we can note an increasing trend to formulate some core research questions in the analysis of microarray data using combinatorial optimization problems and algorithms. Since the problems tend to be of large size, several metaheuristic and exact techniques have been recently proposed *(10–15)*. This chapter illustrates their practical relevance for gene selection problems in microarray data analysis.

The chapter is organized as follows. **Section 2** presents the gene ordering problem, which we address with a metaheuristic method (a memetic algorithm). **Section 3** describes different models for feature set selection problems. **Section 4** presents computational results for an Alzheimer dataset. The final section contains the conclusions.

2. Gene Ordering Problem

The first problem to be addressed using combinatorial optimization is the *Gene Ordering Problem*. This is a generic name for a number of different problems in unsupervised analysis of microarray data. Informally, given a gene expression matrix, we aim to find a permutation of the genes' positions such that genes with similar expressions profiles tend to be relatively close. These problems are generally addressed in the literature but often they are not well formalized in mathematical terms. Some researchers use constructive *ad hoc* heuristics instead of dealing directly with it as a clearly formalized combinatorial optimization problem. One example of these methodologies is *hierarchical clustering constructive heuristics* such as the popular method by Eisen et al. *(4)* (which provides a rooted tree of gene expression profiles and a visualization of it). However, the solutions obtained by this type of method can have highly co-expressed sets of genes in different clusters unless some optimal rearrangement of the leaves of the tree is performed after the clustering is found.

Another option is a method that fully takes into account the global arrangement of the genes and deals with the inherent NP-hardness of the basic problem. We are given as input a matrix of gene expression values $G = g_{ij}$ with $1 \leq i \leq n$, $1 \leq j \leq m$, where g_{ij} represents the expression level of the gene i in sample j, n is the total number of genes, and m is the total number of samples. The objective of the Gene Ordering problem is to find the *"best"* permutation of the rows (genes) of the matrix G such that genes that are similarly expressed appear relatively close to each other. The "best" refers to a certain objective function that we try to minimize. In turn, we need a measure between the gene expression profiles (across all samples) of a pair of genes (x,y). In the following, we assume that we are using a dissimilarity function D[x,y] that has been chosen beforehand.

Let $\pi = (\pi_1, \pi_2, \ldots, \pi_n)$ be the permutation of the genes' names and $w>0$ an integer number that is a parameter of our method. Let $D[\pi_k, \pi_i]$ be the dissimilarity between the genes in positions k and i; we first calculate the sum:

$$\sum_{i=\max(k-w,1)}^{\min(k+w,n)} \left(w - |k - i| + 1\right).D[\pi_k, \pi_i] \qquad [1]$$

For example, consider a gene in position $k = 5$ in the permutation and suppose $w=2$. The partial sum for this gene will be $D[\pi_5, \pi_3] + 2D[\pi_5, \pi_4] + 2D[\pi_5, \pi_6] + D[\pi_5, \pi_7]$. Note that genes that are closer to the gene in position 5 receive a higher weight in Equation 1.

The objective function for a specific permutation π is given by the sum of Equation 1 calculated for each gene as shown in Equation 2.

$$min \sum_{k=1}^{n} \sum_{i=max(k-w,1)}^{min(k+w,n)} \left(w - |k - i| + 1\right).D[\pi_k, \pi_i] \qquad [2]$$

Finding a permutation π^* that minimizes the objective function Equation 2 is an NP-Hard problem and exact methods are not yet able to provide optimal solutions for problems of the size we deal with in microarray data analysis. However, recent advances in metaheuristics have made possible powerful techniques that give good results in a reasonable amount of time *(9)*. We proposed a memetic algorithm *(16)*, which incorporates a Tabu search *(17)* technique for local search. Our memetic algorithm is described in *(5)*. We have used this memetic algorithm for ordering the genes that appear in **Fig. 19.7** corresponding to the genetic signatures we report for Alzheimer's disease. The next section shows how these genetic signatures are obtained using a gene selection method based on a generalization of the *feature set* problem.

3. Feature Set Problems

We present three problems (the *Min Feature Set*, the *Min (α,β)-Feature Set*, and the *Max Cover (α,β)-k-Feature Set*), and we note that the final objective is always to select a minimum set of features that help to discriminate between classes of samples under study. We also note that the definition of a feature is rather general. In this chapter, we restrict our scope to *one gene corresponds to one feature,* used in conjunction with a well-known method for discretization of gene expression values. As a consequence, in this chapter we use the two terms synonymously; however, a feature can in principle characterize any property of a sample (not necessarily a single gene). Next, we describe the different mathematical models for gene/feature selection.

3.1. Min Feature Set Problem

We explain this mathematical formulation using the following example. Consider a dataset represented by the matrix in **Fig. 19.1**. We can regard it as a tiny microarray dataset, in which each row corresponds to a gene (feature), and each column to an individual experiment (sample). The entries in this matrix are the expression levels of each gene in a given sample (in this example we have only two states, up- or down-regulated, which we assume have been discretized by an independent method). We have 5 genes and 6 samples, which are equally distributed between two class labels (true and false). In this particular case, we aim to find a minimum set of features that discriminate among samples 1, 2, and 3 (label = true) and samples 4, 5, and 6 (label = false).

Before modeling the core problem, it is necessary to define the following terms. An *observation* is any pair of samples with distinct labels. A gene is said to *explain an observation* if it is

	Sample 1	Sample 2	Sample 3	Sample 4	Sample 5	Sample 6
Gene A	0	0	0	0	1	0
Gene B	1	1	1	1	0	1
Gene C	1	1	1	0	1	0
Gene D	1	1	1	1	0	0
Gene E	0	0	1	0	0	0
Label	True	True	True	False	False	False

Fig. 19.1. A very simple example used to illustrate feature set selection models.

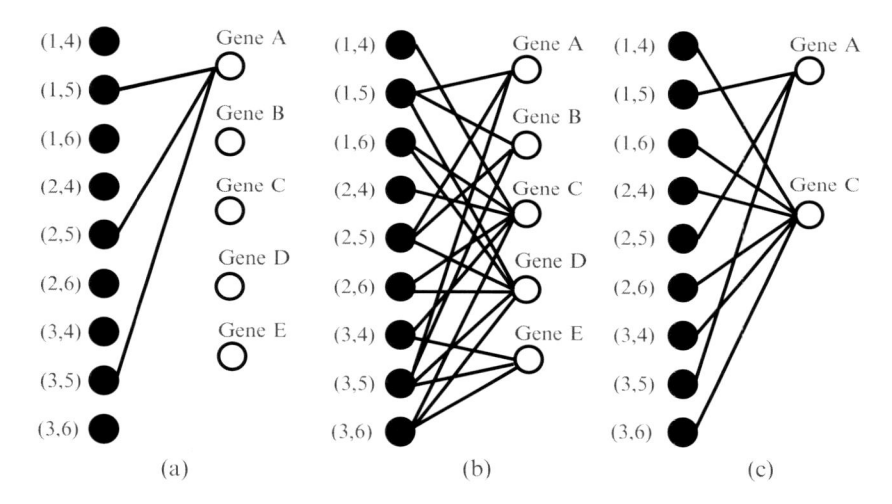

(a)   (b)   (c)

Fig. 19.2. A very simple example used to illustrate feature set selection models. A bipartite graph constructed from an instance of the *Min Feature Set* problem. (**A**) Pairs of samples that belong to different classes (black nodes) and that are discriminated by feature A are linked by an edge. (**B**) The graph created by adding edges for all discriminations. (**C**) A feasible solution of the *Min Feature Set* problem in which two genes are all that is needed to have at least one feature that discriminates any pair of samples.

differentially expressed (has different expression values) for the pair comprising the observation.

The *Min Feature Set* problem can be posed as a graph optimization problem (**Fig. 19.2**), in which each black node represents a relevant observation and each white node represents a gene (feature). An edge links a black node *(p,q)* to a white node *j* if the gene represented by node *j* has distinct expression values for samples *p* and *q* as stated in the matrix. For instance, Gene A is *down-regulated* (has a value of *0*) in sample 1, and is *up-regulated* (has a value of *1*) in sample 5; therefore, there is an edge between node "Gene A" and black node "*(1, 5)*", and we say that *"Gene A covers node (1, 5)"* (**Fig. 19.2A**).

Given the data of **Fig. 19.1**, we can construct the graph as shown in **Fig. 19.2B**. If there was a single gene covering all black nodes, a fact that can be checked by a polynomial time algorithm, this gene would be the simplest explanation that accounts for all the observations. Unfortunately, in microarray data analysis it is highly improbable that even in a very large dataset we will find such a gene. In general, we need a set of genes that, when combined, cover all black nodes. In the present case, Gene A and Gene C together cover all pairs, as shown in **Fig. 19.2C**.

The task of finding the minimal cardinality set of features (genes in our application) that cover all black nodes is defined as the *Min Feature Set* problem. Its decision version is called *k-Feature Set* and was proved NP-Complete more than a decade ago *(18)*. Cotta and Moscato proved in 2003 that the parameterized version of the *k-Feature Set* is W[2]-Complete (where the parameter k is the cardinality of the set) *(19)*. This means that it is unlikely that we will have either a polynomial-time or a fixed-parameter tractable algorithm for this problem. However, due to some properties of this problem, it is possible to apply some *safe reduction rules* (also called *pre-processing rules*) that give an instance of the same problem but of a smaller size. We refer the reader to *(7)* for the description of these reduction rules.

We will show in the next subsection a generalization of this problem by considering the idea of a *multiple explanation* of an observation. Briefly, we want every black node to be covered by several genes that are differentially expressed, and not just a single one. This provides robustness to our selection method and we have also noticed that it helps to uncover subclasses in each of the two classes, if they exist. We may also go further and specify that we want a set that not only differentiates samples with different labels but is also able to explain why two samples have the same label. With these ideas in mind we discuss the *Min (α,β)-Feature set* problem.

3.2. Min (α,β)-Feature Set *Problem*

The *Min (α,β)-Feature Set* problem is a generalization of the *Min Feature Set* problem and was recently introduced by Cotta, Sloper, and Moscato *(20)*. This variant aims to choose a set of features that maximizes the number of differences between two samples with different labels, and also maximizes the similarities between samples with the same label. The parameters *(α,β)* provide a mathematical guarantee of the solution. The parameter α represents the minimum number of genes (features) that must be in a different state in any pair of samples with different labels, and the parameter β represents the minimum number that must be in the same state in any pair of samples that have the same label.

We use the same example shown in **Fig. 19.1** to describe this problem. First, we build a graph (**Fig. 19.3**), in which each black node represents a pair of samples with different labels, each gray

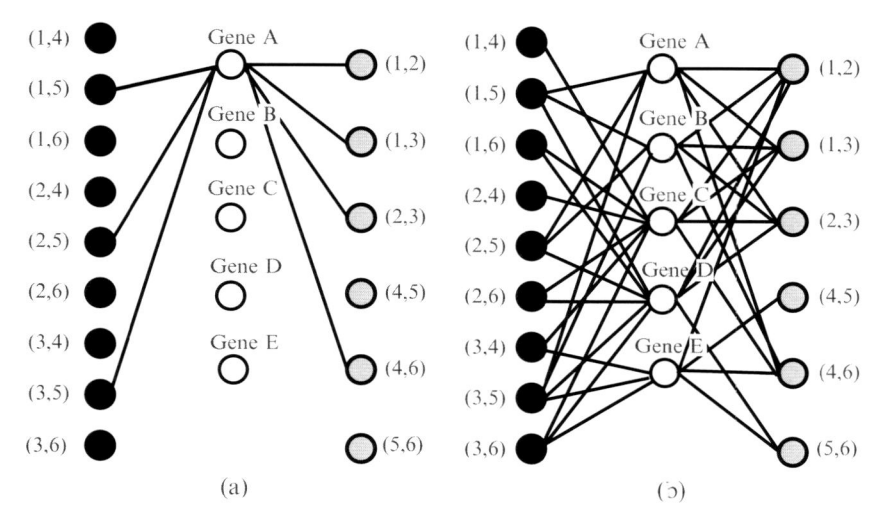

Fig. 19.3. Graph representation for an instance of the *Min (α,β)-Feature Set* problem. (**A**) The pairs of samples that belong to different classes (black nodes) are linked to feature (a) in Fig. 19.2A; however, we now create nodes corresponding to pairs of samples that belong to the same class and these are linked to a feature if that feature does not discriminate between the two samples. (**B**) The resulting graph for the data shown in Fig. 19.1.

node represents a pair of samples with the same label, and each white node represents a gene (feature). An edge links a white node to a black one if the gene helps to explain the observation (they are in a different state). This part of the construction is exactly as for the *Min Feature Set* problem. Analogously, an edge is created between a white and a gray node if the gene is in the same state for that pair of samples. For instance, in **Fig. 19.3A**, Gene A has an edge linking it to the pair of samples *(1, 5)*, and another edge linking the gene to the pair *(1, 3)*, because it is in different states in samples 1 and 5, and in the same state in 1 and 3. After finishing the construction of the graph (**Fig. 19.3B**) we note that the black node *(2, 4)* has degree one. Therefore, the maximum value that α can assume (α_{max}) is 1. By similarity, since the gray node *(4, 5)* also has degree one, the maximum attainable value of β (β_{max}) is also 1. Note also that if we set $\alpha = 1$ and $\beta = 0$, then we have the *Min Feature Set* problem. **Figure 19.4** shows two different optimal solutions for $\alpha = 1$ and $\beta = 1$ for the example.

Next, we present a mathematical formalization of the *Min (α,β)-Feature Set* problem. Let n be the number of genes. The objective function we seek to minimize is Equation 3, in which the variable f_j assumes value *1* if the gene j is selected, and zero otherwise. The constraints discussed before are formalized with linear inequalities. Let $a_{jpq} = 1$ if the feature (gene) j has different states for the samples p and q with different labels; and $a_{jpq} = 0$ otherwise. Each black node has a constraint Equation 4 associated with it, which guarantees that each pair of samples (p,q) is covered at least α times. Similarly, for each gray node we have a

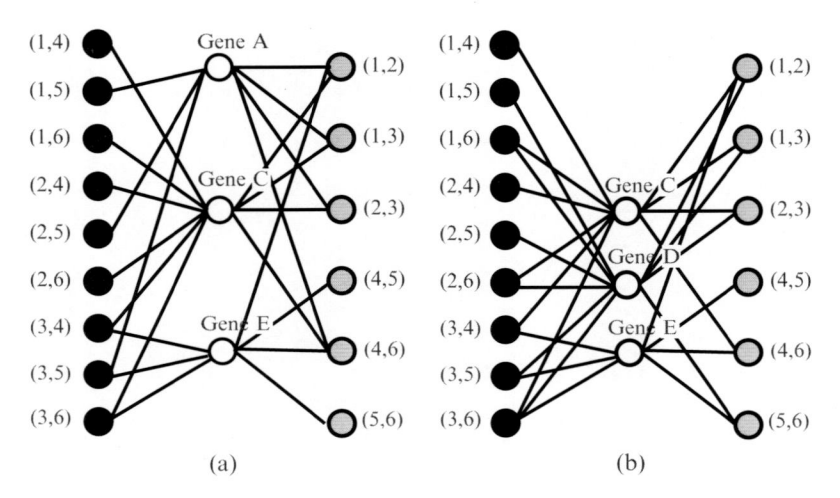

Fig. 19.4. Two optimal solutions for an instance of the *Min* (α = 1, β = 1)-*Feature Set* problem that has as input the matrix displayed in Fig. 19.1. Figures (**A**) and (**B**) show two subgraphs of the graph shown in Fig. 19.3B. These solutions have three features each and (**B**) corresponds to the optimal solution of the *Max Cover* (α = 1, β = 1)-3-*Feature Set* problem.

constraint Equation 5 that guarantees each pair of samples with the same label is covered at least β times, where b_{jpq} is *1* if the feature (gene) j has the same state for the pair (p,q) with labels; and zero otherwise.

$$Min \sum_{j=1}^{n} f_j \qquad [3]$$

$$\sum_{j=1}^{n} a_{jpq} f_j \geq \alpha \; \forall (p,q) label_p \neq label_q \qquad [4]$$

$$\sum_{j=1}^{n} b_{jpq} f_j \geq \beta \; \forall (p,q) label_p = label_q \qquad [5]$$

$$f_j \in \{0,1\} \qquad [6]$$

Since the *Min Feature Set* problem is a special case of the *Min* (α,β)-*Feature Set* problem, its complexity is NP-Hard. However, this problem also has some reduction rules that can help to reduce large instances to more manageable ones. The reduction rules for the *Min* (α,β)-*Feature set* can be found in (7, 20).

3.3. Max Cover (α,β)-k-Feature Set *Problem*

The *Min* (α,β)-*Feature Set* problem can have more than one optimal solution, as shown in **Fig. 19.4**. Among all optimal solutions, with k genes each, we may choose the one that optimizes another objective function. To this end, we define the *Max Cover* (α,β)-k-*Feature Set* problem as follows:

$$Max \sum_{j=1}^{n} c_j f_j \qquad [7]$$

$$\sum_{j=1}^{n} a_{jpq} f_j \geq \alpha \ \forall (p,q) label_p \neq label_q \qquad [8]$$

$$\sum_{j=1}^{n} b_{jpq} f_j \geq \beta \ \forall (p,q) label_p = label_q \qquad [9]$$

$$\sum_{j=1}^{n} f_j = k \qquad [10]$$

$$f_j \in \{0,1\} \qquad [11]$$

Notice that Equation 7 seeks to maximize a weighted sum of features; the weight c_j is equal to the number of pairs of samples that gene j covers. For our example, **Fig. 19.3B**, shows that Gene A covers 7 pairs of samples (black and gray nodes). We can also notice in **Fig. 19.4** that the second solution is the best one according to the objective function Equation 7, since it covers 27 pairs of samples (the first one covers 24 pairs). The constraints Equations 8 and 9 are similar to Equations 4 and 5 from the previous model. The constraint Equation 10 assures that the number of features will be k, this means that if we have solved the *Min* (α,β)-*Feature Set* problem for the same instance, we can search the space of optimal solutions and find the one with maximum coverage, thus applying the techniques in tandem.

To solve the models in Equations 3–6 and 7–11 we use CPLEX (http://www.ilog.com/products/cplex), a mathematical programming commercial software. CPLEX has complete exact methods to solve Integer Programming Models like the ones presented here. Despite the fact that these problems are NP-Hard, and due to the application of reduction rules mentioned before, we have been able to solve large instances in practice using personal computers (*see* **Note 3**).

3.4. Discretization Using an Entropy-Based Heuristic

In the proposed example (**Fig. 19.1**), the gene expression values can only take two values, and therefore can be analyzed via the mathematical models described in **Section 3**. These models have as their basis that each feature can only be in a small number of states. For real microarray datasets, we need to set up appropriate thresholds that will make this discretization useful. In Equations 7 and 8, we used information on the *average* and *standard deviation* of the gene expression values over all genes and samples to create the discretization. Using this approach, genes for which the expression values are located within a small interval may be considered not very relevant as discriminators. Another option is to use a methodology that only uses information from the gene to define a threshold. One example is the entropy-based heuristic introduced by Fayyad and Irani *(21)*. Fayyad and Irani's method evaluates the amount of information that can be gained when a set of samples S are divided into two non-empty sets, S_1 and S_2. It also excludes genes that cannot provide a minimum of information to organize the labels, by applying a test based on the minimum description length principle (MDL).

Consider the example shown in **Fig. 19.5**, which shows the expression values of a gene in nine different samples, each one labeled as either belonging to class C_1 or C_2. Initially, Fayyad and Irani's method identifies the possible partitions of the samples that the gene can induce. The samples are sorted in non-decreasing order of expression values and the possible thresholds are identified (**Fig. 19.5**).

Each identified threshold t_p divides the samples in two sets, the set encoded as 1 (S_1) and the set encoded as 0 (S_2), as shown in **Fig. 19.6**.

Consider $P(C_i,S_j)$ the proportion of samples in S_j that are labelled as C_i. Then we can calculate the *class entropy* as:

$$Ent(S_j) = -\sum_{i=1}^{2} P(C_i,S_j) log_2 \left[P(C_i,S_j) \right] \qquad [12]$$

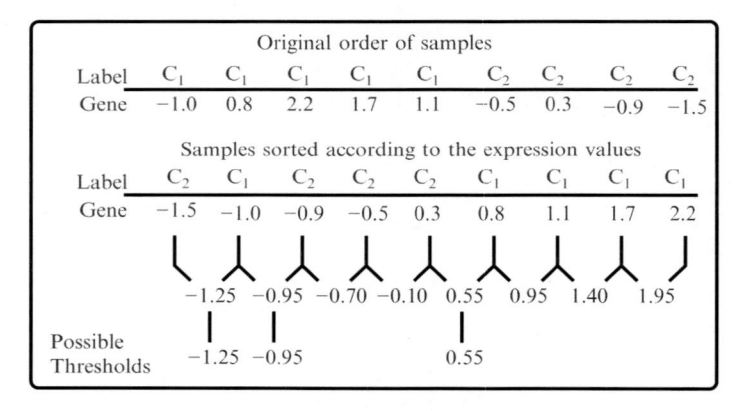

Fig. 19.5. Expression values of a gene in nine samples and the possible thresholds.

	Original order of samples									Initial Entropy
Label	C_1	C_1	C_1	C_1	C_1	C_2	C_2	C_2	C_2	
Gene	−1.0	0.8	2.2	1.7	1.1	−0.5	0.3	−0.9	−1.5	0.991
Possible Thresholds				Induced Discretizations						Entropy
$t1 = -1.25$	1	1	1	1	1	1	1	1	0	0.848
$t2 = -0.95$	0	1	1	1	1	1	1	1	0	0.988
$t3 = 0.55$	0	1	1	1	1	0	0	0	0	0.401

Fig. 19.6. Class information entropy for each possible threshold. The threshold value of 0.55 induces minimum class information entropy.

The *class information entropy* of the partition induced by t_p for gene g, denoted $E(g,t_p,S)$, is given by:

$$E\left(g,t_p,S\right) = \frac{|S_1|}{|S|} Ent\left(S_1\right) + \frac{|S_2|}{|S|} Ent\left(S_2\right) \qquad [13]$$

After evaluating each possible threshold using Equation 13, the one with the minimum class information entropy is chosen (in **Fig. 19.6**, t_3 was chosen). The resulting *information gain* is given by:

$$Gain\left(g,t_p,S\right) = Ent\left(S\right) - E\left(g,t_p,S\right) \qquad [14]$$

Next, the *Minimum Description Length Principle* is applied to evaluate whether the discretization obtained is acceptable. The codification is only acceptable if, and only if:

$$Gain\left(f_j,t_p,S\right) > \frac{log_2(m-1)}{m} + \frac{\Delta\left(f_j,t_p,S\right)}{m} \qquad [15]$$

where m is the total number of samples and:

$$\Delta(f_j, t_p, S) = log_2(3^c-2) - [cEnt(S) - c_1 Ent(S_1) - c_2 Ent(S_2)] \qquad [16]$$

where c is the number of different labels in the original gene; c_1 is the number of different labels in subset S_1 and c_2 is the number of different labels in subset S_2.

If a given gene does not satisfy the MDL criteria given by Equation 15, it is excluded from further consideration.

3.5. Reducing the Effects of the Curse of Dimensionality

As stated by Lottaz et al. (**Chapter 15**), microarray studies suffer from the curse of dimensionality; there are several sets of genes of small cardinalities that are equally able to differentiate samples of different classes, but many of these gene sets might fail as predictors for a sample not present in the dataset. The *MDL* test is useful as initial filtering, removing genes that are not significantly different in both classes. However, this step is not enough and our combinatorial approach is useful for the multivariate gene selection that remains to be done. We have found it useful to produce genetic signatures with large values of α and β. An apparent drawback is that genes in the signature might be false-positives (genes not presenting real predictive power) and noise will be introduced into classification algorithms. However, given the evidence (generally a small number of samples), the approach guarantees that all these possible multi-gene hypotheses of interest will remain in the genetic signature. If, on the contrary, α and β are set too small, the genetic signature obtained may be limited in predictive power and lead to lack of generalization. We also make the observation that our genetic signatures with large values of α and β seem to have good generalization capabilities when used in conjunction with ensembles of weak learners (e.g., *random forest* type of classification algorithms).

Moscato et al. *(8)* make use of other criteria to calibrate the feature selection method in order to find a balance between size of a molecular signature and its discriminative power, reducing the effect of the curse of dimensionality. This does not eliminate the need for test sets (composed of different samples) and cross-validation methods, as they are crucial steps to evaluate selected genes quantitatively as predictors. In the next section, we limit ourselves to small dataset examples and set alpha and beta close to the maximum values. By doing so we make clear the potential information that these feature selection methods can extract from microarray data.

4. Computational Results

To exemplify the application of the methodologies described in the previous sections, we used a microarray dataset on Alzheimer's disease (AD) introduced by Brown et al. *(6)*. It has been obtained from a coronal hemisection at the level of the hippocampus of two human brains, one taken as a control and one affected by AD. The sections were then sliced according to a grid, thus producing 24 small tri-dimensional units *(voxels)* from each brain. A cDNA microarray was used to determine the gene expression of 2,095 genes.

Figure 19.7 shows the result of our gene ordering algorithm applied to the complete dataset (2,095 genes and 48 samples in

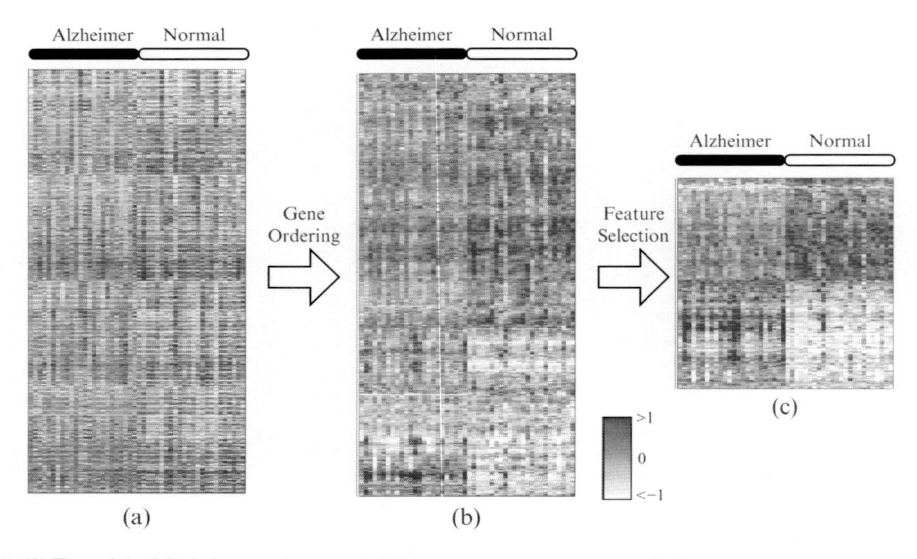

Fig. 19.7. (**A**) The original Alzheimer dataset with 2095 genes and 48 samples. (**B**) The result obtained using the gene ordering problem. (**C**) Solution of *Max Cover ($\alpha = 203$, $\beta = 216$)-464-Feature Set.*

Fig. 19.7A; the same dataset after applying the gene ordering method in **Fig. 19.7B**). The run-time required by our algorithm was 10 minutes. Our ordering already allows identification of some "patterns"; that is, groups of genes that have different expression profiles even if no supervised technique has yet been applied.

We then apply the algorithms used for the *Min (α,β)-Feature Set* and the *Max Cover (α,β)-k-Feature Set* problems as follows. We start applying the entropy-based heuristic to the original dataset to discretize the expression values. It took <2 seconds to run the algorithm on the dataset and according to the *MDL* test, 1,369 genes were excluded, resulting in 726 discretized genes.

For this reduced and now discretized dataset, we compute the maximum attainable values of α and β of any feasible solution to the *Min (α,β)-Feature Set* problem (which we call α_{max} and β_{max}). The values obtained were α_{max} = 203 and β_{max} = 282. We then solved the *Min (α = 203, β = 0)-Feature Set* problem using this dataset, identifying an optimal solution with k = 464 genes. Based in this information, we find the largest value of β such that there is a feasible solution *Min (α = 203, β)-Feature Set* problem and such that the optimal solution has the same number of genes (k = 464). In this case the value of β was 216. Finally, again using the same dataset, we solved the *Max Cover (α = 203, β = 216)-464-Feature Set* problem. The genetic signature can be seen in **Fig. 19.7C**. The run-time required by the gene ordering method for this last figure was 1 minute (*see* **Note 2**).

For a more detailed biological analysis of the results for the same Alzheimer's dataset *see (8)* and for application of this methodology in a Parkinson's dataset and different types of cancer, *see* **Chapter 20** and reference *(7)*, respectively.

5. Conclusions

We presented combinatorial optimization models for two problems with the aim of assisting microarray data analysis. The first problem is a gene ordering problem for an unsupervised analysis of microarray datasets and was addressed with a memetic algorithm. The second is the feature set selection problem, for which the objective is to select groups of genes that discriminate samples in different classes. We presented three different models, for the *Min Feature Set, Min (α,β)-Feature Set*, and *Max Cover (α,β)-k-Feature Set* problems.

All the problems presented in this chapter belong to the NP-hard class yet they can be addressed in practice. For the gene ordering problem, exact methods are not optimal solutions for large instances but the memetic algorithm with a Tabu search embedded gives excellent results in some minutes. The feature set models were solved to optimality using personal computers and were able to extract a large number of differentially expressed genes. In addition, the signatures that the method provides have a mathematical guarantee for robust classification, which is currently not present in other methods for discrimination. For the Alzheimer's disease dataset, we notice that the genes selected cover all pair of samples that belong to different classes and they are discriminated with a guarantee of *at least* 203 genes being differentially expressed.

In **Chapter 20** we apply these methods to a microarray dataset from a mouse model of Parkinson's disease and we provide an in depth analysis of the group of genes selected by our approaches.

6. Notes

1. The purpose of this chapter is to illustrate the application of methods that employ combinatorial optimization techniques for microarray data analysis. However, these are clearly methods that can be applied to other problem domains. We have also tried to communicate the versatility of the core models. It is possible to model other characteristics of interest via selection of other types of weights for the features. In addition, other constraints can be added to the problem formulation, thus allowing the presence/absence of some genes to be part of the molecular signature (as well as other types of mutual exclusion, etc.). Moreover, by looking at genes present in different pathways, a natural extension of this methodology allows the identification of differentially expressed pathways and helps link them with phenotypes of interest.

2. The use of large values of α and β, in conjunction with the *Max Cover* formulation, enables the production of rich molecular signatures with mathematical guarantees of minimum coverage (while being very discriminative and coherent within groups). As a consequence, the differentially expressed genes can be mapped to pathways without losing relevant information.

3. The method is relatively easy to implement and in addition to CPLEX, other commercial packages like XPRESS-MP can be used. In addition, public domain optimization codes exist. The reader may consult the page of A. Neumaier (http://www.mat.univie.ac.at/~neum/glopt/software_g.html).

References

1. Tamayo, P., Ramaswamy, S. (2003) Microarray data analysis: cancer genomics and molecular pattern recognition, in (Ladanyi, M., Gerald, W., eds.), *Expression Profiling of Human Tumors: Diagnostic and Research Applications*. Humana Press, Totowa, NJ.

2. Brazma, A., Vilo, J. (2000) Gene expression data analysis. *FEBS Letts* 480, 17–24.

3. Brown, M. P., Grundy, W. N., Lin, D., et al. (2000) Knowledge-based analysis of microarray gene expression data by using support vector machines. *Proc Natl Acad Sci U S A* 97, 62–267.

4. Eisen, M., Spellman, P., Brown, P., et al. (1998) Cluster analysis and display of genome-wide expression patterns. *Proc Natl Acad Sci U S A* 95, 14863–14868.

5. Moscato, P., Mendes, A., Berretta, R. Benchmarking (2007) a memetic algorithm for ordering microarray data. *BioSystems* 88 (I-2), 56–75.

6. Brown, V., Ossadtchi, A., Khan, A., et al. (2002) High-throughput imaging of brain gene expression. *Genome Res* 12, 244–254.

7. Berretta R., Mendes, A., Moscato, P. (2005) Integer programming models and algorithms for molecular classification of cancer from microarray data. Proceedings of the 28th Australasian Computer Science Conference, in (V. Estivill-Castro, ed.), *Conferences in Research and Practice in Information Technology* 38, 361–370.

8. Moscato P., Berretta R., Hourani M., et al. (2005) Genes related with Alzheimer's disease: a comparison of evolutionary search, statistical and integer programming approaches. Proceedings of EvoBIO2005: 3rd European Workshop on Evolutionary Bioinformatics, in (Rothlauf, F., et al. eds.), *Lecture Notes in Computer Science* 3449, 84–94.

9. Pardalos, P. M., Resende, M. G. C. (2002) *Handbook of Applied Optimization*. Oxford University Press, New York.

10. Sun, M., Xiong, M. (2003) A mathematical programming approach for gene selection and tissue classification. *Bioinformatics* 19, 1243–1251.

11. Merz, P. (2003) Analysis of gene expression profiles: an application of memetic algorithms to the minimum sum-of-squares clustering problem. *BioSystems* 72, 99–109.

12. Lee, S., Kim, Y., Moon, B. (2003) Finding the optimal gene order in displaying microarray data. Proceedings of GECCO2003: Genetic and Evolutionary Computation Conference, in (Cantu-Paz, E., et al., eds), *Lecture Notes in Computer Science* 2724, 2215–2226.

13. Cotta, C., Moscato, P. (2003) A memetic-aided approach to hierarchical clustering from distance matrices: application to phylogeny and gene expression clustering. *Biosystems* 72, 75–97.

14. Greenberg, H., Hart, W., Lancia, G. (2004) Opportunities for combinatorial optimization in computational biology. *INFORMS J Comput* 16, 211–231.

15. Rizzi, R., Bafna, V., Istrail, S., et al. (2002) Practical algorithms and fixed-parameter tractability for the single individual SNP haplotyping problem. Proc. 2nd Annual Workshop on Algorithms in Bioinformatics (WABI), in (Guigo, R., Gusfield, D., eds.), *Lecture Notes in Computer Science* 2452, 29–43.

16. Moscato, P., Cotta, C. (2003) A gentle introduction to memetic algorithms. in (Glover, F., Kochenberger, G., eds.), *Handbook of Metaheuristics*. Kluwer Academic Publishers, Boston.

17. Glover, F., Laguna, M. (1997) *Tabu Search*. Kluwer Academic Publishers, Norwell, MA.

18. Davies, S., Russell, S. (1994) *NP*-completeness of searches for smallest possible feature sets, in (Greiner, R., Subramanian, D., eds.), *AAAI Symposium on Intelligent Relevance*. New Orleans, AAAI Press.

19. Cotta, C., Moscato, P. (2003) The k-FEATURE SET Problem is $W[2]$-Complete. *J Comput Syst Sci* 67, 686–690.

20. Cotta, C., Sloper, C., Moscato, P. (2004) Evolutionary search of thresholds for robust feature selection: application to the analysis of microarray data, in (Raidl, G., et al. eds.), *Applications of Evolutionary Computing Lecture Notes in Computer Science*. 3005, 21–30.

21. Fayyad, U., Irani, K. (1993) Multi-interval discretization of continuous-valued attributes for classification learning. *Proceedings of the 13th International Joint Conference on Artificial Intelligence*. 1022–1029.

Chapter 20

Genetic Signatures for a Rodent Model of Parkinson's Disease Using Combinatorial Optimization Methods

Mou'ath Hourani, Regina Berretta, Alexandre Mendes, and Pablo Moscato

Abstract

This chapter illustrates the use of the combinatorial optimization models presented in **Chapter 19** for the Feature Set selection and Gene Ordering problems to find genetic signatures for diseases using microarray data. We demonstrate the quality of this approach by using a microarray dataset from a mouse model of Parkinson's disease. The results are accompanied by a description of the currently known molecular functions and biological processes of the genes in the signatures.

Key words: Parkinson's disease, combinatorial optimization, gene selection, microarray data analysis, feature selection, gene ordering.

1. Introduction

Parkinson's disease (PD) is the second most common progressive neurodegenerative disorder, after Alzheimer's disease (AD). It is characterized by four main symptoms, caused by the lack of the neurotransmitter dopamine: shaking or rhythmic movement, stiffness of the limbs, slowness of movement, and poor balance and coordination. It usually occurs later in life, with an average age of onset of 62.4 years for patients in the United States. In PD, for reasons that are not fully explained yet, dopaminergic neurons in the region of the brain called *substantia nigra* begin to malfunction and eventually die. This causes a decrease in the amount of available dopamine, which transports signals to the parts of the brain that control movement initiation and coordination.

Jonathan M. Keith (ed.), *Bioinformatics, Volume II: Structure, Function and Applications, vol. 453*
© 2008 Humana Press, a part of Springer Science+Business Media, Totowa, NJ
Book doi: 10.1007/978-1-60327-429-6 Springerprotocols.com

Even though the mechanism that triggers the dopaminergic neurons' death in PD is not entirely understood, most scientists believe that genetic factors play an important role. Indeed, it is estimated that about 15–25% of PD patients have a relative that also has the disease. Several "PD-related genes" have been reported and there is conclusive evidence of a genetic link *(1)*. Research into the genetic basis of this disease is currently being strongly supported in developed countries, as they have an aging population and PD will surely impact on their public health systems.

This chapter uses the combinatorial optimization models presented in **Chapter 19** (*Min (α,β)-Feature Set* problem, *Max Cover (α,β)-k-Feature Set* problem, and *Gene Ordering* problem) to identify different gene expression profiles in a dataset introduced by Brown et al. *(2)* which is available in the public domain (http://labs.pharmacology.ucla.edu/smithlab/genome_multiplex). The dataset contains the expression levels of 7,139 genes (and ESTs) in 80 samples. The samples were taken from 40 different regions of the brains of a PD-affected rodent and a control, which is labeled "normal." Detailed information about the procedure used to extract the brain tissue samples and how the microarray instance was generated can be accessed via the authors' web site and reference *(2)*.

This chapter is organized as follows. **Section 2** describes the main points of the Min (α,β)-Feature Set problem. **Section 3** describes the PD microarray dataset. **Sections 4** and **5** present computational results and conclusions, respectively.

2. The *Min (α,β)-Feature Set* and the *Max Cover (α,β)-k-Feature Set* Problems

The Min (α,β)-Feature Set problem is a generalization of the Min Feature Set problem, and it has been introduced with the aim of identifying subsets of features with robust discriminative power *(3)*. It can have more than one optimal solution and, in this case, we may choose the solution that maximizes the number of pairs of examples covered. These mathematical models are described in detail in **Chapter 19**.

These new models have been used to analyze microarray data from an AD study *(4)*, and for the molecular classification of different types of cancer *(5, 6)*. We have also written an illustrative paper on its application to political science in a study that led to rules predicting the 2004 U.S. presidential election's outcome, 2 months in advance of the actual voting, only based on historical information from previous elections *(7)*. As mentioned, both models are described in **Chapter 19**, but one point that is worth emphasizing again is that the mathematical models cannot use gene expression values directly as input variables. They must first be discretized into a small set of classes. The method chosen to perform this discretization is based on the class-entropy of each gene and is also used in **Chapter 19**. We will refer to elements of the entropy thresholding procedure throughout this chapter.

3. Parkinson's Disease Microarray Data

The microarray data are based on a PD model obtained by the administration of toxic doses of methamphetamine to the C57BL/6J strain of mice. The methamphetamine doses gradually destroy the dopaminergic nerve, in a process resembling the neurodegenerative process observed in PD. The brains from a control and from methamphetamine-treated mice were cut into 40 cubic *voxels* each (10 coronal sections, each cut in four voxels, two for each hemisphere of the brain) and then analyzed, resulting in a microarray with 7,139 genes (and ESTs) identifiable by their accession numbers. We removed from the analysis all genes with more than 30% of missing values, reducing the size of the dataset to 7,035 genes. The missing values of the remaining genes were imputed using the LLSImpute software *(8)*. From an initial literature survey, we found some genes that were present in this dataset and which were previously linked to PD (**Table 20.1**).

The genetic profiles of the genes in the table are depicted in **Fig. 20.1** *(bottom)*, as well as 35 genes reportedly linked to PD in reference *(1)* (*see* **Fig. 20.1** *(top)*). We note that some of the gene-expression profiles in **Fig. 20.1** present only a very moderate

Table 20.1
Genes reportedly linked to Parkinson's disease and also present in the 7,035-gene microarray dataset introduced in Brown et al. *(2)*

Symbol	Accession #	Symbol	Accession #	Symbol	Accession #
APP	AA110872	IDE	AA111232	RNF19	AI326766
CCK	AI322505	MAPT	AA028410	SERPINA3g	AA210495
CSNK2A1	AA259439	PARK7	AA108826	SERPINA3g	W83447
GSTO1	AI430895	PARP1	AA032357	SERPINA3k	AA241936
HFE	AA217236	PINK1	AA023263	SNCA	W41663
HFE	W11889	PINK1	AA259352	UCHL1	AA285969

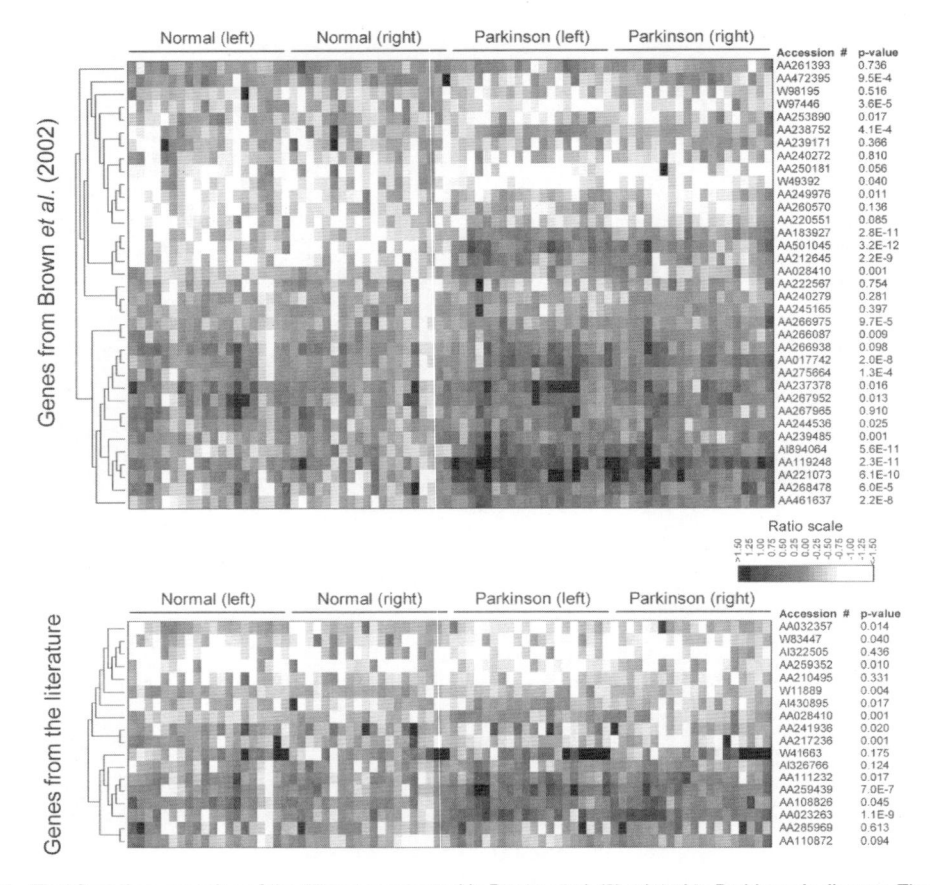

Fig. 20.1. *(Top)* Genetic expression of the 35 genes reported in Brown et al. **(2)** related to Parkinson's disease. The experiments were divided into Normal/Parkinson and left/right hemisphere of the brain. Only a few differences are noticeable between the groups of normal and Parkinson's samples. *(Bottom)* Eighteen genes from our Parkinson's disease survey that are also presented in the dataset introduced in reference **(2)**. They do not show a clear distinct activation profile between the types of tissue either. Beside each gene, we show the *p*-value quantifying the degree to which the gene expression levels in Parkinson-affected and normal tissues are distinct, calculated using the Wilcoxon-Mann-Whitney test.

difference between the normal and PD affected brains. This is also reflected by the corresponding *p*-values, which were calculated using the Wilcoxon-Mann-Whitney test (a non-parametric statistical test).

4. Computational Results

This section presents genetic signatures for PD using several choices of the parameters α and β. First, in this chapter all tissue samples from the same mouse were considered to belong to the same class, no matter what region of the brain they were extracted from. This follows the approach of reference *(2)* in order to facilitate a comparison of the two approaches.

Figure 20.2A presents an image of the original dataset with 7,035 genes, ordered using the memetic algorithm described in **Chapter 19**. It is clear that the brain of the affected rodent has a more uniform gene expression pattern across the different regions of the brain (corresponding to columns in the array). Using the entropy-based heuristic described in **Chapter 19**, we selected 1,984 genes as informative and discretized their gene expression values into two categories (under- or over-expressed). Next, we present a series of genetic signatures obtained using algorithms for the *Min (α,β)-Feature Set* and *Max Cover (α,β)-k-Feature Set* problems and compare characteristics of the solutions obtained for different choices of α and β.

4.1. Signatures for (α,β) = (α_{max},β_{max}) and (α,β) = ($\alpha_{max},\beta_{maximal}$)

It is very simple to identify the maximum values of α and β for which a feasible solution is available. To establish α_{max} first compute, for every pair of samples, the number of genes that are differentially expressed (over-expressed in one sample and under-expressed in the other). Clearly, for one particular pair of samples this number is at its minimum, and α_{max} is that particular number. Analogously, to find β_{max} compute the number of genes that are similarly expressed for every pair of samples belonging to the same class (and the pair of samples that have the minimum value determines β_{max}).

The genetic signature for (α,β) = (α_{max},β_{max}) contains the largest number of genes in all our tests. Solutions with such a large number of genes are interesting because they increase the chances of uncovering large networks of co-expressed genes, as well as complex genetic pathways. For our 1,984-genes dataset, the α_{max} and β_{max} values are equal to 623 and 847, respectively. We have solved to optimality a *Min (623,847)-Feature Set* problem, obtaining a solution with 1,660 genes. Finally, we solve a *Max Cover (623,847)-1,660-Feature Set* problem to obtain the group of 1,660 genes that maximize the number of pairs of examples covered.

Another possible setting for α and β, explained in **Chapter 19**, is $\alpha = \alpha_{max}$ and β is maximal, in which the procedure to find $\beta_{maximal}$ is as follows. Initially, we set $\alpha = \alpha_{max}$ and solve the *Min ($\alpha_{max},0$)-Feature Set* problem. For the PD data, this resulted in 1,502 genes. Then we gradually increase the value of β until the number of genes in the optimal solution exceeds 1,502. The largest value of β that does not increase the number of genes is its maximal value. It maximizes the similarities within classes while keeping the same number of genes in the optimal Feature Set as when this requirement is not enforced ($\beta = 0$). The value of $\beta_{maximal}$ when we have $\alpha = \alpha_{max}$ is 676 and the result for the *Max Cover*

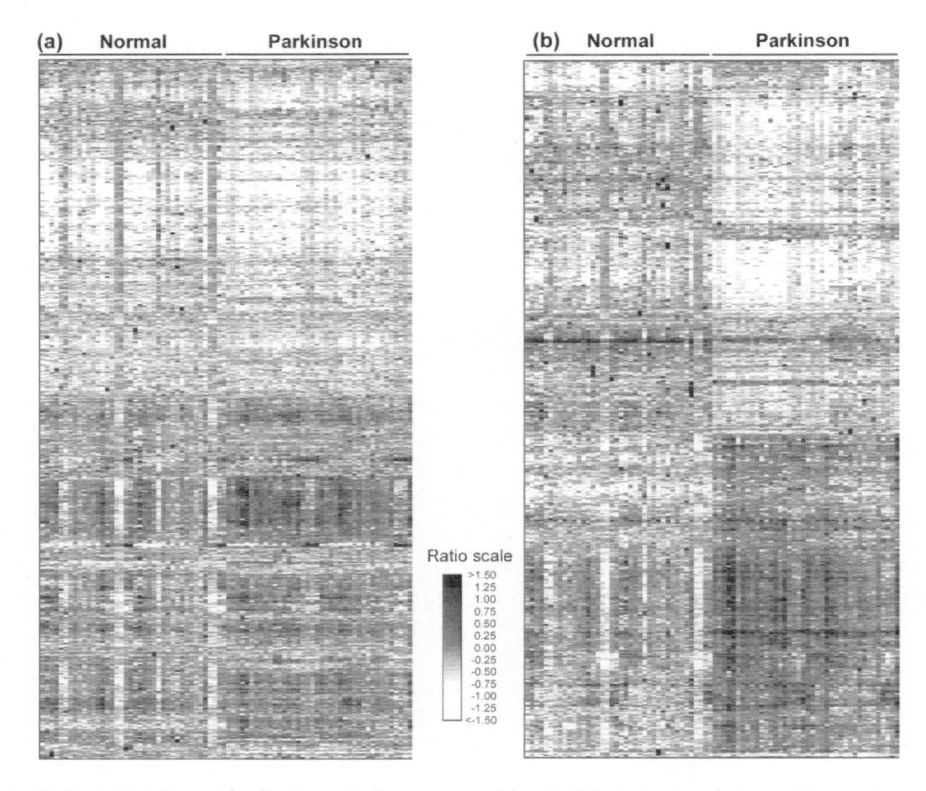

Fig. 20.2. (**A**) Complete dataset for Parkinson's disease, containing 7,035 genes sampled over 80 experiments (tissue samples were extracted from 40 voxels from the normal mouse brain and 40 voxels from the mouse model of the disease). (**B**) Genetic signature using $(\alpha,\beta) = (\alpha_{max},\beta_{maximal}) = (623,676)$. The signature contains 1,502 genes, 158 genes less than with $(\alpha,\beta) = (\alpha_{max},\beta_{max})$ and 482 less than those selected by the entropy criterion alone.

(623,676)-1,502-Feature Set problem is shown in **Fig. 20.2B**. The result for $(\alpha,\beta) = (\alpha_{max},\beta_{max})$ is not shown because visually it is very similar to that using $(\alpha,\beta) = (\alpha_{max},\beta_{maximal})$: one has only 158 genes less than the other.

4.2. Signatures for $\alpha = \beta \approx k$

In some cases, biologists are interested in smaller genetic signatures with genes differentially expressed across all—or the majority—of the samples from each class. Such genes typically have *class information entropy* either equal to zero or very close to zero. Traditionally, they are uncovered by calculating the *p*-values for all genes using an appropriate statistical test, such as the Wilcoxon-Mann-Whitney, and then selecting the lowest ones. However, even though genes with low *p*-values have distinctive expression profiles for the two classes of samples, *there is no* a priori *guarantee on their capability to discriminate* between all pairs of samples *that belong to different classes*. The *Min* (α,β)-*Feature Set* approach provides such a mathematical guarantee. If the final goal of the genetic signature is to build a classifier, this guarantee can be highly beneficial for robust classification.

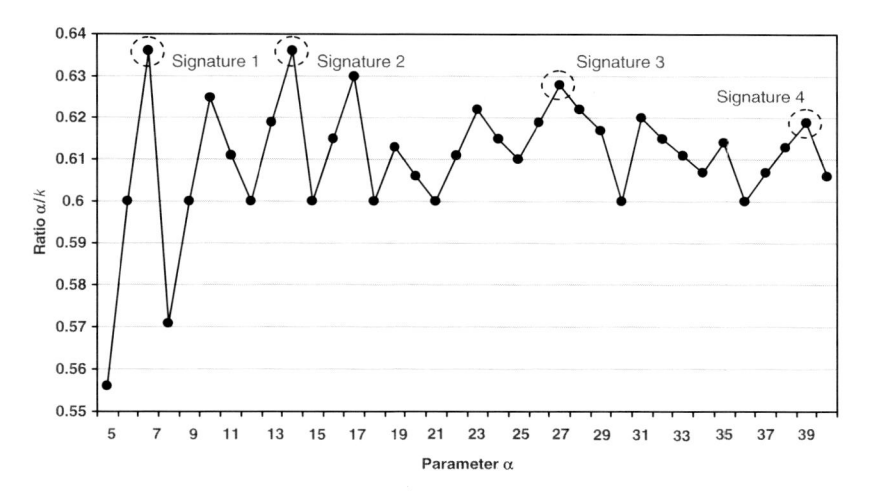

Fig. 20.3. The graph shows the ratio α/k for different values of α. Four points are circled, corresponding to the genetic signatures when $\alpha = \beta = 7, 14, 27$, and 39, since the goal is to maximize the ratio α/k. The values of α and β indicate different levels of robustness, i.e., α expresses the minimum number of genes that should be differentially expressed in any pair of samples from different classes. Analogously, β expresses the minimum number of genes that should be similarly expressed in any pair of samples from the same class. The four corresponding signatures are shown in Fig. 20.4A–D.

In a genetic signature composed only of genes with class information entropy zero, every gene is an ideal discriminator of the observed data. A signature containing k genes with class information entropy zero is an optimal solution for the Min (α,β)-Feature Set problem, where $\alpha = \beta = k$. In the dataset we are studying there is not a single gene with class information entropy equal to zero; thus, it is not possible to obtain such a signature. An alternative approach would be to find genetic signatures where $\alpha = \beta < k$, and the proportion α/k is maximized (or "locally" maximized). **Figure 20.3** shows how such a signature can be selected.

The graph depicted in **Fig. 20.3** has a maximum of 0.636, when $\alpha = \beta = 14$, resulting in k = 22 genes. For comparison purposes, we also give a picture for other locally maximal signatures in **Fig. 20.4A–D**. The genetic signature in **Fig. 20.4A** contains 11 genes and the expression profile differences between the normal and PD samples are easily noticed. In **Fig. 20.4B**, the number of genes increases to 22. There is a clear differentiation between the normal and the PD samples.

The signatures in **Fig. 20.4C** and **D** have 43 and 63 genes, respectively, most of them with very low p-values (the highest is just 5.6E-4). There is no straightforward way to determine which signature is the best. Actually, the intersections between them are quite large. There are nine genes that appear in all four signatures, 11 that appear in three, 13 that appear in two and 44 that appear only once, giving a total of 77 different genes in the union

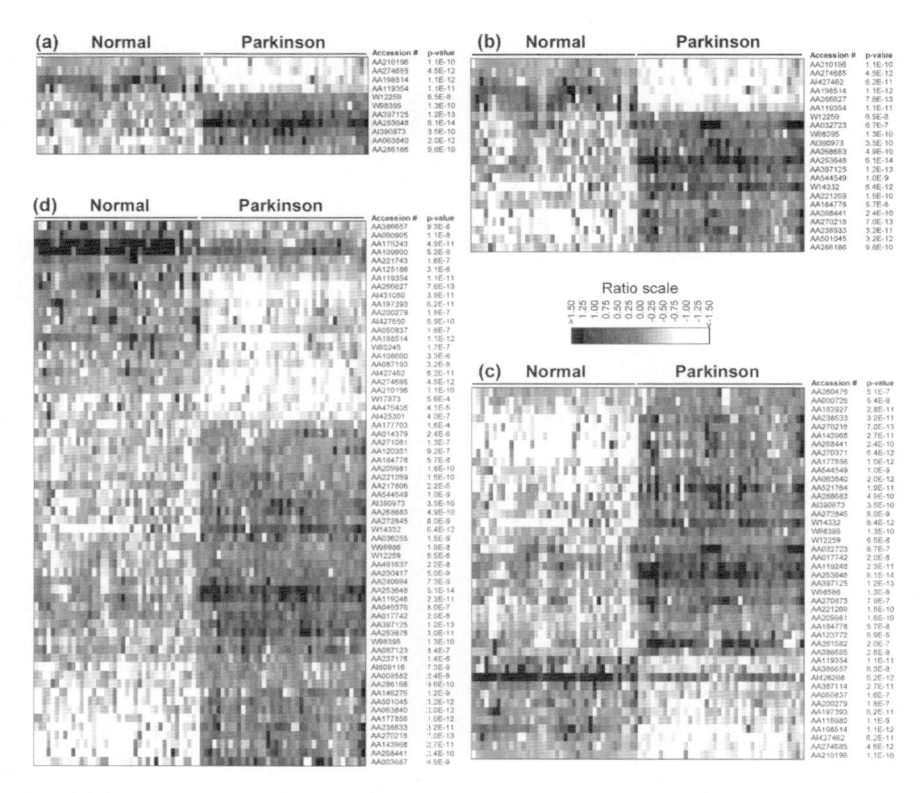

Fig. 20.4. Genetic signatures associated with the four points evidenced in Fig. 20.3. (**A**) Signature 1: $(\alpha,\beta) = (7,7)$ with 11 genes. (**B**) Signature 2: $(\alpha,\beta) = (14,14)$ with 22 genes. (**C**) Signature 3: $(\alpha,\beta) = (27,27)$ with 43 genes. (**D**) Signature 4: $(\alpha,\beta) = (39,39)$ with 63 genes. The four signatures present very distinctive expression profiles for the normal and the Parkinson's samples and the associated p-values are shown beside each gene to corroborate this assertion.

of the four signatures. Of these, 44 have some known biological function and are reported in **Table 20.2**.

Next, we present an analysis of the genes in **Table 20.2** that uses information about pathways and biological processes to support potential links with PD. Eleven genes present in **Table 20.2** belong to at least one known pathway. They are CSNK2A1 (PD pathway); GSK3B (AD, presenilin pathway); GRB2 and CAPN6 (Huntington's disease pathway); SLC1A2 (inotropic glutamate receptor pathway); CATNS (cadherin signaling pathway); FGF1 and PPP2R5D (FGF signaling pathway); CFLAR (apoptosis signaling pathway); MTA2 (p53 signaling pathway); and CPOX (heme biosynthesis pathway).

The first gene, CSNK2A1, has already been reported as linked to PD *(9)* and is also listed in **Table 20.1**. GSK3B, GRB2, and CAPN6 are involved in the pathways of other neurodegenerative diseases—AD and Huntington's disease—also indicating a role in neurodegeneration processes. Indeed, GSK3B was recently shown to regulate neuronal death in Chen et al. *(10)*.

Table 20.2
List containing the 44 genes present in the four signatures depicted in Fig. 20.4 that have a known biological function or belong to a known pathway

Symbol	Accession #	Symbol	Accession #	Symbol	Accession #
AAK1	AA036255	**GSK3B**	AA143968	**PCTK1**	AA230417
CAPN6	AA260476	**GSK3B**	AA270371	**PEX19**	AA050837
CATNS	AA008582	HBA-A1	AA109900	PON2	W98586
CD1D1	AA397125	HDAC5	AA017742	PPARBP	AA544549
CFLAR	AI425301	HSD3B4	AA274685	PPP2R4	AA087123
CPOX	AA108600	KIFAP3	AA268683	PPP2R5D	AA271081
CSNK2A1	AA272845	LCN2	AA087193	RNF39	AA386685
DDR1	W98395	LNPEP	AA217806	SLC1A2	AA238533
EXTL3	W99986	MIC2L1	AA253678	SNX27	AA177856
FGF1	AA261582	MRPS25	AA501045	STK35	AI431080
FMO5	AA237178	MTA2	AA461637	STK39	AA177703
FXR2H	AA119248	MTMR6	AA184778	TMPO	AA118680
GDI1	AA060905	NEIL3	AI427462	UBE2I	AA120351
GFI1B	AA266627	OGT	AA146275	VIPR2	AA000726
GRB2	AA183927	PCBD	W14332		

For the genes marked in boldface, there is already a reported link with Parkinson's disease or we can support a potential link via other neurodegenerative diseases or brain-related processes.

SLC1A2 belongs to the inotropic glutamate receptor pathway and plastic alterations in the synaptic efficacy of inotropic glutamatergic receptors reportedly contribute to motor dysfunction in PD (11). CATNS belongs to the cadherin signaling pathway, and in 2004, Uemura et al. (12) reported that a cadherin family member, namely N-cadherin, regulates the synaptic activity of neurons. The genes FGF1 and PPP2R5D are in the FGF signaling pathway, and Corso et al. (13) found that fibroblast growth factor (FGF) receptors contribute to the degeneration of the substantia nigra neurons in PD. Even though the next two genes, CFLAR and MTA2, are in different pathways, they are both linked to apoptosis, or cell death. Also, MTA2 promotes the deacetylation of p53, modulating cell growth arrest and apoptosis. Recent studies have shown a relation between p53 and brain degeneration (14).

Apart from the 11 genes that belong to known pathways, 33 others have well-established molecular functions and/or biological processes that link them to brain processes and neurodegenerative diseases. mRNA transcription regulation is the biological function of three genes in **Table 20.2**. A recent work reported that D1—a dopamine receptor—mRNA transcription regulation changes after chronic levodopa (L-dopa) treatment, linking it with PD *(15)*. Even though no causal relationship was initially established, further investigation is necessary to check whether GFI1B, HDAC5, and PCBD could also be responsible for D1 regulation. Moreover, PCBD also plays a role in pterin metabolism. Pterin cofactors are required for tyrosine hydroxylase (TH) activity, which converts tyrosine into L-dopa. The final conversion of L-dopa to dopamine is controlled by an enzyme called dopa-decarboxylase *(16)*. The relation between pterin metabolism and PD is well-known, since fibroblasts supplemented with pterin cofactors were found to produce L-dopa *(17)*. Moving from PD to AD, the PON2 polymorphism might be a risk factor associated to late-onset of AD, as it prevents the degeneration of neuronal cells associated with atherosclerosis *(18)*.

Another interesting gene is GDI1, which regulates neurotransmitter release and is responsible for X-linked non-specific mental retardation *(19)*. The genes PEX19, PPARBP, and HSD3B4 participate in lipid and cholesterol metabolism. Differences in movement of cholesterol between different cellular compartments of the central nervous system (CNS) and across the blood–brain barrier to the plasma have been detected in mice with one form of neurodegenerative disease (Niemann-Pick type C). Cholesterol metabolism plays an important role in AD *(20)*. Other neurodegenerative diseases, like amyotrophic lateral sclerosis (ALS) and PD, are currently being studied for possible links with lipid metabolites and pathways *(21)*. Three genes, NEIL3, LCN2, and MIC2L1, are related to DNA repair, oncogenesis, and tumor suppression, respectively. Recent data suggest that the maintenance of DNA integrity is critical in neurodegenerative processes *(22)*. On the other hand, LCN2 is an oncogene involved in cell cycle regulation. There is a strong connection between cell cycle and neurodegenerative diseases. In AD, for example, neurons normally protected from apoptotic and mitotic stimuli lose this characteristic. This change creates a quasi-mitotic status, with excessive neuronal mitochondria (created to produce energy for the cell division) producing free radicals. The resulting oxidative damage finally triggers an apoptotic stimulus, destroying the neuron *(23)*. Additionally, tumor suppressor genes, such as p53, p27, and WT1, have been linked to neurodegenerative processes *(14, 24, 25)*. However, even though MIC2L1 is also a tumor suppressor, there is no reported link yet between it and PD or other neurodegenerative diseases. As mentioned, apoptosis signaling plays an important role in neurodegeneration. This issue is relevant to another gene in the PD signatures, FXR2H;

its biological function is related to induction of apoptosis. Also, it is one of the main genes responsible for fragile X mental retardation, playing a role in synaptic growth, structure, and long-term plasticity *(1)*. Another relevant biological function is T-cell mediated immunity, associated to the genes CD1D1 and TMPO. It has been known for many years that individuals with PD present immune system alterations. Actually, PD patients exhibit a lower frequency of infections and cancer, suggesting a stimulation of the immune system. This hypothesis is supported by the observation of T-cell activation leading to the production of interferon-γ in PD *(26)*.

G-protein plays a variety of roles in numerous neuronal functions. Moreover, the interaction between G-protein–coupled receptor kinases (GRKs) and β-arrestins regulates physiological responsiveness to psychostimulants, indicating a potential involvement in brain disorders, such as addiction, PD, and schizophrenia *(27)*. G-protein–mediated signaling is present as the biological function of genes SNX27 and VIPR2. The next gene, MTMR6, has a biological function of protein phosphatase, which represents the final pathway in the action of transmitters and hormones at neuronal level *(28)*. Also, there is a clear link between protein phosphorylation and PD, since it regulates NMDA receptors, contributing to the pathogenesis of motor dysfunction in PD subjects *(29)*. MRPS25 is a mitochondrial ribosomal protein and mitochondria proteins play a decisive role in apoptosis by regulating the release of cytochrome c and other pro-apoptotic factors *(30, 31)*.

KIFAP3 is known to regulate neurite outgrowth, a structure developed when neurons begin to aggregate. Neurites are protrusions of cytoplasm that will eventually become axons and dendrites. More specifically, KIFAP3 regulates the axonal transport of membranous components necessary for neurite extension *(32)*. Another important gene is PCTK1, which interacts with p35 in the regulation of Cdk5, playing a role in the central nervous system development *(33)*.

The gene OGT has been recently linked to AD *(34)*. It promotes τ-glycosylation, increasing the phosphorylation of the microtubule associated protein τ, which is one of the pathological hallmarks of AD. Finally, DDR1 plays a variety of biological roles. First, it signals the initiation and/or maintenance of granule cell axon extension in the developing cerebellum *(35)*. It is also a transcriptional target for p53. Inhibition of DDR1 function resulted in increased cell apoptosis in response to genotoxic stress induced by p53 *(36)*.

For the other genes in **Table 20.2**, it was not possible to find a link with neurodegenerative diseases or other brain-related processes. Summarizing, from the 77 distinct genes depicted in **Fig. 20.4**, 44 belong to known pathways or have known biological function. For 33 of them, we could support a potential link with neurodegenerative diseases or known processes.

5. Conclusions

This chapter presented an application of the (α,β)-k-Feature Set approach described in **Chapter 19**. A microarray dataset with 7,035 genes and 80 experiments, sampled from an experimental mouse model of PD and a control, was used to test the methodology. The tests included different configurations for the parameters α and β, and images of the resulting genetic signatures were presented. Finally, for a set containing four smaller genetic signatures, we identified potential links with neurodegenerative diseases or neurological processes for the majority of the genes.

The class-entropy filtering alone removed about 72% of the original genes from the analysis. However, the existence of almost 2,000 genes that are sufficiently differentially expressed in both brains to pass the entropy criterion, indicates that this PD model is characterized by a strong change in the overall brain's genetic activity. More experimentation would be required to see if these changes are also observed in other cases. The tests with very small values of α and β brought some interesting results, with remarkably distinct gene expression profiles across the two types of samples and also a strong homogeneity in the expression profile of the PD-associated samples.

One of the main characteristics of the (α,β)-k-Feature Set approach—and its greatest strength—is its flexibility in terms of how to set the parameters. It allows the biologist to obtain genetic signatures with mathematical guarantees not provided by other methods. The biologist can then use signatures with different sizes and degrees of robustness, and adjust the method to better suit his or her needs.

References

1. Feany, M. (2004) New genetic insights into Parkinson's disease. *NEJM* 351, 1937–1940.
2. Brown, V., Ossadtchi, A., Khan, A., et al. (2002) Multiplex three-dimensional brain gene expression mapping in a mouse model of Parkinson's disease. *Genome Res* 12, 868–884.
3. Cotta, C., Sloper, C., Moscato, P. (2004) Evolutionary search of thresholds for robust feature selection: application to microarray data, in *EvoBIO2004*. Proceedings of the 2nd European Workshop in Evolutionary Computation and Bioinformatics, in (Raidl, G., et al., eds.), *Lecture Notes in Computer Science* 3005, 31–40.
4. Moscato, P., Berretta, R., Hourani, M., et al. (2005a) Genes related with Alzheimer's disease: a comparison of evolutionary search, statistical and integer programming approaches, in *EvoBIO2005*. Proceedings of the 3rd European Workshop on Evolutionary Bioinformatics, in (Rothlauf, F.,

et al., eds.), *Lecture Notes in Computer Science* 3449, 84–94.

5. Berretta, R., Mendes, A., Moscato, P. (2005) Integer programming models and algorithms for molecular classification of cancer from microarray data. ACSC2005: Proceedings of the 28th Australasian Computer Science Conference, in (Estivill-Castro, V., ed.), *Conferences in Research and Practice in Information Technology* 38, 361–370.

6. Cotta, C., Langston, M., Moscato, P. (2006) Combinatorial and algorithmic issues for microarray data analysis, in (Gonzalez, T., ed.), *Handbook of Approximation Algorithms and Metaheuristics.* Chapman & Hall/CRC, Boca Raton, FL.

7. Moscato, P., Mathieson, L., Mendes, A., et al. (2005b) The electronic primaries: Predicting the U.S. presidency using feature selection with safe data reduction. *ACSC2005:* Proceedings of the 28th Australasian Computer Science Conference, in (Estivill-Castro, V., ed.), *Conferences in Research and Practice in Information Technology* 38, 371–380.

8. Kim, H., Golub, G., Park, H. (2004) Missing value estimation for DNA microarray gene expression data: local least squares imputation. *Bioinformatics* 21, 187–198.

9. Lee, G., Tanaka, M., Park, K., et al. (2004) Casein kinase II-mediated phosphorylation regulates alpha-synuclein/synphilin-1 interaction and inclusion body formation. *J Biol Chem* 279, 6834–6839.

10. Chen, G., Bower, K. A., Ma, C., et al. (2004) Glycogen synthase kinase 3beta (GSK3beta) mediates 6-hydroxydopamine-induced neuronal death. *FASEB J* 18, 1162–1164.

11. Bibbiani, F., Oh, J. D., Kielaite, A., et al. (2005) Combined blockade of AMPA and NMDA glutamate receptors reduces levodopa-induced motor complications in animal models of PD. *Exp Neurol* 196, 422–429.

12. Uemura, K., Kuzuya, A., Shimohama, S. (2004) Protein trafficking and Alzheimer's disease. *Curr Alzheimer Res* 1, 1–10.

13. Corso, T. D., Torres, G., Goulah, C., et al. (2005) Transfection of tyrosine kinase deleted FGF receptor-1 into rat brain substantia nigra reduces the number of tyrosine hydroxylase expressing neurons and decreases concentration levels of striatal dopamine. *Mol Brain Res* 139, 361–366.

14. Jacobs, W., Walsh, G., Miller, F. (2004) Neuronal survival and p73/p63/p53: a family affair. *Neuroscientist* 10, 443–455.

15. Albert, I., Gulgoni, C., Hakansson, K., et al. (2005) Increased D1 dopamine receptor signaling in levodopa-induced dyskinesia. *Ann Neurol* 57, 17–26.

16. Widner, B., Leblhuber, F., Fuchs, D. (2002) Increased neopterin production and tryptophan degradation in advanced Parkinson's disease. *J Neural Trans* 109, 181–189.

17. Wolff, J., Fisher, LJ., Xu, L., et al. (1989) Grafting fibroblasts genetically modified to produce L-dopa in a rat model of Parkinson disease. *Proc Natl Acad Sci U S A* 86, 9011–9014.

18. Shi, J., Zhang, S., Tang, M., et al. (2004) Possible association between Cys311Ser polymorphism of paraoxonase 2 gene and late-onset Alzheimer's disease in Chinese. *Mol Brain Res* 120, 201–204.

19. D'Adamo P.. Menegon, A., Lo Nigro, C., et al. (1998) Mutations in GDI1 are responsible for X-linked non-specific mental retardation. *Nat Genet* 19, 134–139.

20. Kolsch, H., Heun, R. (2003) Polymorphisms in genes of proteins involved in cholesterol metabolism: evidence for Alzheimer's disease? *Neurobiol Lipids* 2, 5–7.

21. Mattson, M. (2004) Metal-catalyzed disruption of membrane protein and lipid signaling in the pathogenesis of neurodegenerative disorders. *Ann NY Acad Sci* 1012, 37–50.

22. Francisconi, S., Codenotti, M., Ferrari-Toninelli, G., et al. (2005) Preservation of DNA integrity and neuronal degeneration. *Brain Res Rev* 48, 347–351.

23. Raina, A., Zhu, X., Rottkamp, C., et al. (2000) Cyclin toward dementia: cell cycle abnormalities and abortive oncogenesis in Alzheimer disease. *J Neurosci Res* 61, 128–133.

24. Ogawa, O., Lee, H., Zhu, X., et al. (2003) Increased p27, an essential component of cell cycle control, in Alzheimer's disease. *Aging Cell* 2, 105–110.

25. Lovell, M., Xie, C., Xiong, S., et al. (2003) Wilms' tumor suppressor (WT1) is a mediator of neuronal degeneration associated with the pathogenesis of Alzheimer's disease. *Brain Res* 983, 84–96.

26. Czlonkowska, A., Kurkowska-Jastrzebska, I., Czlonkowski, A., et al. (2002) Immune processes in the pathogenesis of Parkinson's disease: a potential role for microglia and nitric oxide. *Med Sci Mon* 8, RA165–RA177.

27. Gainetdinov, R., Premont, R., Bohn, L., et al. (2004) Desensitization of G protein-coupled receptors and neuronal functions. *Annu Rev Neurosci* 27, 107–144.

28. Magnoni, M., Govoni, S., Battaini, F., et al. (1991) The aging brain: protein

phosphorylation as a target of changes in neuronal function. *Life Sci* 48, 373–385.

29. Chase, T., Oh, J. (2000) Striatal mechanisms and pathogenesis of parkinsonian signs and motor complications. *Ann Neurol* 47, S122–S129.

30. Nicholson, D., Thornberry, N. (2003) Apoptosis: life and death decisions. *Science* 299, 214–215.

31. Lindholma, D., Eriksson, O., Korhonen, L. (2004) Mitochondrial proteins in neuronal degeneration. *Biochem Biophys Res Commun* 321, 753–758.

32. Takeda, S., Yamazaki, H., Seog, D., et al. (2000) Kinesin superfamily protein 3 (kif3) motor transports fodrin-associating vesicles important for neurite building. *J Cell Biol* 148, 1255–1266.

33. Cheng, K., Li, Z., Fu, W., et al. (2002) Pctaire1 interacts with p35 and is a novel substrate for Cdk5/p35. *J Biol Chem* 277, 31988–31993.

34. Robertson, L., Moya, K., Breen, K. (2004) The potential role of tau protein O-glycosylation in Alzheimer's disease. *J Alzheimer's Dis* 6, 489–495.

35. Bhatt, R., Tomoda, T., Fang, Y., et al. (2000) Discoidin domain receptor 1 functions in axon extension of cerebellar granule neurons. *Gene Dev* 14, 2216–2228.

36. Ongusaha, P., Kim, J., Fang, L., et al. (2003) p53 induction and activation of DDR1 kinase counteract p53-mediated apoptosis and influence p53 regulation through a positive feedback loop. *EMBO J* 22, 1289–1301.

Section IV

Analytical and Computational Methods

Chapter 21

Developing Fixed-Parameter Algorithms to Solve Combinatorially Explosive Biological Problems

Falk Hüffner, Rolf Niedermeier, and Sebastian Wernicke

Abstract

Fixed-parameter algorithms can efficiently find optimal solutions to some computationally hard (NP-hard) problems. This chapter surveys five main practical techniques to develop such algorithms. Each technique is circumstantiated by case studies of applications to biological problems. It also presents other known bioinformatics-related applications and gives pointers to experimental results.

Key words: Computationally hard problems, combinatorial explosions, discrete problems, fixed-parameter tractability, optimal solutions.

1. Introduction

Many problems that emerge in bioinformatics require vast amounts of computer time to be solved optimally. A typical example is the following: Given a series of n experiments of which some pairs have conflicting results (i.e., at least one must have been faulty), identify a minimum-size subset of experiments to eliminate so that no conflict remains. This problem is notoriously difficult to solve. For this and many other problems, the root of these difficulties can be identified as their *NP-hardness*, which implies a combinatorial explosion in the solution space that apparently cannot be easily avoided *(1)*. Thus, whenever a problem is proven to be NP-hard, it is common to employ heuristic algorithms, approximation algorithms, or attempt to sidestep the problem whenever large instances need to be solved. All the concrete case studies we provide in this survey deal with NP-hard problems.

Jonathan M. Keith (ed.), *Bioinformatics, Volume II: Structure, Function and Applications. vol. 453*
© 2008 Humana Press, a part of Springer Science+Business Media, Totowa, NJ
Book doi: 10.1007/978-1-60327-429-6 Springerprotocols.com

Often, however, it is not only the size of an instance that makes a problem hard to solve, but rather its structure. The concept of fixed-parameter tractability (FPT) reflects this observation and renders it more precise by measuring structural hardness by a so-called *parameter*, which is typically a non-negative integer variable denoted k. This generalizes the concept of "easy special cases" that are known for virtually all NP-hard problems: Whenever the parameter k turns out to be small, a fixed-parameter algorithm is going to be provably efficient while guaranteeing the optimality of the solution obtained. For instance, our fault identification problem can be solved quickly whenever the number of faulty experiments is small—an assumption easy to make in practice, since otherwise the results would not be worth much anyway. A particular appeal of FPT is that the parameter can be chosen from basically boundless possibilities. For instance, we might in our example choose the maximum number of conflicts for a single experiment or the size of the largest group of pairwise conflicting experiments to be the parameter. This makes FPT a many-pronged attack that can be adapted to different practical applications of one problem. Note, however, that not all parameters need to lead to efficient algorithms; in fact, FPT provides tools to classify parameters as "not helpful," meaning that we cannot expect efficient solvability even when the parameter is small.

Fixed-parameter algorithms have by now facilitated many success stories in bioinformatics. Several techniques have emerged as being applicable to large classes of problems. In the main part of this work, we present five of these techniques, namely kernelization (**Section 3.1**), depth-bounded search trees (**Section 3.2**), tree decompositions of graphs (**Section 3.3**), color-coding (**Section 3.4**), and iterative compression (**Section 3.5**). We start each section by giving an overview of basic concepts and ideas, followed by one to three detailed case studies concerning practically relevant bioinformatics problems. Finally, we survey known applications, implementations, and experimental results, thereby highlighting the strengths and fields of applicability of each technique.

2. Preliminaries

Before discussing the main techniques, we continue with a crash course of computational complexity theory and a formal definition for the concept of fixed-parameter tractability. Furthermore, some terms from graph theory are introduced, and we present our running example problem Vertex Cover.

2.1. Computational Complexity Theory

A core topic of computational complexity theory is the evaluation and comparison of different algorithms for a problem *(2)*. Since most algorithms are designed to work with inputs of arbitrary length, the efficiency (or *complexity*) of an algorithm is not stated just for a single input *(instance)* or a collection of inputs, but as a function that relates the input length n to the number of steps that are required to execute the algorithm. Since instances of the same size might take different amounts of time, the *worst-case* running time is considered. This figure is given in an asymptotic sense; the standard way for this being the *big-O notation*: we say that $f(n) = O(g(n))$ when $f(n)/g(n)$ is upper-bounded by a positive constant in the limit for large n *(3)*.

Determining the computational complexity of problems (meaning the best possible asymptotic running time of an algorithm for them) is a key issue in theoretical computer science. Of central importance herein is to distinguish between problems that can be solved efficiently and those that presumably cannot. To this end, theoretical computer scientists have coined the notions of *polynomial-time solvable* on the one hand and *NP-hard* on the other *(1)*. In this sense, polynomial-time solvability has become a synonym for efficient solvability. This means that for a size-n input instance of a problem, an optimal solution can be computed in $O(n^c)$ time, where c is some positive constant. By way of contrast, the (unproven, yet widely believed) working hypothesis of theoretical computer science is that NP-hard problems cannot be solved in $O(n^c)$ time for any constant c. More specifically, typical running times for NP-hard problems are of the form $O(c^n)$ for some constant $c > 1$, or even worse; that is, we have an exponential growth of the number of computation steps.

As there are thousands of practically important NP-hard optimization problems, and the number is continuously growing *(4)*, several approaches have been developed that try to circumvent the assumed computational intractability of NP-hard problems. One such approach is based on polynomial-time approximation algorithms, where one gives up seeking optimal solutions in order to have efficient algorithms *(5, 6)*. Another commonly employed strategy is that of heuristics, where one gives up any provable performance guarantees concerning running time and/or solution quality by developing algorithms that "usually" behave well in "most" practical applications *(7)*.

2.2. Parameterized Complexity

For many applications the compromises inherent to approximation algorithms and heuristics are not satisfactory. Fixed-parameter algorithms can provide an alternative by providing optimal solutions with useful running time guarantees *(8)*. The core concept is formalized as follows.

Definition 1: An instance of a parameterized problem consists of a problem instance I and a parameter k. A parameterized problem is fixed-parameter tractable if it can be solved in $f(k) \times |I|^{O(1)}$ time, where f is a computable function solely depending on the parameter k, not on the input size $|I|$.

For NP-hard problems, $f(k)$ will of course not be polynomial—since otherwise we would have an overall polynomial-time algorithm—but typically be exponential like 2^k. Clearly, fixed-parameter tractability captures the notion of "efficient for small parameter values": for any constant k, we obtain a polynomial-time algorithm. Moreover, the exponent of the polynomial must be independent of k, which means that the combinatorial explosion is *completely confined to the parameter*.

As an example, consider again the identification of k faulty experiments among n experiments. We can solve this problem in $O(2^n)$ time by trying all possible subsets. However, this is not feasible for $n > 40$. In contrast, a fixed-parameter algorithm with running time $O(2^k \times n)$ exists, which allows the problem to be solved even for $n = 1,000$, if $k < 20$ (as will be discussed in **Section 3.1.3**, instances can even be solved for much larger values of k in practice by an extension of this approach).

Note that—as parameterized complexity theory points out—there are problems that are probably not fixed-parameter tractable.

Two recent monographs are available on parameterized complexity, one focusing on theoretical foundations *(9)*, and one focusing on techniques and algorithms *(10)*, the latter also being the focus of our considerations here.

2.3. Graph Theory

Many of the problems we deal with in this chapter are from graph theory *(11)*. A graph $G = (V,E)$ is given by a set of *vertices V* and a set of *edges E*, where every edge $\{v,w\}$ is an undirected connection of two vertices v and w. Throughout this chapter, we use n to denote the number of vertices and m to denote the number of edges. For a set of vertices $V' \subseteq V$, the *induced sub-graph* $G[V']$ is the graph $(V', \{\{v,w\} \in E | v,w \in V'\})$; that is, the graph G restricted to the vertices in V'.

It is not hard to see that we can formalize our introductory example problem of recognizing faulty experiments as a graph problem: vertices correspond to experiments, and edges correspond to pairs of conflicting experiments. Thus, we need to chose a small set of vertices (the experiments to eliminate) so that each edge is incident to at least one chosen vertex. This is formally known as the NP-hard Vertex Cover problem, which serves as a running example for several techniques in this work.

Vertex Cover
Input: A graph $G = (V,E)$ and a nonnegative integer k.
Task: Find a subset of vertices $C \subseteq V$ with k or fewer vertices such that each edge in E has at least one of its endpoints in C.

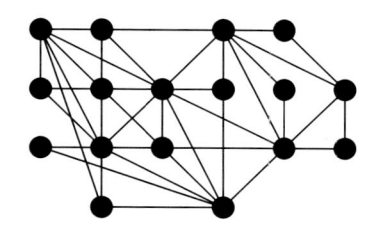

Fig. 21.1. A graph with a size-8 Vertex Cover (cover vertices are marked black).

The problem is illustrated in **Fig. 21.1**. Vertex cover can be considered the *Drosophila* of fixed-parameter research in that many initial discoveries that influenced the whole field originated from studies of this single problem.

3. Fixed-Parameter Tractability Techniques

3.1. Kernelization: Data Reduction with Guaranteed Effectiveness

The idea of data reduction is to quickly pre-solve those parts of a given problem instance that are (relatively) easy to cope with, shrinking it to those parts that form the "really hard" core of the problem. Computationally expensive algorithms need then only be applied to this core. In some practical scenarios, data reduction may even reduce instances of a seemingly hard problem to triviality. Once an effective (and efficient) reduction rule has been found, it is useful in virtually any problem-solving context, whether it be heuristic, approximative, or exact.

This section introduces the concept of *kernelizations*, that is, polynomial-time data reduction with guaranteed effectiveness. These are closely connected to and emerge within the FPT framework.

3.1.1. Basic Concepts

Today, there are many examples of combinatorial problems that would not be solvable without employing heuristic data reduction and preprocessing algorithms. For example, commercial solvers for hard combinatorial problems such as the integer linear program solver CPLEX heavily rely on data-reducing preprocessors for their efficiency *(12)*. Obviously, many practitioners are aware of the concept of data reduction in general. The reason why they should also consider FPT in this context is that fixed-parameter theory provides a way to use data reduction rules not only in a heuristic way, but with guaranteed performance using so-called *kernelizations*. For a reduced instance, these guarantee an upper bound on its size that solely depends on the parameter value. To render a precise definition of this:

Definition 2 (8, 10): Let I be an instance of a parameterized problem with given parameter k. A reduction to a problem kernel (or kernelization) is a polynomial-time algorithm that replaces I by a new instance I′ and k by a new parameter $k′ \leq k$ such that—independently of the size of I—the size of I′ is guaranteed to only depend on some function in k. Furthermore, the new instance I′ must have a solution with respect to the new parameter k′ if and only if I has a solution with respect to the original parameter k.

Kernelizations can help to understand the practical effectiveness of some data reduction rules and, conversely, the quest for kernelizations can lead to new and powerful data reduction rules based on deep structural insights.

Intriguingly, there is a close connection between fixed-parameter tractable problems and those problems for which there exists a problem kernel—they are exactly the same. Unfortunately, the running time of a fixed-parameter algorithm directly obtained from a kernelization is usually not practical and, in the other direction, there exists no constructive scheme for developing data reduction rules for a fixed-parameter tractable problem. Hence, the main use of this equivalence is to establish the fixed-parameter tractability or amenability to kernelization of a problem; it is also useful for showing that we need not search any further (e.g., if a problem is known to be fixed-parameter intractable, we do not need to look for a kernelization).

3.1.2. Case Study

This section first illustrates the concept of kernelization by a simple example concerning the Vertex Cover problem. It then shows how a generalization of this method leads to a very effective kernelization.

3.1.2.1. A Simple Kernelization for Vertex Cover

Consider our running example, Vertex Cover. In order to cover an edge in the graph, one of its two endpoints *must* be in the Vertex Cover. If one of these is a degree-1 vertex (i.e., it has exactly one neighbor), then the other endpoint has the potential to cover more edges than this degree-1 vertex, leading to a first data reduction rule.

Reduction Rule VC1: For degree-1 vertices, put their neighboring vertex into the cover.

Here, "put into the cover" means adding the vertex to the solution set and removing it and its incident edges from the instance. Note that this reduction rule assumes that we are only looking for *one* optimal solution to the Vertex Cover instance we are trying to solve; there may exist other minimum Vertex Covers that do include the reduced degree-1 vertex. (There exist suitable reduction rules when it is of interest to enumerate *all* Vertex Covers of a given graph, but these are beyond the scope of this work. For instance, Damaschke *(13)* suggests the notion of a *full kernel* that contains all solutions in a compressed form and thereby allows enumeration of them. This is an emerging field of research, and not many full kernels are known.)

After applying Rule VC1, we can further do the following in the fixed-parameter setting where we ask for a Vertex Cover of size at most k.

> *Reduction Rule VC2*: If there is a vertex of degree at least $k + 1$, put this vertex into the cover.

The reason this rule is correct is that if we did not take v into the cover, then we would have to take every single one of its $k+1$ neighbors into the cover in order to cover all edges incident to v. This is not possible because the maximum allowed size of the cover is k.

After exhaustively performing Rules VC1 and VC2, no vertex in the remaining graph has a degree higher than k, meaning that at most k edges can be covered by choosing an additional vertex into the cover. Since the solution set may be no larger than k, the remaining graph can have at most k^2 edges if it is to have a solution. Clearly, we can assume without loss of generality that there are no isolated vertices (i.e., vertices with no incident edges) in a given instance. In conjunction with Rule VC1, this means that every vertex has degree at least two. Hence, the remaining graph can contain at most k^2 vertices.

Abstractly speaking, what we have done so far is the following: After applying two *polynomial-time* data reduction rules to an instance of Vertex Cover, we arrived at a reduced instance whose size can *solely be expressed in terms of the parameter k*. Hence, considering Definition 2, we have found a kernelization for Vertex Cover.

3.1.2.2. Improving Effectiveness by Generalization: Crown Reductions

Besides the simple-to-achieve size-k^2 problem kernel for Vertex Cover we have just discussed, there are several more advanced kernelization techniques for Vertex Cover. The more advanced methods generally feature two important improvements: First, they do not require the parameter k to be stated explicitly beforehand (contrary to Rule VC2); that is, they are *parameter-independent*. Second, they improve the upper bounds on the kernel size to being linear in k. We explore one of these advanced methods in more detail here, namely the so-called *crown reduction* rules *(14, 15)*.

Crown reduction rules are a prominent example for an advanced kernelization for Vertex Cover; they constitute a generalization of the elimination of degree-1 vertices we have seen in Rule VC1.

A *crown* in a graph consists of an independent set I (i.e., no two vertices in I are connected by an edge) and a set H containing all vertices adjacent to I. In order for $I \cup H$ to be a crown,

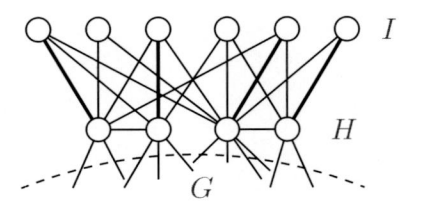

Fig. 21.2. A graph G with a crown $I \cup H$. Note how the thick edges constitute a maximum matching of size $|H|$ in the bipartite graph induced by the edges between I and H.

there has to exist a size-$|H|$ maximum bipartite matching in the bipartite graph induced by the edges between I and H, that is, one in which every element of H is matched. (*See*, e.g., *(3)* for an introduction on bipartite matching and the involved algorithmics.) An example for a crown structure is given in **Fig. 21.2**.

If there is a crown $I \cup H$ in the input graph G, then we need *at least* $|H|$ vertices to cover all edges in the crown. However, since all edges in the crown can be covered by taking *at most* $|H|$ vertices into the cover (as I is an independent set), there is a minimum-size Vertex Cover for G that contains all the vertices in H and none of the vertices in I. We may thus delete any given crown $I \cup H$ from G, reducing k by $|H|$. In a sense, the degree-1 vertices we took care of in Rule VC1 are the simplest crowns.

Two issues remain to be dealt with, namely how to find crowns efficiently and giving an upper bound on the size of the problem kernel that can be obtained via crown reductions. It turns out that finding crowns can be achieved in polynomial time *(15)*. The size of the thus reduced instance is upper-bounded via the following theorem (a proof of which can be found in *(14)*):

Theorem 1: A graph that is crown-free and has a Vertex Cover of size at most k can contain at most 3k vertices.

In this way, by generalizing Rule VC1—which by itself does not constitute a kernelization—we have obtained an efficient kernelization that does not require an explicitly stated value for the parameter k and yields a small kernel for Vertex Cover. (Another example for a data reduction for Vertex Cover that is based on schemes that generalize to arbitrarily large graph substructures is given by Chen et al. *(16)*.) Even smaller kernels of size upper-bounded by $2k$ are attainable with other methods, for example using the so-called "Nemhauser-Trotter kernelization" (*see (10)* for details).

3.1.3. Applications and Implementations

Since many initial and influential discoveries concerning FPT were made from studies of Vertex Cover, it comes as no surprise that the experimental field is more advanced for this problem than for others from the realm of fixed-parameter tractability. Abu-Khzam et al. *(14)* studied various kernelization methods for Vertex Cover and their practical performance both with respect to time as well as with respect to the resulting kernel size. For bioinformatics-related networks derived from protein databases and microarray data, they found that crown reductions turn out to be very fast to compute in practice and can be as effective as approaches with a better worst-case bound of $2k$ (e.g., the Nemhauser-Trotter reduction). Therefore, Abu-Khzam et al. recommend always using crown reduction as a general preprocessing step when solving Vertex Cover before attempting other, more costly, reduction methods.

Solving Vertex Cover is of relevance to many bioinformatics-related scenarios such as microarray data analysis *(17)* and the computation of multiple sequence alignments *(18)*. Besides solving instances of Vertex Cover, another important application of Vertex Cover kernelizations is that of searching maximum cliques (i.e., maximum-size fully connected sub-graphs) in a graph. Here, use is made of the fact that an n-vertex graph has a maximum clique of size $(n-k)$ if and only if its complement graph has a size-k minimum Vertex Cover (i.e., the "edgewise inverse" graph that contains exactly those edges the original graph does not contain.) Details and experimental results for this with applications to computational biology are given, e.g., by Abu-Khzam et al. *(19)*. State-of-the art implementations of fixed-parameter algorithms and kernelizations for Vertex Cover enable finding cliques and dense sub-graphs consisting of 200 or more vertices (e.g., *see (17)*) in biological networks such as they appear in the analysis of microarray data.

Another biologically relevant clustering problem where kernelizations have been successfully implemented is the Clique Cover problem. Here, the task is to cover all edges of a graph using at most k cliques (these may overlap). Using data reduction, Gramm et al. *(20)* showed instances with a solution size of up to $k=250$ to be solvable in practice. Finally, a further example for a clustering problem in which kernelizations have proved to be quite useful in practice is the Cluster Editing problem that we discuss in more detail in **Section 3.2.2.2** *(21)*.

3.2. Depth-Bounded Search Trees

Once the data reductions we have discussed in the previous section have been applied to a problem instance, we are left with the "really hard" problem kernel to be solved. A standard way to explore the huge search space related to optimally solving a computationally hard problem is to perform a systematic exhaustive search. This can be organized in a tree-like fashion, which is the subject of this section.

3.2.1. Basic Concepts

Search trees algorithms—also known as backtracking algorithms, branching algorithms, or splitting algorithms—certainly are no new idea and have extensively been used in the design of exact algorithms (e.g., *see (3, 22)*). The main contribution of fixed-parameter theory to search tree algorithms is the consideration of search trees whose *depth is bounded from above by the parameter.* Combined with insights on how to find useful—and possibly non-obvious—parameters, this can lead to search trees that are much smaller than those of naive brute-force searches. For example, a very naive search tree approach for solving Vertex Cover is to just take one vertex and branch into two cases: either this vertex is in the Vertex Cover or not. For an n-vertex graph, this leads to a search tree of size $O(2^n)$. As we outline in this section, we can do much better than that and obtain a search tree whose depth is upper-bounded by k, giving a size bound of $O(2^k)$ (extending what we discuss here, there are even better search trees of size $O(1.28^k)$ possible). Since usually $k \ll n$, this can draw the problem into the zone of feasibility even for large graphs (as long as k is small).

Besides depth-bounding, fixed-parameter theory provides additional means to provably improve the speed of search tree exploration, particularly by interleaving this exploration with kernelizations, that is, data reduction is applied to partially solved instances during the exploration.

3.2.2. Case Studies

Starting with our running example Vertex Cover, this section introduces the concept of depth-bounded search trees by three case studies.

3.2.2.1. Vertex Cover Revisited

For many search tree algorithms, the basic idea is to find a small subset of the input instance in polynomial time such that at least one element of this subset *must* be part of an optimal solution to the problem. In the case of Vertex Cover, the simplest such subset is any two vertices that are connected by an edge. By definition of the problem, one of these two vertices *must* be part of a solution or the respective edge would not be covered. Thus, a simple search-tree algorithm to solve Vertex Cover on a graph G proceeds by picking an arbitrary edge $e = \{v,w\}$ and recursively searching for a Vertex Cover of size k-1 both in G-v and G-w. (For a vertex $v \in V$, we define G-v to be the graph G with both the vertex v and the edges incident to v removed.) That is, the algorithm *branches* into two sub-cases knowing one of them must lead to a solution of size at most k—provided that it exists.

As shown in **Fig. 21.3**, the recursive calls of the simple Vertex Cover algorithm can be visualized as a tree structure. Because the depth of the recursion is upper-bounded by the parameter value and we always branch into two sub-cases, the number of cases that are considered by this tree—its size, so to say—is upper-bounded by $O(2^k)$. Note how this size is independent of the size of the input instance and only depends on the value of the parameter k.

The currently "best" search trees for Vertex Cover are of worst-case size $O(1.28^k)$ *(16)* and mainly achieved by elaborate

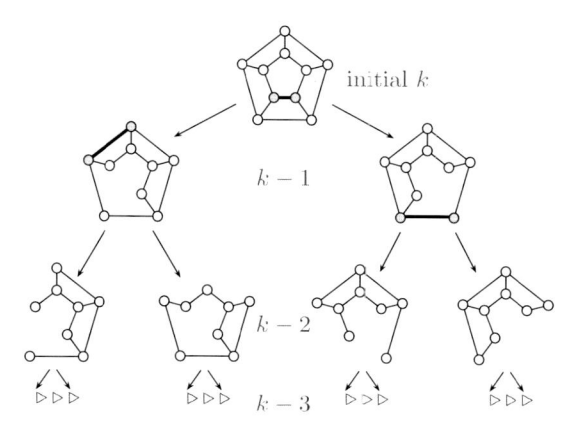

Fig. 21.3. Simple search tree for finding a Vertex Cover of size at most k in a given graph. The size of the tree is upper-bounded by $O(2^k)$.

case distinctions. However, for practical applications it is always concrete implementation and testing that has to decide whether the administrative overhead caused by distinguishing more and more cases pays off. A simpler algorithm with slightly worse search tree size bounds may be preferable.

3.2.2.2. The Cluster Editing Problem

For Vertex Cover, we have found a depth-bounded search tree by observing that at least one endpoint of any given edge *must* be part of the cover. A somewhat similar approach can be used to derive a depth-bounded search tree for the following clustering problem:

> Cluster Editing
> *Input*: A graph $G = (V,E)$ and a nonnegative integer k.
> *Task*: Find whether we can modify G to consist of disjoint cliques (i.e., fully connected components) by adding or deleting at most k edges.

An illustration for this problem is given in **Fig. 21.4**. Cluster Editing has important applications in graph-modeled data clustering, e.g., to cluster microarray data: Here, the individual datapoints are represented as vertices. Edges connect two vertices if these have similar expression profiles. The underlying assumption is that the datapoints will form dense clusters in the constructed graph that are only sparsely connected to each other. Thus, by adding and removing only a few edges, we can reveal the underlying correlation structure in the form of disjoint cliques.

Similar to Vertex Cover, a search tree for Cluster Editing can be obtained by noting that the desired graph of disjoint cliques forbids a certain structure: If two vertices are connected by an edge, then their neighborhoods must be the same. Hence, whenever we encounter two connected vertices u and v in the input

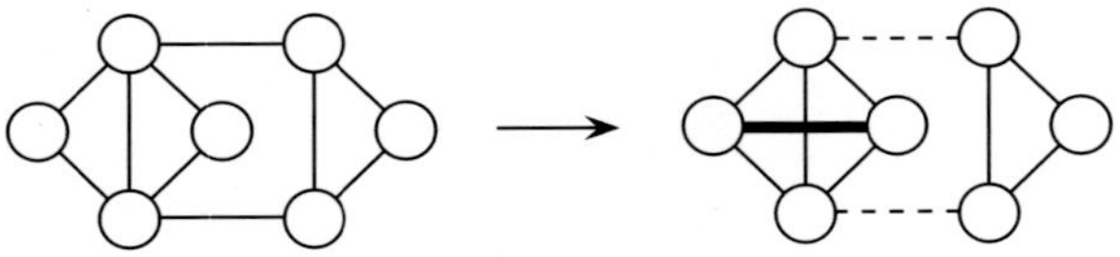

Fig. 21.4. Illustration for the Cluster Editing problem: By removing two edges from and adding one edge to the graph on the left (i.e., $k = 3$), we can obtain a graph that consists of two disjoint cliques.

graph G that are connected by an edge and where one vertex, say u, has a neighbor w that is not connected to v, we are compelled to do one of three things: Either remove the edge $\{u,v\}$, or connect v with w, or remove the edge $\{u,w\}$. Note that each such modification counts with respect to the parameter k. Therefore, exhaustively branching into three cases for at most k forbidden substructures, we obtain a search tree of size $O(3^k)$ to solve Cluster Editing. The currently best branching scheme has a search-tree size of $O(1.92^k)$ *(23)*; interestingly, it was derived by using computer-aided algorithm design. Experimental results for Cluster Editing are reported in *(21)*.

3.2.2.3. The Center String Problem

The Center String problem is also known as Consensus String or Closest String.

> Center String
> *Input*: A set of k length-l strings $s_1,\ldots,s_k$ and a non-negative integer d.
> *Task*: Find a *center string* s that satisfies $d_H(s,s_i) \leq d$ for all $i = 1,\ldots,k$.

Here, $d_H(s,s_i)$ denotes the Hamming distance between two strings s and s_i, that is, the number of positions where s and s_i differ. Note that there are at least two immediately obvious parameterizations of this problem. The first is given by choosing the "distance parameter" d and the second is given by the number k of input strings. Both parameters are reasonably small in various applications; we refer to Gramm et al. *(24)* for more details. Here, we focus on the parameter d.

One application scenario in which this problem appears is in *primer design* where we try to find a small DNA sequence called *primer* that binds to a set of (longer) target DNA sequences as a starting point for replication of these sequences. How well the primer binds to a sequence is mostly determined by the number of positions in that sequence that hybridize to it. Although often done by hand, Stojanovic et al. *(25)* proposed a computational approach for finding a well-binding primer of length l. First, the target sequences are aligned, that is, as many matching positions within the sequences as possible are grouped into columns. Then, a "sliding window" of length l is moved over this alignment,

```
...GGTGAG ATCTATAGAAGT TGAATGC...
...GGTGGA ATCTACAGTAAC GGATTGT...
...GGCGAG ATCTACAGAAGT GGAATGC...
...GGCGAG ATCTATAGAGAT GGAATGC...
...GGCAAG ATCTATAGAAGT GGAATGC...
```

closest string: ATCTACAGAAAT

primer candidate: TAGATGTCTTTA

Fig. 21.5. Illustration to show how DNA primer design can be achieved by solving Center String instances on length-λ windows of aligned DNA sequences. (Note that the primer candidate is not the center string sought after but its nucleotide-wise complement.)

giving a Center String problem for each window position. **Figure 21.5** illustrates this (*see (26)* for details).

In the remainder of this case study, we sketch a fixed-parameter search tree algorithm for Center String due to Gramm et al. *(24)*, the parameter being the distance d. Unlike for Vertex Cover and Cluster Editing, the central challenge lies in even *finding* a depth-bounded search tree, which is all but obvious at a first glance. Once found, however, the derivation of the upper bound for the search tree size is straightforward. The underlying algorithm is very simple to implement, and we are not aware of a better one. The main idea behind the algorithm is to maintain a *candidate string* $\acute{S}$ for the center string and compare it to the strings $s_1, \ldots, s_k$. If $\acute{S}$ differs from some s_i in more than d positions, then we know that $\acute{S}$ needs to be modified in at least one of these positions to match the character that s_i has there. Consider the following observation:

> *Observation 1:* Let d be a non-negative integer. If two strings s_i and s_j have a Hamming distance greater than 2d, then there is no string that has a Hamming distance of at most d to both of s_i and s_j.

This means that s_i is allowed to differ from $\acute{S}$ in at most $2d$ positions. Hence, among any $d + 1$ of those positions where s_i differs from $\acute{S}$, at least one must be modified to match s_i. This can be used to obtain a search tree that solves Center String.

We start with a string from $\{s_1, \ldots, s_k\}$ as the candidate string $\acute{S}$, knowing that a center string can differ from it in at most d positions. If $\acute{S}$ already is a valid center string, we are done. Otherwise, there exists a string s_i that differs from $\acute{S}$ in more than d positions, but less than $2d$. Choosing any $d+1$ of these positions, we branch into $(d + 1)$ sub-cases, each sub-case modifying a position in $\acute{S}$ to match s_i. This position cannot be changed any more further down in the search tree (otherwise, it would not have made sense to make it match s_i at that position). Hence, the depth of the search tree is upper-bounded by d, for if we were to go deeper down in the tree, then $\hat{S}$ would differ in more than d positions from the original string we started with. Thus, Center String can be solved by exploring a search tree of size $O((d+1)^d)$ *(24)*. Combining data reduction with this search tree, we arrive at the following.

Theorem 2: Center String can be solved in $O(k \times l + k \times d \times (d+1)^d)$ steps.

It might seem as if this result is purely of theoretical interest—after all, the term $(d+1)^d$ becomes prohibitively large already for, say, $d = 15$. However, two things are to be noted in this respect: First, for one of the main applications of Center String, primer design, d is very small (most often <4). Second, empirical analysis reveals that when the algorithm is applied to real-world and random instances, it often beats the proven upper bound by far, solving many real-world instances in less than a second. The algorithm is also faster than an Integer Linear Programming formulation of Center String when the input consists of many strings and l is small *(24)*.

Unfortunately, many variants of Center String—roughly speaking, these deal with finding a matching *sub*-string and distinguish between strings to which the center is supposed to be close and to which it should be distant—are known to be intractable from a fixed-parameter point of view *(27–29)*.

3.3.1. Applications and Implementations

In combination with data reduction, the use of depth-bounded search trees has proved itself quite useful in practice, for example, allowing one to find Vertex Covers of more than 10,000 vertices in some dense graphs of biological origin *(19)*. It should also be noted that search trees trivially allow for a parallel implementation: When branching into sub-cases, each process in a parallel setting can further explore one of these branches with no additional communication required. Cheetham et al. *(18)* were the first to practically demonstrate this with their parallel Vertex Cover solver; they achieve a near-optimum (i.e., linear with the number of processors employed) speedup on multiprocessor systems, solving instances with $k \geq 400$ in mere hours. Recent research even indicates that in some cases, parallelizing may yield a super-linear speedup because the branches that lead to a solution are explored earlier than in a sequential setting *(19)*.

Besides in fixed-parameter theory, search tree algorithms are studied extensively in the area of artificial intelligence and heuristic state space search. There, the key to speed-ups are *admissible heuristic evaluation functions*, which quickly give a lower bound on the distance to the goal. The reason that admissible heuristics are rarely considered by the FPT community in their works is that they typically cannot improve the asymptotic running time (see, e.g., *(31)* for a counterexample). Still, the speed-ups obtained in practice can be quite pronounced, as demonstrated for Vertex Cover *(30)*.

As with kernelizations, algorithmic developments outside the fixed-parameter setting can make use of the insights that have been gained in the development of depth-bounded search trees in a fixed-parameter setting. A recent example for this is the Minimum Quartet Inconsistency problem arising in the construction of evolutionary trees. Here, an algorithm that uses depth-bounded search trees was developed by Gramm and Niedermeier *(31)*. Their insight was recently used by Wu et al. *(32)* to develop a (non-parameterized) faster algorithm for this problem.

In conclusion, depth-bounded search trees with clever branching rules are certainly one of the first approaches to try when solving fixed-parameter tractable problems in practice.

3.4. Tree Decompositions of Graphs

Many NP-hard graph problems become computationally feasible when they are restricted to graphs without cycles; that is, trees or collections of trees (forests). Trees, however, form a very limited class of graphs that often do not suffice as a model in real-life applications. Hence, as a compromise between general graphs and trees, one might want to look at "tree-like" graphs. This likeness is formalized by the concept of *tree decompositions* of graphs. Introduced by Robertson and Seymour about 20 years ago, tree decompositions nowadays play a central role in algorithmic graph theory *(11)*. In this section, we survey some important aspects of tree decompositions and their algorithmic use with respect to computational biology and FPT.

3.4.1. Basic Concepts

There is a very helpful and intuitively appealing characterization of tree decompositions in terms of a *robber–cop game* in a graph *(33)*: A robber stands on a graph vertex and, at any time, he can run at arbitrary speed to any other vertex of the graph as long as there is a path connecting both. He is not permitted to run through a cop, though. A cop, at any time, either stands at a graph vertex or is in a helicopter (i.e., he is above the game board). The goal is to land a helicopter on the vertex occupied by the robber. Note that, due to the use of helicopters, cops are not constrained to move along graph edges. The robber can see a helicopter approaching its landing vertex and he may run to a new vertex before the helicopter actually lands. Thus, for a set of cops the goal is to occupy all vertices adjacent to the robber's vertex and land one more remaining helicopter on the robber's vertex itself. The *treewidth* of the graph is the minimum number of cops needed to catch a robber minus one (observe that if the graph is a tree, two cops suffice and trees hence have a treewidth of one) and a corresponding *tree decomposition* is a tree structure that provides the cops with a scheme to catch the robber. The tree decomposition indicates bottlenecks in the graph and thus reveals an underlying scaffold-like structure that can be exploited algorithmically.

Formally, tree decompositions and treewidth center around the following somewhat technical definition; **Fig. 21.6** shows a graph together with an optimal tree decomposition of width two.

Definition 3: Let G = (V,E) be a graph. A tree decomposition of G is a pair $<\{X_i | i \in I\}, T>$ where each X_i is a subset of V, called a *bag*, and T is a tree with the elements of I as nodes. The following three properties must hold:

1. $\bigcup_{i \in I} X_i = V$;
2. for every edge $\{u,v\} \in E$, there is an $i \in I$ such that $\{u,v\} \subset X_i$; and
3. for all $i,j,k \in I$, if j lies on the path between i and k in T, then $X_i \cap X_k \subset X_j$.

The width of $<\{X_i | i \in I\}, T>$ equals $\max\{|X_i| | i \in I\} - 1$. The treewidth of G is the minimum k such that G has a tree decomposition of width k.

The third condition of the definition is often called the *consistency property*. It is important in dynamic programming, the main algorithmic tool when solving problems on graphs of bounded treewidth. An equivalent formulation of this property is to demand that for every graph vertex v, all bags containing v form a connected sub-tree.

For trees, the bags of a corresponding tree decomposition are simply the two-element vertex sets formed by the edges of the tree. In the definition, the subtraction of 1 thus ensures that trees have a treewidth of 1. In contrast, a clique of n vertices has

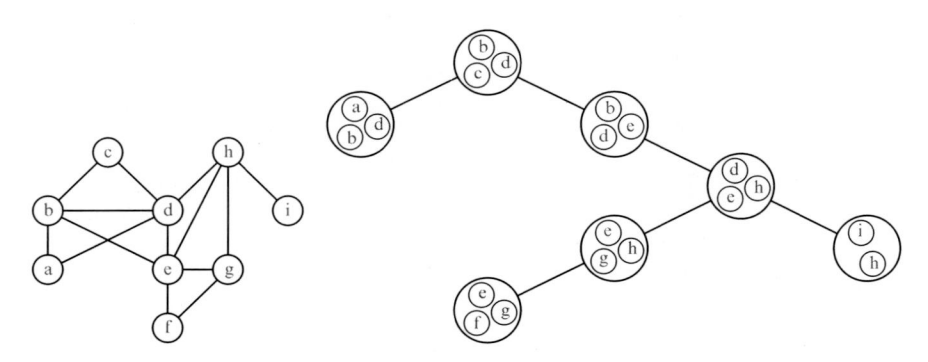

Fig. 21.6. A graph together with a tree decomposition of width 2. Observe that—as demanded by the consistency property—each graph vertex induces a sub-tree in the decomposition tree.

treewidth n-1. The corresponding tree decomposition trivially consists of one bag containing all graph vertices; in fact, no tree decomposition with smaller width is attainable. More generally, it is known that every complete sub-graph of a graph G is completely "contained" in a bag of G's tree decomposition.

Tree decompositions of graphs are connected to another central concept in algorithmic graph theory: *graph separators* are vertex sets whose removal from the graph separates the graph into two or more connected components. Each bag of a tree decomposition forms a separator of the corresponding graph.

Given a graph, determining its treewidth is an NP-hard problem itself. However, several tools and heuristics exist that construct tree decompositions *(34)*, and for some graphs that appear in practice, computing a tree decomposition is easy. Here, we concentrate on the algorithmic use of tree decompositions, assuming that they are provided to us.

3.4.2. Case Study

Typically, tree decomposition based algorithms proceed according to the following two-stage scheme:

1. Find a tree decomposition of bounded width for the input graph.

2. Solve the problem by dynamic programming on the tree decomposition, starting from the leaves.

Intuitively speaking, the decomposition tree provides us with sort of a scaffold that allows an efficient and consistent processing through the graph in order to solve the given problem. Note that this scaffold leads to optimal solutions even when the utilized tree decompositions are not optimal; however, the algorithm will run slower and consume more memory in that case.

To exemplify dynamic programming on tree decompositions, we make use of our running example Vertex Cover and sketch a fixed-parameter dynamic programming algorithm for Vertex Cover with respect to the parameter treewidth.

> *Theorem 3*: For a graph G with a given width-ω tree decomposition $<\{X_i | i \in I\}, T>$, an optimal Vertex Cover can be computed in $O(2^\omega \times \omega \times |I|)$ steps.

The basic idea of the algorithm is to examine for each bag X_i all of the at most $2^{|X_i|}$ possibilities to obtain a Vertex Cover for the sub-graph $G[X_i]$. This information is stored in tables A_i, $i \in I$. Adjacent tables are updated in a bottom-up process starting at the leaves of the decomposition tree. Each bag of the tree decomposition thus has a table associated with it. During this updating process it is guaranteed that the "local" solutions for each sub-graph associated with a bag of the tree decomposition are combined into a "globally optimal" solution for the overall graph G. (We omit several technical details here; these can be found in *(10)*.)

Observe that the following points of Definition 3 guarantee the validity of the preceding approach.

1. The first condition in Definition 3, that is, $\bigcup_{i \in I} X_i = V$, makes sure that every graph vertex is taken into account during the computation.

2. The second condition in Definition 3, that is, "$\forall e \in E \,\exists\, i \in I: e \in X_i$", makes sure that all edges can be treated and thus will be covered.

3. The third condition in Definition 3 guarantees the consistency of the dynamic programming, since information concerning a particular vertex v is only propagated between neighbored bags that both contain v.

One thing to keep in mind for a practical application is that storing dynamic programming tables requires memory space that grows exponentially in the treewidth. Hence, even for "small" treewidths, say, between 10 and 20, the computer program may run out of memory and break down.

3.4.3. Applications and Implementations

Tree decomposition based algorithms are a valuable alternative whenever the underlying graphs have small treewidth. As a rule of thumb, the typical border of practical feasibility lies somewhere below a treewidth of 20 for the underlying graph. Successful implementations for solving Vertex Cover with tree decomposition approaches have been reported by Alber et al. *(35)* and Betzler et al. *(36)*.

Another recent practical application of tree decompositions was given by Xu et al. *(37)*, who proposed a tree decomposition based algorithm to solve two problems encountered in protein structure prediction, namely the prediction of backbone structure and side-chain prediction. To this end, they modeled these two problems as a graph labeling problem and showed that the resulting graphs have a very small treewidth in practice, allowing the problems to be solved efficiently.

Besides taking an input graph, computing a tree decomposition for it, and hoping that the resulting tree decomposition has a small treewidth, there have also been cases in which a problem is modeled as a graph problem such that it can be *proved* that the resulting graphs have a tree decomposition with small treewidth that can efficiently be found. As an example, Song et al. *(38)* used a so-called conformational graph to specify the consensus sequence-structure of an RNA family. They demonstrated that the treewidth of this graph is basically determined by the structural elements that appear in the RNA. More precisely, they show that if there is a bounded number of crossing stems, say k, in a pseudoknot structure, then the resulting graph has treewidth $(2 + k)$. Since the number of crossing stems is usually small, this yields a fast algorithm for searching RNA secondary structures.

A further strong example concerning tree decomposition–based algorithms are probabilistic inference networks, which play a vital role in several decision support systems in medical, agricultural, and other applications *(39, 40)*. Here, tree decomposition based dynamic programming is a key solving method.

3.5. Color-Coding

The color-coding method due to Alon et al. *(41)* is a general method for finding small patterns in graphs. It is clearly not as generally applicable as data reduction or search trees, but can lead to very efficient algorithms for certain problems. In its simplest form, it can solve the Minimum Weight Path problem, which asks for the cheapest path of length k in a graph. This has been employed with protein–protein interaction networks to find signaling pathways *(42, 43)* and to evaluate pathway similarity queries *(44)*.

3.5.1. Basic Concepts

Naïvely trying all roughly n^k possibilities of finding a small structure of k vertices within a graph of n vertices quickly leads to a combinatorial explosion, making this approach infeasible even for rather small input graphs of a few hundred vertices. The central idea of color-coding is to randomly color each vertex of the graph with one of k colors and to "hope" that all vertices in the sub-graph searched for obtain different colors (it becomes *colorful*). When this happens, the task of finding the sub-graph is greatly simplified: It can be solved by dynamic programming in a running time in which the exponential part depends only on the size k of the substructure searched for, as opposed to the $O(n^k)$ running time of the naïve approach.

Of course, most of the time the target structure is not actually colorful. Therefore, we have to repeat the process of randomly coloring and then searching (called *trial*) many times with a fresh coloring until with sufficiently high probability at least once our target structure is colorful. Since the number of trials also depends only on k (albeit exponentially), this algorithm has a fixed-parameter running time.

3.5.2. Case Study

Formally stated, the problem we consider is the following:

> Minimum Weight Path
> *Input:* An undirected edge-weighted graph G and a non-negative integer k.
> *Task:* Find a simple length-k path in G that minimizes the sum over its edge weights.

This problem is well known to be NP-hard *(1)*. What makes the problem hard is the requirement of *simple* paths, that is, paths in which no vertex may occur more than once (otherwise, it is easily solved by traversing a minimum-weight edge k-1 times). Given a fixed coloring of vertices, finding the minimum-weight colorful path is accomplished by dynamic programming: assume that for some $i < k$ we have computed a value $W(v,S)$ for every vertex $v \in V$ and cardinality-i subset S of vertex colors that denotes the minimum weight of a path that uses each color in S exactly once and ends in v. Clearly, this path is simple because no color is used more than once. We can now use this to compute the values $W(v,S)$ for all cardinality-$(i+1)$ subsets S and vertices $v \in V$, because a colorful length-$(i+1)$ path that ends in a vertex $v \in V$ can be composed of a colorful length-i path that does not use the color of v and ends in a neighbor of v.

More precisely, we let:

$$W(v,S) = \min_{c = \{u,v\} \in E}(W(u,S \setminus \{\text{color}(v)\}) + w(e)). \qquad [1]$$

See **Fig. 21.7** for an example. It is easy to verify that the dynamic programming takes $O(2^k m)$ time. Whenever the minimum-weight length-k path in the input graph is colored with k colors (i.e., every vertex has a different color), then it is found. The problem, of course, is that the coloring of the input graph is random and hence many coloring *trials* have to be performed to ensure that the minimum-weight path is found with a high probability. More precisely, the probability of any length-k path (including the one with minimum weight) being colorful in a single trial is:

$$P_c = \frac{k!}{k^k} > \sqrt{2\pi k}e^{-k} \qquad [2]$$

because there are k^k ways to arbitrarily color k vertices with k colors and $k!$ ways to color them such that no color is used more than once. Using t trials, a path of length k is found with probability $1-(1-P_c)^t$. To ensure that a colorful path is found with a probability greater than $1\text{-}\varepsilon$ (for some $0 < \varepsilon \le 1$), at least:

$$t(\varepsilon) = \left\lceil \frac{\ln \varepsilon}{\ln(1 - P_c)} \right\rceil = |\ln \varepsilon|.O(e^k) \qquad [3]$$

trials are therefore needed, which bounds the overall running time by $2^{O(k)} \times n^{O(1)}$. Although the result is only correct with a certain probability, the user can specify any desired error probability, say

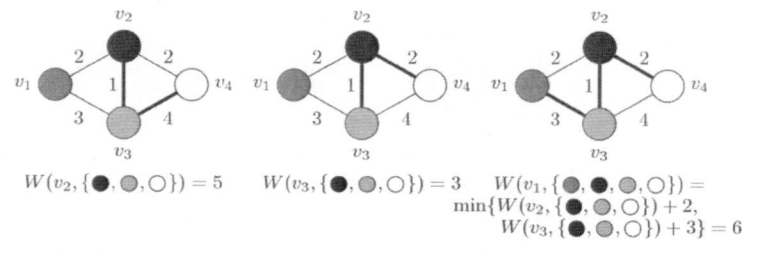

$W(v_2, \{\bullet,\ominus,\bigcirc\}) = 5 \qquad W(v_3, \{\bullet,\ominus,\bigcirc\}) = 3 \qquad W(v_1, \{\bullet,\bullet,\ominus,\bigcirc\}) = \min\{W(v_2, \{\bullet,\ominus,\bigcirc\}) + 2, W(v_3, \{\bullet,\ominus,\bigcirc\}) + 3\} = 6$

Fig. 21.7. Example for solving Minimum Weight Path using the color-coding technique. Here, using (1) a new table entry *(right)* is calculated using two already known entries *(left and middle)*.

0.1%, and even very low error probabilities do not incur excessive extra running time costs.

3.5.3. Applications and Implementations

Protein interaction networks represent proteins by nodes and mutual protein–protein interaction probabilities by weighted edges. They are a valuable source of information for understanding the functional organization of the proteome. Scott et al. *(43)* demonstrated that *high-scoring simple paths* in the network constitute plausible candidates for linear signal transduction pathways; *simple* meaning that no vertex occurs more than once, and *high-scoring* meaning that the product of edge weights is maximized. (To match the definition of Minimum Weight Path, one works with the *weight* $w(e)$: $= -\log p(e)$ of an edge e with interaction probability $p(e)$ between e's endpoints, such that the goal is to minimize the sum of weights for the edges of a path.)

The currently most efficient implementation based on color-coding *(42)* is capable of finding optimal paths of length up to 13 in seconds within the yeast protein interaction network, which contains about 4,500 vertices. A variant with better theoretical running time has recently been suggested *(45)*, but to the best of our knowledge has not yet been implemented.

A particularly appealing aspect of the color-coding method is that it can be easily adapted to many practically relevant variations of the problem formulation:

- The set of vertices where a path can start and end can be restricted (e.g., to force it to start in a membrane protein and end in a transcription factor *(43)*).

- Not only the minimum-weight path can be sought after but rather a collection of low-weight paths (typically, one demands that these paths must differ in a certain amount of vertices to ensure that they are diverse and not small modifications of the global minimum-weight path).

- Recently, it has been demonstrated that pathway queries to a network, that is, the task of finding a pathway in a network that is as similar as possible to a query pathway, can be handled with color-coding *(44)*.

Besides path problems, color-coding has also been used to give algorithms that find small trees in graphs *(43)* and for graph packing problems *(45)*, in which the goal is to find many disjoint copies of a pattern in a graph. No application outside the realm of small patterns in graphs is currently known.

3.6. Iterative Compression

Of the techniques we survey, iterative compression is by far the youngest, appearing first in a work by Reed, Smith, and Vetta in 2004 *(46)*. Although it is perhaps not quite as generally applicable as data reduction or search trees, it appears to be useful for solving a wide range of problems and has already led to significant breakthroughs in showing fixed-parameter tractability results.

The central concept of iterative compression is to employ a so-called *compression routine*.

3.6.1. Basic Concepts

Definition 4: A compression routine is an algorithm that, given a problem instance and a solution of size k, either calculates a smaller solution or proves that the given solution is of minimum size.

Using this routine, one finds an optimal solution to a problem by inductively building up the problem structure and iteratively compressing intermediate solutions. The point is that if the compression routine is a fixed-parameter algorithm with respect to the parameter k, then so is the whole algorithm.

The main strength of iterative compression is that it allows one to see the problem from a different angle. For the compression routine, we do not only have the problem instance as input, but also a solution, which carries valuable structural information on the input. Also, the compression routine does not need to find an optimal solution at once, but only any better solution. Therefore, the design of a compression routine can often be simpler than designing a complete fixed-parameter algorithm.

However, although the mode of use of the compression routine is usually straightforward, finding the compression routine itself is not—even when we already know a problem to be fixed-parameter tractable, it is not clear that a compression routine with interesting running time exists.

3.6.2. Case Studies

The showcase for iterative compression is the Minimum Fragment Removal problem, also known as Graph Bipartization (**Fig. 21.8**). This problem appears in the context of SNP haplotyping *(47)*. When analyzing DNA fragments obtained by shotgun sequencing, it is initially unknown to which of the two chromosome copies of a diploid organism a fragment belongs. However, we can determine for some pairs of fragments that they cannot belong to the same

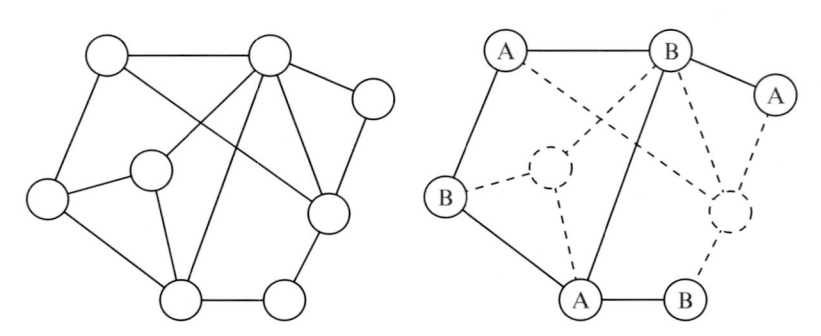

Fig. 21.8. A Minimum Fragment Removal instance *(left)*, and an optimal solution *(right)*: When deleting two fragments *(dashed)*, the remaining fragments can be allocated to the two chromosome copies **(A and B)** such that no conflicting fragments get the same assignment.

chromosome copy since they contain conflicting information at some SNP locus. Using this information, it is straightforward to reconstruct the chromosome assignment. We can model this as a graph problem, in which the fragments are the vertices and a conflict is represented as an edge. The task is then to color the vertices with two colors such that no vertices with the same color are connected by an edge.

The problem gets difficult in the presence of errors such as parasite DNA fragments, which randomly conflict with other fragments. In this scenario, we ask for the least number of fragments to remove such that we can get a consistent fragment assignment. Using the number of fragments k to be removed as a parameter is a natural approach, since the result is only meaningful for small k anyway.

Iterative compression provided the first fixed-parameter algorithm for Minimum Fragment Removal with this parameter *(46)*. We sketch how to apply this to finding an optimal solution (a *removal set*) for Minimum Fragment Removal: starting with an empty graph G' and an empty fragment removal set C', we add one vertex v at a time from the original graph to both G' (including edges to vertices already in G') and the removal set. Then C' is still a valid removal set. Although C' may not be optimal (it can be too large by 1), we can find an optimal removal set for G' by applying the compression routine to G' and C'. Since eventually $G' = G$, we obtain an optimal removal set for G.

The compression routine itself works by examining a number of vertex cuts in an auxiliary graph (i.e., a set of vertices whose deletion makes the graph disconnected), a task that can be accomplished in polynomial time by maximum flow techniques. We refer to the literature for details *(46, 48)*. The running time of the complete algorithm is $O(3^k \times mn)$.

3.6.3. Applications and Implementations

Nearly all of the currently known iterative compression algorithms solve *feedback set problems* in graphs; that is, problems in which one wishes to destroy certain cycles in a graph by deleting at most k vertices or edges *(49)* for a survey on feedback set problems). Probably the most prominent among them is Feedback Vertex Set, which also has applications for genetic linkage analysis *(50)*. However, the currently known iterative compression algorithms *(51, 52)* exhibit a rather bad combinatorial explosion; a randomized fixed-parameter algorithm from *(53)* so far appears to be better for practical applications.

Although thus no convincing practical results for iterative compression-based algorithms for Feedback Vertex Set are currently known, first experimental results for iterative compression-based algorithms for Minimum Fragment Removal are quite encouraging. An implementation, improved by heuristic speedups, can solve all problems from a testbed based on human genome data within minutes, whereas established methods are only able to solve about half of the instances within reasonable time *(48)*.

3.7. Conclusion

We surveyed several techniques for developing efficient fixed-parameter algorithms for computationally hard (biological) problems. A broader perspective is given in the recent monograph *(10)*. Since many biologically relevant computational problems appear to "carry small parameters," we firmly believe that there will continue to be a strong interaction between parameterized complexity analysis and algorithmic bioinformatics. Since the theory of fixed-parameter algorithmics is still very young (basically starting in the 1990s), the whole field is still vividly developing, offering many tools and challenges with potentially high impact, particularly in the field of computational (molecular) biology.

4. Notes

1. Parameter choice: Fixed-parameter algorithms are the better the smaller the parameter value is. Hence, try to find parameterizations with small parameter values.

2. A natural way to find significant problem parameterizations is to identify parameters that measure the distance from tractable (i.e., efficiently solvable) problem instances.

3. Avoid exponential-space algorithms when there are alternatives, since in practice this usually turns out to be more harmful than exponential time.

4. Always start by designing data reduction rules. They are helpful in combination with basically any algorithmic technique you might try later.

5. Use data reduction not only as a preprocessing step. In search tree algorithms, branching can produce new opportunities for data reduction. Therefore, apply data reductions repeatedly during the course of the whole algorithm.

6. In search tree algorithms, try to avoid complicated case distinctions when a simpler search strategy yields almost the same running time bounds.

7. Use high-level programming languages that allow for quicker implementation of ideas and are less error prone. For exponential-time algorithms, algorithmic improvements can often lead to a speed-up by several orders of magnitude, dwarfing the speedup of 2–10 gained by choice of programming language.

8. Try to augment search tree algorithms with admissible heuristic evaluation functions (*see* **Section 3.2.3**).

9. Many fixed-parameter algorithms can be easily adapted to run on parallel machines.

10. Do not be afraid of bad upper bounds for fixed-parameter algorithms—the analysis is worst-case and often much too pessimistic.

Acknowledgments

This work was supported by the Deutsche Forschungsgemeinschaft, Emmy Noether research group PIAF (fixed-parameter algorithms), NI 369/4 (Falk Hüffner), and the Deutsche Telekom Stiftung (Sebastian Wernicke).

References

1. Garey, M. R., Johnson, D. S. (1979) *Computers and Intractability: A Guide to the Theory of NP-Completeness*. Freeman, New York.

2. Papadimitriou, C. H. (1994) *Computational Complexity*. Addison-Wesley, Reading, Mass.

3. Cormen, T. H., Leiserson, C. E., Rivest, R. L., et al. (2001) *Introduction to Algorithms*, 2nd ed. MIT Press, Cambridge, MA.

4. Papadimitriou, C. H. (1997) NP-completeness: a retrospective. *Proc. 24th International Colloquium on Automata, Languages and Programming (ICALP'97)* Degano, P., Gorrieri, R., Marchetti-Spaccamela, A., (eds). Springer, New York.

5. Ausiello, G., Crescenzi, P., Gambosi, G., et al. (1999) *Complexity and Approximation: Combinatorial Optimization Problems and Their Approximability Properties*. Springer, New York.

6. Vazirani, V. V. (2001) *Approximation Algorithms*. Springer, New York.

7. Michalewicz, Z., Fogel, D. B. (2004) *How to Solve It: Modern Heuristics*, 2nd ed. Springer, New York.

8. Downey, R. G., Fellows, M. R. (1999) *Parameterized Complexity*. Springer, New York.

9. Flum, J., Grohe, M. (2006) *Parameterized Complexity Theory*. Springer, New York.

10. Niedermeier, R. (2006) *Invitation to Fixed-Parameter Algorithms*. Oxford University Press, New York.

11. Diestel, R. (2005) *Graph Theory*, 3rd ed. Springer, New York.

12. Bixby, R. E. (2002) Solving real-world linear programs: a decade and more of progress. *Operations Res* 50, 3–15.

13. Damaschke, P. (2006) Parameterized enumeration, transversals, and imperfect phylogeny reconstruction. *Theoret Comp Sci* 351, 337–350.

14. Abu-Khzam, F. N., Collins, R. L., Fellows, M. R., et al. (2004) Kernelization algorithms for the Vertex Cover problem: theory and experiments. *Proc. 6th Workshop on Algorithm Engineering and Experiments (ALENEX '04)*. SIAM, Bangkok, Thailand.

15. Chor, B., Fellows, M. R., Juedes, D. W. (2004) Linear kernels in linear time, or how to save k colors in $O(n^2)$ steps. *Proc. 30th International Workshop on Graph-Theoretic Concepts in Computer Science (WG'04)*. Springer, New York.

16. Chen, J., Kanj, I. A., Xia, G. (2006) Improved parameterized upper bounds for Vertex Cover. *Proc. 31st International Symposium on Mathematical Foundations of Computer Science (MFCS '06)*, volume 4162 of *LNCS*. Springer, New York.

17. Chesler, E. J., Lu, L., Shou, S., et al. (2005) Complex trait analysis of gene expression uncovers polygenic and pleiotropic networks that modulate nervous system function. *Nat Genet* 37, 233–242.

18. Cheetham, J., Dehne, F. K. H. A., Rau-Chaplin, A., et al. (2003) Solving large FPT problems on coarse-grained parallel machines. *J Comput Syst Sci* 67, 691–706.

19. Abu-Khzam, F. N., Langston, M. A., Shanbhag, P., et al. (2006) Scalable parallel algorithms for FPT problems. *Algorithmica* 45, 269–284.

20. Gramm, J., Guo, J., Hüffner, F., et al. (2006) Data reduction, exact, and heuristic algorithms for clique cover. *ACM J Exp Algorith.*

21. Dehne, F. K. H. A., Langston, M. A., Luo, X., et al. (2006) The cluster editing problem: Implementations and experiments. *Proc. 2nd International Workshop on Parameterized and Exact Computation (IWPEC'06).* Springer, New York.

22. Skiena, S. S. (1998) *The Algorithm Design Manual.* Springer-Verlag, New York.

23. Gramm, J., Guo, J., Hüffner, F., et al. (2004) Automated generation of search tree algorithms for hard graph modification problems. *Algorithmica* 39, 321–347.

24. Gramm, J., Niedermeier, R., Rossmanith, P. (2003) Fixed-parameter algorithms for closest string and related problems. *Algorithmica* 37, 25–42.

25. Stojanovic, N., Florea, L., Riemer, C., et al. (1999) Comparison of five methods for finding conserved sequences in multiple alignments of gene regulatory regions. *Nucl Acids Res* 27, 3899–3910.

26. Gramm, J. (2003) *Fixed-Parameter Algorithms for the Consensus Analysis of Genomic Sequences.* PhD thesis, Universität Tübingen, Germany.

27. Fellows, M. R., Gramm, J., Niedermeier, R. (2006) On the parameterized intractability of motif search problems. *Combinatorica* 26, 141–167.

28. Gramm, J., Guo, J., Niedermeier, R. (2006) Parameterized intractability of distinguishing substring selection. *Theory Comput Syst* 39, 545–560.

29. Marx, D. (2005) The closest substring problem with small distances. *Proc. of the 46th Annual IEEE Symposium on Foundations of Computer Science (FOCS '05).* Springer, New York.

30. Felner, A., Korf, R. E., Hanan, S. (2004) Additive pattern database heuristics. *J Artif Intell Res* 21, 1–39.

31. Gramm, J., Niedermeier, R. (2003) A fixed-parameter algorithm for minimum quartet inconsistency. *J Comput Syst Sci* 67, 723–741.

32. Wu, G., You, J.-H., Lin, G. (2005) A lookahead branch-and-bound algorithm for the maximum quartet consistency problem. *Proc. 5th Workshop on Algorithms in Bioinformatics (WABI '05).* Springer, New York.

33. Bodlaender, H. L. (1998) A partial k-arboretum of graphs with bounded treewidth. *Theoretical Computer Science* 209, 1–45.

34. Bodlaender, H. L. (2005) Discovering treewidth. *Proc. 31st Conference on Current Trends in Theory and Practice of Computer Science (SOFSEM'05).* Springer, New York.

35. Alber, J., Dorn, F., Niedermeier, R. (2005) Empirical evaluation of a tree decomposition based algorithm for Vertex Cover on planar graphs. *Discrete Appl Math* 145, 219–231.

36. Betzler, N., Niedermeier, R., Uhlmann, J. (2006) Tree decompositions of graphs: Saving memory in dynamic programming. *Discrete Opt* 3, 220–229.

37. Xu, J., Jiao, F., Berger, B. (2005) A tree-decomposition approach to protein structure prediction. *Proc. 4th International IEEE Computer Society Computational Systems Bioinformatics Conference (CSB 2005).* IEEE Computer Society, Los Alamitos, CA.

38. Song, Y., Liu, C., Malmberg, R. L., et al L. (2005) Tree decomposition based fast search of RNA structures including pseudoknots in genomes. *Proc. 4th International IEEE Computer Society Computational Systems Bioinformatics Conference (CSB 2005).* IEEE Computer Society, Los Alamitos, CA.

39. Jensen, F. V. (2001) *Bayesian Networks and Decision Graphs.* Springer, New York.

40. Bachoore, E., Bodlaender, H. L. (2006) *Weighted Treewidth: Algorithmic Techniques and Results.* Technical Report UU-CS-2006-013. University of Utrecht, Utrecht, The Ntherlands.

41. Alon, N., Yuster, R., Zwick, U. (1995) Color-coding. *JACM* 42, 844–856.

42. Hüffner, F., Wernicke, S., Zichner, T. (2007). Algorithm engineering for color-coding to facilitate signaling pathway detection, in *Advances in Bioinformatics and Computational Biology.* World Scientific, Hackensack, NJ, in press.

43. Scott, J., Ideker, T., Karp, R. M., et al. (2006) Efficient algorithms for detecting signaling pathways in protein interaction networks. *J Comput Biol* 13, 133–144.

44. Shlomi, T., Segal, D., Ruppin, E., et al. (2006) QPath: a method for querying

pathways in a protein–protein interaction network. *BMC Bioinformat* 7, 199.

45. Kneis, J., Mölle, D., Richter, S., et al. (2006) Divide-and-color. *Proc. 32nd International Workshop on Graph-Theoretic Concepts in Computer Science (WG '06)*, Springer, New York, in press.

46. Reed, B., Smith, K., Vetta, A. (2004) Finding odd cycle transversals. *Oper Res Letts* 32, 299–301.

47. Panconesi, A., Sozio, M. (2004) Fast hare: A fast heuristic for single individual SNP haplotype reconstruction. *Proc. 4th Workshop on Algorithms in Bioinformatics (WABI '04)*, volume 3240 of *LNCS*. Springer, New York.

48. Hüffner, F. (2005) Algorithm engineering for optimal graph bipartization. *Proc. 4th International Workshop on Efficient and Experimental Algorithms (WEA '05)*. Springer, New York.

49. Festa, P., Pardalos, P. M., Resende, M. G. C. (1999) Feedback set problems, in (Du, D. Z.,

Pardalos, P. M., eds.), *Handbook of Combinatorial Optimization, vol.* A. Kluwer, Dordrecht, The Netherlands.

50. Becker, A., Geiger, D., Schäffer, A. (1998) Automatic selection of loop breakers for genetic linkage analysis. *Hum Genet* 48, 49–60.

51. Dehne, F. K. H. A., Fellows, M. R., Langston, M. A., et al. (2005) An $O(2^{O(k)}n^3)$ FPT algorithm for the undirected feedback vertex set problem. *Proc. 11th International Computing and Combinatorics Conference (COCOON'05)*. Springer, New York.

52. Guo, J., Gramm, J., Hüffner, F., et al. (2005) Improved fixed-parameter algorithms for two feedback set problems. *Proc. 9th Workshop on Algorithms and Data Structures (WADS'05)*. Springer, New York.

53. Becker, A., Bar-Yehuda, R., Geiger, D. (2000) Randomized algorithms for the loop cutset problem. *J Artif Intell Res* 12, 219–234.

Chapter 22

Clustering

Geoffrey J. McLachlan, Richard W. Bean, and Shu-Kay Ng

Abstract

Clustering techniques are used to arrange genes in some natural way, that is, to organize genes into groups or clusters with similar behavior across relevant tissue samples (or cell lines). These techniques can also be applied to tissues rather than genes. Methods such as hierarchical agglomerative clustering, k-means clustering, the self-organizing map, and model-based methods have been used. This chapter focuses on mixtures of normals to provide a model-based clustering of tissue samples (gene signatures) and gene profiles.

Key words: Clustering of tissue samples, clustering of gene profiles, hierarchical agglomerative methods, partitional methods, k-means, model-based methods, normal mixture models, mixtures of factor analyzers, mixtures of linear mixed-effects models.

1. Introduction

DNA microarray technology, first described in the mid-1990s, is a method to perform experiments on thousands of gene fragments in parallel. Its widespread use has led to a huge growth in the amount of expression data available. A variety of multivariate analysis methods have been used to explore these data for relationships among the genes and tissue samples. Cluster analysis has been one of the most frequently used methods for these purposes. It has demonstrated its utility in the elucidation of unknown gene function, the validation of gene discoveries, and the interpretation of biological processes; see (1–2) for examples.

The main goal of microarray analysis of many diseases, in particular of unclassified cancer, is to identify as yet unclassified cancer subtypes for subsequent validation and prediction, and ultimately

Jonathan M. Keith (ed.), *Bioinformatics, Volume II: Structure, Function and Applications, vol. 453*
© 2008 Humana Press, a part of Springer Science + Business Media, Totowa, NJ
Book doi: 10.1007/978-1-60327-429-6 Springerprotocols.com

to develop individualized prognosis and therapy. Limiting factors include the difficulties of tissue acquisition and expense of microarray experiments. Thus, often microarray studies attempt to perform a cluster analysis of a small number of tumor samples on the basis of a large number of genes, and can result in gene-to-sample ratios of approximately 100-fold.

Many researchers have explored the use of clustering techniques to arrange genes in some natural order; that is, to organize genes into clusters with similar behavior across relevant tissue samples (or cell lines). Although a cluster does not automatically correspond to a pathway, it is a reasonable approximation that genes in the same cluster have something to do with each other or are directly involved in the same pathway.

It can be seen there are two distinct but related clustering problems with microarray data. One problem concerns the clustering of the tissues on the basis of the genes; the other concerns the clustering of the genes on the basis of the tissues. This duality in cluster analysis is quite common. In the present context of microarray data, one may be interested in grouping tissues (patients) with similar expression values or in grouping genes on patients with similar types of tumors or similar survival rates.

One of the difficulties of clustering is that the notion of a cluster is vague. A useful way to think about the different clustering procedures is in terms of the shape of the clusters produced *(3)*. The majority of the existing clustering methods assume that a similarity measure or metric is known *a priori*; often the Euclidean metric is used. However, clearly, it would be more appropriate to use a metric that depends on the shape of the clusters. As pointed out by *(4)*, the difficulty is that the shape of the clusters is not known until the clusters have been found, and the clusters cannot be effectively identified unless the shapes are known.

Before we proceed to consider the clustering of microarray data, we give a brief account of clustering in a general context. For a more detailed account of cluster analysis, the reader is referred to the many books that either consider or are devoted exclusively to this topic; for example *(5–9)* and *(10)*. A recent review article on clustering is *(11)*.

1.1. Brief Review of Some Clustering Methods

Cluster analysis is concerned with grouping a number (n) of entities into a smaller number (g) of groups on the basis of observations measured on some variables associated with each entity. We let $y_j = (y_{1j}, \ldots, y_{pj})^T$ be the observation or feature vector containing the values of p measurements $y_{1j}, \ldots, y_{pj}$ made on the jth entity ($j = 1, \ldots, n$) to be clustered. These data can be organized as a matrix:

$$\Upsilon_{p \times n} = ((y_{vj})); \qquad [1]$$

that is, the jth column of $\Upsilon_{p \times n}$ is the obervation vector y_j.

In discriminant analysis (supervised learning), the data are classified with respect to g known classes and the intent is to form a classifier or prediction rule on the basis of these classified data for assigning an unclassified entity to one of the g classes on the basis of its feature vector. In contrast to discriminant analysis, in cluster analysis (unsupervised learning), there is no prior information on the group structure of the data or, in the case in which it is known that the population consists of a number of classes, there are no data of known origin with respect to the classes. The clustering problem falls into two main categories, which overlap to some extent (12):

1. What is the best way of dividing the entities into a given number of groups, where there is no implication that the resulting groups are in any sense a natural division of the data? This is sometimes called dissection or segmentation.

2. What is the best way to find a natural subdivision of the entities into groups? Here by natural clusters, it is meant that the clusters can be described as continuous regions of the feature space containing a relatively high density of points, separated from other such regions by regions containing a relatively low density of points (5). It is therefore intended that natural clusters possess the two intuitive qualities of internal cohesion and external isolation (13).

Sometimes the distinction between the search for naturally occurring clusters as in 2 and other groupings as in 1 is stressed; see, for example (14). However, often it is not made, particularly as most methods for finding natural clusters are also useful for segmenting the data. Essentially, all methods of cluster analysis attempt to imitate what the eye and brain do so well in $p = 2$ dimensions. For example, in the scatter plot (**Fig. 22.1**) of the expression values of two smooth muscle related genes on ten tumors and ten normal tissues from the colon cancer data of (15), it is very easy to detect the presence of two clusters of equal size without making the meaning of the term cluster explicit.

Clustering methods can be categorized broadly as being hierarchical or nonhierarchical. With a method in the former category, every cluster obtained at any stage is a merger or split of clusters obtained at the previous stage. Hierarchical methods can be implemented in a so-called agglomerative manner (bottom-up), starting with $g = n$ clusters or in a divisive manner (top-down), starting with the n entities to be clustered as a single cluster. In practice, divisive methods can be computationally prohibitive unless the sample size n is very small. For instance, there are $2^{(n-1)}-1$ ways of making the first subdivision. Hence hierarchical methods are usually implemented in an agglomerative manner, as to be discussed further in the next section. In (16), a hybrid clustering method was proposed that combines

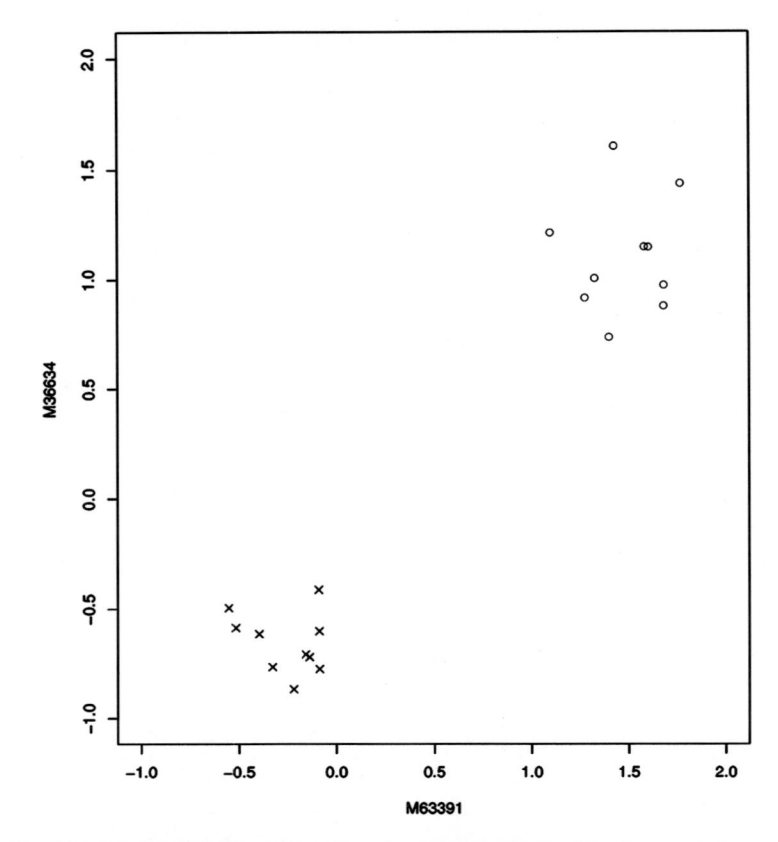

Fig. 22.1. Scatter plot of the expression values of two genes on 10 colon cancer tumors (x) and 10 normal tissues (o).

the strengths of bottom-up hierarchical clustering with that of top-down clustering. The first method is good at identifying small clusters, but not large ones; the strengths are reversed for top-down clustering.

One of the most popular nonhierarchical methods of clustering is k-means, where k refers to the number of clusters to be imposed on the data. It seeks to find $k = g$ clusters that minimize the sum of the squared Euclidean distances between each observation y_j and its respective cluster mean; that is, it seeks to minimize the trace of W, tr W, where:

$$W = \sum_{i=1}^{g} \sum_{j=1}^{n} z_{ij}(y_j - \bar{y}_i)(y_j - \bar{y}_i)^T \qquad [2]$$

is the pooled within-cluster sums of squares and products matrix, and:

$$\overline{y}_i = \sum_{j=1}^{n} y_j \Big/ \sum_{j=1}^{n} z_{ij} \qquad\qquad [3]$$

is the sample mean of the ith cluster. Here z_{ij} is a zero-one indicator variable that is one or zero, according as y_j belongs or does not belong to the ith cluster ($i = 1, \ldots, g$; $j = 1, \ldots, n$). It is impossible to consider all partitions of the n observations into g clusters unless n is very small, since the number of such partitions with non-empty clusters is the Stirling number of the second kind:

$$\frac{1}{n!} \sum_{j=1}^{n} (-1)^{(n-1)} \binom{n}{i} n^g, \qquad\qquad [4]$$

which can be approximated by $g^n / g!$; see (8). In practice, k-means is therefore implemented by iteratively moving points between clusters so as to minimize tr W. In its simplest form, each observation y_j is assigned to the cluster with the nearest centre (sample mean) and then the center of the cluster is updated before moving on to the next observation. Often the centers are estimated initially by selecting k points at random from the sample to be clustered.

Other partitioning methods have been developed, including k-methods (8), which is similar to k-means, but constrains each cluster center to be one of the observations y_j. The self-organizing map (17) is similar to k-means, but the cluster centers are constrained to lie on a (two-dimensional) lattice. It is well known that k-means tends to lead to spherical clusters since it is predicated on normal clusters with (equal) spherical covariance matrices. One way to achieve elliptical clusters is to seek clusters that minimize the determinant of W, $|W|$, rather than its trace, as in (18); see also (19) who derived this criterion under certain assumptions of normality for the clusters.

In the absence of any prior knowledge of the metric, it is reasonable to adopt a clustering procedure that is invariant under affine transformations of the data; that is, invariant under transformations of the data of the form:

$$y \rightarrow Cy + a \qquad\qquad [5]$$

where C is a nonsingular matrix. If the clustering of a procedure is invariant under [5] for only diagonal C, then it is invariant under change of measuring units but not rotations. However, as commented upon in (20), this form of invariance is more compelling than affine invariance. The clustering produced by minimization of $|W|$ is affine invariant.

In the statistical and pattern recognition literature in recent times, attention has been focused on model-based clustering via mixtures of normal densities. With this approach, each observation vector y_j is assumed to have a g-component normal mixture density:

$$f(y_j; \Psi) = \sum_{i=1}^{g} \pi_i \phi(y_j; \mu_i, \Sigma_i), \qquad [6]$$

where $\phi(y; \mu_i, \Sigma_i)$ denotes the p-variate normal density function with mean μ_i and covariance matrix Σ_i, and the π_i denote the mixing proportions, which are non-negative and sum to one. Here the vector Ψ of unknown parameters consists of the mixing proportions π_i, the elements of the component means μ_i, and the distinct elements of the component-covariance matrix Σ_i, and it can be estimated by its maximum likelihood estimate calculated via the EM algorithm (*see (21–22)*). This approach gives a probabilistic clustering defined in terms of the estimated posterior probabilities of component membership $\tau_i(y_j; \hat{\Psi})$, where $\tau_i(y_j; \Psi)$ denotes the posterior probability that the jth feature vector with observed value y_j belongs to the ith component of the mixture ($i = 1, …, g$; $j = 1, …, n$). Using Bayes' theorem, it can be expressed as:

$$\tau_i(y_j; \Psi) = \frac{\pi_i \phi(y_j; \mu_i, \Sigma_i)}{\sum_{h=1}^{g} \pi_h \phi(y_j; \mu_h, \Sigma_h)}. \qquad [7]$$

It can be seen that with this approach, we can have a "soft" clustering, whereby each observation may partly belong to more than one cluster. An outright clustering can be obtained by assigning y_j to the component to which it has the greatest estimated posterior probability of belonging. The number of components g in the normal mixture model Equation 6 has to be specified in advance (*see* **Note 1**).

As noted in *(23)*, "Clustering methods based on such mixture models allow estimation and hypothesis testing within the framework of standard statistical theory." Previously, Marriott *(12)* (page 70) had noted that the mixture likelihood-based approach "is about the only clustering technique that is entirely satisfactory from the mathematical point of view. It assumes a well-defined mathematical model, investigates it by well-established statistical techniques, and provides a test of significance for the results." One potential drawback with this approach is that normality is assumed for the cluster distributions. However, this assumption would appear to be reasonable for the clustering of microarray data after appropriate normalization.

One attractive feature of adopting mixture models with elliptically symmetric components such as the normal or its more robust version in the form of the t density *(22)* is that the implied clustering is invariant under affine transformations in Equation 5. Also, in the case in which the components of the mixture correspond to externally defined sub-populations, the unknown parameter vector Ψ can be estimated consistently by a sequence of roots of the likelihood equation. Note that this is not the case if a criterion such as minimizing $|W|$ is used.

In the preceding, we have focused exclusively on methods that are applicable for the clustering of the observations and variables considered separately; that is, in the context of clustering microarray data, methods that would be suitable for clustering the tissue samples and genes considered separately rather than simultaneously. Pollard and van der Laan *(24)* proposed a statistical framework for two-way clustering; *see* also *(25)* and the references therein for earlier approaches to this problem. More recently, *(26)* reported some results on two-way clustering (bi-clustering) of tissues and genes. In their work, they obtained similar results to those obtained when the tissues and the genes were clustered separately.

2. Methods

Although biological experiments vary considerably in their design, the data generated by microarray experiments can be viewed as a matrix of expression levels. For M microarray experiments (corresponding to M tissue samples), where we measure the expression levels of N genes in each experiment, the results can be represented by an $N \times M$ matrix. For each tissue, we can consider the expression levels of the N genes, called its *expression signature*. Conversely, for each gene, we can consider its expression levels across the different tissue samples, called its *expression profile*. The M tissue samples might correspond to each of M different patients or, say, to samples from a single patient taken at M different time points. The $N \times M$ matrix is portrayed in **Fig. 22.2**, where each sample represents a separate microarray experiment and generates a set of N expression levels, one for each gene.

Against the preceding background of clustering methods in a general context as given in the previous section, we now consider their application to microarray data, concentrating on a model-based approach using normal mixtures. However, first we consider the application of hierarchical agglomerative methods, given their extensive use for this purpose in bioinformatics.

2.1. Clustering of Tissues: Hierarchical Methods

For the clustering of the tissue samples, the microarray data portrayed in **Fig. 22.2** are in the form of the matrix in Equation 1 with $n = M$ and $p = N$, and the observation vector y_j corresponds to the expression signature for the jth tissue sample. In statistics, it is usual to refer to the entirety of the tissue samples as the sample, whereas the biologists tend to refer to each individual expression signature as a sample, and we follow this practice here.

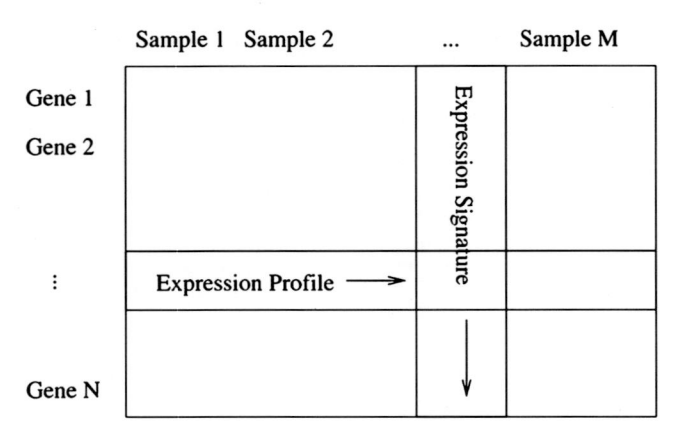

Fig. 22.2. Gene expression data from *M* microarray experiments represented as a matrix of expression levels with the *N* rows corresponding to the *N* genes and the *M* columns to the *M* tissue samples.

The commonly used hierarchical agglomerative methods can be applied directly to this matrix to cluster the tissue samples, since they can be implemented by consideration of the matrix of proximities, or equivalently, the distances, between each pair of observations. Thus, they require only $O(n^2)$ or at worst $O(n^3)$ calculations, where $n = M$ and the number M of tissue samples is limited usually to <100. The situation would be different with the clustering of the genes, as then $n = N$ and the number N of genes could be in the tens of thousands.

In order to compute the pairwise distances between the observations, one needs to select an appropriate distance metric. Metrics that are used include Euclidean distance and the Pearson correlation coefficient, although the latter is equivalent to the former if the observations have been normalized beforehand to have zero means and unit variances. Having selected a distance measure for the observations, there is a need to specify a linkage metric between clusters. Some commonly used metrics include single linkage, complete linkage, average linkage, and centroid linkage. With single linkage, the distance between two clusters is defined by the distance between the two nearest observations (one from each cluster), whereas with complete linkage, the cluster distance is defined in terms of the distance between the two most distant observations (one from each cluster). Average linkage is defined in terms of the average of the $n_1 n_2$ distances between all possible pairs of observations (one from each cluster), where n_1 and n_2 denote the number of observations in the two clusters in question. For centroid linkage, the distance between two clusters is the distance between the cluster centroids (sample means). Another commonly used method is Ward's procedure *(27)*, which joins clusters so as to minimize the within-cluster variance (the trace of *W*). Lance and Williams *(28)* have presented a simple linear system of equations as a unifying framework for these different linkage measures. Eisen et al. *(2)* were the first to apply cluster analysis to microarray data, using average linkage with a correlation-based metric. The nested clusters produced by a hierarchical method of clustering can be portrayed in a tree diagram, in which the extremities (usually shown at the bottom) represent the individual observations, and the branching of the tree gives the order of joining together. The height at which clusters of points are joined corresponds to the distance between the clusters. However, it is not clear in general how to choose the number of clusters.

To illustrate hierarchical agglomerative clustering, we use nested polygons in **Fig. 22.3** to show the clusters obtained by applying it to six bivariate points, using single-linkage with Euclidean distance as the distance measure. It can be seen that the cluster of Observations 5 and 6 is considerably closer to the cluster of 1 and 2 than Observation 4 is.

There is no reason why the clusters should be hierarchical for microarray data. It is true that if there is a clear, unequivocal grouping, with little or no overlap between the groups, any method will reach this grouping. However, as pointed out by *(12)*,

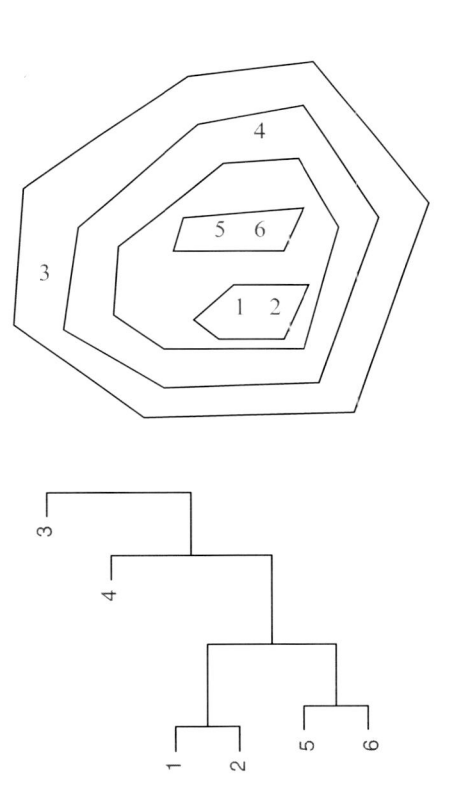

Fig. 22.3. An illustrative example of hierarchical agglomerative clustering.

"hierarchical methods are not primarily adapted to finding groups." For instance, if the division into $g = 2$ groups given by some hierarchical method is optimum with respect to some criterion, then the subsequent division into $g = 3$ groups is unlikely to be so. This is due to the restriction that one of the groups must be the same in both the $g = 2$ and $g = 3$ clusterings. As explained by *(12)*, this restriction is not a natural one to impose if the purpose is to find a natural grouping of the data. In the sequel, we therefore focus on nonhierarchical methods of clustering. As advocated by *(12)* (page 67), "it is better to consider the clustering problem *ab initio*, without imposing any conditions."

2.2. Clustering of Tissues: Normal Mixtures

More recently, increasing attention is being given to model-based methods of clustering of microarray data *(29–32)*, among others. However, the normal mixture model [6] cannot be directly fitted to the tissue samples if the number of genes p used in the expression signature is large. This is because the component-covariance matrices Σ_i are highly parameterized with ½ $p(p + 1)$ distinct elements each. A simple way of proceeding in the clustering of high-dimensional data would be to take the component-covariance matrices Σ_i to be diagonal. However, this leads to clusters whose axes are aligned with those of the feature space, whereas in practice the clusters are of arbitrary orientation. For instance, taking the Σ_i to be a common multiple of the identity matrix leads to a soft-version of k-means, which produces spherical clusters.

Banfield and Raftery *(33)* introduced a parameterization of the component-covariance matrix Σ_i based on a variant of the standard spectral decomposition of $\Sigma_i (i = 1, \ldots g)$. However, if p is large relative to the sample size n, it may not be possible to use this decomposition to infer an appropriate model for the component-covariance matrices. Even if it were possible, the results may not be reliable due to potential problems with near-singular estimates of the component-covariance matrices when p is large relative to n.

Hence, in fitting normal mixture models with unrestricted component-covariance matrices to high-dimensional data, we need to consider first some form of dimension reduction and/or some form of regularization. A common approach to reducing the number of dimensions is to perform a principal component analysis (PCA). However, the latter provides only a global linear model for the representation of the data in a lower-dimensional subspace. Thus, it has limited scope in revealing group structure in a data set. A global nonlinear approach can be obtained by postulating a finite mixture of linear (factor) sub-models for the distribution of the full observation vector y_j given a relatively small number of (unobservable) factors. That is, we can provide a local dimensionality reduction method by a mixture of factor analyzers model, which is given by Equation 6 by imposing on the component-covariance matrix Σ_i, the constraint:

$$\Sigma_i = B_i B_i^T + D_i \qquad (i = 1,...,g), \qquad [8]$$

where B_i is a $p \times q$ matrix of factor loadings and D_i is a diagonal matrix $(i = 1, ..., g)$. We can think of the use of this mixture of factor analyzers model as being purely a method of regularization. However, in the present context, it might be possible to make a case for it being a reasonable model for the correlation structure between the genes. This model implies that the latter can be explained by the linear dependence of the genes on a small number of latent (unobservable variables) specific to each component.

The EMMIX-GENE program of *(31)* has been designed for the clustering of tissue samples via mixtures of factor analyzers. In practice we may wish to work with a subset of the available genes, particularly as the fitting of a mixture of factor analyzers will involve a considerable amount of computation time for an extremely large number of genes. Indeed, the simultaneous use of too many genes in the cluster analysis may serve only to create noise that masks the effect of a smaller number of genes. Also, the intent of the cluster analysis may not be to produce a clustering of the tissues on the basis of all the available genes, but rather to discover and study different clusterings of the tissues corresponding to different subsets of the genes *(24, 34)*. As explained in *(35)*, the tissues (cell lines or biological samples) may cluster according to cell or tissue type (e.g., cancerous or healthy) or according to cancer type (e.g., breast cancer or melanoma). However, the same samples may cluster differently according to other cellular characteristics, such as progression through the cell cycle, drug metabolism, mutation, growth rate, or interferon response, all of which have a genetic basis.

Therefore, the EMMIX-GENE procedure has two optional steps before the final step of clustering the tissues. The first step considers the selection of a subset of relevant genes from the available set of genes by screening the genes on an individual basis to eliminate those that are of little use in clustering the tissue samples. The usefulness of a given gene to the clustering process can be assessed formally by a test of the null hypothesis that it has a single component normal distribution over the tissue samples (*see* **Note 2**). Even after this step has been completed, there may still be too many genes remaining. Thus, there is a second step in EMMIX-GENE in which the retained gene profiles are clustered (after standardization) into a number of groups on the basis of Euclidean distance so that genes with similar profiles are put into the same group. In general, care has to be taken with the scaling of variables before clustering of the observations, as the nature of the variables can be intrinsically different. In the present context the variables (gene expressions) are measured on the same scale. Also, as noted, the clustering of the observations (tissues) via normal mixture models is invariant under changes in scale and location. The clustering of the tissue samples can be carried out on the basis of the groups considered individually (using some or all of the genes within a group) or collectively. For the latter, we can replace each group by a representative (a metagene) such as the sample mean as in the EMMIX-GENE procedure.

To illustrate this approach, we applied the EMMIX-GENE procedure to the colon cancer data of *(15)*. It consists of $n = 2000$ genes and $p = 62$ columns denoting 40 tumors and 22 normal tissues. After applying the selection step to this set, there were 446 genes remaining in the set. The remaining genes were then clustered into 20 groups, which were ranked on the basis of $-2\log\lambda$, where λ is the likelihood ratio statistic for testing $g = 1$ vs. $g = 2$ components in the mixture model. The heat map of the second ranked group G_2 is shown in **Fig. 22.4**. The clustering of the tissues on the basis of the 24 genes in G_2 resulted in a partition of the tissues in which one cluster contains 37 tumors *(1–29, 31–32, 34–35, 37–40)* and three normals (48, 58, 60), and the other cluster contains three tumors (30, 33, 36) and 19 normals *(41–47, 49–57, 59, 61–62)*. This corresponds to an error rate of six out of 62 tissues compared to the "true" classification given in *(15)*. (This is why here we examine the heat map of G_2 instead of G_1.) For further details about the results of the tissue clustering procedure on this dataset, *see (31)*.

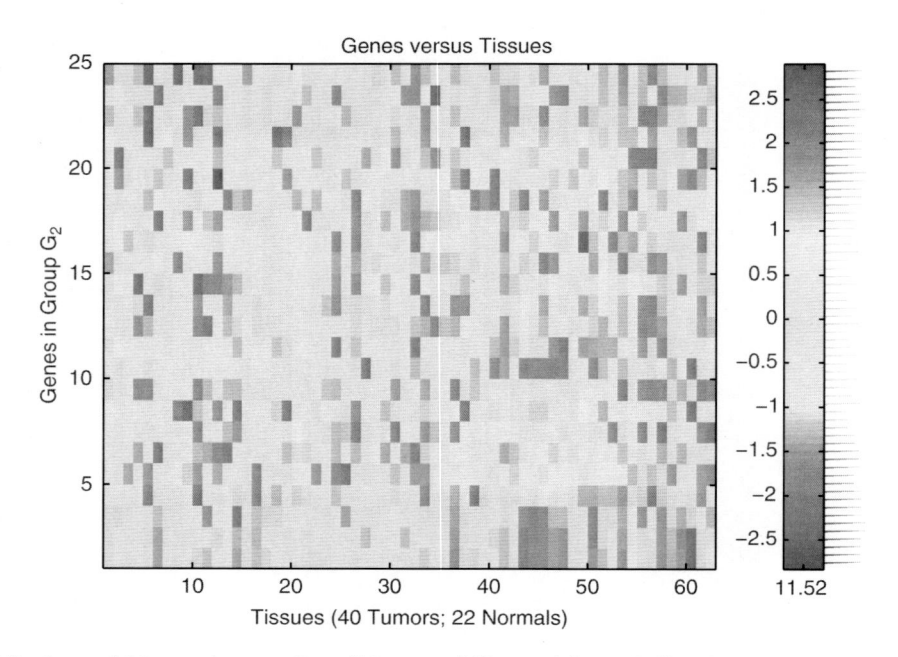

Fig. 22.4. Heat map of 24 genes in group G_2 on 40 tumor and 22 normal tissues in Alon data.

**2.3. Clustering
of Gene Profiles**

In order to cluster gene profiles, it might seem possible just to interchange rows and columns in the data matrix in Equation 1. However, with most applications of cluster analysis in practice it is assumed that:

1. There are no replications on any particular entity specifically identified as such.

2. All the observations on the entities are independent of one another.

These assumptions should hold for the clustering of the tissue samples, although the tissue samples have been known to be correlated for different tissues due to flawed experimental conditions. However, condition *(2)* will not hold for the clustering of gene profiles, since not all the genes are independently distributed, and condition *(1)* will generally not hold either as the gene profiles may be measured over time or on technical replicates. Although this correlated structure can be incorporated into the normal mixture model in Equation 6 by appropriate specification of the component-covariance matrices Σ_i, it is difficult to fit the model under such specifications. For example, the M-step (the maximization step of the EM algorithm) may not exist in closed form. Accordingly, we now consider the EMMIX-WIRE model of Ng et al. *(36)*, who adopt conditionally a mixture of linear mixed models to specify this correlation structure among the tissue samples and allow for correlations among the genes. It also enables covariate information to be incorporated into the clustering process.

For a gene microarray experiment with repeated measurements, we have for the *j*th gene ($j = 1,\ldots,n$), when $n = N$, a feature vector (profile vector) $y_j = (y_{1j}^{\ T}, \ldots, y_{tj}^{\ T})^T$, where *t* is the number of distinct tissues in the experiment and:

$$y_{lj} = (y_{l1j},\ldots,y_{lrj})^T \qquad (l = 1,\ldots,t)$$

contains the *r* replications on the *j*th gene from the *l*th tissue. Note that here, the *r* replications can also be time points. The dimension *p* of the profile vector y_j is equal to the number of microarray experiments, $M = rt$. Conditional on its membership of the *i*th component of the mixture, the EMMIX-WIRE procedure assumes that y_j follows a linear mixed-effects model (LMM):

$$y_j = X\beta_i + Ub_{ij} + Vc_i + \varepsilon_{ij}, \tag{9}$$

where the elements of β_i (a *t*-dimensional vector) are fixed effects (unknown constants) ($i = 1,\ldots,g$). In [9], b_{ij} (a q_b-dimensional vector) and c_i (a q_c-dimensional vector) represent the unobservable gene- and tissue-specific random effects, respectively, conditional on membership of the *i*th cluster. These random effects represent the variation due to the heterogeneity of genes and tissues (corresponding to $b_i = (b_{i1}^{\ T}, \ldots, b_{in}^{\ T})^T$ and c_i, respectively).

The random effects b_{ij} and c_i, and the measurement error vector ε_{ij} are assumed to be mutually independent. In Equation 9, X, U, and V are known design matrices of the corresponding fixed or random effects. The dimensions q_b and q_c of the random effects terms b_{ij} and c_i are determined by the design matrices U and V which, along with X and H, specify the experimental design to be adopted.

With the LMM, the distributions of b_{ij} and c_i are taken, respectively, to be multivariate normal $N_{q_b}(0, q_b I_{q_b})$ and $N_{q_c}(0, q_c I_{q_c})$, where I_{q_b} and I_{q_c} are identity matrices with dimensions being specified by the subscripts. The measurement error vector ε_{ij} is also taken to be multivariate normal $N_p(0, A_i)$, where $A_i = \mathrm{diag}(H\varphi_i)$ is a diagonal matrix constructed from the vector $(H\varphi_i)$ with $\varphi_i = (\sigma_{i1}^2, \ldots, \sigma_{iq_c}^2)^T$ and H is a known $p \times q_e$ zero-one design matrix. That is, we allow the ith component-variance to be different among the p microarray experiments.

The vector ψ of unknown parameters can be obtained by maximum likelihood via the EM algorithm, proceeding conditionally on the tissue-specific random effects c_i. The E- and M-steps can be implemented in closed form. In particular, an approximation to the E-step by carrying out time-consuming Monte Carlo methods is not required. A probabilistic or outright clustering of the genes into g components can be obtained, based on the estimated posterior probabilities of component membership given the profile vectors and the estimated tissue-specific random effects $\hat{c}_i (i = 1,\ldots,g)$.

To illustrate this method, we report here an example from (36) who used it to cluster some time course data from the yeast cell-cycle study of (37). The data consist of the expression levels of 612 genes for the yeast cells at $p = 18$ time points. With reference to [9], the design matrix was taken be an 18×2 matrix with the $(l + 1)$th row ($l = 0,\ldots,17$):

$$(\cos(2\pi(7l)/\omega + \Phi) \quad \sin(2\pi(7l)/\omega + \Phi)), \qquad [10]$$

where the period of the cell cycle ω was taken to be 53 and the phase offset Φ was set to zero. The design matrices for the random effects parts were specified as $U = 1_{18}$ and $V = I_{18}$. That is, it is assumed that there exist random gene effects b_{ij} with $q_b = 1$ and random temporal effects $c_i = (c_{i1}, \ldots c_{iq_c})^T$ with $q_c = p = 18$. The latter introduce dependence among expression levels within the same cluster obtained at the same time point. Also, $H = 1_{18}$ and $\phi_i = \sigma_i^2$ ($q_c = 1$) so that the component variances are common among the $p = 18$ experiments.

The number of components g was determined using the Bayesian information criterion (BIC) for model selection. It indicated here that there are twelve clusters. The clustering results for $g = 12$ are given in **Fig. 22.5**, where the expression profiles for genes in each cluster are presented.

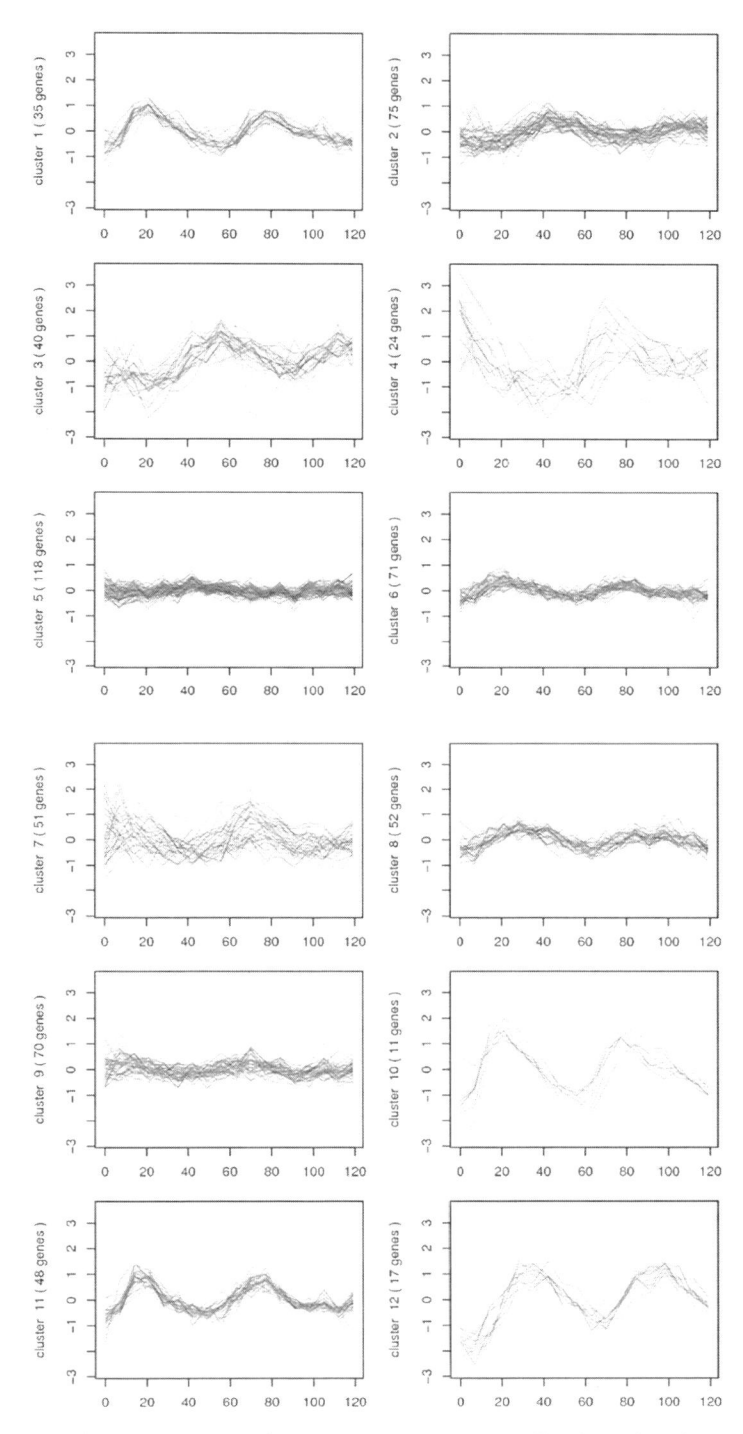

Fig. 22.5. Clustering results for Spellman yeast cell cycle data. For all the plots, the x-axis is the time point and the y-axis is the gene-expression level.

3. Notes

1. For both procedures, as with other partitional clustering methods, the number of clusters g needs to be specified at the outset. As both procedures are model-based, we can make a choice as to an appropriate value of g by consideration of the likelihood function. In the absence of any prior information as to the number of clusters present in the data, we monitor the increase in the log likelihood function as the value of g increases. At any stage, the choice of $g = g_o$ versus $g = g_1$, for instance $g_1 = g_o + 1$, can be made by either performing the likelihood ratio test or by using some information-based criterion, such as BIC (Bayesian information criterion). Unfortunately, regularity conditions do not hold for the likelihood ratio test statistic λ to have its usual null distribution of chi-squared with degrees of freedom equal to the difference d in the number of parameters for $g = g_1$ and $g = g_o$ components in the mixture models. One way to proceed is to use a resampling approach as in *(38)*. Alternatively, one can apply BIC, which leads to the selection of $g = g_1$ over $g = g_o$ if $-2\log \lambda$ is greater than $d \log(n)$. The value of d is obvious in applications of EMMIX-GENE, but is not so clear with applications of EMMIX-WIRE, due to the presence of random effects terms *(36)*.

2. The most time-consuming step of the three steps is the gene selection step of EMMIX-GENE. This step is slower than the others as a mixture of two normals is fitted for each gene, instead of a multivariate normal being fitted to a group of genes or a metagene simultaneously. A faster but ad hoc selection step is to make the decision for each gene on the basis of the interquartile range of the gene expression values over the tissues.

References

1. Alizadeh, A., Eisen, M. B., Davis, R. E., et al. (2000) Distinct types of diffuse large B-cell lymphoma identified by gene expression profiling. *Nature* 403, 503–511.

2. Eisen, M. B., Spellman, P. T., Brown, P. O., et al. (1998) Cluster analysis and display of genome-wide expression patterns. *Proc Natl Acad Sci USA* 95, 14863–14868.

3. Reilly, C., Wang, C., Rutherford, R. (2005) A rapid method for the comparison of cluster analyses. *Stat Sin* 15, 19–33.

4. Coleman, D., Dong, X. P., Hardin, J., et al. (1999) Some computational issues in cluster analysis with no a priori metric. *Comput Statist Data Anal* 31, 1–11.

5. Everitt, B. S. (1993) *Cluster Analysis*, 3rd ed. Edward Arnold, London.

6. Hartigan, J. A. (1975a) *Clustering Algorithms*. Wiley, New York.

7. Hastie, T., Tibshirani, R. J., Friedman, J. H. (2001) *The Elements of Statistical Learning*. Springer-Verlag, New York.

8. Kaufman, L., Rousseeuw, P. (1990) *Finding Groups in Data: An Introduction to Cluster Analysis*. Wiley, New York.

9. Ripley, B. D. (1996) *Pattern Recognition and Neural Networks*. Cambridge University Press, Cambridge, UK.

10. Seber, G. A. F. (1984) *Multivariate Observations*. Wiley, New York.

11. Kettenring, J. R. (2006) The practice of cluster analysis. *J. Classification* 23, 3–30.

12. Marriott, F. H. C. (1974) *The Interpretation of Multiple Observations*. Academic Press, London.

13. Cormack, R. M. (1971) A review of classification (with discussion). *J R Statist Soc A* 134, 321–367.

14. Hand, D. J., Heard, N. A. (2005) Finding groups in gene expression data. *J Biomed Biotech* 2005, 215–225.

15. Alon, U., Barkai, N., Notterman, D. A., et al. (1999) Broad patterns of gene expression revealed by clustering analysis of tumor and normal colon tissues probed by oligonucleotide arrays. *Proc Natl Acad Sci USA* 96, 6745–6750.

16. Chipman, H., Tibshirani, R. (2006) Hybrid hierarchical clustering with applications to microarray data. *Biostatistics* 7, 286–301.

17. Kohonen, T. (1989) *Self-organization and Associative Memory*, 3rd ed. Springer-Verlag, Berlin.

18. Friedman, H. P., Rubin, J. (1967) On some invariant criteria for grouping data. *J Amer Statist Assoc* 62, 1159–1178.

19. Scott, A. J., Symons, M. J. (1971) Clustering methods based on likelihood ratio criteria. *Biometrics* 27, 387–397.

20. Hartigan, J. A. (1975b) Statistical theory in clustering. *J. Classification* 2, 63–76.

21. McLachlan, G. J., Basford, K. E. (1988) *Mixture Models: Inference and Applications to Clustering*. Marcel Dekker, New York.

22. McLachlan, G. J., Peel, D. (2000) *Finite Mixture Models*. Wiley, New York.

23. Aitkin, M., Anderson, D., Hinde, J. (1981) Statistical modelling of data on teaching styles (with discussion). *J Roy Stat Soc A* 144, 419–461.

24. Pollard, K. S., van der Laan, M. J. (2002) Statistical inference for simultaneous clustering of gene expression data. *Math Biosci* 176, 99–121.

25. Getz, G., Levine, E., Domany, E. (2000) Coupled two-way clustering analysis of gene microarray data. *Cell Biol* 97, 12079–12084.

26. Ambroise, C., Govaert, G. (2006) Model based hierarchical clustering. Unpublished manuscript.

27. Ward, J. H. (1963). Hierarchical grouping to optimize an objective function. *J Am Stat Assoc* 58, 236–244.

28. Lance, G. N., Williams, W. T. (1967) A generalized theory of classificatory sorting strategies: I. Hierarchical systems. *Comp J* 9, 373–380.

29. Ghosh, D., Chinnaiyan, A. M. (2002) Mixture modelling of gene expression data from microarray experiments. *Bioinformatics* 18, 275–286.

30. Yeung, K. Y., Fraley, C., Murua, A., et al. (2001) Model-based clustering and data transformations for gene expression data. *Bioinformatics* 17, 977–987.

31. McLachlan, G. J., Bean, R. W., Peel, D. (2002) A mixture model-based approach to the clustering of microarray expression data. *Bioinformatics* 18, 413–422.

32. Medvedovic, M., Sivaganesan, S. (2002) Bayesian infinite mixture model based clustering of gene expression profiles. *Bioinformatics* 18, 1194–1206.

33. Banfield, J. D., Raftery, A. E. (1993) Model-based Gaussian and non-Gaussian clustering. *Biometrics* 49, 803–821.

34. Friedman, J. H., Meulman, J. J. (2004) Clustering objects on subsets of attributes (with discussion). *J Roy Stat Soc B* 66, 815–849.

35. Belitskaya-Levy, I. (2006) A generalized clustering problem, with application to DNA microarrays. *Statist Appl Genetics Mol Biol* 5, Article 2.

36. Ng, S. K., McLachlan G. J., Wang, K., et al. (2006) A mixture model with random-effects components for clustering correlated gene-expression profiles. *Bioinformatics* 22, 1745–1752.

37. Spellman, P., Sherlock, G., Zhang, M. Q., et al. (1998) Comprehensive identification of cell cycle-regulated genes of the yeast Saccharomyces cerevisiae by microarray hybridization. *Mol Biol Cell* 9, 3273–3297.

38. McLachlan, G. J. (1987) On bootstrapping the likelihood ratio test statistic for the number of components in a normal mixture. *Appl Statist* 36, 318–324.

Chapter 23

Visualization

Falk Schreiber

Abstract

Visualization is a powerful method to present and analyze a large amount of data. It is increasingly important in bioinformatics and is used for exploring different types of molecular biological data, such as structural information, high throughput data, and biochemical networks. This chapter gives a brief introduction to visualization methods for bioinformatics and presents two commonly used techniques in detail: heatmaps and force-directed network layout.

Key words: visualization, data exploration, heatmaps, force-directed layout, graph drawing.

1. Introduction

Visualization is the transformation of data, information, or knowledge into a visual form, such as images and diagrams. It uses the human ability to take in a large amount of data in a visual form and to detect trends and patterns in pictures easily. Visualization is a helpful method to analyze and explore data or communicate information, a fact expressed by the common proverb "A picture speaks a thousand words." The visual representation of data or knowledge is not a particularly modern technique, but rather is as old as human society. Rock engravings and cave images can be seen as an early form of visual communication between humans. Molecular biological information has also been represented visually for a long time. Well-known examples are illustrations in books, such as molecular structures (e.g., DNA) or biological processes (e.g., cell cycle, metabolic pathways). Most of the other chapters in this book use visualizations to illustrate concepts or present information.

Jonathan M. Keith (ed.), *Bioinformatics, Volume II: Structure, Function and Applications vol. 453*
© 2008 Humana Press, a part of Springer Science + Business Media, Totowa, NJ
Book doi: 10.1007/978-1-60327-429-6 Springerprotocols.com

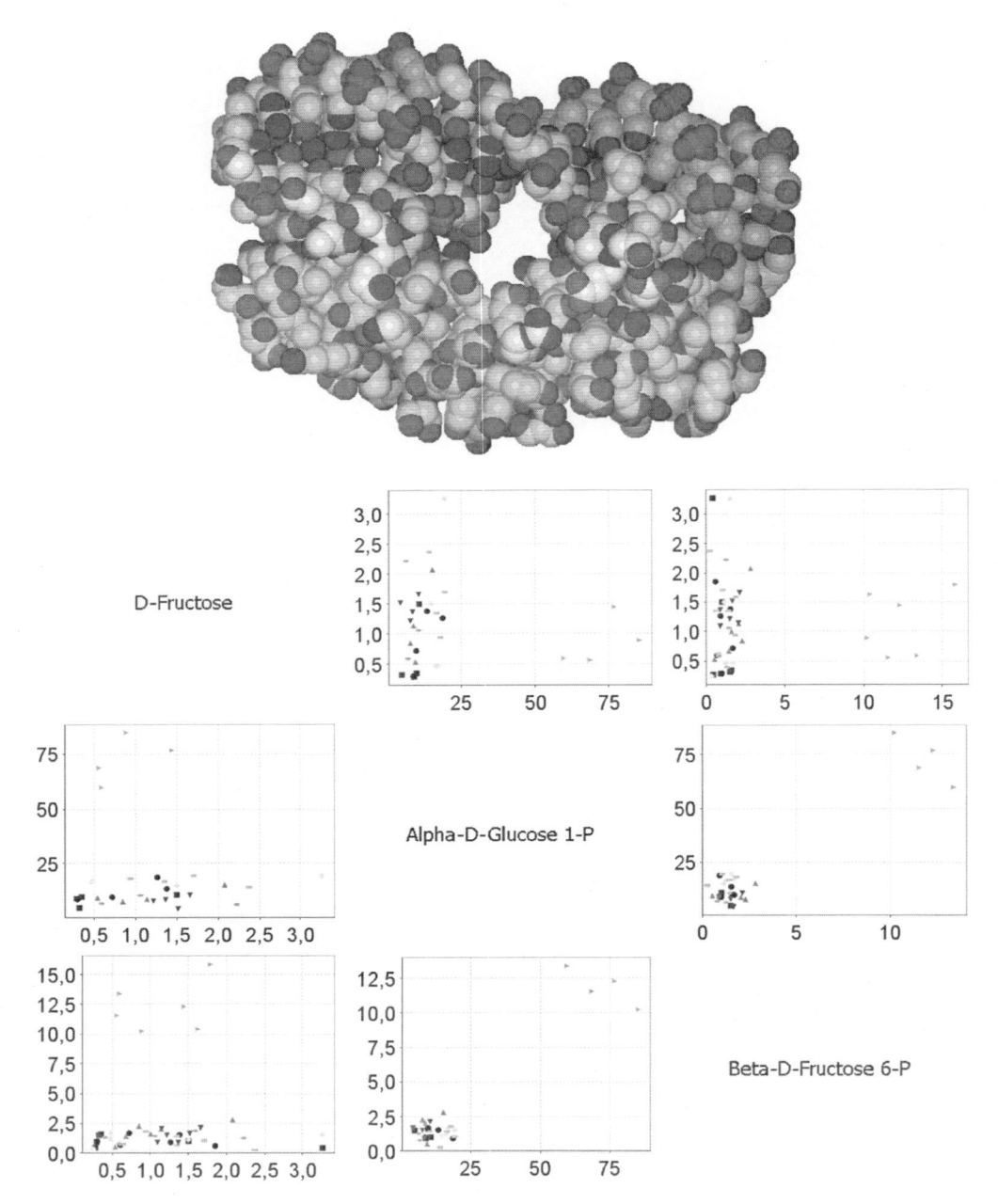

Fig. 23.1. Examples of visualizations in bioinformatics: (top) three-dimensional structure of a molecule (produced with Molw PDB Viewer *(13)*), (bottom) scatter-plot matrix of metabolite profiling data of different lines of an organism (produced with VANTED *(14)*), (next page bottom) line-graph of time series data of the concentration of a metabolite (produced with VANTED *(14)*), and (next page top) layout of a metabolic pathway (produced with BioPath *(15)*).

Nowadays visualization is an increasingly important method in bioinformatics to present very diverse information. Structural information of molecules can be shown in 2D (e.g., secondary structure of proteins or structural formulae of substances) and 3D

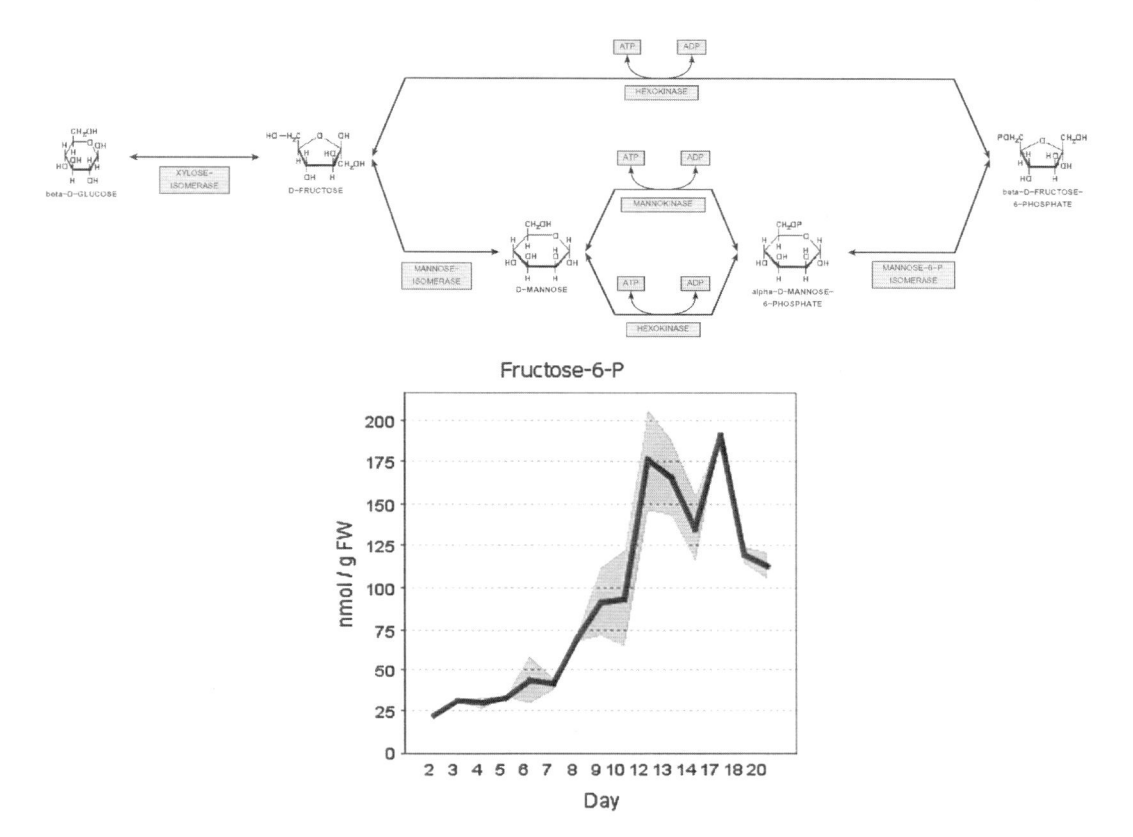

Fig. 23.1. (continued)

space *(1, 2)*. Genome and sequence annotation is often displayed in linear or circular representations with additional annotations *(3, 4)*. Expression and metabolite profiles are high-dimensional data that can be visualized with techniques such as bar-charts, line-graphs, scatter-plot matrices *(5)*, parallel coordinates *(6)*, heatmaps *(7)*, and tree-maps *(8)*. There are several methods to visualize hierarchical structures (e.g., phylogenetic trees) *(9, 10)* and biochemical networks (e.g., metabolic pathways) *(11, 12)*. Typical examples of visualizations in bioinformatics are shown in **Fig. 23.1**.

This chapter presents two particularly important techniques in detail: *heatmaps* and *force-directed network layout*.

Heatmaps are a standard method to visualize and analyze large-scale data obtained by the high throughput technologies discussed in previous chapters. These technologies lead to an ever-increasing amount of molecular-biological data, deliver a snapshot of the system under investigation, and allow the comparison of a biological system under different conditions or in different developmental stages. Examples include gene expression data *(16)*, protein data *(17)*, and the quantification of metabolite concentrations *(18)*. A typical visualization of such data using a heatmap is shown in **Fig. 23.2**.

Force-directed network layout is the main method used to visualize biological networks. Biological networks are important in bioinformatics. Biological processes form complex networks such as metabolic pathways, gene regulatory networks, and protein–protein interaction networks. Furthermore, the data obtained by high throughput methods and biological networks are closely related. There are two common ways to interpret experimental data: *(1)* as a biological network and *(2)* in the context of an underlying biological network. A typical example of the first interpretation is the analysis of interactomics data, for example, data from two-hybrid experiments *(20)*. The result of these experiments is information as to whether proteins interact pairwise with each other or not. Taking many different protein pairs into account, a protein–protein interaction network can be directly derived. An example of the second interpretation is the analysis of metabolomics data, such as data from mass spectrometry based metabolome analysis *(18)*. These experiments give, for example, time series data for different metabolites, which can be mapped onto metabolic networks and then analyzed within the network context. A visualization of a network using force-directed layout is shown in **Fig. 23.3**.

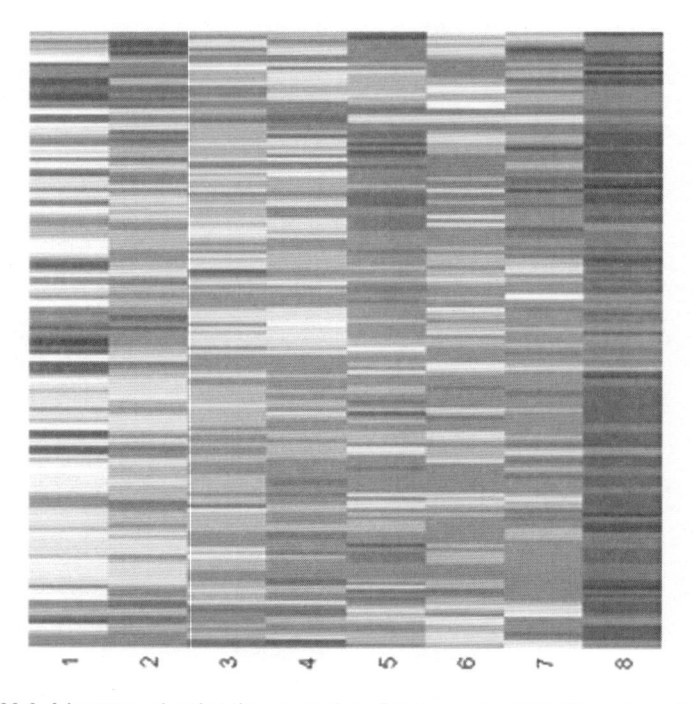

Fig. 23.2. A heatmap showing the expression of genes under eight different conditions *(19)*. Each line represents a gene.

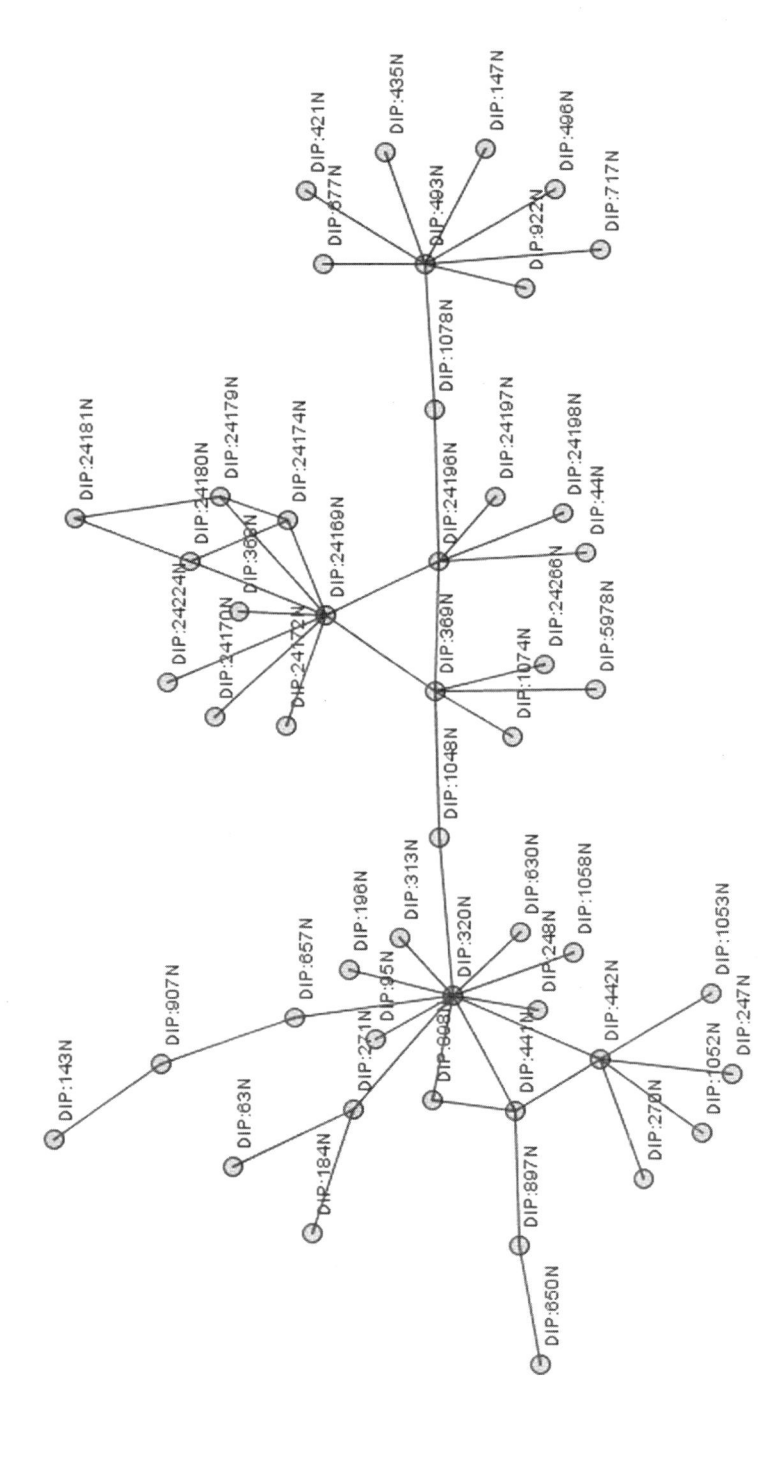

Fig. 23.3. A picture of a protein–protein interaction network based on the force-directed layout method (produced with CentiBiN (21)).

2. Methods

2.1. Heatmaps

High throughput data are often represented by a two-dimensional (2D) matrix M. Usually the rows represent the measured entities (e.g., expression of genes) and the columns represent the different samples (e.g., different time points, environmental conditions, or genetically modified lines of an organism). To show patterns in the data it is often useful to rearrange the rows and/or columns of the matrix so that similar rows (columns) are close to each other, for example, to place genes that have similar expression patterns close together.

A heatmap is a 2D, colored grid of the same size as the matrix M in which the color of each place is determined by the corresponding value of the matrix, as shown in **Fig. 23.4**.

For a given matrix M, the algorithm to produce a heatmap is as follows:

1. (Optional) Rearrange the rows of the matrix as follows: Compute a distance matrix containing the distance between each pair of rows (consider each row as a vector). There are several possible distance measures (e.g., Euclidean distance, Manhattan distance, correlation coefficient). Based on the distance matrix either rearrange the rows directly such that neighboring rows have only a small distance, or compute a hierarchical clustering (using one of the various methods available, such as complete linkage or single linkage). Rearrange the rows such that a crossing-free drawing of the tree representing the hierarchical clustering is obtained and similar rows are close together. Details of this rearranging step and several variations can be found in **Chapter 19** and in *(7, 22, 23)*.

2. (Also optional) Rearrange the columns of the matrix similarly.

3. Use a color scheme such that the distances between the colors represents the distances between the values of the elements of the matrix M (*see* **Note 1**). Assign to each matrix element its color and compute a grid visualization and (optional) dendrogram(s) displaying the hierarchical clustering(s) for rows/columns as shown in **Fig. 23.4**.

Free software to produce such visualizations is, for example, the R programming package *(19)*.

2.2. Force-Directed Network Layout

Biological networks are commonly represented as graphs. A graph $G = (V,E)$ consists of a set of vertices $V = \{v_1,\ldots, v_n\}$ representing the biological objects (e.g., proteins) and a set of edges $E \subseteq \{(v_i,v_j) \mid v_i,v_j \in V\}$ representing the interactions between the biological objects (e.g., interactions between proteins). To visualize a graph, a layout has to be computed, that is, coordinates for

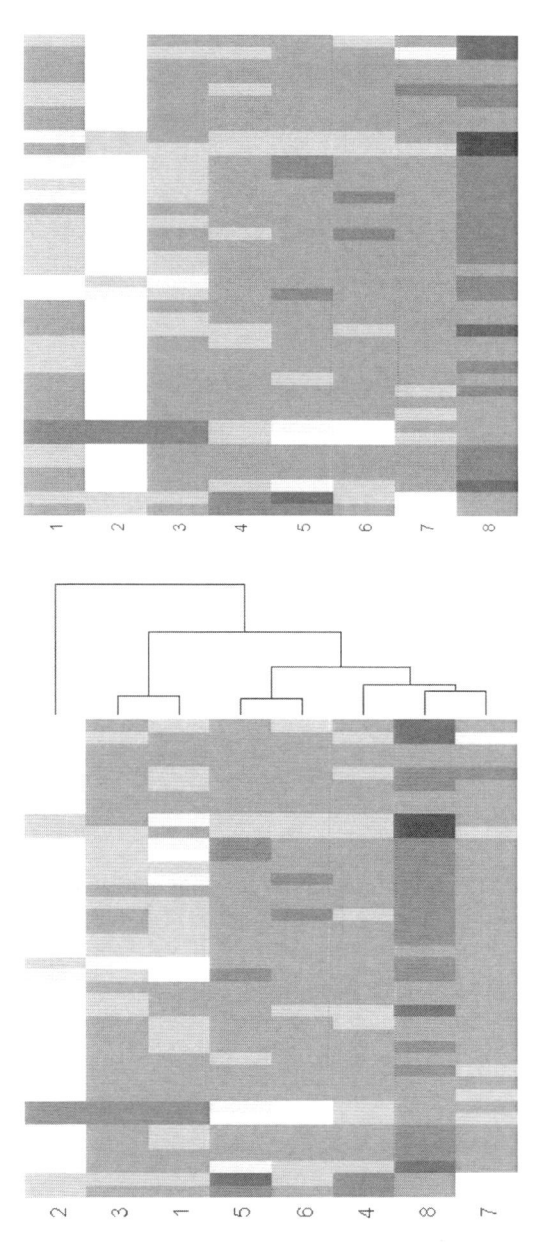

Fig. 23.4. *(Top)* A heatmap of the data set in the given order (x-axis: conditions, y-axis: genes). *(Bottom)* Rearrangement of columns (conditions) and dendrogram showing a hierarchical clustering of the different conditions (conditions with similarly expressed genes are close together *(19)*.

the vertices and curves for the edges. In the following we present the force-directed graph layout approach usually applied to biological networks.

A force-directed layout method uses a physical analogy to draw graphs by simulating a system of physical forces defined on the graph. It produces a drawing that represents a locally minimal energy configuration of the physical system. Such layout methods are popular as they are easy to understand and implement, and give good visualization results. In general, force-directed layout methods consist of two parts: *(1)* a system of forces defined by the vertices and edges, and *(2)* a method to find positions for the vertices (representing the final layout of the graph) such that for each vertex the total force is zero *(24)*. There are several frequently used varieties of force-directed methods *(25–28)*.

Here we use a force model that interprets vertices as mutually repulsive "particles" and edges as "springs" connecting the particles. This results in attractive forces between adjacent vertices and repulsive forces between non-adjacent vertices. To find a locally minimal energy configuration iterative numerical analysis is used. In the final drawing, the vertices are connected by straight lines.

For a given graph $G = (V,E)$ the algorithm to compute a layout $l(G)$ is as follows (*see* also **Fig. 23.5**):

1. Place all vertices on random positions. This gives an initial layout $l_0(G)$ (*see* **Note 2**).

2. Repeat the following steps (Steps 3–4) until a stop criterion (e.g., number of iterations, quality of current layout) is reached.

3. For the current layout $l_i(G)$ compute for each vertex $v \in V$ the force $F(v) = \sum_{(u,v) \in E} f_a(u,v) + \sum_{(u,v) \in V \times V} f_r(u,v)$, which is the sum of all attractive forces f_a and all repulsive forces f_r affecting v. For 2D or 3D drawings these force vectors consist of two (x,y) or three (x,y,z) components, respectively. For example, for the x component the forces f_a and f_r are defined as:

$$f_a(u,v) = c_1 * (d(u,v) - l) * \frac{x(v) - x(u)}{d(u,v)} \text{ and}$$

$$f_r(u,v) = \frac{c_2}{d(u,v)^2} * \frac{x(v) - x(u)}{d(u,v)},$$

respectively, where l is the optimal distance between any pair of adjacent vertices, $d(u,v)$ is the current distance between the vertices u and v, $x(v)$ is the x-coordinate of vertex v, and c_1, c_2 are positive constants (*see* **Note 3**). The other components are similarly defined.

4. Move each vertex in the direction of $F(v)$ to produce a new layout $l_{i+1}(G)$ (*see* **Note 4**).

Free software packages to produce such network layouts are, for example, JUNG *(29)* and Gravisto *(30)*.

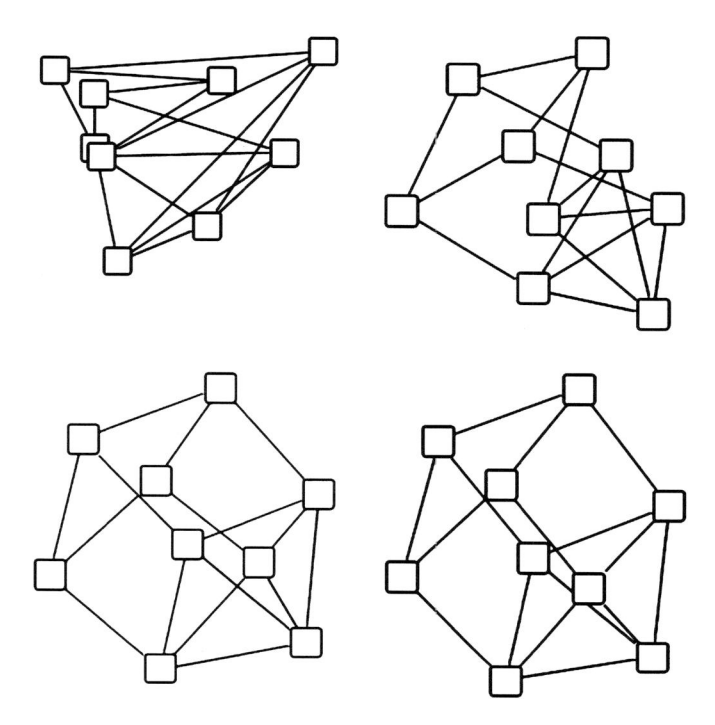

Fig. 23.5. Visualization of a graph at different steps of the force-directed layout (from top left clock-wise): initial layout, after 10, 25, and 100 iterations, respectively.

3. Notes

1. Do not use a red-green color scheme as quite a number of people are red-green colorblind and therefore unable to interpret the visualization.

2. The initial positions of the vertices should not be on a line. An alternative to a random placement is to use a given initial layout that then can be improved by the force-directed layout algorithm.

3. The parameters l, c_1, and c_2 greatly affect the final drawing. A good way to find appropriate values for a specific graph is to interactively change them during the run of the algorithm until appropriate interim results are obtained.

4. It is possible to dampen the force $F(v)$ with an increasing number of iterations to allow large movements of vertices in the beginning and only small movements close to the end of the algorithm. This can help to avoid "oscillation effects" in which vertices repeatedly jump between two positions.

References

1. Tsai, C. S. (2003) *Introduction to Computational Biochemistry*. Wiley-Liss, New York.

2. Can, T., Wang, Y., Wang, Y.-F., et al. (2003) FPV: fast protein visualization using Java3D. *Bioinformatics* 19, 913–922.

3. Helt, G. A., Lewis, S., Loraine, A. E., et al. (1998) BioViews: Java-based tools for genomic data visualization. *Genome Res* 8, 291–305.

4. Kerkhoven, R., van Enckevort, F. H. J., Boekhorst, J., et al. (2004) Visualization for genomics: the microbial genome viewer. *Bioinformatics* 20, 1812–1814.

5. Andrews, D. F. (1972) Plots of high-dimensional data. *Biometrics* 29, 125–136.

6. Inselberg, A., Dimsdale, B. (1990) Parallel coordinates: a tool for visualizing multi-dimensional geometry. *Proc Visual '90*, 361–370.

7. Eisen, M. B., Spellman, P. T., Brown, P. O., et al. (1998). Cluster analysis and display of genome-wide expression patterns. *Proc Natl Acad Sci U S A* 95, 14863–14868.

8. Baehrecke, E. H., Dang, N., Babaria, K., et al. (2004) Visualization and analysis of microarray and gene ontology data with treemaps. *BMC Bioinformatics* 5, 84.

9. Hughes, T., Hyun, Y., Liberles, D. (2004) Visualising very large phylogenetic trees in three dimensional hyperbolic space. *BMC Bioinformatics* 5, 48.

10. Rost U., Bornberg-Bauer, E. (2002) TreeWiz: interactive exploration of huge trees. *Bioinformatics* 18, 109–114.

11. Schreiber, F. (1992) High quality visualization of biochemical pathways in BioPath. *In Silico Biol* 2, 59–73.

12. Sirava, M., Schäfer, T., Eiglsperger, M., et al. (2002) BioMiner: modeling, analyzing, and visualizing biochemical pathways and networks. *Bioinformatics* 18, S219–S230.

13. Molw PDB Viewer 4.0. http://www.cris.com/~Molimage/.

14. Junker, B. H., Klukas, C., Schreiber, F. (2006) VANTED: a system for advanced data analysis and visualization in the context of biological networks. *BMC Bioinformatics* 7, 109.

15. Forster, M., Pick, A., Raitner, M., et al. (2002) The System Architecture of the BioPath system. *In Silico Biol* 2, 415–426.

16. Duggan, D., Bittner, B., Chen, Y., et al. (1999) Expression profiling using cDNA microarrays. *Nat Genet* 21, 11–19.

17. MacBeath, G. (2002) Protein microarrays and proteomics. *Nat Genet* 32, 526–532.

18. Villas-Boas, S. G., Mas, S., Akesson, M., et al. (2005) Mass spectrometry in metabolome analysis. *Mass Spectrom Rev* 24, 613–646.

19. R. http://www.r-project.org/.

20. Ito, T., Chiba, T., Ozawa, R., et al. (2001) A comprehensive two-hybrid analysis to explore the yeast protein interactome. *Proc Natl Acad Sci U S A* 98, 4569–4574.

21. CentiBin. http://centibin.ipk-gatersleben.de/.

22. Bar-Joseph, Z., Gifford, D. K., Jaakkola, T. S. (2001) Fast optimal leaf ordering for hierarchical clustering. *Bioinformatics* 17, S22–S29.

23. Biedl, T., Brejová, B., Demaine, E. D., et al. (2001) Optimal arrangement of leaves in the tree representing hierarchical clustering of gene expression data. Technical report 2001-14, Dept. of Computer Science, University of Waterloo.

24. Di Battista, G., Eades, P., Tamassia, R., et al. (1999) *Graph Drawing*. Prentice-Hall, Upper Saddle River, New Jersey.

25. Eades, P. (1984) A heuristic for graph drawing. *Congressus Numerantium* 42, 149–160.

26. Fruchterman, T., Reingold, E. (1991) Graph drawing by force-directed placement. *Software Pract Exper* 21, 1129–1164.

27. Kamada, T., Kawai, S. (1989) An algorithm for drawing general undirected graphs. *Inf Proc Letts* 31, 7–15.

28. Sugiyama, K., Misue, K. (1995) Graph drawing by magnetic spring model. *J Vis Lang Comput* 6, 217–231.

29. JUNG. http://jung.sourceforge.net/.

30. Gravisto. http://www.gravisto.org/.

Chapter 24

Constructing Computational Pipelines

Mark Halling-Brown and Adrian J. Shepherd

Abstract

Many bioinformatics tasks involve creating a computational pipeline from existing software components and algorithms. The traditional approach is to glue components together using scripts written in a programming language such as Perl. However, a new, more powerful approach is emerging that promises to revolutionise the way bioinformaticians create applications from existing components, an approach based on the concept of the scientific workflow.

Scientific workflows are created in graphical environments known as workflow management systems. They have many benefits over traditional program scripts, including speed of development, portability, and their suitability for developing complex, distributed applications. This chapter explains how to design and implement bioinformatics workflows using free, Open Source software tools, such as the Taverna workflow management system. We also demonstrate how new and existing tools can be deployed as Web services so that they can be easily integrated into novel computational pipelines using the scientific workflow paradigm.

Key words: Computational pipeline, scientific workflow, workflow management system, Web service, applications server.

1. Introduction

When a bioinformatician needs to solve a new task, a common approach is to construct a computational pipeline from existing software components and algorithms. Data flows from one component to the next in a user-defined sequence.

A simple example is the following pipeline for generating sequence profiles:

Given a genome identifier, repeat the following until finished:

1. Retrieve next protein-coding nucleic acid sequence belonging to the given genome from an appropriate data resource

Jonathan M. Keith (ed.), *Bioinformatics, Volume II: Structure, Function and Applications vol. 453*
© 2008 Humana Press, a part of Springer Science+Business Media, Totowa, NJ
Book doi: 10.1007/978-1-60327-429-6 Springerprotocols.com

(e.g., the GenBank database using the BioPerl Bio::DB:: GenBank module).

2. Generate an amino acid sequence for the current nucleic acid sequence using the BioJava Translate DNA module.

3. Find amino acid sequences that closely match the current sequence (e.g., by scanning it against the SWISS-PROT database using PSI-BLAST).

4. Align the sequence (e.g., using the Emma interface to ClustalW at the EBI) against a set of its homologs.

5. Output the resulting multiple alignment.

Even this simple pipeline raises a number of issues:

- Where are the components located (locally or remotely) and how can one access them?

- Are external components available in a reliable and consistent manner?

- How easy is it to integrate a given component into the pipeline? (i.e., is it only available via a point-and-click Web interface?)

- Is the output from component x in the format required by component y?

- How can the parallel nature of the pipeline (i.e., the fact that it is possible to process multiple sequences simultaneously) be exploited?

This chapter is about approaches to constructing computational pipelines in the bioinformatics domain. The focus here is on the future, rather than the past. There is an ongoing revolution that is transforming the way bioinformaticians construct computational pipelines. A completely new and powerful approach is now possible compared to what was feasible just a few years ago.

We begin by contrasting the two approaches, old and new. Under the current (old) regime, bioinformaticians typically develop novel computational pipelines by combining existing software components using scripts (written, for example, in the Perl programming language). Since existing components are rarely designed with the needs of application developers in mind, integrating them into new applications is often difficult, and may involve extensive coding to handle idiosyncratic data formats and enable scripts to automatically communicate with a given component via its Web interface (a technique known as "screen scraping"). To manage a given combination of internal and external resources, the bioinformatician may have to acquire considerable expertise and undertake more non-trivial coding (i.e., using Perl modules such as the CGI for interacting with Web-based applications, using the DBI for interacting with databases, etc.).

Under the new, emerging regime, bioinformaticians construct novel applications in a graphical environment known as a workflow management system (WMS) that enables components (represented by boxes) to be joined together (using arrows) as required. Many of the components themselves are likely to be dedicated Web services that have well-defined interfaces and make data available in standardised formats (e.g., XML). Combining them may require no coding at all. The WMS is designed to handle combinations of local and external resources in a seamless manner, so no coding by the bioinformatician is required; again, choices can be made directly using the workflow software interface.

The workflow paradigm offers several advantages over its predecessor:

- WMSs facilitate the rapid development of applications via an easy-to-use graphical interface.

- Workflows are typically much easier to understand than scripts because they are good at hiding low-level implementation details.

- Most WMSs have an array of features that makes it much easier to build applications that exploit distributed components and multiple hosts in an efficient manner.

- Workflows provide a platform-independent interface to components, thereby promoting the development of reusable, distributed, general-purpose scientific tools, rather than custom-made scientific applications.

Since bioinformatics applications typically involve integrating multiple tools and handling large sets of data stored in multiple formats, workflows are of particular utility in this field. Indeed, within the wider scientific community, bioinformatics has been one of the first fields in which significant numbers of practitioners have begun to use WMSs.

Before moving on to the practical issues involved in designing and implementing a workflow using a WMS, we briefly introduce two important concepts that enable us to categorise different kinds of workflow—workflow patterns, and workflow models. These have a direct bearing on the choices we make when designing a workflow.

1.1. Workflow Patterns   In order to understand the structure of an arbitrary workflow (i.e., the graph connecting the set of components in the workflow), it is useful to be aware of a set of core workflow patterns, each of which exhibits a distinct type of relationship between its constituent components *(1)*. The simplest is the sequence pattern in which successive components form a chain, with the execution of each component (except the first) dependent on the execution of its predecessor. The sequence pattern is the basic ingredient of a simple linear pipeline. A variation on the standard, strictly linear topology, which has intermediate storage nodes linked to specific pipeline components, is illustrated in **Fig. 24.1**.

Whereas in a sequence pattern components are executed serially, in a parallel pattern multiple components are executed concurrently. The move from serial to parallel execution can greatly reduce the total execution time of the workflow application, but is only possible for components that do not depend upon each other. **Figure 24.2** illustrates a parallelized workflow that exploits the independence of the majority of its components. In bioinformatics, parallel workflows are often appropriate when it is necessary to apply multiple components to the same set of data. One example amongst many is the prediction of antigenic peptides in protein sequences using multiple prediction methods *(2)*.

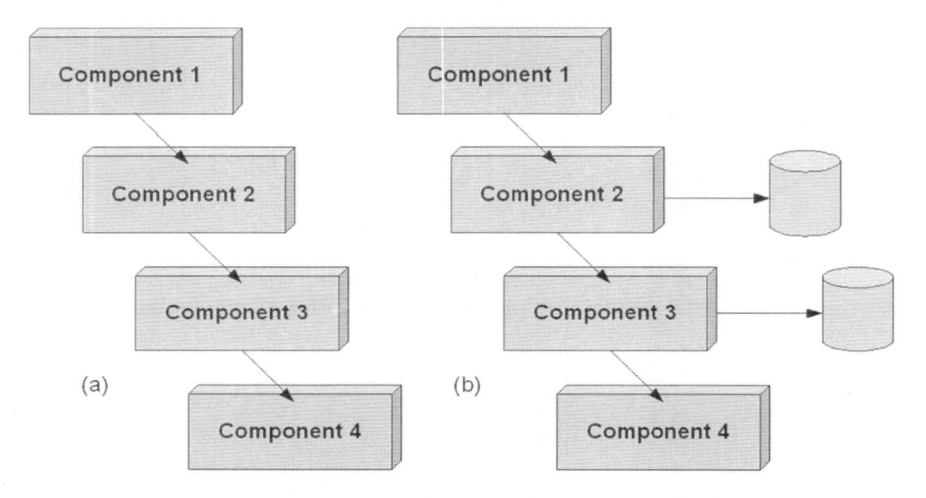

Fig. 24.1. (**A**) A simple linear pipeline. (**B**) The same pipeline with intermediate storage.

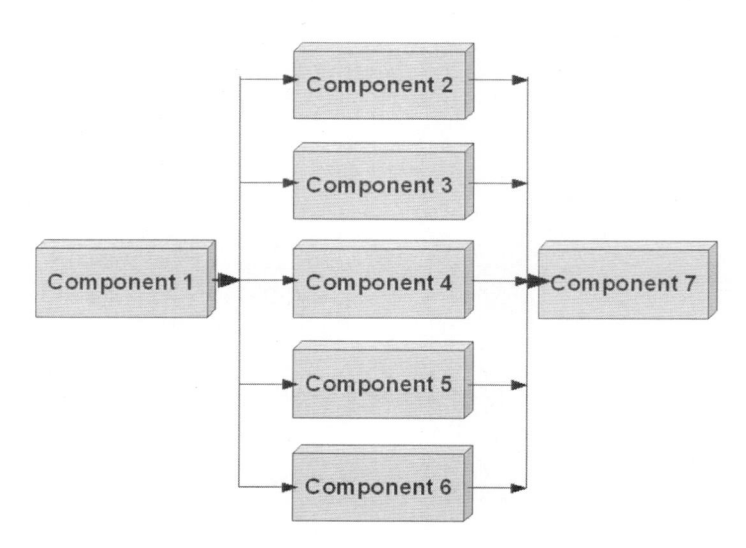

Fig. 24.2. A highly parallelized workflow.

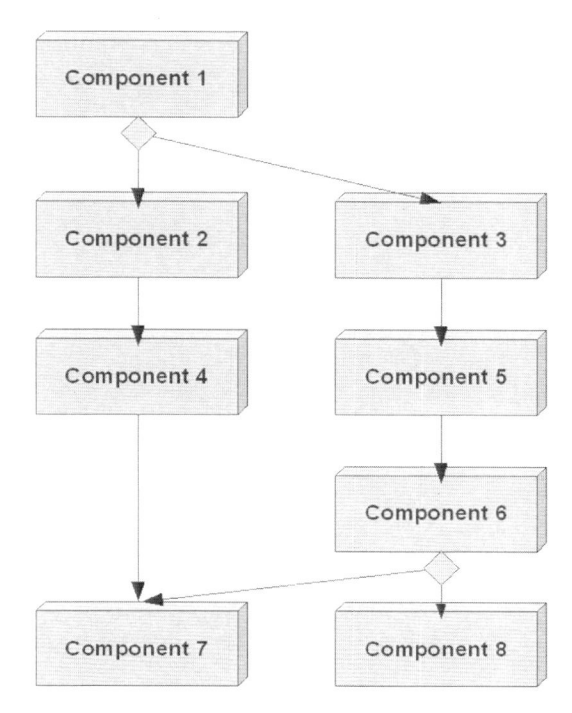

Fig. 24.3. A workflow that incorporates decision logic.

A so-called choice pattern adds decision logic to the workflow structure. The flow of data through a workflow containing such a pattern is decided at run-time and depends conditionally on the output from specific components. For example, in **Fig. 24.3** the direction of data flow after component 1 depends on the output from that component (i.e., on whether its output satisfies, or fails to satisfy, the conditions specified by the workflow logic).

Iterative, or cyclical patterns permit a block of two or more components to be executed repeatedly. This is achieved using a loop, as illustrated in **Fig. 24.4**. Iteration occurs frequently in bioinformatics applications, for example, looping through a list of sequences or sites on a sequence.

By combining different workflow patterns in a single workflow design, it is possible to construct applications that carry out tasks of arbitrary complexity, and to construct highly efficient, parallelized solutions where resources and the dependency structure of the task allow.

1.2. Workflow Models

There is an important distinction between concrete and abstract workflow models. Whereas a concrete (executable, or static) workflow specifies the resources that are allocated to each task, an abstract workflow leaves resources unspecified. An abstract workflow provides a high-level, highly portable description of an

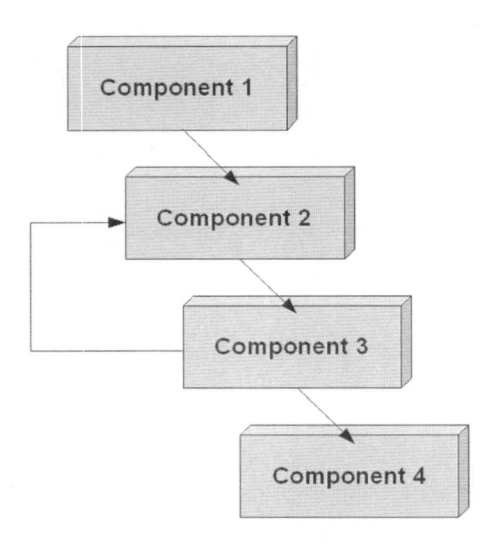

Fig. 24.4. A workflow with an iterative loop between components 2 and 3.

application that ignores implementation details. Given a WMS with appropriate mechanisms for identifying and evaluating resources, decisions about resource allocation can be automated and deferred until run-time. Abstract workflows are particularly appropriate in environments where the availability of resources is highly changeable, exemplified by the "Grid".

> "GRID: Integrated, collaborative use of high-end computers, networks, databases, and scientific instruments owned and managed by multiple organisations." -http://globus.org.

In cases where the user needs to retain control over at least part of the resource allocation process, it is possible to generate a concrete workflow from an abstract workflow prior to execution time. Indeed, it is possible to generate multiple concrete workflows from a single abstract workflow, each using a different set of resources. Most WMSs (e.g., ICENI, Taverna, Triana) support both concrete and abstract models.

2. Workflow Management Systems and Web Services

2.1. Workflow Management Systems

A WMS is software that enables a user to design a workflow, coordinate its components within a distributed environment, execute the workflow application, and monitor the flow of data when the application is running. A typical WMS contains five main parts:

- The design (or composition) section provides a graphical or language-based interface for modeling the desired flow of data using available components and computational resources.

- The resource information (or information retrieval) section is concerned with gathering and presenting information about available resources and execution platforms.

- The scheduling section is concerned with the mapping and management of resources, and the coordination of workflow processes and activities.

- The fault management section is concerned with handling faults that occur during the execution of a workflow. It can be used to automatically switch the flow of work to an alternative resource when a default resource fails (i.e., when an external server goes down).

- The data transfer and management section is concerned with ensuring that data is available in the locations where it is needed during the execution of the workflow.

In the remainder of this section we discuss each of these major WMS elements in turn.

2.1.1. Workflow Design

The workflow design section of a WMS provides a high-level view with which the user constructs a workflow. Different WMSs offer contrasting types of interface for designing workflows, of which the two most important are graphical (graph-based) modeling interfaces and language-based modeling interfaces.

Graphical interfaces provide much easier routes to designing workflows for the general user than language-based interfaces. A graphical interface allows the user to drag and drop components onto the workflow composition area and connect them together by clicking or dragging lines. **Figure 24.5** shows the workflow composition interface of the Taverna WMS. In this simple example, there is a three-component workflow in the workflow diagram window (lower left) that retrieves a sequence given a sequence identifier as input. Additional services can be dragged from the advanced model explorer (top left) for incorporation into the workflow.

Different WMSs take different approaches to the graphical modeling of workflows. For example, some exploit established formalisms, notably Petri Nets or UML (Unified Modeling Language), whereas others take their own, unique approach. This variety of approaches means that graphical models of workflows cannot be readily exchanged between different management systems. Perhaps more importantly, the expressiveness of different approaches varies considerably, with some systems supporting only the sequence and parallel workflow patterns described in the previous section.

Language-based interfaces to WMSs typically involve the use of a mark-up language based on XML (Extensible Markup Language), for example BPEL4WS (Business Process Execution Language for Web Services). Although most users are unlikely to want to write workflows directly using such a language (not least

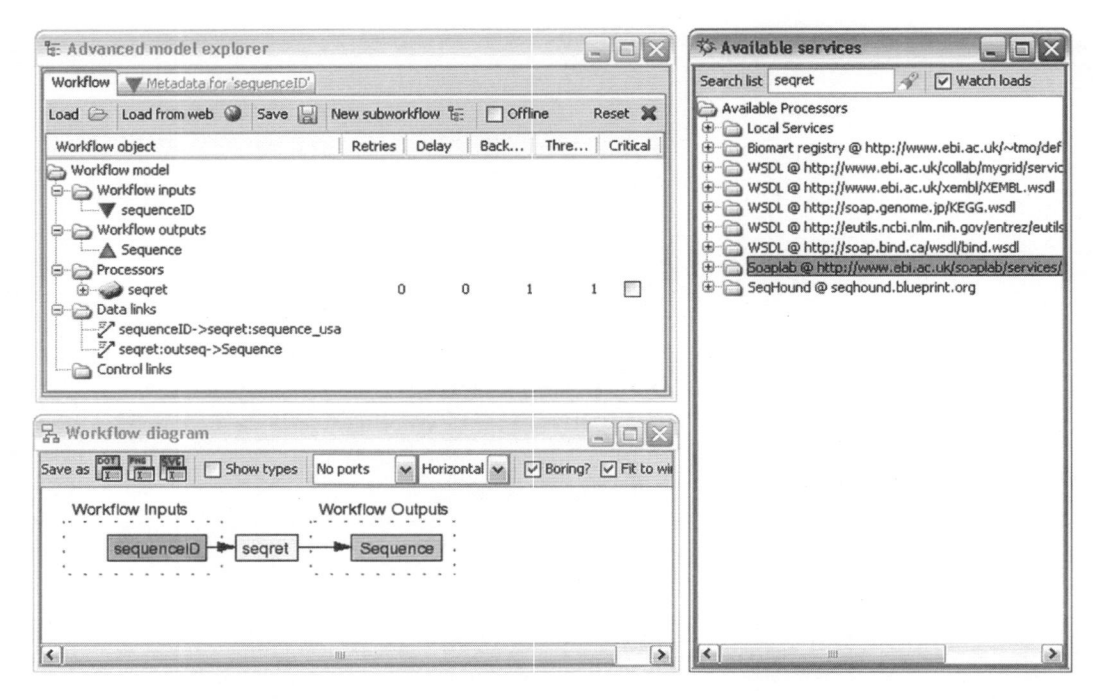

Fig. 24.5. Taverna's Workflow Composition Interface.

because of the complexity of the associated syntax), workflow languages are important because they specify workflows in a convenient form for undertaking validation, manipulation, and interchange. In practice, most WMSs offer the best of both worlds in that language-based workflow definitions can be generated automatically from a graphically designed workflow. Workflow languages provide a potential mechanism for porting a workflow from one WMS to another, although not all WMSs support the same workflow languages.

2.1.2. Resource Information

Before we can add components to a workflow, we need to know what components are available and some essential information about each component: its name, what it does, what input it requires, and what output it produces. Some Web services provide further information about their resource and platform requirements. In addition, some WMSs store historical information about the performance of components and resources that can be used as the basis for future design and scheduling decisions (e.g., the ICENI performance repository system).

What information is available and how one goes about retrieving it depends on the application and the WMS being used. Web applications that exploit Web service protocols describe their properties and capabilities in an unambiguous, computer-interpretable form using a core set of mark-up language constructs that WMSs understand (*see* **Section 2.3**. for details). Information about the availability and status of Grid resources, on the other hand, is available from various sources. For example, the Monitoring and Discovery System (MDS) provides information about CPU type and load, operating system and memory size. The MDS is used by several WMSs, including GridAnt.

An increasing number of bioinformatics components are being made publicly available in a form that enables ordinary users to incorporate them directly into their workflows. Some key examples are the Web services made available by large organizations, such as the European Bioinformatics Institute (EBI) and National Center for Biotechnology Information (NCBI). Several lists of bioinformatics-related Web services are available online (e.g., the MyGrid project page at twiki.mygrid.org.uk/twiki/bin/view/Bioinformatics/BioinformaticsWebServices). *See* **Note 1** for details on how to create your own Web Services and deploy them in Taverna.

2.1.3. Workflow Scheduling and Data Management

One of the major advantages of using a sophisticated WMS to develop computational pipelines is its ability to automatically assign tasks to different resources and schedule the execution of workflow components by taking into account the resource requirements of different components, the nature and availability of different resources, as well as any explicit requirements specified by the users *(3)*. Developing effective scheduling strategies is one of the biggest challenges for the creators of WMSs. Scheduling tasks in workflows that exploit decision logic and/or iteration (choice and/or iterative patterns) poses particular problems, as decisions made at run-time (regarding the direction of flow in a choice pattern, or the number of iterations in an iterative pattern) determine what constitutes an effective strategy. This is a key reason why WMSs have been quicker to support sequence and parallel workflow patterns than choice and iterative patterns.

Different WMSs approach the scheduling problem in radically different ways, with some opting for a centralized approach (e.g., Taverna, ICENI, GridAnt), others for a hierarchical or a decentralized approach (e.g., Triana). In a centralised approach, there is a single scheduler that controls every task in the workflow. Hierarchical and decentralized approaches have multiple schedulers at different resources, but differ in that a decentralized approach does not have a central scheduler. Although details about the pros and cons of different scheduling strategies lie outside the scope of this chapter (*see (4)* for a useful summary), it is worth pointing out that, given a complex workflow and multiple distributed resources, the effectiveness of different WMSs in devising an appropriate schedule is likely to vary a great deal. However, given the speed at which WMS software is being refined, it is impossible to recommend particular systems at this juncture.

2.1.4. Fault Management

There are a number of reasons why the execution of a workflow can fail. Typical causes are: resource overload (running out of memory or disk space), network failure, and (often unexplained) difficulties with an external service or resource. Since failures of one sort or another are highly likely to occur from time to time, many WMSs incorporate mechanisms for preventing or recovering from such failures where possible.

Within workflow environments, failure handling techniques can be directed toward individual tasks or the whole workflow (5). The commonest approach to task-level fault tolerance is known as the retry technique, whereby a service that fails is simply re-invoked. This approach is supported by most WMSs (e.g., Triana, ICENI, Taverna). Other approaches include the alternate resource and checkpoint/restart techniques (which involve switching a failed task to a different resource) and the replication technique (a brute-force technique which runs the same task simultaneously on multiple hosts in the expectation that at least one will execute successfully).

Whereas task-level fault tolerance is about individual components and resources, workflow-level fault tolerance is about designing the workflow in a way that makes it robust against certain anticipated failures. One obvious approach is to provide alternative routes by which the execution of the workflow can proceed given the failure of one or more components. In the case where no alternative route is available, user-specified error handling can be built into the workflow so that the workflow terminates elegantly (and stores intermediate data) even when failure cannot be (currently) avoided.

2.1.5. Data Transfer and Management

Bioinformatics applications often require large quantities of data to be processed. Given a workflow that combines resources from different sites, data will typically have to be moved around between locations. Things can quickly get complicated; input data for each remote component may have be staged at the remote site, whilst the output from each component may require replication (or transformation) and transferral to another location.

There is a key distinction between WMSs that require the user to specify the intermediate transfer of data (e.g., GridAnt) and those that automate the process (e.g., ICENI, Taverna, Triana, and Kepler). For a concise summary of the pros and cons of different automatic data transfer strategies, *see (4)*.

2.2. Current Workflow Management Systems

In recent years, the world of scientific workflow management has undergone a radical transformation, with many new WMSs being introduced, and many existing systems being substantially upgraded. Information about more than thirty WMSs is currently available online at the Scientific Workflow Survey site (www.extreme.indiana.edu/swf-survey/). Here we concentrate on four

Table 24.1
Workflow management system characteristics

Name	Prerequisite	Integration	Design systems	Availability
Taverna	Java 1.4+	Web Services	Language/ Graph	LGPL
Triana	Grid Application Toolkit (GAT)	GAT (Web Service)	Graph	Apache License
ICENI	Globus Toolkit/J2EE Application Server	Globus/Web Services	Language/Graph	ICENI Open Source License
GridAnt	Apache Ant/Globus Toolkit	Globus	Language	GTPL

of the most widely used scientific WMSs: Taverna, Triana, ICENI, and GridAnt. The main focus here is on the unique characteristics of each system. Some of the key design and integration characteristics of these four WMSs are summarized in **Table 24.1**.

2.2.1. Taverna

Taverna (6), developed by the myGrid (7) project, is a WMS specifically designed for bioinformaticians developing Grid-based applications. Taverna provides two approaches to constructing workflows: a graphical user interface and a workflow language called Scufl (Simple Conceptual Unified Flow Language) with an XML representation XScufl. Unlike most workflow languages (e.g., BEPL4WS), Scufl supports an implicit iteration strategy; a service that expects an input I is executed iteratively when invoked with an input list I1, I2, I3, etc. Taverna supports multithreading for iterative processes, and is thus particularly well-suited to applications that are to be executed on a computer cluster or farm. Services are invoked and data managed by the FreeFluo workflow enactor.

Taverna supports several different types of service. In addition to standard WSDL-based Web services, Taverna is capable of constructing workflows using both Soaplab, the SOAP-based Analysis Web Service developed at the EBI (http://www.ebi.ac.uk/soaplab/), and local services written in Java. Service discovery is possible via the Feta Service Search plug-in, which supports searches by function or by input/output parameters as well as by name.

The Taverna workbench can be downloaded from the Taverna project web site (taverna.sourceforge.net).

2.2.2. Triana

Triana is a WMS developed by the GridLab project (8). Recently, Triana has been redesigned and integrated with Grids via the GridLab GAT (Grid Application Toolkit) interface (9). In Triana, tasks can be grouped and then saved to file for later use. This is a highly flexible approach, as a grouped task can be used in the same way as any other task, promoting a modular approach to application development.

Initially developed to support the GEO 600 gravitational wave detection project (10), Triana incorporates tools that support the analysis of one-dimensional data sets such as signal processing algorithms, mathematical functions and visualization tools (e.g., the 2D plotting tool XMGrace).

The Triana user interface is disconnected from the underlying system. Clients can log into a remote Triana Controlling Service (TCS), build and run a workflow remotely and visualize the results locally using a PC, PDA, or other device. Clients can log onto and off from the TCS at will without stopping the execution of the workflow.

Triana can be downloaded from the Triana project web site (http://www.trianacode.org/index.html).

2.2.3. ICENI

The ICENI (Imperial College e-Science Networked Infrastructure) (11) WMS is arguably the most sophisticated system discussed in this chapter. ICENI can take into account quality of service metrics specified by the user, so that the same workflow can be tailored to the requirements and expectations of different users, for example by optimizing for execution time rather than expenditure, or vice versa. ICENI implements several scheduling algorithms that aim to find near-optimal mappings of components onto resources given the current user requirements. As part of its overall scheduling strategy, ICENI takes into account historical information (regarding the past performance of different component/resource combinations) stored in a performance repository system. ICENI can be downloaded from the ICENI pages on the London e-Science Centre web site (http://www.lesc.imperial.ac.uk/iceni/).

2.2.4. GridAnt

GridAnt (12, 13) developed by Argonne National Laboratory, is a WMS that is built upon the popular Apache Ant (14) build tool for Java. What distinguishes GridAnt from the other WMSs discussed here is its simplicity. GridAnt supports only the concrete workflow model and language-based workflow design. There are no built-in sophisticated strategies for scheduling, fault tolerance and data transfer—all must be defined or directed by the user. However, GridAnt is a convenient tool for developing Grid applications that have simple process flows (sequence and parallel patterns) and which are specifically designed to work with the Globus Toolkit.

The Java CoG Kit GridAnt module can be downloaded from the Java CoG Kit web site (wiki.cogkit.org/index.php/Main_Page).

2.3. Web Services

A Web service is a Web-based application designed to support interoperable interaction using the eXtensible Mark-up Language (XML), Simple Object Access Protocol (SOAP), Web Services Description Language (WSDL), and Universal Description, Discovery, and Integration (UDDI) open standards over

Table 24.2
Web Services Protocol Stack

Orchestration Flow language	BPEL WSFL	Integration
Security Reliability Transaction	WS-Security WS-RM(Reliability) WS-Transactions	Quality of service
Service description Service discovery	WSDL UDDI	Discovery and description
Message packaging State transfer Remote procedure calls	SOAP REST XML-RPC	Meassaging
Hypertext transfer Mail transfer File transfer Blocks exchange	HTTP SMTP FTP BEEP	Transport

an Internet protocol backbone. XML is used to tag the data, SOAP is used to transfer the data, WSDL is used for describing available services, and UDDI is used in the discovery of available services. **Table 24.2** shows a full list of the standards that make up Web services. (Software systems developed using servlets, JSP, or JWIG are interactive services, where the client is a human being using a browser. However the main topic of this section will be application-to-application Web services, where the client is itself a program.)

Web services provide a loose coupling between an application that uses the Web service and the Web service itself. A Web service may undergo changes without disrupting the workflows using that Web service provided the interface to the service remains the same. By communicating via XML, Web services can be incorporated into distributed applications running on disparate systems without the need for customized code (*see* **Notes 1** and **2**). All of the standards and protocols which are utilised by Web services are open and in text-based formats, making it easy for developers to comprehend. Web services utilize HyperText Transfer Protocol (HTTP) over Transmission Control Protocol (TCP) and can therefore work through existing firewall security measures. **Figure 24.6** provides a schematic overview of how a simple Web service works.

2.3.1. Application Servers

To deploy your own Web service you need to install and run an application server. An application server provides middleware services for security, state maintenance, along with data access and persistence.

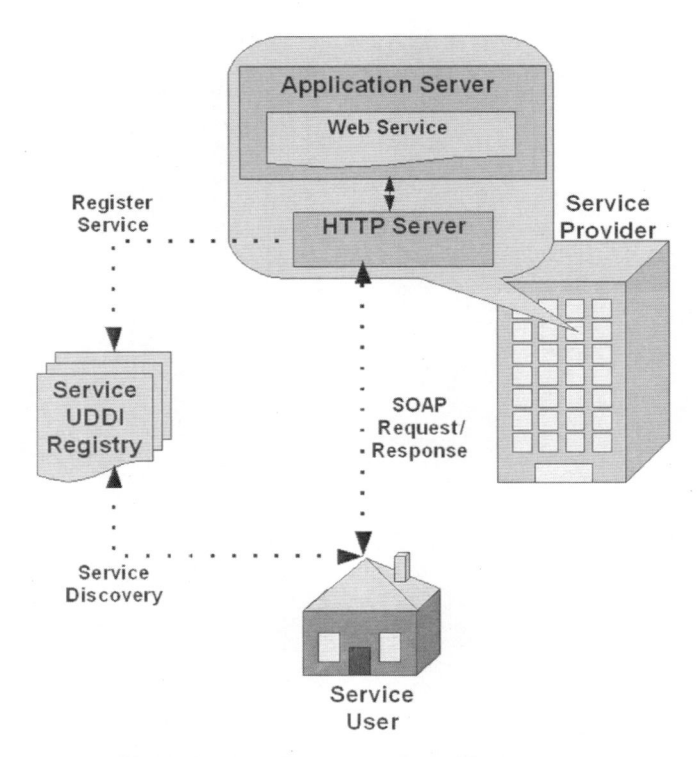

Fig. 24.6. The workings of a simple Web service. Interactions between the service user, the service provider and the registry.

J2EE provides a framework for developing and deploying Web services on an application server. Most free application servers are J2EE compliant. The J2EE specification defines numerous API services and multiple application programming models for application development and integration. These include technologies such as Java Messaging Service (JMS), J2EE Connector Architecture (JCA), Java API for XML Messaging (JAXM), Java API for XML Processing (JAXP), Java API for XML Registries (JAXR), and Java API for XML-based RPC (JAX-RPC).

WebSphere (IBM) and Weblogic (BEA) are the better known commercial J2EE application servers, whereas JBoss is a non-commercial Open Source J2EE-compliant application server. Because it is Java-based, JBoss can be used on any operating system that supports Java.

An alternative approach is to combine two Apache products: Tomcat and AXIS. Tomcat is a free, Open Source implementation of Java Servlet and JavaServer Pages (JSP) technologies that includes its own HTTP server. By adding AXIS, Tomcat can be used as a Web service server. **Figure 24.7** shows the relationships between the client, Tomcat and AXIS.

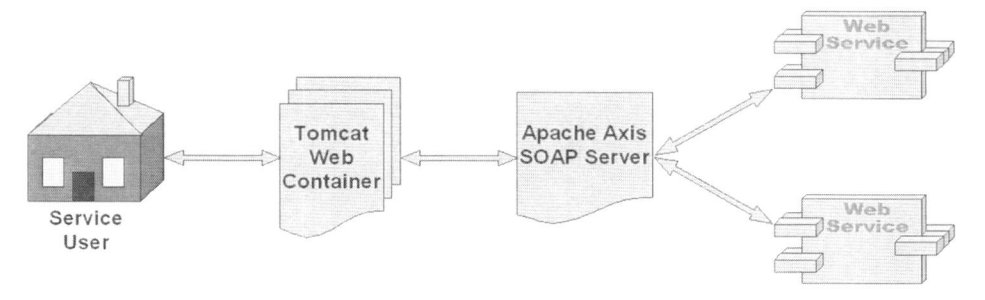

Fig. 24.7. The relationship between the service user, Apache Tomcat, Apache AXIS and Web services.

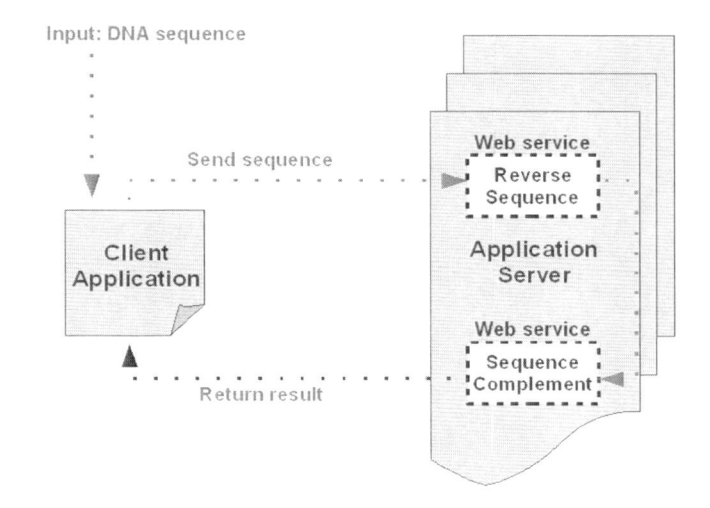

Fig. 24.8. Graphical representation of a DNA reverse complement workflow.

3. Notes

1. Creating your own Web service: This section will explain how to create, deploy and use Web services within a scientific workflow. As an example, we develop a workflow that takes a DNA sequence as input and returns its reverse complement. For present purposes we assume that the workflow is to be developed in a Microsoft Windows environment. However, all of the operations described below are easily transferable to a Linux system, and all tools and software are available for both operating systems.

   A graphical summary of the workflow is presented in **Fig. 24.8**.

   a. Configuration: To begin we need to select a Java environment, a Java IDE (if desired) and an application server with a Web service container. There are several possible

Java environments, including: J2SE JDK from Sun (java. sun.com); Blackdown JDK (www.blackdown.org); and BEA WebLogic JRockit (www.bea.com).

Here we use the Sun JDK, as other products are generally designed for the development of software with specialised requirements, for example JRockit provides excellent support for threading and garbage management, facilities that are not needed by the application we are developing.

At the time of writing, the most recent version of the Sun JDK is JDK 5.0, which can be downloaded from http:// java.sun.com/j2se/1.5.0/download.jsp. Installation is straightforward, although it is important to check that the $JAVA_HOME variable has been set and that the $JAVA_HOME/bin folder has been added to the system path.

Our chosen application server is the Sun Application Server SE 8, which is widely used, and is distributed with the J2EE SDK. It can be downloaded from http://java. sun.com/j2ee/1.4/download.html. During installation, ensure that the box labelled "Add bin directory to PATH" is selected. On our own system, the server was set to run on the default port 8080.

An Integrated Development Environment (IDE) provides tools for organizing source code, for compiling and linking source code, and for running and debugging the compiled programs. There are a variety of IDEs available for Java, including: NetBeans (www.netbeans.org); JBuilder (Borland); and Eclipse (www.eclipse.org). We chose Eclipse as it is freely available and provides excellent facilities for integrating third-party tools and plug-ins. Eclipse can be downloaded from http://www.eclipse. org/downloads/.

b. Developing Java Components: The starting point for developing a J2EE Web service is the so-called Service Endpoint Interface (SEI). An SEI is a Java interface that declares the methods that a client can invoke on the service. For each Java class which we wish to publish as a Web service, we need to define the implementation class and an SEI.

For the simple workflow we are developing here, we need to define two Java classes that we have named ReverseString.java and ComplementString.java. ReverseString. java takes a DNA sequence as input and returns that sequence reversed; ComplementString.java takes a DNA sequence as an input and returns its complement. (The code for these two classes can be found in the info box in the following.) We placed the two Java files in separate folders (C:\apps\reverse and C:\apps\complement) and created a build sub-folder in each (C:\apps\reverse\build and C:\apps\complement\build). The files were compiled and the output written to their respective build folders.

```
Reverse String Endpoint Interface
package reversestring;
import java.rmi.Remote;
import java.rmi.RemoteException;
public interface ReverseStringIF extends Remote {
    public String reverseIt(String s) throws Remote-
    Exception;
}

Reverse String Implementation

package reversestring;
public class ReverseStringImpl implements Revers-
eStringIF {
  public String reverseIt(String s) {
    return new StringBuffer(s).reverse();
}
}

Complement String Endpoint Interface
package complement;
import java.rmi.Remote;
import java.rmi.RemoteException;
public interface ComplementIF extends Remote {
    public  String  complement(String  s)  throws
    RemoteException;
}

Complement String Implementation
package complement;
public class ComplementStringImpl implements Com-
plementIF {
public String complement(String s) {
    char [] temp = new char[s.length()];
    for (int i=0; i<s.length(); i++) {
        if(s.charAt(i) == 'A') temp[i] = 'T';
        if(s.charAt(i) == 'G') temp[i] = 'C';
        if(s.charAt(i) == 'C') temp[i] = 'G';
        if(s.charAt(i) == 'T') temp[i] = 'A';
    }
    return new String(temp);
  }
}
```

c. Creating the Configuration/WSDL/Mapping Files: To generate the WSDL file for each Java class, a configuration file (typically named config.xml) must be created and passed to the wscompile tool (An example configuration file can be found in the info box below).

```
<?xml version="1.0" encoding="UTF-8"?>
<configuration xmlns="http://java.sun.com/xml/ns/
jax-rpc/ri/config">
  <service name="ReverseAString"
  targetNamespace="http://localhost:8080"
    typeNamespace="http://localhost:8080" packageNa
    me="reversestring">
    <interface name="reversestring.ReverseStringIF"/>
  </service>
</configuration>
```

By running wscompile, a WSDL and a mapping file is created in the relevant build directory.

```
wscompile -define -mapping build&backslash;mapping.
xml -d build -nd build - classpath build config.xml
```

    d. Packaging and Deploying the Web Services: The Web services can be packaged and deployed using the deploy-tool tool, which comes with J2EE. This tool creates the deployment descriptor and a .war file, and can be used to directly deploy the Web service. The following information box gives a breakdown of the actions that are required to package and deploy a Web service using deploytool.

```
 1) Start a New Web Component wizard by selecting
    File→New→Web Component and then click Next.
 2) Select the button labelled Create New Stand-Alone
    WAR Module.
 3) In the WAR File field, click Browse and navigate
    to your folder eg. C:/complement
 4) In the File Name field, enter a Service Name.
 5) Click Create Module File and then click Edit
    Contents.
 6) In the tree under Available Files, locate your
    directory.
 7) Select the build subdirectory then click Add
    and then click OK.
 8) In the Context Root field, enter/complement-jaxrpc
    then click Next.
 9) Select the Web Services Endpoint radio button
    and then click Next.
10) In the WSDL File combo box, select your WSDL
    file.
11) In the Mapping File combo box, select your
    mapping file and then click next.
```

```
12) In the Service Endpoint Implementation combo
    box, select complement. ComplementStringImpl
    and then click next.
13) In the Service Endpoint Interface combo box,
    select complement.complementIF.
14) In the Namespace combo box, select http://
    localhost:8080.
15) In the Local Part combo box, select complemen-
    tIFPort.
16) Click Next and the click Finish.
```

Specifying the Endpoint Address
```
 1) To access this service, we will specify a
    service endpoint address URL: http://local-
    host:8080/complement-jaxrpc/complement
 2) In deploytool, select your Service Name in the tree.
 3) In the tree, select ComplementStringImpl.
 4) Select the Aliases tab.
 5) In the Component Aliases table, add /component.
 6) In the Endpoint tab, select component for the End-
    point Address in the Sun-specific Settings frame.
 7) Select File→Save.
```

Deploying the Service
```
 1) In deploytool tree, select your Service Name.
 2) Select Tools→Deploy.
```

It is possible to confirm that the service has been deployed correctly by looking at the WSDL file at the following URL, http://localhost:8080/complement-jaxrpc/complement?WSDL.

2. Deploying a Web service in Taverna: Once the Web service has been deployed it can be integrated into a workflow using a WMS that supports Web services. For example, Taverna allows the user to add a Web service by following the instructions in the info box below.

```
In the "Available Services" Window:
1)  Right-click on "Available Processors" and select
    "Add new WSDL scavenger".
2)  Insert your WSDL address, e.g. http://localhost:8080/
    complement-jaxrpc/complement?WSDL
3)  Click OK
```

References

1. van der Aalst, W., van Hee, K. (2002) *Workflow Management: Models, Methods, and Systems.* MIT Press, Cambridge, MA.

2. Halling-Brown, M. D., Cohen, J., Darlington, J., et al. (2006) Constructing an antigenic peptide prediction pipeline within ICENI II. *BMC Bioinformatics,* submitted.

3. Hamscher, V., Schwiegelshohn, U., Streit, A., et al. (2000) *Evaluation of Job-Scheduling Strategies for Grid Computing.* Springer-Verlag, Heidelberg, Germany.

4. Yu, J., Buyya, R (2005) *A Taxonomy of Workflow Management Systems for Grid Computing . Technical Report GRIDS-TR-2005-1,Grid Computing and Distributed Systems Laboratory.* University of Melbourne.

5. Hwang, W. H., Huang, S. Y. (2003) Estimation in capture–recapture models when covariates are subject to measurement errors. Biometrics 59, 1113–1122.

6. Oinn, T., Addis, M., Ferris, J., et al. (2004) Taverna: a tool for the composition and enactment of bioinformatics workflows. Bioinformatics 20, 3045–3054.

7. Stevens, R. D., Robinson, A. J., Goble, C. A. (2003) myGrid: personalised bioinformatics on the information grid. Bioinformatics 19, i302–i304.

8. Taylor, I., Shields, M., Wang, I. (2003) *Resource Management of Triana P2P Services. Grid Resource Management.* Kluwer, Netherlands.

9. Allen, G., Davis, K., Dolkas, K. N., et al. (2003) *Enabling Applications on the Grid-A GridLab Overview.* SAGE, London.

10. The GEO600 Team (1994) *GEO600: Proposal for a 600 m Laser-Interferometric Gravitational Wave Antenna.* MPQ Report 190.

11. Furmento, N., Lee, W., Mayer, A., et al. (2002) *ICENI: An Open Grid Service Architecture Implemented with Jini.* Super-Computing, Baltimore.

12. von Laszewski, G., Amin, A., Hategan, M., et al. (2004) *GridAnt:A ClientControllable Grid Workflow System.* IEEE CS Press, Los Alamitos, CA.

13. Amin K, von Laszewski G. (2003) GridAnt: A Grid Workflow System. Manual. http://www.globus.org/cog/projects/gridant/gridant-manual.pdf

14. http://ant.apache.org/.

Chapter 25

Text Mining

Andrew B. Clegg and Adrian J. Shepherd

Abstract

One of the fastest-growing fields in bioinformatics is text mining: the application of natural language processing techniques to problems of knowledge management and discovery, using large collections of biological or biomedical text such as MEDLINE. The techniques used in text mining range from the very simple (e.g., the inference of relationships between genes from frequent proximity in documents) to the complex and computationally intensive (e.g., the analysis of sentence structures with parsers in order to extract facts about protein–protein interactions from statements in the text).

This chapter presents a general introduction to some of the key principles and challenges of natural language processing, and introduces some of the tools available to end-users and developers. A case study describes the construction and testing of a simple tool designed to tackle a task that is crucial to almost any application of text mining in bioinformatics—identifying gene/protein names in text and mapping them onto records in an external database.

Key words: Text mining, natural language processing, part-of-speech tagging, named entity recognition, parsing, information retrieval, information extraction.

1. Introduction

The biological sciences are very much knowledge-driven, with facts recorded in the text of scientific journals being the gold standard of accepted truth. However, the exponential growth in the amount of available biological literature *(1)* and the ever-increasing fragmentation of the life sciences into more and more specialized disciplines do not bode well for the ability of researchers to find and keep track of all the information relevant to their fields. The MEDLINE database *(2)* has added between 1,500 and 3,500 bibliographic records per working day since 2002 *(3)*, around three quarters of which have abstracts, and that now comprises about

Jonathan M. Keith (ed.), *Bioinformatics, Volume II: Structure, Function and Applications, vol. 453*
© 2008 Humana Press, a part of Springer Science + Business Media, Totowa, NJ
Book doi: 10.1007/978-1-60327-429-6 Springerprotocols.com

14 million entries. However, this is only one source of textual biological knowledge, along with full-text electronic journals and textbooks, patent filings, internal lab and project reports, comments fields in biomolecular databases, and topic-specific curated repositories such as The Interactive Fly *(4, 5)*. The development of natural language processing (NLP) techniques tailored to the biological domain has been driven by the resulting problems of literature overload that face scientists today.

It is easy to conceive of situations where a biologist's workload might be lessened by suitably targeted text-mining tools. One commonly cited scenario involves the construction of a descriptive picture of the causal relationships between a set of genes that are differentially expressed in a disease state, as revealed by microarray experiments for example. Although it is often possible to make some headway by clustering these genes according to their functional annotations, a literature trawl will be necessary in all but the simplest cases in order to identify the specific regulatory pathways involved *(6)*. Another task for which text-mining techniques have been seen as a potential boon is the curation of model organism databases. Curators at these projects spend many hundreds of biologist-hours each week scanning the newly published literature for discoveries about genes or proteins in the organism of interest, in order to enter the results into publicly searchable databases. Much research has been focused on developing NLP approaches to find the appropriate parts of the appropriate articles and extract candidate annotations to present to the curators *(7, 8)*. More generally, the problem of integrating structured databases with unstructured textual resources, through terminological indexing or document classification for example, is an ever-present challenge in industrial-scale biology.

1.1. What Makes Biological Text Mining Hard?

Regardless of the particular application or the approach taken, there are certain issues that are common to almost any natural language project in molecular biology, and as a result the practical section of this chapter focuses on these. The issues in question relate to the nomenclature of biomolecular entities. Although there is a canonical list of gene and protein names and symbols for each organism that has been the subject of genomic studies, maintained by organizations such as HUGO *(9, 10)*, FlyBase *(11, 12)* and SGD *(13, 14)*, along with a set of guidelines for naming new genes and proteins and referring to old ones, the extent to which these lists and rules are actually adhered to varies considerably from species to species.

As a result, some genes and proteins in many species have a surprisingly large array of names and symbols (**Fig. 25.1**), some of which can be written in several different ways once variations in punctuation, spacing, abbreviation, and orthography (Greek letters, Roman numerals, etc.) are taken into account. Although

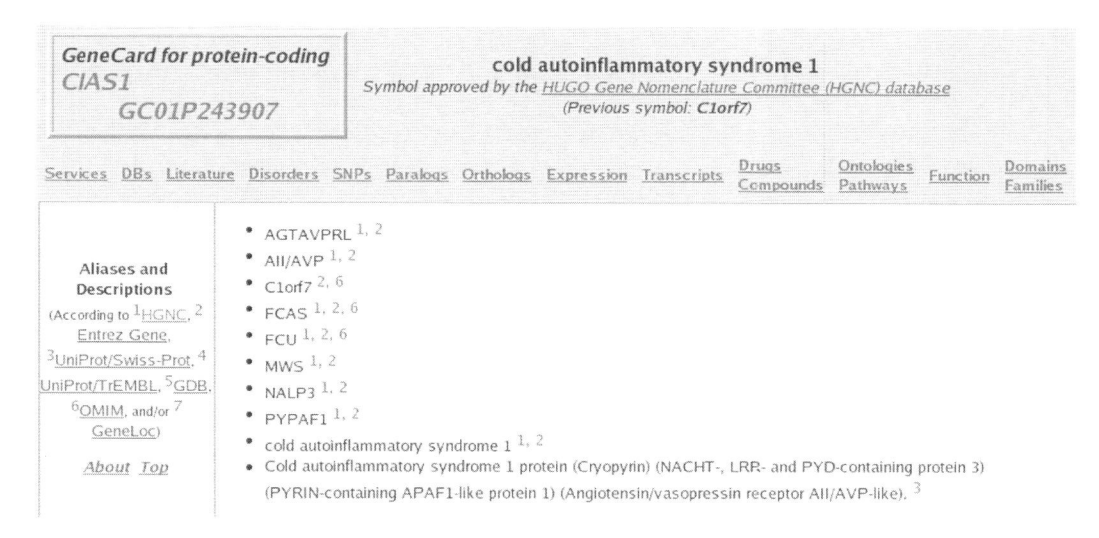

Fig. 25.1. Part of the GeneCards record for the human gene known most often as *CIAS1*, showing a plethora of names and symbols reflecting different aspects of function, structure, and pathology.

some resources, such as Gene Ontology *(15, 16)* and GeneCards *(17)* compile alias lists for the entities they cover, it is doubtful that such lists will ever be comprehensive enough to account for the variations in actual usage in the literature. Furthermore, even if a particular entity has only a small, tractable list of aliases, it is very common for some of those aliases to overlap with names for other genes or proteins in the same or different species, diseases or syndromes (*Angelman syndrome, brain tumor—see* **Note 1**), other biological or general scientific terms (*CELL, gluon*) or common English words (*CLOCK, for*).

In addition to these basic terminological hurdles and related problems involving names and abbreviations for chemicals, cell lines, laboratory procedures, and so on, any attempt to process biological text at a more linguistically aware level faces additional challenges. Most successful tools in mainstream computational linguistics are based on rules derived by hand or learned by statistical methods from large amounts of manually annotated data. However, equivalent quantities of annotated data are simply not available yet in the molecular biology domain. While a linguist can mark up part-of-speech (POS) tags (common noun, present tense verb, etc.) or phrase structure trees on text drawn from a corpus of newspaper articles or telephone conversations without any domain-specific background knowledge, the same is not true of biological corpora, in which the skills of both linguists and biologists are required. Enough manpower has been thrown at the POS annotation problem to produce decent amounts of this kind of data over the last few years, and projects such as GENIA *(18, 19)* and BioIE *(20, 21)* are beginning to redress the balance as far as syntactic annotation goes. The volume of grammatical structure data available is still more suited to evaluation than retraining, however, so various groups are testing mainstream language processing algorithms on biological text as-is, and using *ad hoc* methods where possible to detect and/or correct errors *(22–25)*.

2. Existing Tools for End Users

Despite these difficulties, various complete packages are available that enable end users in the life sciences to perform text mining without having any detailed knowledge of the computational processes involved. Most of these come in the form of Web applications, and many are free to use, at least for academic research.

2.1. Information Retrieval–Based Systems

One broad subset of the available software uses algorithms based on information retrieval (IR) to find relevant documents, clusters of documents, or connections between documents. IR is the branch of computer science that underlies search engines *(26)*; although the systems discussed here go significantly beyond the likes of Google *(27)* in semantic richness, and the ultimate output is not simply a list of documents, but a set of putative associations inferred from the subject matter of those documents, the term indexing techniques and statistical relevance measures employed are similar.

One of the original and most influential systems in this category was Arrowsmith *(28, 29)* (since re-implemented in a more user-friendly form at *(30)*), which is based on the idea of "complementary but disjoint literatures." Put simply, the idea is that if two biomedical topics (e.g., two genes, or a drug and a disease) are not discussed together in MEDLINE, there may yet be a set of keywords or phrases that appear in the search results for both that provide possible mechanisms by which the two may be related.

The original version of Arrowsmith also supports the ability to start with one topic and speculatively explore first directly related, and then indirectly related terms (and thus possible new hypotheses). This approach is also taken by BITOLA *(31, 32)*, a similar system designed to help mine potential gene–disease relationships from their linking topics and shared chromosomal locations. Both single- and dual-topic queries are supported by Manjal *(33)*, which also supports multiple topic queries and the graphical representation of results for easy navigation. All of these systems support the filtering of topic lists by semantic types such as Chemicals & Drugs or Amino Acid Sequence.

Another influential system with a markedly different approach to Arrowsmith is PubGene *(34, 35)*. This system constructs networks of genes by inferring a relationship whenever two genes are mentioned in the same MEDLINE abstract. The associations are ordered by frequency of co-occurrence, giving the user the ability to explore the networks visually and give preference to those associations that are supported by more instances. A similar approach is taken by the DRAGON suite of tools *(36)*, which allow visualization of co-occurrence networks in various areas of biology, including specialized query forms for *Streptomyces* and *Arabidopsis thaliana*. A more traditional take on IR is demonstrated by BioEx *(37)*, which allows results to be filtered by Medical Subject Headings (MeSH) terms, and also provides an alternative query form for finding definitions of biomedical terms.

Perhaps the most sophisticated example of IR in the biological domain is provided by Textpresso *(38, 39)*, an advanced search engine for articles about *Caenorhabditis elegans*. As well as keywords, the user can search by various ontology terms describing entities and events at the molecular or physiological level, and various categories that collect together descriptive terms referring to spatial or temporal relationships, positive and negative effects, and so on. Also, by allowing the user to restrict query matching to sentences rather than whole documents, queries can be built up that rival information extraction approaches (see the following) for richness and specificity without requiring much linguistic processing.

2.2. Information Extraction–Based Systems

Whereas information retrieval works at the level of documents, information extraction (IE) works at the level of individual pieces of information within documents—entities and their attributes, facts about them, and the events and relationships in which they are involved *(40)*. The preceding systems blur the distinctions between traditional IR and IE somewhat, for example, by using a synonym list (thesaurus) to relate several distinct keywords to the same entity, mapping keywords to ontological categories, or matching queries to sentences rather than whole documents; however, the whole-document/extracted-fact distinction remains useful.

The PubGene approach is extended and taken further into the realms of IE by ChiliBot *(41, 42)*, which has similar user inputs (a set of gene or protein names, along with other arbitrary keywords) and outputs (a network of relationships). However, linguistic analysis of individual sentences in each MEDLINE abstract is used to classify relationships into categories (e.g., inhibitory or stimulatory), thus allowing more specificity and richness than PubGene. A similar but more narrowly focused approach is taken by GIFT *(43, 44)*, which is designed to extract and categorize gene-regulation relationships from literature on *Drosophila melanogaster*. The resulting networks can be presented in graphical or tabular form.

Another example of a system that crosses the boundaries between IR and IE is EBIMed *(45)*, which initially retrieves MEDLINE abstracts according to a relevance-ranked keyword search, and then identifies the names of genes and proteins, Gene Ontology terms, and drugs and species by thesaurus lookup. These steps are similar to those taken by Textpresso. However, the results are presented to the user not document by document or excerpt by excerpt, but by pairwise associations between the entities identified in the previous step, as detected by regular expressions designed to flexibly match search terms and interaction keywords. Importantly, the entities are linked back to the corresponding databases at the EBI, allowing users to drill down to abstracts or sentences containing names of the entities of interest, or through to the database records of the entities themselves—a facility that is not supported by many current applications.

All the applications discussed so far have been Web-based; a different solution is presented by BioRAT *(46, 47)*, which is a downloadable Java desktop application. As in EBIMed, the user first specifies a search query, for which BioRAT finds and downloads matching documents from PubMed *(2)*—plus full article text where possible—or Google. Alternatively, a local document collection can be supplied. Then the user can construct extraction queries based on customizable keyword lists and templates, allowing BioRAT to detect entities, relationships, and events mentioned in the text and tabulate them. A similar approach is taken by POSBIOTM/W *(48)*, although in this case the user has less control over the design of queries.

2.3. Integrated Text Mining

Although the systems discussed in the preceding section are intended to provide standalone text mining capabilities, there are others that integrate the output of text-mining algorithms into broader knowledge management and discovery platforms. One such system is STRING *(49, 50)*, which enables users to interactively explore networks of interacting genes and proteins in which the network linkages are based on various sources of evidence. As well as literature co-occurrence, these also include pathway databases, microarray datasets, assay results, genomic context, and phylogenies. The results are displayed graphically, and the user can click through to any piece of supporting evidence, for example, a database record or MEDLINE abstract.

Another way to integrate text-mining functionality is to link NLP components to other more traditional bioinformatics components in a scientific workflow framework (*see* **Chapter 24** of this book). Two such pilot systems based on the web services model are described in *(51)*, which integrate information extraction with the results of BLAST *(52)* searches and SWISS-PROT *(53)* queries, respectively, in order to facilitate investigations of genetic diseases. The information extraction software used in these studies is described in *(54)*, although in principle any program which is scriptable—that is, which can be run as a batch job without interactive control—could be embedded in a computational pipeline like this.

2.4 Commercial Packages

In addition to the freely available systems already discussed, various commercial text-mining packages for the biological domain are on the market. These are available from companies such as Alma Bioinformatica *(55)*, Ariadne Genomics *(56)*, Autonomy *(57)*, Exergen Biosciences *(58)*, IBM *(59)*, LION bioscience *(60)*, Linguamatics *(61)*, PubGene's commercial spin-off *(62)*, SAS *(63)*, SPSS *(64)*, Stratagene *(65)*, and TEMIS *(66)*. Such solutions are primarily marketed to pharmaceutical and biotechnology companies.

3. Components for Developers

Although the systems described so far cover a wide variety of general purpose and specialist requirements, the unique characteristics of many bioinformatics projects necessitate the development of bespoke solutions. Also, it is frequently the case that algorithms tuned for one specific domain will outperform their general purpose rivals. Fortunately, there are various standalone programs and libraries available that encapsulate functional building blocks of text-mining systems and are available for free, some of which have been developed from the ground up for bioinformatics applications or retrained on biological data. Since there are hundreds of NLP components and libraries available from the broader computational linguistics community, this section gives preference to those that are of particular utility or interest to bioinformatics developers.

3.1. Sentence Boundary Detection

A simple and yet crucial foundation for many tasks is the splitting of text into individual sentences. It has been shown that even when simply inferring biological relationships from co-occurrences of entity names, operating at the level of individual sentences gives the best tradeoff between precision and recall scores (*see (67)*, and **Note 2**). At the other end of the scale, analysis of text using a sophisticated grammatical parser will be difficult if not totally impossible if the boundaries between the sentences are not accurately determined, as the parser will be hampered in its attempts to fit syntactic tree or graph structures to the sequences of words. Although one might think sentence boundaries in English are obvious, there are many situations in which a full stop does not indicate the end of a sentence, for example, or the first word in a new sentence does not begin with a capital letter (e.g., gene names in certain species). Sentence boundary detection functionality is often included in more sophisticated tools (e.g., MedPost

(68, 69)); alternatively, the LingPipe suite (70) includes a bio-medically trained sentence splitter in its Java API.

3.2. Part-of-Speech Tagging

POS tagging is useful as it provides evidence for decisions made by other components. For example, a named entity recognizer might filter out candidate gene names that do not end in nouns, to reduce false-positives, or an IE algorithm might examine sequences of POS tags around keywords in order to decide whether a regular expression matches. Incorrect POS tags can severely hamper the accuracy achievable by syntactic parsers (22). Although almost any POS tagger can be retrained on any available annotated text (e.g., from GENIA, BioIE, or MedTag (71, 72)), several taggers are available that come pretrained on biological or biomedical corpora, including MedPost, the GENIA tagger (73, 74), the Biosfier tagger (75), and ETIQ (76).

3.3. Named Entity Recognition

Named entity recognition refers to the process of tagging entity names where they appear in text—usually just genes and proteins, but sometimes a wider variety of entities such as cell lines and types, drugs and other small molecules, diseases, etc. There is a wide selection of named entity recognizers available with different strengths, weaknesses, entity classes and modes of operation, and the interested reader is advised to compare (or perhaps combine) several. Popular examples include LingPipe and ABNER (77, 78), both of which can be invoked separately or used as Java APIs; GAPSCORE (79, 80), which can be accessed remotely as a Web service from several programming languages; POSBIOTM/NER, which can also be used as a web application (81, 82); and NLProt (83, 84), KeX/PROPER (85, 86), GeneTaggerCRF (75), and ABGene (87, 88), which are command line–based.

Note, however, that with the exception of NLProt, none of these tools link back from the name of an entity to the corresponding accession number in an appropriate database. This means that they are incapable of actually identifying or disambiguating the entities whose names they find; this functionality must be supplied by the user (see **Section 3.6**). Their utility comes from their ability to detect novel variations on entity names that are not present verbatim in the databases, and also to discriminate (albeit imperfectly) between true entity names and other strings that they might resemble, based on textual context.

3.4. Sentence Structure Extraction

Unless the objective is to treat each sentence or document as a "bag of words," in which connections are simply inferred from the presence or frequency of certain terms, some syntactic (grammatical) processing will be necessary to extract the semantics (meaning) of the text. One approach is to define regular expressions over keywords (or parts of words), entity names, and POS tags, so that queries can bind tightly to the parts of a sentence that are required for a particular fact to be expressed, and more loosely to those parts that are less necessary or subject to more variability of expression. This can be accomplished in the Corpus Workbench query processor (*see (89, 90)* and **Note 3**), which also deals elegantly with the linear phrase (or "chunk") boundaries that can be predicted by the latest version of the GENIA tagger.

Such analyses are still, however, somewhat "shallow"; that is, they group words into phrases but they do not model the relationships between words or phrases. Consider for example the sentence "the C26 element alone, unlike other NF-kappa B-binding elements, could not be activated by p40tax." There is a long-distance (negative) relationship between p40tax and C26 via the verb activated, which skips over the intervening phrase. A parser can produce data structures representing these relationships, which may be in the form of phrase structure trees, like those produced by the Charniak parser (*see (91, 92)* and **Note 4**), or word dependency graphs, like those generated by the Link Grammar parser (*see (93, 94)* and **Note 5**). The potential of such data structures for enhancing biomedical information extraction has become the focus of much research.

3.5. Indexing and Searching Documents

The resource of choice for open-source search engine development is Lucene *(95, 96)*, a highly capable Java (*see* **Note 6**) toolkit from the Apache Software Foundation, makers of the popular Apache Web server. The LingPipe distribution contains the Lucene libraries and comes with a tutorial on using them to store, index, and retrieve MEDLINE abstracts. Also included is a similar tutorial on populating a traditional SQL database with MEDLINE, and using LingPipe's named entity recognizer to collect entity occurrence and co-occurrence frequencies.

3.6. Case Study: Dictionary-Based Gene/ Protein Name Identification

In order to illustrate some of the challenges involved in text mining and propose a potential solution to the named entity problem that has a bearing on most natural language projects in bioinformatics, we present a simple, high throughput named entity recognition and identification protocol. Inspired by (97–100), this algorithm is designed for and tested on the human genome, requires no training data, and unlike most of the named entity tools described in the preceding, provides a genomic database accession number for each hit (*see* **Note 7**).

The principles behind this approach are twofold. First, although dictionary-based methods have serious coverage limitations for most organisms, the kinds of nomenclature variations that lead to false-negative results are, to some extent, predictable and automatically reproducible. Second, if broad coverage results in a large number of false-positives—acronyms, words, or phrases that resemble genuine entity names—an existing named entity recognizer with high precision (or an ensemble of several) can be employed later as a filter in order to prune the set of candidate hits. Our experimental protocol therefore proceeds as follows:

1. Construct a human gene/protein name dictionary from several databases.

2. Generate plausible lexical and orthographic variations of each name.

3. Compile a "keyword tree" data structure to hold the dictionary in memory.

4. Search a collection of sentences for names occurring in the keyword tree.

5. Discard hits that do not agree with a third-party statistical named entity recognizer.

3.6.1. Collating the Dictionary

The dictionary collation task can be broken down into the following discrete steps:

1. Collect human gene/protein symbols and full names from UniProt (101), GeneCards, and Ensembl (102) (*see* **Note 8**).

2. Join lists by Ensembl ID and remove redundant names.

3. Collapse names to lower case.

4. Remove most punctuation (*see* **Note 9**).

5. Replace all white space with logical token boundaries.

6. Insert token boundaries between runs of letters and digits, and around the remaining punctuation.

7. Check for and remove redundant names again.

The initial redundant list of unprocessed records from the three databases gave us 20,988 entities, of which all but 34 had at least one name, and 92,052 name-entity pairs. The normalization process ensured that, for example, the names "NF Kappa-B", "NF Kappa B", or "NF-kappa-B" would all lead to the same normalized form "nf kappa b"—in which the space now represents a logical token boundary rather than an actual ASCII space character.

3.6.2. Generating the Name Variants

Our name-variant generator is deterministic and rule based, unlike the probabilistic algorithm described in (99), and has two components: an orthographic rewriter, and a lexical rewriter. Both are applied recursively to each normalized name in the dictionary until no new variations are generated. The lexical rewriter runs first, taking the following steps:

1. For each token in the name:

   a. If the token is in the stopword list (*see* **Note 10**), generate a copy of the name with this token left out.

   b. Add a new name to the dictionary if it is not already present.

   For a name like "nacht lrr and pyd containing protein 3", for example, the resulting variations are "nacht lrr pyd containing protein 3", "nacht lrr and pyd containing 3," and "nacht lrr pyd containing 3". Then the orthographic rewriter runs:

2. For each token in the name:

    a. If the token is numeric and between 1 and 20:

      i. Generate a copy of the name.

      ii. Replace this token in copy with a Roman numeral equivalent.

    b. Otherwise, if the token is a Roman numeral between I and XX:

      i. Generate a copy of the name.

      ii. Replace this token in copy with numeric equivalent.

    c. Otherwise, if the token is a single letter with a Greek equivalent:

      i. Generate a first copy of the name.

      ii. Replace this token in the first copy with spelled-out Greek letter (e.g., "alpha").

      iii. Generate a second copy of the name.

      iv. Replace this token in the second copy with an SGML-encoded Greek character (e.g., "&agr;" *see* **Note 11**).

    d. Otherwise, if the token is a spelled-out Greek letter:

      i. Generate a copy of the name.

      ii. Replace this token in the copy with a single-letter equivalent.

    e. Add new name(s) to dictionary if not already present.

Thus, an original name such as "ATP1B", having been normalized to the three-token sequence "atp 1 b", would be expanded by the orthographic rewriter to the following set of names: "atp 1 b", "atp 1 beta", "atp 1 &bgr;", "atp i b", "atp i beta", "atp i &bgr;" (*see* **Note 12**).

Once again, redundant names for each entity were pruned after this process, as were names that consisted of a single character, names with more than 10 tokens, and names that exactly matched a list of around 5,000 common English words. This left 354,635 distinct names in the dictionary, of which 17,217 (4.9%) were ambiguous (i.e., referring to more than one entity). On average, each entity had 18.1 names, and each name referred to 1.1 entities.

3.6.3. Compiling the KeyWord Tree

In order to use the fast, scalable Aho-Corasick dictionary search algorithm (*see* **Note 13**), the dictionary must be compiled into a data structure called a keyword tree (*see* **Note 14**). This takes several minutes and a considerable amount of memory (*see* **Note 15**), but allows the corpus to be searched at a speed proportional only to the corpus size and not to the number of entities in the dictionary.

3.6.4. Searching the Corpus

Once the keyword tree is built, the corpus can be searched as follows:

1. For each sentence in the corpus:
   a. Collapse to lower case.
   b. Remove punctuation, as in **Section 3.6.1**.
   c. Retokenize as in **Section 3.6.1**.
   d. For each letter in the sentence:
      i. Pass the letter to the AhoCorasick algorithm.
      ii. the algorithm signals a hit ending here:
         —Store the start and end points and the content of the hit.
   e. For each hit in this sentence:
      i. Remove the hit if it is subsumed by another hit (*see* **Note 16**).

3.6.5. Filtering the Results

Like all purely dictionary-based methods, our approach makes no attempt to filter false-positives out of its results. However, it is relatively simple to employ a third-party named entity recognizer based on word and sentence features as a filter. We used ABNER as it has a well-documented Java API and good published performance scores, although any of the systems described in **Section 3.3**—or an ensemble of several—could be used instead.

We used the filter in two modes; results for each are given in the following. In the *exact* mode, all hits generated by our dictionary-based method were required to exactly match hits generated by ABNER. This is equivalent to running ABNER first, and then searching for the candidate entities it discovers in the dictionary. In the *inexact* mode, we only required that hits from the dictionary search overlapped with hits generated by ABNER in order to be retained. This is a looser criterion that is more forgiving of boundary detection errors.

3.6.6. Evaluation

In order to evaluate the effectiveness of our technique, we derived a test dataset from the non–species-specific GENETAG-05 corpus (the named entity portion of MedTag) by selecting sentences that came from journals related to human biology or medicine (*see* **Note 17**), or which contained the strings "human", "patient" or "child." This yielded 3,306 sentences. We compared the output of our software against this gold standard after enabling each major feature, calculating the precision and recall score each time (*see* **Note 18**), as well as the proportion of true-positives for which the name refers to more than one entity. The results are presented in **Table 25.1**; F-measure is the harmonic mean of precision and recall (*see* **Note 19**) and is designed to penalize systems in which the two scores are grossly out of balance.

Table 25.1
Precision, recall, and F-measure measured after implementing each feature, on a set of 3,306 human-related sentences from GENETAG-05

Features in use	P	R	F
Just normalization (tokens and punctuation)	23.1	31.0	26.4
+ Filtering of unsuitable names	40.4	30.9	35.0
+ Case insensitivity	35.6	41.4	38.3
+ Numeral substitution (Roman-Arabic)	35.6	42.0	38.7
+ Letter substitution (Roman-Greek)	35.6	42.4	38.7
+ Uninformative word removal	32.4	43.5	37.1
+ ABNER filtering (exact)	92.6	19.6	32.3
+ ABNER filtering (inexact)	58.6	37.9	46.0
ABNER baseline performance	82.3	81.4	81.9

All values are percentages.

3.6.7. Making Sense of the Results

These results illustrate several important points. As with many NLP tasks, there is a natural trade-off between precision and recall. This trade-off is most strikingly apparent in the difference between requiring exact matches and allowing inexact matches when using ABNER as a filter.

The ABNER baseline performance is included to demonstrate what a non-dictionary-based named entity recognizer is capable of, but it must be noted again that the output of ABNER does not include links back to genomic database identifiers so it cannot provide any information on *which* gene or protein has been tagged each time. The very low recall score when allowing only exact matches between ABNER and our algorithm makes it clear how few of the entities tagged by ABNER are *exactly* present in even an expanded dictionary. By contrast, our algorithm makes many more errors, but provides a link to a genomic database (in this case, an Ensembl record) for each entity it finds.

Because the test set was generated in a partly automated manner, we manually examined a random subset of 300 names in the test set that were missing from the dictionary, in order to determine the cause of their omission. Although many were due to genuine deficiencies in the dictionary, a considerable number (33, 11.0%) were occurrences of plural, family, group, or otherwise generic names that we did not set out to cover (*see* **Note 20**). An even larger subset (44, 14.7%) were genes or proteins from other organisms that happened to be mentioned in a human-relevant sentence, but as many as 20 (6.7%) seemed to be annotator errors (*see* **Note 21**). Taken together, these figures suggest that the effective false-negative (recall error) rates may be around a third lower than those reported in **Table 25.1**.

Our prototype system was originally designed for a bibliometric investigation into research coverage of the human genome, and thus was built with a gene's-eye view of human biology. However, the results drew attention to the inadequacy of considering names for genes and for their protein products to be interchangeable, a shortcut also taken by many (perhaps most) named entity recognition projects. This practice disregards the fact that biomedically important proteins are often hetero-oligomers or other complexes composed of the products of several genes (*see* **Note 22**).

3.6.8. Dealing with Ambiguity

There are two kinds of ambiguity that must be considered when detecting biological entities of a specific category. The first is ambiguity between multiple entities in that category, for example, gene acronyms that stand for more than one gene; and the second is ambiguity between entities of that category and strings of text that might represent other classes of entity or any other words or phrases. Our algorithm itself does not make any attempt to deal explicitly with either kind of ambiguity. The latter kind of ambiguity is tackled by the use of ABNER as a filter, as a given string may or may not be tagged as an entity name by ABNER depending on its context in the sentence. The former kind needs a more subtle approach. We briefly discuss two algorithms that could potentially be applied to either of these problems, and one that only addresses the issue of ambiguous names in a gene/protein dictionary.

The first approach is used in EBIMed, described in (103), and is geared specifically toward the disambiguation of abbreviations, but is not restricted to entities of any particular class. It makes a distinction between local and global abbreviations; local abbreviations are those whose long forms are given before they are used, and these are resolved with the help of an automatically extracted dictionary, whereas global abbreviations, which do not have explicitly stated long forms, require the use of statistical term-frequency measures. A very different approach is proposed in (104) for ambiguous biomedical terms from the Unified Medical Language System (UMLS) (105), and makes use of the fact that UMLS comes with various hand-curated tables of related concepts. For each possible sense (corresponding concept) of an ambiguous term, the algorithm counts the number of related concepts that are referred to by terms in the textual neighborhood of the ambiguous term, and the sense with the highest count wins. Given the availability of various curated resources in the molecular biology domain, that list of already-known relationships between genes, proteins, and other entities or concepts, it is easy to see how one might adapt this approach.

Finally, the simple but mostly effective method used in *(97)* is to look elsewhere in the same abstract for other, unambiguous aliases for the entities referred to by each ambiguous name. In the rare cases, in which this does not resolve the ambiguity—that is, more than one of the senses of the ambiguous name seem to be referred to by other names—the ambiguous name is discarded.

3.6.9. Concluding Remarks

An important lesson from this worked example is that although naïve dictionary-matching methods perform comparatively poorly, a few simple variant-generation methods can improve recall by around 50%, and these methods in conjunction with inexact filtering from a high performance named entity recognizer can roughly double overall performance (F-measure). The potential is great for improving on these early gains with a more comprehensive dictionary (*see* **Note 23**), and more variant generation rules (*see* **Note 24**).

Although the precision, recall, and F-measure scores for our dictionary algorithm are considerably less than those achieved by ABNER on the same data, the advantages of using an algorithm that returns an actual database identifier are substantial. Furthermore, it must be noted that many of the existing text-mining systems described in **Section 2** use simple exact-matching techniques that can only be expected to perform about as well as the lowest scores reported in the preceding. In a real-world application based on our method, one might want to consider a best of both worlds approach. A set of high-confidence annotations could first be made based on inexact agreement between the dictionary method and ABNER or another named entity recognizer, and then augmented with a set of lower-confidence annotations obtained by finding (via fuzzy string matching) the closest name in the database to each of the remaining tags in the recognizer's output.

4. Notes

1. In the human genome, names like these usually refer to heritable diseases or cancers caused by mutations in the gene. The situation is not so simple with other organisms; *brain tumor* is a *D. melanogaster* gene that results in a tumor-like growth in the developing larva's brain that is lethal before adulthood (106), and is therefore somewhat unlike the class of human cancers with which it shares its name.

2. *Precision* is the proportion of reported instances that are correct, and *recall* is the proportion of true instances that were found, where an instance is just an occurrence of a name in a named entity task, for example, or perhaps a pair of interacting entities in an information extraction task. Typically, it is straightforward to increase either of these scores at the expense of the other, for most NLP algorithms; the real challenge is increasing one while the other also rises, or at least stays roughly the same.

3. An example of using Corpus Workbench to query biological text is in *(43)*.

4. Although the Charniak parser was originally designed for and trained on non-specialist English, a modified version with its POS component trained on biomedical text *(23)* is available from the same web site.

5. Although the Link Grammar was originally designed for non-specialist English, a method for augmenting its vocabulary automatically with a biomedical lexicon is described in (107).

6. In addition, Lucene is currently being ported to C; see the cited web site for details.

7. For cases in which the same name refers to several distinct entities (homonyms), this method returns all of the associated accession numbers.

8. We treated genes and their protein products as interchangeable for the purposes of our investigation (a shortcut that is common in many named entity recognition systems), although we discuss some of the drawbacks of this approach in **Section 3.6.7**.

9. We retained "&" and ";" as these are necessary for understanding SGML escape sequences, which sometimes appear in MEDLINE to represent special characters—e.g. "&agr;" for "α". We also kept "+" as it is useful for distinguishing positive ions, which are common in protein names, from similar-looking sequences of characters. We decided to remove all instances of "-" though, as it is much more commonly used as a hyphen or dash, which is less selective.

10. This is a list of uninformative and optional words that occur in gene and protein names, such as "the", "of", and "a", as well as generic terms such as "gene", "protein", "precursor", "molecule", etc.

11. Note that this is a one-way translation as there are no SGML character codes in our dictionary, although the reverse translation could be implemented straightforwardly.

12. The letter "i" is excluded from the Roman to Greek letter conversion to avoid unnecessary and unhelpful generation of the "iota" character every time a digit "1" was processed.

13. A thorough description of the algorithm and its associated data structure is given in (108). Rather than re-implement it ourselves, we adapted a Java version available at (109).

14. We added a token boundary character at the beginning and end of each entity name so that they would only match when aligned with token boundaries in the corpus; this means that the name "octn2" will be matched in the string "octn2-mediated", for example, but the name "car" will not be matched in the string "carrier".

15. Our non-optimized prototype implementation requires up to 768 Mb of heap for the Java virtual machine to complete the task happily.

16. This step ensures that, for example, the string "OB receptor" does not also register a hit for "OB".

17. We looked for journal names containing the strings "Hum" or "Child", or "Clin" or "Med" but not "Vet"—the journal names are supplied in standard abbreviated format.

18. Due to the existence of multiple acceptable alternative annotations for many of the names in GENETAG-05, this calculation was slightly more complicated than usual. The formula for recall is:

$$\frac{\# \text{ true positives}}{\# \text{ true positives } + \# \text{ false negatives}}$$

and the formula for precision is:

$$\frac{\# \text{ true positives}}{\# \text{ true positives } + \# \text{ false positives}}$$

but in this case we had to define a true positive as any predicted name that matches any one of the alternative annotations for a given instance of an entity in GENETAG-05, a false-positive as any predicted name that matches none of the annotations for any entity in GENETAG-05, and a false-negative as a GENETAG-05 entity with no annotations predicted by our algorithm. Also, since GENETAG-05 does not contain any kind of instance identifier, we had to pre-process the annotation lists supplied to group annotations together based on overlaps, in order to determine which groups of annotations referred to the same instance of the same entity. Finally, an additional level of complexity was added by the fact that our dictionary algorithm will generate multiple hits for different entities in which several entities share the same name. For scoring purposes, each of these ambiguous hits was treated as a single (true- or false-) positive.

19. It is calculated as follows:

$$\frac{2 \times P \times R}{P + R}$$

20. These included cases such as "E1 genes", "anti-viral proteins", "MHC class II promoters" and "blood hemoglobin".

21. Many of these were non–gene/protein entities that had been tagged in error, such as "statin" (a class of drug), "platelet activating factor" (a phospholipid), and "immunoperoxidase" (a lab technique). Also present were several sequence features that broke the annotators' own guidelines on what should be included, for example "DXS52" (a sequence tagged site), "Alu" (a class of common repeat sequences), and "HS40" (a regulatory element).

22. For example, "fibrinogen," "nuclear factor (NF)-kappaB", "casein", and the immunoglobulin family.

23. An obvious way to improve on the dictionary we used would be to include gene/protein families as distinct meta-entities, as these are often mentioned collectively by authors, for example "MAPK" or "human E2F". Another option would be to include names of domains (e.g., "SH3" and "PH") and complexes of multiple gene products as discussed in **Section 3.6.7**.

24. For example, automatic acronym generation, reordering of words in long names, and prepending "h" (for "human") to short symbols, as this is a common abbreviation in the literature.

Acknowledgments

This work was supported by the Biotechnology and Biological Sciences Research Council and AstraZeneca. The authors thank Mark Halling-Brown for supplying the dictionary and A. G. McDowell for implementing (and advising on) the Aho-Corasick algorithm.

References

1. Cohen, K. B., Hunter, L. (2004) Natural language processing and systems biology, in (Dubitzky, W., Azuaje, F., eds.), *Artificial Intelligence Methods and Tools for Systems Biology*. Kluwer, Dordrecht.

2. MEDLINE via PubMed, http://www.pubmed.org/

3. MEDLINE Fact Sheet, http://www.nlm.nih.gov/pubs/factsheets/medline.html

4. Brody, T. (1999) The Interactive Fly: gene networks, development and the Internet. *Trends Genet* 15, 333–334.

5. The Interactive Fly, http://flybase.bio.indiana.edu/allieddata/lk/interactivefly/aimain/1aahome.htm

6. Shatkay, H., Edwards, S., Wilbur, W. J., et al. (2000) Genes, themes, and microarrays: Using information retrieval for large-scale gene analysis, in (Bourne, P., Gribskov, M., Altman, R., et al., eds.), *Proceedings of the Eighth International Conference on Intelligent Systems for Molecular Biology*. AAAI Press, Menlo Park, CA.

7. Hersh, W., Bhupatiraju, R. T. (2003) Of mice and men (and rats and fruit flies): the TREC genomics track, in (Brown, E., Hersh, W., and Valencia, A., eds.), *ACM SIGIR'03 Workshop on Text Analysis and Search for Bioinformatics: Participant Notebook*. Association for Computing Machinery, Toronto, Canada.

8. Hirschman, L., Yeh, A., Blaschke, C., et al. (2005) Overview of BioCreAtIvE: critical assessment of information extraction for biology. *BMC Bioinformatics* 6:S1.

9. Wain, H. M., Bruford, E. A., Lovering, R. C., et al. (2002) Guidelines for human gene nomenclature. *Genomics* 79, 464–470.

10. HUGO Gene Nomenclature Committee, http://www.gene.ucl.ac.uk/nomenclature/

11. Drysdale, R. A., Crosby, M. A., The FlyBase Consortium. (2005) FlyBase: genes and gene models. *Nucl Acids Res* 33, D390–D395.

12. FlyBase: A Database of the *Drosophila* genome, http://flybase.bio.indiana.edu/

13. Cherry, J. M. (1995) Genetic nomenclature guide. Saccharomyces cerevisiae. in Trends Genetics Nomenclature Guide, *Trends Genetics*, p. 11–12.

14. *Saccharomyces* Genome Database, http://www.yeastgenome.org/

15. Ashburner, M., Ball, C. A., Blake, J. A., et al. (2000) Gene ontology: tool for the unification of biology. The Gene Ontology Consortium. *Nat Genet* 25, 25–29.

16. Gene ontology, http://www.geneontology.org

17. Rebhan, M., Chalifa-Caspi, V., Prilusky, J., et al. (1997), GeneCards: encyclopedia for genes, proteins and diseases. http://bioinformatics.weizmann.ac.il/cards

18. Kim, J.D., Ohta, T., Tateisi, Y., et al. (2003) GENIA corpus—a semantically annotated corpus for biotextmining. *Bioinformatics* 19, i180–i182.

19. The GENIA Project, http://www.tsujii.is.s.u-tokyo.ac.jp/~genia/

20. Kulick, S., Bies, A., Liberman, M., et al. (2004) Integrated annotation for biomedical information extraction, in (Hirschman, L., Pustejovsky, J., eds.), *HLTNAACL 2004 Workshop: BioLINK 2004, Linking Biological Literature, Ontologies and Databases*. Association for Computational Linguistics, Boston.

21. Mining the Bibliome, http://bioie.ldc.upenn.edu/

22. Clegg, A. B., Shepherd, A. J. (2005) Evaluating and integrating treebank parsers on a biomedical corpus, in (Jansche, M., ed.), *Association for Computational Linguistics Workshop on Software CDROM*. Association for Computational Linguistics, Ann Arbor, MI.

23. Lease, M., Charniak, E. (2005) Parsing biomedical literature, in (Dale, R., Wong, K.-F., Su, J., et al., eds.), *Proceedings of the Second International Joint Conference on Natural Language Processing (IJCNLP'05)*. Jeju Island, Korea.

24. Wermter, J., Fluck, J., Stroetgen, J., et al. (2005) Recognizing noun phrases in biomedical text: an evaluation of lab prototypes and commercial chunker, in (Hahn, U., and Valanaa A. eds.), *Proceedings of the First International Symposium on Semantic Mining in Biomedicine*. Hinxton, UK.

25. Grover, C., Lapata, M., Lascarides, A. (2005) A comparison of parsing technologies for the biomedical domain. *Nat Language Engin* 11, 27–65.

26. van Rijsbergen, C. J. (1979) *Information Retrieval*, 2nd ed. Butterworths, London.

27. Google, http://www.google.com/

28. Smalheiser, N. R., Swanson, D. R. (1998) Using ARROWSMITH: a computer-assisted approach to formulating and assessing scientific hypotheses. *Comput Methods Progr Biomed* 57, 149–153.

29. Arrowsmith 3.0, http://kiwi.uchicago.edu/

30. Arrowsmith @ University of Illinois at Chicago, http://arrowsmith.psych.uic.edu/arrowsmith_uic/index.html

31. Hristovski, D., Peterlin, B., Mitchell, J. A., et al. (2003) Improving literature based discovery support by genetic knowledge integration. *Stud Health Technol Informat* 95, 68–73.

32. BITOLA, http://www.mf.unilj.si/bitola/

33. Manjal, http://sulu.infoscience.uiowa.edu/Manjal.html

34. Jenssen, T.-K., Lægreid, A., Komorowski, J., et al. (2001) A literature network of human genes for high-throughput analysis of gene expression. *Nat Genet* 28, 21–28.

35. PubGene, http://www.pubgene.org/

36. DRAGON Genome Explorer, http://research.i2r.astar.edu.sg/DRAGON/

37. BioEx, http://monkey.dbmi.columbia.edu/Biology/

38. Müller, H.-M., Kenny, E. E., Sternberg, P. W. (2004) Textpresso: an ontology-based information retrieval and extraction system for biological literature. *PLoS Biology* 2(11).

39. Textpresso, http://www.textpresso.org/

40. NIST Message Understanding Conference web archive, http://www.itl.nist.gov/iaui/894.02/related_projects/muc/

41. Chen, H., Sharp, B. M. (2004) Content-rich biological network constructed by mining PubMed abstracts. *BMC Bioinformatics* 5:147.

42. ChiliBot, http://www.chilibot.net/index.html

43. Domedel-Puig, N., Wernisch, L. (2005) Applying GIFT, a Gene Interactions Finder in Text, to fly literature. *Bioinformatics* 21, 3582–3583.

44. Gene Interactions Finder in Text, http://gift.cryst.bbk.ac.uk/gift/

45. EBIMed, http://www.ebi.ac.uk/Rebholz-srv/ebimed/index.jsp

46. Corney, D. P. A., Buxton, B. F., Langdon, W. B., et al. (2004) Biorat: extracting biological information from full-length papers. *Bioinformatics* 20, 3206–3213.

47. BioRAT: a Biological Research Assistant for Text Mining, http://bioinf.cs.ucl.ac.uk/biorat/

48. POStech Biological Text-Mining System, http://isoft.postech.ac.kr/Research/Bio/bio.html

49. von Mering, C., Jensen, L. J., Snel, B., et al. (2005) STRING: known and predicted protein-protein associations, integrated and transferred across organisms. *Nucl Acids Res* 33, D433–D437.

50. STRING—Search Tool for the Retrieval of Interacting Genes/Proteins, http://string.embl.de/

51. Gaizauskas, R., Davis, N., Demetriou, G., et al. (2004) Integrating biomedical text mining services into a distributed workflow environment, in *Proceedings of the UK e-Science All Hands Meeting*. Nottingham, UK.

52. Altschul, S. F., Gish, W., Miller, W., et al. (1990) Basic local alignment search tool. *J Mol Biol* 215, 403–410.

53. Boeckmann, B., Bairoch, A., Apweiler, R., et al. (2003) The SWISS-PROT protein knowledge base and its supplement TrEMBL in 2003. *Nucl Acids Res* 31, 365–370.

54. Gaizauskas, R., Hepple, M., Davis, N., et al. (2003) Ambit: Acquiring medical and biological information from text, in *Proceedings of the UK e-Science All Hands Meeting*, Nottingham, UK.

55. Alma Bioinformatica, http://www.almabioinfo.com/

56. Ariadne Genomics, http://www.ariadnegenomics.com/

57. Autonomy, http://www.autonomy.com/

58. Exergen Biosciences, http://www.exergenbio.com/

59. IBM, http://www.ibm.com/

60. LION bioscience, http://www.lionbioscience.com/

61. Linguamatics, http://www.linguamatics.com/

62. PubGene, http://www.pubgene.com/

63. SAS, http://www.sas.com/

64. SPSS, http://www.spss.com/

65. Stratagene, http://www.stratagene.com/

66. TEMIS, http://www.temis-group.com/

67. Ding, J., Berleant, D., Nettleton, D., et al. (2002) Mining MEDLINE: abstracts, sentences, or phrases? in *Proceedings of the 7th Pacific Symposium on Biocomputing*. World Scientific Publishing, Lihue, HI.

68. Smith, L., Rindflesch, T., Wilbur, W. J. (2004) MedPost: a part-of-speech tagger for biomedical text. *Bioinformatics* 20, 2320–2321.

69. Medpost ftp site, ftp://ftp.ncbi.nlm.nih.gov/pub/lsmith/MedPost/ medpost.tar.gz

70. LingPipe, http://alias-i.com/lingpipe/

71. Smith, L. H., Tanabe, L., Rindflesch, T., et al. (2005) MedTag: a collection of biomedical annotations, in (Bozanis, P., and Houstis, E. N., eds.), *Proceedings of the ACLISMB Workshop on Linking Biological Literature, Ontologies and Databases: Mining Biological Semantics*. Association for Computational Linguistics, Detroit.

72. MedTag, ftp://ftp.ncbi.nlm.nih.gov/pub/lsmith/MedTag

73. Tsuruoka, Y., Tateishi, Y., Kim, J.-D., et al. (2005) Developing a robust part-of-speech tagger for biomedical text, in *Advances in Informatics: 10th Panhellenic Conference on Informatics*. Springer-Verlag, Volos, Greece.

74. GENIA Tagger, http://www-tsujii.is.s.u-tokyo.ac.jp/GENIA/tagger/

75. Biosfier Software Distribution, http://www.cis.upenn.edu/datamining/software_dist/biosfier/

76. ETIQ, http://www.lri.fr/ia/Genomics/formulaire_ETIQ.html

77. Settles, B. (2005) ABNER: an open source tool for automatically tagging genes, proteins and other entity names in text. *Bioinformatics* 21, 3191–3192.

78. ABNER: A Biomedical Named Entity Recognizer, http://www.cs.wisc.edu/~bsettles/abner/

79. Chang, J. T., Schtze, H., Altman, R. B. (2004) GAPSCORE: finding gene and protein names one word at a time. *Bioinformatics* 20, 216–225.

80. Gene and Protein Name Server, http://bionlp.stanford.edu/gapscore/

81. Song, Y., Kim, E., Lee, G. G., et al. (2005) POSBIOTM-NER: a trainable biomedical named-entity recognition system. *Bioinformatics* 21, 2794–2796.

82. POStech Biological Text-Mining System, http://isoft.postech.ac.kr/Research/BioNER/POSBIOTM/NER/main.html

83. Mika, S., Rost, B. (2004) Protein names precisely peeled off free text. *Bioinformatics* 20, i241–i247.

84. NLProt, http://cubic.bioc.columbia.edu/services/nlprot/

85. Fukuda, K., Tsunoda, T., Tamura, A., et al. (1998) Toward information extraction: Identifying protein names from biological papers, in *Proceedings of the Pacific Symposium on Biocomputing (PSB'98)*, Hawaii.

86. KeX, http://www.hgc.jp/service/tooldoc/KeX/intro.html

87. Tanabe, L., Wilbur, W. J. (2002) Tagging gene and protein names in biomedical text. *Bioinformatics* 18, 1124–1132.

88. ABGene, ftp://ftp.ncbi.nlm.nih.gov/pub/tanabe/

89. Christ, O. (1994) A modular and flexible architecture for an integrated corpus query system, in *Proceedings of the Third Conference on Computational Lexicography and Text Research (COMPLEX'94)*, Budapest.

90. IMS Corpus Workbench, http://www.ims.uni-stuttgart.de/projekte/CorpusWorkbench/

91. Charniak, E. (2000) A maximum-entropy-inspired parser, in *Proceedings of the first conference on North American chapter of the Association for Computational Linguistics*, Morgan Kaufmann Publishers, San Francisco.

92. BLLIP Resources, http://www.cog.brown.edu/Research/nlp/resources.html

93. Sleator, D., Temperley, D. (1993) Parsing English with a link grammar, in *Proceedings of the Third International Workshop on Parsing Technologies*, Tilburg, Netherlands.

94. Link Grammar, http://www.link.cs.cmu.edu/link/

95. Hatcher, E., Gospodnetić, O. (2004) *Lucene in Action*. Manning Publications, Greenwich, CT.

96. Lucene, http://lucene.apache.org/

97. Cohen, A. M. (2005) Unsupervised gene/protein named entity normalization using automatically extracted dictionaries, in *Proceedings of the ACL-ISMB Workshop on Linking Biological Literature, Ontologies and Databases: Mining Biological Semantics*, Association for Computational Linguistics, Detroit.

98. Tsuruoka, Y., Tsujii, J. (2003) Boosting precision and recall of dictionary-based protein name recognition, in (Ananiadou, S., Tsujii, J., eds.), *Proceedings of the ACL 2003 Workshop on Natural Language Processing in Biomedicine*. Association for Computational Linguistics, Sapporo, Japan.

99. Tsuruoka, Y., Tsujii, J. (2003) Probabilistic term variant generator for biomedical terms, in *Proceedings of the 26th Annual International ACM SIGIR Conference*, Association for Computing Machinery, Toronto, Canada.

100. Fundel, K., Güttler, D., Zimmer, R., et al. (2005) A simple approach for protein name identification: prospects and limits. *BMC Bioinformatics* 6(Suppl 1):S15.

101. Apweiler, R., Bairoch, A., Wu, C., et al. (2004) UniProt: the Universal Protein knowledgebase. *Nucl Acids Res* 32, D115–D119.

102. Hubbard, T., Andrews, D., Caccamo, M., et al. (2005) Ensembl 2005. *Nucl Acids Res* 33, D447–D453.

103. Gaudan, S., Kirsch, H., Rebholz-Schuhmann, D. (2005) Resolving abbreviations to their senses in Medline. *Bioinformatics* 21, 3658–3664.

104. Widdows, D., Peters, S., Cederberg, S., et al. (2003) Unsupervised monolingual and bilingual word-sense disambiguation of medical documents using UMLS, in (Ananiadou, S., Tsujii, J., eds.), *Proceedings of the ACL 2003 Workshop on Natural Language Processing in Biomedicine*. Association for Computational Linguistics, Sapporo, Japan.

105. The Unified Medical Language System, http://www.nlm.nih.gov/research/umls/

106. Arama, E., Dickman, D., Kimchie, Z., et al. (2000) Mutations in the β-propeller domain of the *Drosophila brain tumor (brat)* protein induce neoplasm in the larval brain. *Oncogene* 19, 3706–3716.

107. Svolovits, P. (2003) Adding a medical lexicon to an English parser, in (Musen, M., ed.), *Proceedings of the AMIA 2003 Annual Symposium*. American Medical Informatics Association, Bethesda, MD.

108. Gusfield, D. (1997) *Algorithms on Strings, Trees and Sequences: Computer Science and Computational Biology*, Cambridge University Press, Cambridge, UK.

109. Notes from A. G. McDowell, http://www.mcdowella.demon.co.uk/programs.html

INDEX